Lecture Notes in Computer   6661

Commenced Publication in 1973
Founding and Former Series Editors:
Gerhard Goos, Juris Hartmanis, and Jan va

Raffaele Giancarlo   Giovanni Manzini (Eds.)

Combinatorial Pattern Matching

22nd Annual Symposium, CPM 2011
Palermo, Italy, June 27-29, 2011
Proceedings

 Springer

Volume Editors

Raffaele Giancarlo
University of Palermo, Department of Mathematics
Via Archirafi 34, 90123 Palermo, Italy
E-mail: raffaele@math.unipa.it

Giovanni Manzini
University of 'Piemonte Orientale', Department of Computer Science
Viale T. Michel 11, 15121 Alessandria, Italy
E-mail: manzini@mfn.unipmn.it

ISSN 0302-9743   e-ISSN 1611-3349
ISBN 978-3-642-21457-8   e-ISBN 978-3-642-21458-5
DOI 10.1007/978-3-642-21458-5
Springer Heidelberg Dordrecht London New York

Library of Congress Control Number: 2011928675

CR Subject Classification (1998): F.2, I.5, H.3.3, J.3, I.4.2, E.4, G.2.1, E.1

LNCS Sublibrary: SL 1 – Theoretical Computer Science and General Issues

Typesetting: Camera-ready by author, data conversion by Scientific Publishing Services, Chennai, India

Printed on acid-free paper

Springer is part of Springer Science+Business Media (www.springer.com)

Preface

The papers contained in this volume were presented at the 22nd Annual Symposium on Combinatorial Pattern Matching (CPM 2011) held in Mondello (Palermo), Italy, during June 27–29, 2011.

All the papers presented at the conference are original research contributions. We received 70 submissions from 20 countries; each paper was reviewed by at least three reviewers. The whole submission and review process was carried out with the invaluable help of the EasyChair conference system.

The committee decided to accept 36 papers. The program also included three invited talks by Nello Cristianini from the University of Bristol, UK, Gadi Landau from the University of Haifa, Israel, and Martin Vingron from the Max Planck Institute for Molecular Genetics, Berlin, Germany.

The objective of the annual CPM meetings is to provide an international forum for research in combinatorial pattern matching and related applications. It addresses issues of searching and matching strings and more complicated patterns such as trees, regular expressions, graphs, point sets, and arrays. The goal is to derive non-trivial combinatorial properties of such structures and to exploit these properties in order to either achieve superior performance for the corresponding computational problems or pinpoint conditions under which searches cannot be performed efficiently. The meeting also deals with problems in computational biology, data compression and data mining, coding, information retrieval, natural language processing, and pattern recognition.

The Annual Symposium on Combinatorial Pattern Matching started in 1990, and has since taken place every year. Previous CPM meetings were held in Paris, London, Tucson, Padova, Asilomar, Helsinki, Laguna Beach, Aarhus, Piscataway, Warwick, Montreal, Jerusalem, Fukuoka, Morelia, Istanbul, Jeju Island, Barcelona, London, Ontario, Pisa, Lille, and New York.

Starting from the third meeting, proceedings of all meetings have been published in the LNCS series, volumes 644, 684, 807, 937, 1075, 1264, 1448, 1645, 1848, 2089, 2373, 2676, 3109, 3537, 4009, 4580, 5029, 5577, and 6129.

Selected papers from the first meeting appeared in volume 92 of *Theoretical Computer Science*, from the 11th meeting in volume 2 of *Journal of Discrete Algorithms*, from the 12th meeting in volume 146 of *Discrete Applied Mathematics*, from the 14th meeting in volume 3 of *Journal of Discrete Algorithms*, from the 15th meeting in volume 368 of *Theoretical Computer Science*, from the 16th meeting in volume 5 of *Journal of Discrete Algorithms*, from the 19th meeting in volume 410 of *Theoretical Computer Science*, and from the 20th meeting in volume 9 of *Journal of Discrete Algorithms*.

For this year, a special issue of *Theoretical Computer Science* is already planned for expanded versions of selected extended abstracts presented at the symposium.

Special thanks are due to the members of the Program Committee who worked very hard to ensure the timely review of all the submitted manuscripts, and participated in stimulating discussions that led to the selection of the papers for the conference.

April 2011   Raffaele Giancarlo
                                                                Giovanni Manzini

Best Student Paper Award

This year the Program Committee Co-chairs and the Local Organizing Committee sponsored a Best Student Paper Award. The award was reserved for papers authored solely by PhD students or by researchers in their first year of a Post-Doc assignment.

Among the 70 submissions received by the Program Committee, five of them were eligible for the award. The committee decided unanimously to assign the award to the paper:

Succincter Text Indexing with Wildcards

Chris Thachuk

Department of Computer Science,
University of British Columbia, Vancouver, Canada

We study the problem of indexing text with wildcard positions, motivated by the challenge of aligning sequencing data to large genomes that contain millions of single nucleotide polymorphisms (SNPs) —positions known to differ between individuals. SNPs modeled as wildcards can lead to more informed and biologically relevant alignments. We improve the space complexity of previous approaches by giving a succinct index requiring $(2 + o(1))n \log \sigma + O(n) + O(d \log n) + O(k \log k)$ bits for a text of length n over an alphabet of size σ containing d groups of k wildcards. The new index is particularly favorable for larger alphabets and comparable for smaller alphabets, such as DNA. A key to the space reduction is a result we give showing how any compressed suffix array can be supplemented with auxiliary data structures occupying $O(n) + O(d \log \frac{n}{d})$ bits to also support efficient dictionary matching queries. We present a new query algorithm for our wildcard index that greatly reduces the query working space to $O(dm + m \log n)$ bits, where m is the length of the query. We note that compared to previous results this reduces the working space by two orders of magnitude when aligning short read data to the human genome.

Organization

Program Committee

Alexandr Andoni	Microsoft Research SVC, USA
Mikhail Atallah	Purdue University, USA
Jérémy Barbay	University of Chile, Chile
Frédérique Bassino	Université Paris 13, France
Anne Bergeron	Université du Québec, Montreal, Quebec
Raphaël Clifford	University of Bristol, UK
Aldo de Luca	University of Naples "Federico II", Italy
Chiara Epifanio	University of Palermo, Italy
Johannes Fischer	Karlsruhe Institute of Technology, Germany
Travis Gagie	Aalto University, Finland
Raffaele Giancarlo	University of Palermo, Italy (Co-chair)
Danny Hermelin	Max Planck Institute for Informatics, Germany
Wing-Kai Hon	National Tsing Hua University, Taiwan
Juha Kärkkäinen	University of Helsinki, Finland
Giosué Lo Bosco	University of Palermo, Italy
Stefano Lonardi	University of California Riverside, USA
Sabrina Mantaci	University of Palermo, Italy
Giovanni Manzini	University of Piemonte Or., Italy (Co-chair)
Burkhard Morgenstern	University of Göttingen, Germany
J. Ian Munro	University of Waterloo, Canada
Veli Mäkinen	University of Helsinki, Finland
Joong Chae Na	Sejong University, South Korea
Christian Pedersen	Aarhus University, Denmark
Wojciech Plandowski	University of Warsaw, Poland
Simon J. Puglisi	RMIT, Australia
Rajeev Raman	University of Leicester, UK
Mireille Régnier	INRIA-Saclay, France
Kunihiko Sadakane	National Institute of Informatics, Japan
David Sankoff	University of Ottawa, Canada
Giorgio Satta	University of Padova, Italy
Srinivasa Rao Satti	Seoul National University, South Korea
Roded Sharan	TelAviv University, Israel
William F. Smyth	McMaster University, Canada
Peter Stadler	University of Leipzig, Germany
Gabriel Valiente	Technical University of Catalonia, Spain
Susana Vinga	Inesc-ID, Portugal
Siu-Ming Yiu	University of Hong Kong, Hong Kong
Michal Ziv-Ukelson	Ben-Gurion University of the Negev, Israel

Steering Committee

Alberto Apostolico   University of Padova, Italy, and Georgia
       Institute of Technology, USA
Maxime Crochemore   Université Paris-Est, France, and King's
       College London, UK
Zvi Galil   Georgia Institute of Technology, USA

Organizing Committee

Chiara Epifanio   University of Palermo, Italy
Raffaele Giancarlo   University of Palermo, Italy
Giosué Lo Bosco   University of Palermo, Italy
Sabrina Mantaci   University of Palermo, Italy

Web and Publications Committee

Fabio Bellavia   University of Palermo, Italy
Alessio Langiu   University of Palermo, Italy
Carmen Lupascu   University of Palermo, Italy
Giovanna Rosone   University of Palermo, Italy
Luca Pinello   Harvard University, USA
Filippo Utro   IBM T.J. Watson Research Center, USA

External Referees

Atkins, Leon
Bankevich, Anton
Belazzougui, Djamal
Belcaid, Mahdi
Blin, Guillaume
Bouvel, Mathilde
Brejova, Bronislava
Bruckner, Sharon
Bucci, Michelangelo
Béal, Marie-Pierre
Camacho, Philippe
Canovas, Rodrigo
Carpi, Arturo
Claude, Francisco
Clément, Julien
Culpepper, Shane
Davoodi, Pooya
De Luca, Alessandro

Duma, Denisa
Dvorkin, Mikhail
Elberfeld, Michael
Elloumi, Mourad
Farzan, Arash
Flamm, Christoph
Francisco, Alexandre
Fraser, Robert
Gamzu, Iftah
Giambruno, Laura
Gog, Simon
Gotthilf, Zvi
Grabowski, Szymon
He, Meng
Jalsenius, Markus
Jansson, Jesper
Jiang, Shuai
Kaltenbach, Hans-Michael

Karhumaki, Juhani
Kim, Sung-Ryul
Kopelowitz, Tsvi
Kreft, Sebastian
Kubica, Marcin
Kufleitner, Manfred
Kulikov, Alexander
Kuruppu, Shanika
Lee, Inbok
Liptak, Zsuzsanna
Ma, Jian
Mnich, Matthias
Mosig, Axel
Mozes, Shay
Nekrich, Yakov
Nicaud, Cyril
Nielsen, Jesper
Nikolenko, Sergey
Noé, Laurent
Parida, Laxmi
Park, Heejin
Pinello, Luca
Pinhas, Tamar
Polishko, Anton
Ponty, Yann
Popa, Alexandru
Poulalhon, Dominique
Radoszewski, Jakub
Restivo, Antonio
Russo, Luis
Rytter, Wojciech

Sach, Benjamin
Salmela, Leena
Sand, Andreas
Sanders, Peter
Schmiedl, Christina
Sciortino, Marinella
Segev, Danny
Silverbush, Dana
Sim, Jeong Seop
Simonsen, Martin
Simpson, Jamie
Sirotkin, Alexander
Sirén, Jouni
Speck, Jochen
Tarhio, Jorma
Tataru, Paula
Thankachan, Sharma V.
Toivonen, Jarkko
Ustaoglu, Berkant
Vaglica, Roberto
Valenzuela, Daniel
Verbin, Elad
Verzotto, Davide
Vialette, Stéphane
Vyahhi, Nikolay
Välimäki, Niko
Walen, Tomasz
Zakov, Shay
Zaroda, Artur
Zizza, Rosalba

Sponsoring Institutions

University of Palermo
University of Piemonte Orientale

Table of Contents

Algorithms on Grammar-Compressed Strings

Gad M. Landau[*]

Department of Computer Science, University of Haifa, Israel
Department of Computer Science and Engineering, NYU-Poly, Brooklyn, NY, USA
landau@cs.haifa.ac.il

Grammar based compression, where one replaces a long string by a small context-free grammar that generates the string, is a simple and powerful paradigm that captures many of the popular compression schemes, including the Lempel-Ziv family, Run-Length Encoding, Byte-Pair Encoding, Sequitur and Re-Pair.

Let S be a string of length N given as a grammar $G(S)$ of size n. The random access problem is to compactly represent $G(S)$ while supporting fast random access queries. That is, given an index i, report $S[i]$ without decompressing S. We will first present a linear space representations of $G(S)$ that supports $O(\log N)$ random access time. This representation extends to efficiently support substring decompression. Namely, we can decompress any substring $S[i]...S[j]$ in the same complexity as a random access query and additional $O(j-i)$ time.

Once we obtain an efficient substring decompression method, it can then serve as a basis for a compressed version of classical pattern matching. Namely, we can take any black-box (uncompressed) approximate pattern matching algorithm and turn it into a corresponding algorithm over grammar compressed strings. We will then focus on a specific algorithm for computing the edit distance of two grammar-compressed strings. This algorithm requires $O(nN)$ time and uses the compression in a more complicated way (i.e., not through random access queries)

---

[*] Partially supported by the National Science Foundation Award 0904246, Israel Science Foundation grant 347/09, Yahoo, Grant No. 2008217 from the United States-Israel Binational Science Foundation (BSF) and DFG.

R. Giancarlo and G. Manzini (Eds.): CPM 2011, LNCS 6661, p. 1, 2011.

Automatic Discovery of Patterns in Media Content

Nello Cristianini

Intelligent Systems Laboratory
University of Bristol
nello@support-vector.net

Abstract. The strong trend towards the automation of many aspects of scientific enquiry and scholarship has started to affect also the social sciences and even the humanities. Several recent articles have demonstrated the application of pattern analysis techniques to the discovery of non-trivial relations in various datasets that have relevance for social and human sciences, and some have even heralded the advent of "Computational Social Sciences" and "Culturomics". In this review article I survey the results obtained over the past 5 years at the Intelligent Systems Laboratory in Bristol, in the area of automating the analysis of news media content. This endeavor, which we approach by combining pattern recognition, data mining and language technologies, is traditionally a part of the social sciences, and is normally performed by human researchers on small sets of data. The analysis of news content is of crucial importance due to the central role that the global news system plays in shaping public opinion, markets and culture. It is today possible to access freely online a large part of global news, and to devise automated methods for large scale constant monitoring of patterns in content. The results presented in this survey show how the automatic analysis of millions of documents in dozens of different languages can detect non-trivial macro-patterns that could not be observed at a smaller scale, and how the social sciences can benefit from closer interaction with the pattern analysis, artificial intelligence and text mining research communities.

1   Introduction

As an increasing number of research tasks are automated, the nature itself of scientific investigation is evolving, with a shift from a hypothesis-driven to a data-driven model of enquiry [15,31,4,21,23]. While this trend is most visible in the biological and physical sciences, it is also increasingly affecting areas of scholarship that seemed to be beyond the reach of automated methods, such as the humanities and the social sciences [27,38,30] .

While a hypothesis-driven approach to science involves careful design of experiments, often to deliberately discriminate between two competing hypotheses, in a data-driven approach the collection of vast amounts of data precedes the formulation of any hypotheses. So in genomics, for example, the sequencing of complete genomes is seen as the starting point for investigations whose details

R. Giancarlo and G. Manzini (Eds.): CPM 2011, LNCS 6661, pp. 2–13, 2011.
© Springer-Verlag Berlin Heidelberg 2011

are not known at the time of sequencing. Data mining and automated pattern analysis are obviously central to this new scientific method, as are data management aspects relative to the use of massive datasets. Statistics, data structures and algorithms are the language of this new way of doing science [11,12].

One qualitative difference between this approach and previous ones is its focus on "exhaustive" data: genomes, proteomes, transcriptomes are words denoting the full set of all genes, proteins and transcribed RNA that are present in an organism. Many more omics have been proposed, from metabolomics to interactomics [14]. In biology there have also been references to the bibliome, intended as the full set of published literature that is relevant to a given area of study [22]. Each of these approaches implies the use of automated means of data collection and analysis.

In the social and human sciences, the collection of vast datasets was often motivated by commercial applications. Social networks have been charted (as a side effect of email usage or social networking websites), user behavior data have been collected (as part of marketing efforts) and media content has been digitized (as part of the current business model, which involves the offer of free access to content, in return for advertising revenue) [27,38,9,8].

All of this data is ripe for social and human science research, and indeed this has already started. An investigation of 5% of all books ever written was published last year in Science and heralded as the start of "culturomics" an analogous to all the omics that have appeared in the biological sciences [30] . This investigation answered questions such as the evolution of spelling conventions, and the first use of certain words, among other things. Similarly, studies based on very large social networks have appeared [27]. It is not only the number of such data-driven studies that is increasing, so is also the complexity of the relations that can be detected by automated means. Accessing aspects of content or style, even in multilingual text, is now possible by advanced algorithms, developed for various applications in data mining, pattern recognition, and web technologies.

In this paper we will review a series of results that have been obtained over the past 5 years at the Intelligent Systems Laboratory at the University of Bristol, involving the analysis of vast amounts of news content. We call the contents of all the media "the mediasphere" and while it is now conceivable to chart all of it, at the present we are analyzing a large portion of it, which we believe to be representative only for Europe and United States. This exposition is not intended to present novel results (as all results discussed here are the fruit of various collaborations within the ISL and have been published elsewhere), nor to cover the entire emerging field of automated media analysis. Other projects are under way, with similar goals to ours, most notably the Lydia project [29,6], and the European Media Monitor [36].

The studies presented here involve the analysis of large corpora (sometimes in the order of millions of documents) and a diverse set of techniques, to answer questions such as: is there a gender bias in the coverage of news, and does it depend on the topic? Is the writing style related to the topic? Is there any pattern in the way the leading european outlets select the stories they publish every day? Can we automatically detect new memes emerging in the news stream? Can we relate the content of textual streams to objective quantities, such as rainfall or flu-levels in the real world? These questions traditionally fall within the remit of the social sciences, and are addressed by methods such as "coding" (annotation of articles by human analysts), followed by statistical analysis of the results. The size of these studies is necessarily limited to articles appeared in a few outlets and weeks, or to a select topic (as an example of such style of analysis see [1]). We attack this problem by gathering large amounts of data, with a dedicated software infrastructure, and deploying various algorithms for pattern analysis, language technologies and data mining. Where the quality of information extracted from each article may not be as high as that of human analysts, we can apply our methods to millions of documents, therefore extracting large scale patterns and trends that would not be accessible by conventional methods.

As a demonstration of a diversity of approaches, in this article we will address the analysis of multilingual data, to reconstruct some aspects of the EU media content [18], the detection of relations between gender bias and topic[2], the analysis of style: how readability and sentiment are related to topic and outlet [19], and the detection of events in textual streams (both from traditional and from social media) [34,26].

This article starts by surveying our data management infrastructure, NOAM (News Outlets Analysis and Monitoring)[17], and then briefly describing some of the experiments that we have mentioned above. As much more remains to be done, we will devote the Conclusions to discuss the road ahead, and the implications of this general line of research for scientific method in general.[1]

2   Data Acquisition and Management

In order for us to perform the experiments described in the next section, it was necessary to create a scalable infrastructure for data gathering and management. The need to access very large amounts of data also introduced constraints on our analysis methods. Our system is called NOAM and is described in [17]. We cannot describe here the details of our infrastructure, besides mentioning that it currently is based on 5 dedicated servers, and centered around a MySQL database. All news-items stored in our system can be annotated by various software modules, that can apply tags describing different aspects of the content. These modules operate nearly independently, gradually improving the annotation of existing content, but also generating new content as in the case of the machine translation module. Wherever possible we built our modules around open source software, as is the case for machine translation, information extraction, and support vector machines, while all the rest had to be developed within the group.

---

[1] We invite the readers to access the articles and Demo websites mentioned in this article, via our unified project website: http://mediapatterns.enm.bristol.ac.uk.

Our system currently monitors the contents of about 1500 outlets of various types (newspaper, magazine, broadcast), in 22 European languages, and from 193 different countries. English language news outlets represent the main part of this set (498), the rest being formed by news outlets in different languages, or sources of different type, such as press releases or blogs. The system checks multiple times per day for any new content by reading the news feeds of these online outlets, and stores in the database the basic information: title, description, URL; as well as some tags inherited from tags hand-assigned to the feeds list (e.g.: location, language, media type). Once in the system, these news items are processed, augmented and annotated by our software modules.

The full textual content of the articles is retrieved, starting from the URL provided by the RSS feed, by a "**scraper** module", and stored in the same database. If the document is not in english, a module based on the **Statistical-Machine-Translation** (SMT) package Moses [25] is used to generate another news-item, in English. We trained the SMT module using data from Europarl [24] and JRC-Acquis Corpora, so to cover all languages of the EU [37]. Various **topic classifiers** based on 1-class and 2-class Support Vector Machines [33,10,7] have been trained on Reuters [28] and New York Times [32] corpora, producing modules that can operate on English language text to assign topic labels, such as: "crime", "business", "science", "disasters", and so on [37]. The system also extracts **named entities** (people, locations, organizations) by using a module built around the open source tool Gate [13], augmented with various adaptations to deal with co-reference resolution [2,19] . In this way, a database of people is generated, along with attributes such as their gender and domain tags (e.g.: "Barack Obama, male, politics"). Information about the **style of writing** can be extracted by using standard metrics, including tools to measure the readability of a text as a function of word and sentence length [20]; and tools to measure the "sentiment" value of adjectives [16], which we have incorporated into larger pipelines aimed at assessing the degree of "linguistic subjectivity" of a text (details in [2,19]). All English language articles are also indexed by a suffix tree based on words, and statistical annotation is maintained about the frequency of all **word n-grams**, so that surprising changes in frequency are detected, and interesting n-grams are discovered [34]. **Clustering** is used to link together all articles, from all outlets, that cover the same events. The resulting cluster is called a "story". Based on this software infrastructure, we performed various experiments aimed at detecting and understanding patterns in media content. Overall, we have analysed 30 million online documents. A separate pipeline has also been used to gather **twitter** content, so to experiment with it as a means to predict quantities such as flu-levels and rainfall in the UK [26]. The term we use for this task is 'nowcasting' as it infers the present, rather than the future. Many other modules are currently being developed, for user modeling, information extraction, image analysis, and more.

3 Experiments: Patterns in Content

As an illustration of the style of enquiry that we described above, we will present here some of the patterns we have detected in the global news content. They were all published in various conferences and journals, so that we will omit most technical details, focusing instead on the findings.

3.1 News Coverage in the EU

A key question in media content analysis is: what makes news? In other words: out of all the many stories that could be covered every day, how do outlets pick the few stories they will feature in their main pages or feeds?

We selected the leading news outlets from each EU country (the top 10 outlets by web traffic - as measured by Alexa ranking - where available) that also had a RSS feed, and collected all articles from their main-feed (roughly corresponding to first page in newspapers), over a period of 6 months. This resulted into 1.3M news items, in 22 different languages. We machine-translated all into English, making sure to remove all untranslated words that might survive this process. We then clustered all articles into "stories" (sets of nearby articles in the bag-of-words representation), and represented each news outlet as a "bag of stories". The question we asked was: is there any stable pattern in the stories that each outlet chooses to cover?

The answer emerged by just linking outlets that share many stories (as measured by chi-square) and splitting the resulting network into communities (by applying a threshold to the chi-square statistic). The outlet-communities were found to be formed mostly by outlets from the same country, suggesting that the key differentiator among European outlets in the choice of stories to cover is still nationality. Once this had been established, we merged all outlets of each country into single nodes (regarding them as super outlets) and we repeated the exercise. What emerged now is a network showing the similarity among EU countries in their coverage of news stories (see Figure 1).

We compared this network with three other networks, obtained by using trade-relations data, Eurovision voting patterns, and geographic proximity. Each of these was found to be significantly correlated to the media-content similarity network. Finally, we also measured the deviation of each country from the "average EU content", and ranked all 27 countries by that measure. This was shown to be statistically correlated to both the year of accession to the EU and whether they use a national currency or the Euro. All the results in this section have been reproduced from the article [18], where the technical details can be found.

What is remarkable of this type of analysis, is that the macroscopic pattern we detected emerges from multiple small independent choices made by thousands of editors every day, and yet it reveals a clear structure in the european mediasphere. These results demonstrate how millions of documents in dozens of languages can be processed by machine, to reveal macro-patterns at the continental level that could never be observed by conventional methods of analysis.

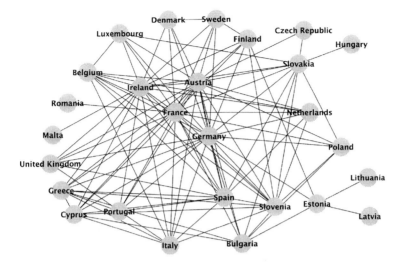

Fig. 1. This network represents the similarities among EU countries in the choice of stories they cover in their leading newspapers. It was generated by analysing 1M documents in 23 different languages, from about 200 newspapers, representing the 27 countries of the EU.

3.2   Gender, Style and Topic

Gender. Of the top 100 richest people in the world in 2011, 90 are men and 10 are women[2]. We call this the M/F ratio, and it is one of the most obvious biases in many areas of society. What is less obvious is how this bias changes with topic, or area of life.

The M/F ratio is higher in some domains and lower in others: of the top 50 richest athletes, all 50 are male[3]; of the top 100 "celebrities" 35 are females[4]. Of the 10 richest fashion models, 10 are females[5].

We asked the question: does media attention follow similar patterns? Of the 1000 most mentioned people, how many are men and how many are women? Does this ratio also change with topic? Answering this question implies being able to reliably detect people and their gender, as well as the topic of the articles in which they are mentioned. By doing this on large quantities of text, one can form the charts of the Top 1000 most mentioned people by topic, and measure the gender bias in these lists. We used our infrastructure to do just that, processing 476,528 articles in English language. Our infrastructure included the open source tool Gate, which we used for named entity recognition (NER), although we had

---

[2] http://www.forbes.com/wealth/billionaires/list

[3] http://sportsillustrated.cnn.com/more/specials/fortunate50/

[4] http://www.forbes.com/lists/2010/53/celeb-100-10_The-Celebrity-100.html

[5] http://www.forbes.com/2010/05/12/top-earning-models-business-entertainment-models.html

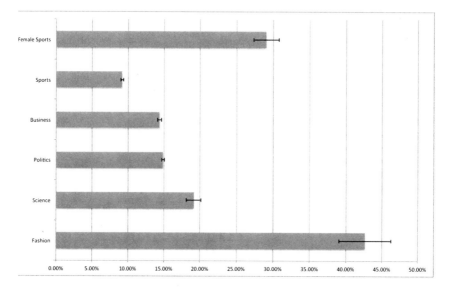

Fig. 2. The percentage of females in the 1000 most popular people, by topic, in 476,528 articles in english language, appeared between 1st Jan 2010 and 30th October 2010

to add various layers to improve the quality of annotation, by leveraging large corpora [2,3]. This was done only on English-language documents. We obtained the results shown in Figure 2.

It is clear that not only the same bias is present also in media coverage, but that it varies with topic in a similar way as the income bias.

Patterns in Style. We were also interested in large scale patterns involving writing style, namely readability and the use of "opinionated" or "judgmental" language. The first quantity is captured by a standard measure, roughly a function of word and sentence length, which has been shown to correlate well with ease of comprehension [20]. In order to capture the second quantity, we use the fraction of all adjectives in the text that are "sentimentally polarized" (that is, adjectives that express a sentiment or a judgment, according to the tool Senti-WordNet [16]). The results showed a clear and significant dependency of style both on topic and on news outlets (See Figure 3 for the relation between outlets and style).

Besides observing reassuring properties, e.g. that children outlets are the most readable, and Op/Ed articles use highly opinionated language, it is interesting to notice that outlets that occupy similar market niches tend to have a similar writing style (e.g., The Sun and The Daily Mirror; or The Times and the New York Times). This kind of analysis can now be performed on a massive scale, revealing properties of outlets that are likely to both shape and reflect the preferences of their readers. These results are reproduced from the articles [2,19], where all the technical details can be found.

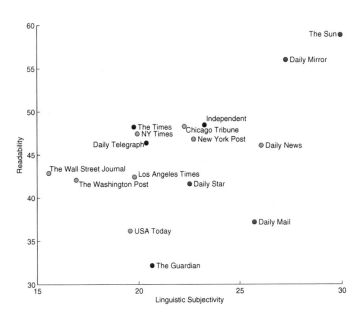

Fig. 3. The leading US and UK newspapers plotted in a "style space", spanned by their readability and "linguistic subjectivity" scores

3.3 Event Discovery in Textual Streams

Another question in media studies concerns the relation between the content of the media and events in the real world. Can we mine the content of large textual streams, such as traditional media or user-generated media, to discover events that are taking place in the real world?

From the point of view of pattern analysis, there are two distinct classes of problems: one is the detection of a specific event in a textual stream (pattern matching), while the other is the detection of "any event" of interest in the same stream (pattern discovery). We have experimented with both settings, developing a tool for event discovery - which we tested on the New York Times corpus [32]- and a tool for the detection of influenza-like-illness - which we tested on the content of Twitter.

The detection of "interesting" events in textual streams naturally depends on what we consider to be an event. In this case, we assumed that any significant change in the statistics of the source generating the data may signal an event, or a change in the state of the world. Under the assumption that the textual stream can be considered as generated by a Markov source (at the level of words), we focused on detecting significant changes in the frequency of word n-grams. The technical challenge is obviously to be able to monitor all possible n-grams, keeping sufficient statistics for each, in order to decide if one of them is significantly changing its frequency. A statistically annotated suffix tree was developed, adapted to deal with the large (and unbounded) alphabet size. While

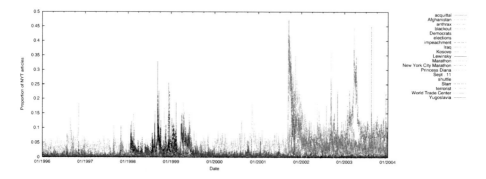

Fig. 4. The most surprising n-grams in a 8-years time series of 1M New York Times articles. Among the highest peaks are n-grams like: world-trade-center; princess-diana; sept. 11; Lewinsky.

the system is now deployed and working on daily news in our system, we have tested it on 8 years of the New York Times corpus, obtaining the results shown in Figure 4.

These results are based on the articles [35,34] where the technical details can be found.

The problem of detecting a specific event in the world, by analysing a textual stream, was addressed by using a supervised machine learning method. A time series of all twitter posts from the UK was collected for several months, along with a time series created by the Health Protection Agency reporting the levels of flu in the country. The idea was to map the text into a vector space representation, and then to use a linear regression algorithm to map words to flu-levels. We used the sparse regression method BoLasso [5], identifying a small subset of words that are highly indicative of the flu levels in the population, and the resulting flu-detector has shown to provide strong correlation with the actual flu levels. The same experiment was also replicated for the prediction of rainfall levels. These results are based on the paper [26] where the technical details can be found.

4 Conclusions

The data revolution of the past decade is changing unexpected aspects of society, business and also science. The data-driven approach to science may mark an important change in scientific method, affecting not only how we do science, but also what we consider as valid or interesting knowledge [11,12]. The humanities and the social sciences are also starting to be affected by this. One particularly important set of data is represented by online news, which now include most leading global news sources. Patterns and trends in the content of this textual stream contain valuable information for media analysts, and social scientists in general.

We have demonstrated how state-of-the-art tools in machine learning, information extraction, and machine translation can today be deployed, to analyze gender bias in news content, readability, subjectivity, and how they all relate to the topic of an article. We have also demonstrated how the publishing choices of news editors in different European countries reflect more general relations between these countries. While it is now possible to ask certain scientific questions to the data, and largely automate the process of answering them, it is also true that these questions are somewhat different from those that were asked before in the same area of scholarship. In fact, the data-driven approach can be seen as complementing the traditional one, rather than competing with it. Currently, we can extract simple information, about entities and events, topics and surprising memes. This, coupled with large amounts of data, can automatically generate social networks, maps, or timelines, or associations between news outlets based on their style, or choice of topics.

The interesting opportunity for pattern analysis researchers is that this frontier stretches our current capabilities, inviting us to collaborate with other scholars, and to work on large scale, noisy and non-trivial data streams. The appeal for social scientists is that this approach has the potential to "extend the boundaries of rigorous quantitative inquiry to a wide array of new phenomena spanning the social sciences and the humanities" as was remarked in [30].

Acknowledgements. The results reviewed and discussed in this study were obtained over several years of collaboration with many students and colleagues from the ISL of the University of Bristol, as well as colleagues from the School of Journalism of Cardiff University. Among others, I would like to thank Ilias Flaounas, Marco Turchi, Tijl De Bie, Omar Ali, Tristan Snowsill, Florent Nicart, Bill Lampos, Elena Hensinger and Justin Lewis. Support for this research has come from the Royal Society (Wolfson Research Merit Award) and the EU (Pascal2 Network, and SMART project).

References

1. Aday, S.: Chasing the bad news: An analysis of 2005 iraq and afghanistan war coverage on nbc and fox news channel. Journal of Communications 60, 144–164 (2010)
2. Ali, O., Flaounas, I., De Bie, T., Mosdell, N., Lewis, J., Cristianini, N.: Automating news content analysis: An application to gender bias and readability. In: Workshop on Applications of Pattern Analysis (WAPA). JMLR: Workshop and Conference Proceedings, Windsor, UK, pp. 36–43 (2010)
3. Ali, O., Cristianini, N.: Information fusion for entity matching in unstructured data. In: Papadopoulos, H., Andreou, A.S., Bramer, M. (eds.) AIAI 2010. IFIP Advances in Information and Communication Technology, vol. 339, pp. 162–169. Springer, Heidelberg (2010)
4. Ariely, D., Berns, G.: Neuromarketing: the hope and hype of neuroimaging in business. Nature Reviews Neuroscience 11, 284–292 (2010)
5. Bach, F.: Bolasso: model consistent lasso estimation through the bootstrap. In: Proceedings of the 25th International Conference on Machine Learning, ICML 2008 (2008)

6. Bautin, M., Ward, C., Patil, A., Skiena, S.: Access: News and blog analysis for the social sciences. In: 19th Int. World Wide Web Conference, WWW 2010 (2010)
7. Chang, C., Lin, C.: LIBSVM : A library for support vector machines. Software available at (2001), http://www.csie.ntu.edu.tw/~cjlin/libsvm
8. Coyle, K.: Mass digitization of books. The Journal of Academic Librarianship 32, 641–645 (2006)
9. Crane, G.: What do you do with a million books? D-Lib Magazine 12 (2006)
10. Cristianini, N., Shawe-Taylor, J.: An Introduction to Support Vector Machines and other Kernel-based learning methods. Cambridge University Press, Cambridge (2000)
11. Cristianini, N.: Scientific method and patterns in data. In: Samalam, V. (ed.) Procs. of the 5th UK BCS Knowledge Discovery and Data Mining Symposium, University of Salford (2009)
12. Cristianini, N.: Are we there yet? Neural Networks (2010)
13. Cunningham, H., Maynard, D., Bontcheva, K., Tablan, V.: Gate: A framework and graphical development environment for robust nlp tools and applications. In: Proc. of the 40th Anniversary Meeting of the Association for Computational Linguistics, Philadelphia, USA, pp. 168–175 (2002)
14. Greenbaum, D., Luscombe, N.M., Janson, R., et al.: Interrelating different types of genomic data, from proteome to secretome: 'oming in on function. Genome Research
15. Editorial: Defining the scientific method. Nature Methods 6, 237 (2009)
16. Esuli, A., Sebastiani, F.: Sentiwordnet: A publicly available lexical resource for opinion mining. In: Proceedings of LREC, pp. 417–422 (2006)
17. Flaounas, I., Ali, O., Turchi, M., Snowsill, T., Nicart, F., Bie, T.D., Cristianini, N.: Noam: News outlets analysis and monitoring system. In: Proceedings of the 2011 ACM SIGMOD International Conference on Management of Data. ACM, New York (2011)
18. Flaounas, I., Turchi, M., Ali, O., Fyson, N., Bie, T.D., Mosdell, N., Lewis, J., Cristianini, N.: The structure of eu mediasphere. PLoS ONE e14243 (2010)
19. Flaounas, I., Ali, O., Bie, T.D., Mosdell, N., Lewis, J., Cristianini, N.: Massive-scale automated analysis of news-content: Topics, style and gender (2011) (submitted for publication)
20. Flesch, R.: A new readability yardstick. Journal of Applied Psychology 32(3), 221–233 (1948)
21. González, M., Barabási, A.L.: Complex networks: From data to models. Nature Physics 3, 224–225 (2007)
22. Grivell, L.: Mining the bibliome: searching for a needle in a haystack? EMBO Reports 3(3), 200–203 (2002), http://www.nature.com/embor/journal/v3/n3/full/embor199.html
23. Janes, K., Yaffe, M.: Data-driven modelling of signal-transduction networks. Nature Reviews Molecular Cell Biology 7, 820–828 (2006)
24. Koehn, P.: Europarl: A parallel corpus for statistical machine translation. In: Machine Translation Summit X, pp. 79–86 (2005)
25. Koehn, P., Hoang, H., et al.: Moses: Open source toolkit for statistical machine translation. In: Annual Meeting-Association for Computational Linguistics ACL 2007, demonstration session, vol. 45 (2007)
26. Lampos, V., De Bie, T., Cristianini, N.: Flu detector-tracking epidemics on twitter. In: Machine Learning and Knowledge Discovery in Databases, pp. 599–602 (2010)

27. Lazer, D., Pentland, A., Adamic, L., Aral, S., Barabási, A., Brewer, D., Christakis, N., Contractor, N., Fowler, J., Gutmann, M., et al.: Computational Social Science. Science 323(5915), 721–723 (2009)
28. Lewis, D., Yang, Y., Rose, T., Li, F.: Rcv1: A new benchmark collection for text categorization research. Journal of Machine Learning Research 5, 361–397 (2004)
29. Lloyd, L., Kechagias, D., Skiena, S.: Lydia: A system for large-scale news analysis. In: Consens, M.P., Navarro, G. (eds.) SPIRE 2005. LNCS, vol. 3772, pp. 161–166. Springer, Heidelberg (2005)
30. Michel, J., Shen, Y., Aiden, A., Veres, A., Gray, M., Pickett, J., Hoiberg, D., Clancy, D., Norvig, P., Orwant, J., et al.: Quantitative Analysis of Culture Using Millions of Digitized Books. Science 331(6014), 176 (2011)
31. Potthast, T.: Paradigm shifts versus fashion shifts? EMBO Reports 10, S42–S45 (2009)
32. Sandhaus, E.: The new york times annotated corpus. In: Linguistic Data Consortium, Philadelphia (2008)
33. Shawe-Taylor, J., Cristianini, N.: Kernel Methods for Pattern Analysis. Cambridge University Press, Cambridge (2004)
34. Snowsill, T., Flaounas, I., Bie, T.D., Cristianini, N.: Detecting events in a million new york times articles. In: Balcátar, J.L., Bonchi, F., Gionis, A., Sebag, M. (eds.) (ECML/PKDD 2010). LNCS (LNAI), vol. 6321, pp. 615–618. Springer, Heidelberg (2010)
35. Snowsill, T., Nicart, F., Stefani, M., de Bie, T., Cristianini, N.: Finding surprising patterns in textual data streams. In: 2nd International Workshop on Cognitive Information Processing, pp. 405–410 (2010)
36. Steinberger, R., Pouliquen, B., der Goot, E.V.: An Introduction to the Europe Media Monitor Family of Applications. In: Information Access in a Multilingual World-Proceedings of the SIGIR 2009 Workshop (SIGIR-CLIR 2009), pp. 1–8 (2009)
37. Turchi, M., Flaounas, I., Ali, O., De Bie, T., Snowsill, T., Cristianini, N.: Found in translation. In: Buntine, W., Grobelnik, M., Mladenić, D., Shawe-Taylor, J. (eds.) ECML PKDD 2009. LNCS, vol. 5782, pp. 746–749. Springer, Heidelberg (2009)
38. Watts, D.: A twenty-first century science. Nature 445(7127), 489 (2007)

Computational Regulatory Genomics

Martin Vingron

Max Planck Institute for Molecular Genetics

Abstract. Genome sequence encodes not only genes but also the regulatory relationships among genes. Thus, the time and spatial patterns of gene expression are also encrypted in the DNA sequence. In order to unravel this other genetic code, regulatory genomics attempts to integrate functional genomics data with sequence data. This talk will summarize several approaches developed in our group, starting with a biophysically motivated method for prediction of transcription factor binding sites. Main applications are the identification of tissue specific transcription factors and the prediction of regulatory changes due to SNPs. Further, the talk will describe some indications that the division of promoters into two classes with high and low CpG contents, respectively, is of functional importance and helps in understanding mammalian promoters. In fact, the two classes of promoters display different features when it comes to binding site usage and tissue specific regulation. The dichotomy is further supported by an analysis of histone modifications in the promoters. Taken together, we interpret this as indication that different regulatory mechanisms govern transcription in these two classes of promoters.

R. Giancarlo and G. Manzini (Eds.): CPM 2011, LNCS 6661, p. 14, 2011.
© Springer-Verlag Berlin Heidelberg 2011

Lempel-Ziv Factorization Revisited

Enno Ohlebusch and Simon Gog

Institute of Theoretical Computer Science, University of Ulm, D-89069 Ulm
{Enno.Ohlebusch,Simon.Gog}@uni-ulm.de

Abstract. For 30 years the Lempel-Ziv factorization of a string has played an important role in data compression, and more recently it was used as the basis of linear time algorithms for the detection of all maximal repetitions (runs) in a string. In this paper, we present two new linear time algorithms: the first one is the fastest and the second is the most space-efficient among all LZ-factorization algorithms known so far.

1 Introduction

Let $S = S[1..n]$ be a string of length n over an alphabet Σ. The LZ-factorization of S is a factorization $S = \omega_1\omega_2 \cdots \omega_m$ such that each ω_k, $1 \leq k \leq m$, is either

(a) a letter $c \in \Sigma$ that does not occur in $\omega_1\omega_2 \cdots \omega_{k-1}$ or
(b) the longest substring of S that occurs at least twice in $\omega_1\omega_2 \cdots \omega_k$.

For example, the LZ-factorization of $S = acaaacatat$ is $\omega_1 = a$, $\omega_2 = c$, $\omega_3 = a$, $\omega_4 = aa$, $\omega_5 = ca$, $\omega_6 = t$, $\omega_7 = at$. The LZ-factorization can be represented by a sequence of pairs $(\mathsf{PrevOcc}_1, \mathsf{LPS}_1), \ldots, (\mathsf{PrevOcc}_m, \mathsf{LPS}_m)$, where in case (a) $\mathsf{PrevOcc}_k = c$ and $\mathsf{LPS}_k = 0$ and in case (b) $\mathsf{PrevOcc}_k$ is a position in $\omega_1\omega_2 \cdots \omega_{k-1}$ at which an occurrence of ω_k starts and $\mathsf{LPS}_k = |\omega_k|$. In our example, the LZ-factorization of $S = acaaacatat$ can be represented by $(a,0), (c,0), (1,1), (3,2), (2,2), (t,0), (7,2)$.

The LZ-factorization plays an important role in data compression (e.g. it is used in `gzip`, `WinZip`, and `PKZIP`). Moreover, it is the basis of linear time algorithms for the detection of all maximal repetitions (runs) in a string [11,14]. The key idea is that maximal repetitions within an element of the factorization have already been detected in its previous occurrence and thus need not be (re-)computed. The space consumption of the LZ-factorization algorithm was identified in [19] as a bottleneck for finding tandem repeats in large DNA sequences. Thus, it is of importance to develop space efficient algorithms.

It is well-known that the LZ-factorization can be computed in linear time with the help of the suffix tree of S [21]. Recently, several algorithms have been developed that use suffix arrays instead [1,4,5,7,6]. The *suffix array* SA of the string S is an array of integers in the range 1 to n specifying the lexicographic ordering of the n suffixes of the string S. To be precise, it satisfies $S_{\mathsf{SA}[1]} < S_{\mathsf{SA}[2]} < \cdots < S_{\mathsf{SA}[n]}$, where S_i denotes the ith suffix $S[i..n]$ of S. A direct linear time construction of the suffix array is possible; see [20] for an overview. Almost

R. Giancarlo and G. Manzini (Eds.): CPM 2011, LNCS 6661, pp. 15–26, 2011.

Algorithm 1. Computation of the LZ-factorization based on LPS and PrevOcc

---

$i \leftarrow 1$
while $i < n$ **do**
    if LPS$[i] = 0$ **then**
        PrevOcc$[i] \leftarrow S[i]$
    output (PrevOcc$[i]$, LPS$[i]$)
    $i \leftarrow i + \max\{1, \text{LPS}[i]\}$

---

all above-mentioned LZ-factorization algorithms additionally use the LCP-array: an array containing the lengths of the longest common prefix between every pair of consecutive suffixes in SA. We use $lcp(u,v)$ to denote the longest common prefix between strings u and v. The LCP-array is an array of integers in the range 1 to $n+1$ such that LCP$[1] = 0$, LCP$[n+1] = 0$, and LCP$[i] = |lcp(S_{\text{SA}[i-1]}, S_{\text{SA}[i]})|$ for $2 \leq i \leq n$. Given the suffix array, the LCP-array can be computed in linear time [13,12].

2   LZ-Factorization by Peak Elimination

Algorithm 1 shows that the Lempel-Ziv factorization can easily be computed from the arrays LPS and PrevOcc, which we define next.

The *longest previous substring* (LPS) is defined by LPS$[1] = 0$ and for k with $2 \leq k \leq n$:

$$\text{LPS}[k] = \max\{\ell : 0 \leq \ell \leq n-k+1; S[k..k+\ell-1] \text{ is a substring of } S[1..k+\ell-2]\}$$

That is, LPS$[k]$ is the length of the longest prefix of S_k that has another occurrence in S starting strictly before position k. If there is no $\ell > 0$ such that $S[k..k+\ell-1]$ is a substring of $S[1..k+\ell-2]$, then LPS$[k] = 0$ because for $\ell = 0$ we have $S[k..k+\ell-1] = \varepsilon$ and the empty string ε is a substring of any other string. We are not only interested in LPS$[k]$ but also in a position $j < k$ at which the longest previous substring occurred. Such a position j will be stored in the array PrevOcc, i.e., PrevOcc$[k] = j$. If there is no such position, i.e., if LPS$[k] = 0$, we set PrevOcc$[k] = \bot$. As an example, the arrays LPS and PrevOcc of $S = acaaacatat$ are depicted (in suffix array order) in Fig. 1.

In order to formulate the important Lemma 1 (see [6, Eqn. (1)]), we define the following two arrays (to deal with boundary cases, we set SA$[0] = 0 = $ SA$[n+1]$).

For any index $1 \leq i \leq n$, let

$$\text{PSV}[i] = \max\{j : 0 \leq j < i \text{ and } \text{SA}[j] < \text{SA}[i]\}$$
$$\text{NSV}[i] = \min\{j : i < j \leq n+1 \text{ and } \text{SA}[j] < \text{SA}[i]\}$$

If there is no index j with $1 \leq j < i$ and SA$[j] <$ SA$[i]$, we have PSV$[i] = 0$ because $0 < i$ and $0 = $ SA$[0] <$ SA$[i]$. Analogously, if there is no index j with

i	SA$[i]$	LCP$[i]$	$S_{\text{SA}[i]}$	PSV$[i]$	NSV$[i]$	LPS$[$SA$[i]]$	PrevOcc$[$SA$[i]]$
0	0		ε				
1	3	0	$aaacatat$	0	3	1	1
2	4	2	$aacatat$	1	3	2	3
3	1	1	$acaaacatat$	0	11	0	$\perp$
4	5	3	$acatat$	3	7	3	1
5	9	1	at	4	6	2	7
6	7	2	$atat$	4	7	1	5
7	2	0	$caaacatat$	3	11	0	$\perp$
8	6	2	$catat$	7	11	2	2
9	10	0	t	8	10	1	8
10	8	1	tat	8	11	0	$\perp$
11	0	0	ε				

Fig. 1. The suffix array of the string $S = acaaacatat$ and additional tables

$i < j \leq n$ and SA$[j] <$ SA$[i]$, we have NSV$[i] = n+1$. Fig. 1 provides an example of the auxiliary arrays PSV and NSV.

Lemma 1. *For every $1 \leq i \leq n$, the following equality holds*

$$\text{LPS}[\text{SA}[i]] = \max\{|\text{lcp}(S_{\text{SA}[\text{PSV}[i]]}, S_{\text{SA}[i]})|, |\text{lcp}(S_{\text{SA}[i]}, S_{\text{SA}[\text{NSV}[i]]})|\}$$

where S_0 is the empty string ε.

In essence, the preceding lemma says that in order to compute LPS$[$SA$[i]]$ for a given index i, it suffices to consider the closest index j with SA$[j] <$ SA$[i]$ preceding i (viz. PSV$[i]$) and the closest index j with SA$[j] <$ SA$[i]$ succeeding i (viz. NSV$[i]$) in the suffix array of S.

We now recall the nice (conceptual) graph representation of SA and LCP introduced in [5]. The graph has $n + 2$ nodes, each of which is labeled with $(i, \text{SA}[i])$; to deal with boundary cases, the graph also contains the nodes labeled with $(0, \text{SA}[0])$ and $(n + 1, \text{SA}[n + 1])$. It is instructive to view each node $(i, \text{SA}[i])$ as a point in the plane $\mathbb{R}^2$. Moreover, consecutive nodes $(i, \text{SA}[i])$ and $(i + 1, \text{SA}[i + 1])$ are connected by an edge labeled with the value LCP$[i + 1]$; see Fig. 2 for an example. In the graph, $|\text{lcp}(S_{\text{SA}[\text{PSV}[i]]}, S_{\text{SA}[i]})|$ is the minimum of the edge labels on the path from node $(\text{PSV}[i], \text{SA}[\text{PSV}[i]])$ to $(i, \text{SA}[i])$. Analogously, $|\text{lcp}(S_{\text{SA}[i]}, S_{\text{SA}[\text{NSV}[i]]})|$ is the minimum of the edge labels on the path from $(i, \text{SA}[i])$ to node $(\text{NSV}[i], \text{SA}[\text{NSV}[i]])$. Now consider a *peak* in the graph, i.e., a node $(i, \text{SA}[i])$ such that SA$[i - 1] <$ SA$[i]$ and SA$[i + 1] <$ SA$[i]$. In this case, $(\text{PSV}[i], \text{SA}[\text{PSV}[i]]) = (i - 1, \text{SA}[i - 1])$ and $(\text{NSV}[i], \text{SA}[\text{NSV}[i]]) = (i+1, \text{SA}[i+1])$. Thus, we have LPS$[SA[i]] = \max\{LCP[i],$ LCP$[i+1]\}$ by Lemma 1 (moreover, PrevOcc$[$SA$[i]]$ can get the value SA$[i-1]$ if LCP$[i] \geq$ LCP$[i+1]$ and the value SA$[i+1]$ if LCP$[i] \leq$ LCP$[i+1]$). That is, the correct value of LPS$[$SA$[i]]$ can be computed as the maximum of the labels of the edges with end point $(i, \text{SA}[i])$. The peak can be eliminated as follows: we delete node $(i, \text{SA}[i])$ and its edges from the graph, and add an edge labeled with $\min(\text{LCP}[i], \text{LCP}[i + 1])$ between

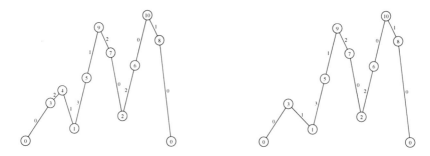

Fig. 2. Left: The graph for the string $S = acaaacatat$. Right: The graph after the elimination of the leftmost peak (yielding $\mathsf{LPS}[4] = 2$ and $\mathsf{PrevOcc}[4] = 3$).

nodes $(i-1, \mathsf{SA}[i-1])$ and $(i+1, \mathsf{SA}[i+1])$. Observe that such a peak elimination is safe because it does not change the values $\mathsf{PSV}[j]$, $\mathsf{NSV}[j]$, $\mathsf{SA}[\mathsf{PSV}[j]]$, and $\mathsf{SA}[\mathsf{NSV}[j]]$ for any index $j \neq i$ with $1 \leq j \leq n$. In the transformed graph, for any index $j \neq i$ with $1 \leq j \leq n$, $\mathsf{LPS}[\mathsf{SA}[j]]$ can again be computed as the maximum of the minimum of the edge labels on the paths from node $(\mathsf{PSV}[j], \mathsf{SA}[\mathsf{PSV}[j]])$ to $(j, \mathsf{SA}[j])$ and the minimum of the edge labels from node $(\mathsf{NSV}[j], \mathsf{SA}[\mathsf{NSV}[j]])$ to $(j, \mathsf{SA}[j])$. This is because the label $\min(\mathsf{LCP}[i], \mathsf{LCP}[i+1])$ of the new edge is the length of the longest common prefix of the suffixes $S_{\mathsf{SA}[i-1]}$ and $S_{\mathsf{SA}[i+1]}$ of S. Hence it is safe to remove peak by peak from the graph in this way. Obviously, the order in which peaks are eliminated is arbitrary. Fig. 2 illustrates the removal of the leftmost peak from the graph.

The algorithm of Crochemore and Ilie [5] scans the arrays SA and LCP from left to right. In other words, the arrays LPS and $\mathsf{PrevOcc}$ are computed by (virtually) building the graph in *suffix array order* and elimination of detected peaks. The algorithm was improved by further considering the case $\mathsf{SA}[i-1] < \mathsf{SA}[i] < \mathsf{SA}[i+1]$ and $\mathsf{LCP}[i] \geq \mathsf{LCP}[i+1]$. In this case $\mathsf{LPS}[\mathsf{SA}[i]] = \mathsf{LCP}[i]$; see [7]. The next corollary stems from [5, Prop. 1].

Corollary 1. *The* LPS *array is a permutation of the array* $\mathsf{LCP}[1..n]$.

3   Ultra-Fast Lempel-Ziv Factorization

All LZ-factorization algorithms in [1,4,5,7,6] except for the first one in [5] and the last two in [4] rely on the suffix array *and* the precomputed LCP-array. According to Corollary 1, if one computes the LZ-factorization via the arrays LPS and $\mathsf{PrevOcc}$, the computation of lcp-values is really necessary. In our opinion, however, it is disadvantageous to *precompute* the LCP-array. Algorithm 2 intermingles the computation of lcp-values with the computation of the arrays LPS and $\mathsf{PrevOcc}$: it computes these arrays by (virtually) building the above-mentioned graph in *text order* and peak elimination. More precisely, in the main

Algorithm 2. Computation of LPS and PrevOcc

Main procedure	Procedure $sop(i, \ell, j)$
for $i \leftarrow 1$ **to** n **do**	**if** $\mathsf{LPS}[i] = \perp$ **then**
$\quad \Phi[\mathsf{SA}[i]] \leftarrow \mathsf{SA}[i-1]$	$\quad \mathsf{LPS}[i] \leftarrow \ell$
$\ell \leftarrow 0$	$\quad \mathsf{PrevOcc}[i] \leftarrow j$
for $i \leftarrow 1$ **to** n **do**	**else**
$\quad j \leftarrow \Phi[i]$	$\quad$ **if** $\mathsf{LPS}[i] < \ell$ **then**
$\quad$ **while** $S[i + \ell] = S[j + \ell]$ **do**	$\quad\quad$ **if** $\mathsf{PrevOcc}[i] > j$ **then**
$\quad\quad \ell \leftarrow \ell + 1$	$\quad\quad\quad sop(\mathsf{PrevOcc}[i], \mathsf{LPS}[i], j)$
$\quad$ **if** $i > j$ **then**	$\quad\quad$ **else**
$\quad\quad sop(i, \ell, j)$	$\quad\quad\quad sop(j, \mathsf{LPS}[i], \mathsf{PrevOcc}[i])$
$\quad$ **else**	$\quad\quad \mathsf{LPS}[i] \leftarrow \ell$
$\quad\quad sop(j, \ell, i)$	$\quad\quad \mathsf{PrevOcc}[i] \leftarrow j$
$\quad \ell \leftarrow max(\ell - 1, 0)$	$\quad$ **else** /* $\mathsf{LPS}[i] \geq \ell$ */
	$\quad\quad$ **if** $\mathsf{PrevOcc}[i] > j$ **then**
	$\quad\quad\quad sop(\mathsf{PrevOcc}[i], \ell, j)$
	$\quad\quad$ **else**
	$\quad\quad\quad sop(j, \ell, \mathsf{PrevOcc}[i])$

procedure it computes lcp-values as in the Φ-algorithm of Kärkkäinen et al. [12].
Suppose that the algorithm has just computed the length ℓ of the longest common prefix of some suffix S_i of S and the suffix S_j that precedes S_i in the suffix array (so among all suffixes of S which are lexicographically smaller than S_i, S_j is the largest). In terms of the graph, this means that the algorithm has just detected an edge with label ℓ between nodes i and j; but of course it does not build the graph. Instead it stores this edge in the arrays LPS and PrevOcc. To be precise, if $i > j$, it stores ℓ in $\mathsf{LPS}[i]$ and j in $\mathsf{PrevOcc}[i]$. Otherwise, if $i < j$, it stores ℓ in $\mathsf{LPS}[j]$ and i in $\mathsf{PrevOcc}[j]$. However, collisions may occur. If, for example $i > j$, it may be the case that the memory cells $\mathsf{LPS}[i]$ and $\mathsf{PrevOcc}[i]$ are occupied already; say $m = \mathsf{LPS}[i]$ and $k = \mathsf{PrevOcc}[i]$. In terms of the graph, this means that a peak has been detected because both k and j are smaller than i. Consequently, the peak is eliminated by a case distinction: either (a) $m < \ell$ or (b) $m \geq \ell$. Let us consider case (a); case (b) is similar. Since $m < \ell$, we set $\mathsf{LPS}[i] \leftarrow \ell$ and $\mathsf{PrevOcc}[i] \leftarrow j$ in accordance with the peak elimination procedure described above. (It should be pointed out that the entries $\mathsf{LPS}[i]$ and $\mathsf{PrevOcc}[i]$ are fixed from this point on.) Moreover, we have to insert a new edge between nodes j and k with label m. As we always store an edge at the index that is the maximum of its node labels, we further distinguish between the cases (i) $k < j$ and (ii) $k > j$. Again, we only consider the first case because the second case can be treated similarly. If the memory cells $\mathsf{LPS}[j]$ and $\mathsf{PrevOcc}[j]$ are still empty (i.e, $\mathsf{LPS}[j] = \perp$ and $\mathsf{PrevOcc}[j] = \perp$), then we store m in $\mathsf{LPS}[j]$ and k in $\mathsf{PrevOcc}[j]$. Otherwise, we have just detected another peak and we eliminate it in the same way as described above. As an example, consider the execution of the main procedure of Algorithm 2 for $S = acaaacatat$ and $i = 4$. Fig. 3 (left) shows

$S[i]$	a	c	a	a	a	c	a	t	a	t
i	1	2	3	4	5	6	7	8	9	10
LPS$[i]$			0	1			0			
PrevOcc$[i]$			0	1			2			

$S[i]$	a	c	a	a	a	c	a	t	a	t
i	1	2	3	4	5	6	7	8	9	10
LPS$[i]$			0	2			0			
PrevOcc$[i]$			0	3			2			

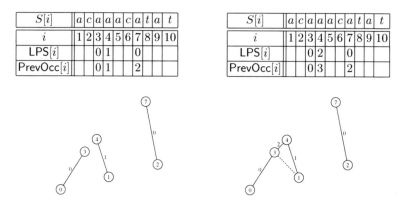

Fig. 3. $i = 4$: Detection and removal of the first peak in the graph

$S[i]$	a	c	a	a	a	c	a	t	a	t
i	1	2	3	4	5	6	7	8	9	10
LPS$[i]$			1	2			0			
PrevOcc$[i]$			1	3			2			

$S[i]$	a	c	a	a	a	c	a	t	a	t
i	1	2	3	4	5	6	7	8	9	10
LPS$[i]$		0	1	2			0			
PrevOcc$[i]$		0	1	3			2			

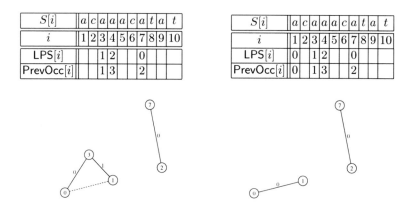

Fig. 4. $i = 4$: Detection and removal of the second peak in the graph

the point of departure. Since $j = \Phi[4] = 3$ (cf. Fig. 1), S_4 is compared with S_3 and the length of the longest common prefix of S_4 and S_3 is 2. This results in the procedure call $sop(4, 2, 3)$ because $i = 4 > 3 = j$. Since LPS$[4] = 1 < 2 = \ell$ and PrevOcc$[4] = 1 < 3 = j$, there is another procedure call $sop(3, 1, 1)$. Fig. 3 (right) shows the situation without the effect of $sop(3, 1, 1)$: peak node 4 is eliminated, i.e., LPS$[4]$ is set to $\ell = 2$ and PrevOcc$[4]$ is set to $j = 3$. Let us turn to the effect of the procedure call $sop(3, 1, 1)$: LPS$[3] = 0 < 1$ and PrevOcc$[3] = 0 < 1$ result in another procedure call $sop(1, 0, 0)$. The left part of Fig. 4 depicts the situation without the effect of $sop(1, 0, 0)$ (peak node 3 is eliminated), while the right part shows the final situation (after the execution of the main procedure for $i = 4$).

Algorithm 2 runs in linear time. This is because the main procedure without the calls to the procedure sop takes $O(n)$ time [12] and an amortized analysis shows that there are at most $2n$ calls to the procedure sop.

4   LZ-Factorization with Succinct Data Structures

Next, we develop a LZ-factorization algorithm that uses as low space as possible. This is of importance because the LZ-factorization is identified in [19] as a bottleneck for finding tandem repeats in large DNA sequences, and a space efficient solution would render the ad hoc solutions of [19] superfluous.

To explain the idea of our second algorithm, we need the following terminology. For a substring ω of S, the ω-interval in the suffix array SA of S is the interval $[lb..rb]$ such that ω is a prefix of $S_{\mathsf{SA}[k]}$ for all $lb \le k \le rb$, but ω is not a prefix of any other suffix of S. Suppose that we have already computed the first $k-1$ factors, say $S[1..j] = \omega_1 \cdots \omega_{k-1}$ and we want to compute the next factor ω_k, which is a prefix of S_{j+1}. To this end, we determine the $S[j+1]$-interval $[lb_1..rb_1]$ in SA and the rank $i = \mathsf{ISA}[j+1]$ of suffix S_{j+1} (where ISA denotes the inverse of the permutation SA). Now $S[j+1]$ has a previous occurrence in S if and only if $\mathsf{PSV}[i]$ or $\mathsf{NSV}[i]$ (or both) lie inside the $S[j+1]$-interval $[lb_1..rb_1]$. If this is the case, $\mathsf{SA}[\mathsf{PSV}[i]]$ or $\mathsf{SA}[\mathsf{NSV}[i]]$ is a previous occurrence of $S[j+1]$. Then we iterate this process: we determine the $S[j+1..j+2]$-interval $[lb_2..rb_2]$ and check whether $\mathsf{PSV}[i]$ or $\mathsf{NSV}[i]$ lie in $[lb_2..rb_2]$, and so on. The maximum ℓ for which the $S[j+1..j+\ell]$-interval $[lb_\ell..rb_\ell]$ contains $\mathsf{PSV}[i]$ or $\mathsf{NSV}[i]$ is the length of ω_k. Moreover, if $\mathsf{PSV}[i]$ ($\mathsf{NSV}[i]$) lies in $[lb_\ell..rb_\ell]$, then $\mathsf{SA}[\mathsf{PSV}[i]]$ ($\mathsf{SA}[\mathsf{NSV}[i]]$) is the start position of a previous occurrence of $S[j+1..j+\ell]$.

For example, suppose that we have already computed the first two factors $\omega_1 = a$ and $\omega_2 = c$ of the LZ-factorization of the string $S = acaaacatat$, and we want to compute the next factor ω_3, a prefix of $S_3 = aaacatat$. We determine the a-interval $[1..6]$ in SA (cf. Fig. 1) and the rank $1 = \mathsf{ISA}[3]$ of suffix S_3. Now a has a previous occurrence at position $\mathsf{SA}[\mathsf{NSV}[1]] = 1$ in S because $\mathsf{NSV}[1] = 3$ lies inside the a-interval $[1..6]$. Since neither $\mathsf{PSV}[1] = 0$ nor $\mathsf{NSV}[1] = 3$ lie inside the aa-interval $[1..1]$, it follows that $\omega_3 = a$.

This approach is similar to the algorithm CPS2 [4], but it is not space-efficient because it uses the suffix array and its inverse. A solution to this problem is to use backward search [8] on the string $T = S^{rev}\$$, where S^{rev} is the reverse of S and $\$$ is a character not occurring in S ($\$$ is smaller than any other character in the alphabet Σ).

Let us recall the backward search technique. In what follows, SA denotes the suffix array of T (not that of S). The Burrows and Wheeler transform transforms T into the string $\mathsf{BWT}[1..n+1]$ defined by $\mathsf{BWT}[i] = T[\mathsf{SA}[i] - 1]$ for all i with $\mathsf{SA}[i] \ne 1$ and $\mathsf{BWT}[i] = \$$ otherwise. In virtually all cases, the Burrows-Wheeler transformed string compresses much better than the original string; see [3]. The permutation LF, defined by $LF(i) = \mathsf{ISA}[\mathsf{SA}[i] - 1]$ for all i with $\mathsf{SA}[i] \ne 1$ and $LF(i) = 1$ otherwise, is called LF-mapping. A compressed full-text index based on a compressed form of the LF-mapping is commonly referred to as FM-index [8]. The LF-mapping can be implemented by $LF(i) = C[c] + Occ(c, i)$ where $c = \mathsf{BWT}[i]$, $C[c]$ is the overall number (of occurrences) of characters in T which are strictly smaller than c, and $Occ(c, i)$ is the number of occurrences of the character c in $\mathsf{BWT}[1..i]$. Ferragina and Manzini [8] showed that it is possible to search a pattern character-by-character backwards in the suffix array SA of

---

Algorithm 3. Construction of the balanced parentheses sequence

$push(0)$ /* SA[0] $= \infty$ */
write '('
for $i \leftarrow 1$ **to** $n + 1$ **do**
  while SA[i] > SA[$top()$] **do**
    $pop()$ and write ')'
  $push(i)$ and write '('
write '))' /* for SA[0] $= \infty$ and SA[$n + 1$] $= \infty$ */

---

string T, without storing SA. Let $c \in \Sigma$ and ω be a substring of T. Given the ω-interval $[i..j]$ in the suffix array SA of T, $backwardSearch(c, [i..j])$ returns the $c\omega$-interval $[C[c] + Occ(c, i-1) + 1 \ldots C[c] + Occ(c, j)]$ if it exists, and $\perp$ otherwise. We use a wavelet tree [10] as a space-efficient implementation of an FM-index. With this implementation both a backward search step and the computation of $LF(i)$ take $\mathcal{O}(\log |\Sigma|)$ time, but it is possible to reduce this to constant time.

Since we work with $T = S^{rev}\$$ instead of S, we have to replace PSV and NSV with the following functions (to deal with boundary cases, we set SA[0] $= \infty =$ SA[$n + 1$]). For any index $1 \leq i \leq n$, we define

$$\mathsf{PGV}(i) = \max\{j : 0 \leq j < i \text{ and } \mathsf{SA}[j] > \mathsf{SA}[i]\}$$
$$\mathsf{NGV}(i) = \min\{j : i < j \leq n+1 \text{ and } \mathsf{SA}[j] > \mathsf{SA}[i]\}$$

The values PGV(i) and NGV(i) can be computed in constant time on a balanced parentheses sequence using $2n + o(n)$ bits that is constructed by Algorithm 3.

– PGV(i) $= rank_{(}(enclose(select_{(}(i))))$
– NGV(i) $= rank_{(}(findclose(select_{(}(i))) + 1$

For space reasons, we refer to [17,9] for details (note that a suffix array does not have equal entries, so the problem here is easier than the general case studied in [17]).

Now we have all ingredients for the space-efficient LZ-factorization algorithm 4, which runs in $\mathcal{O}(n \log |\Sigma|)$ time. Each execution of its outer while-loop computes the next factor of the right-to-left LZ-factorization of S^{rev} (starting at position i') which coincides with the next factor of the left-to-right LZ-factorization of S (starting at $n + 1 - i'$). Note that we do not need the inverse suffix array because the rank of the current suffix can be determined with the help of the LF-mapping. We still need the suffix array to determine the previous occurrence but only once for each factor. That is the reason why it is possible to use a compressed suffix array, which saves even more space.

We would like to point out that the idea of replacing the binary search in the algorithm CPS2 with the backward search in the reverse string was also used in [15] for an alternative Lempel-Ziv parsing (called LZ-End). Furthermore, there is another LZ-factorization algorithm that uses succinct data structures: the online algorithm developed by Okanohara and Sadakane [18]. With the aid of

---

Algorithm 4. LZ-factorization of S based on backward search on $T = S^{rev}\$$

$i \leftarrow n$ /* $|S^{rev}| = n$ */
$j \leftarrow 1$ /* $\$$ appears at index 1 in SA */
while $i > 1$ **do**
  $i' \leftarrow i$
  $[sp..ep] \leftarrow backwardSearch(T[i], [1..n+1])$
  while NGV$[LF(j)] \leq ep$ **or** PGV$[LF(j)] \geq sp$ **do**
    $j \leftarrow LF(j)$
    $[lb..rb] \leftarrow [sp..ep]$
    $i \leftarrow i - 1$
    $[sp..ep] \leftarrow backwardSearch(T[i], [sp..ep])$
  LPS $\leftarrow i' - i$
  if LPS $= 0$ **then**
    PrevOcc $\leftarrow T[i]$
    $i \leftarrow i - 1$
  else if NGV$[j] \leq rb$ **then**
    PrevOcc $\leftarrow n + 1 -$ SA$[$NGV$[j]] - ($LPS $- 1)$
  else /* PGV$[j] \geq lb$ */
    PrevOcc $\leftarrow n + 1 -$ SA$[$PGV$[j]] - ($LPS $- 1)$
  output (PrevOcc, LPS)

---

rank/select operations and range minimum queries, it dynamically maintains succinct representations of the suffix array, the LCP-array, and the BWT. Its worst-case time complexity is $O(n \log^3 n)$.

5   Experimental Results

In our experiments, programs were compiled using g++ version 4.1.2 with options -O3 -DNDEBUG on a 64 bit Linux (Kernel 2.6.16) system equipped with a Dual-Core AMD Opteron processor with 3 GHz and 4GB of RAM. The test cases are described in [4].

We compared our Algorithm 2, called LZ_OG, with the currently fastest algorithms [6]. It should be stressed that the latter algorithms only compute the LPS-array (called LPF in [6]) but *not* the PrevOcc-array. Although this is a severe disadvantage to our Algorithm 2, it still outperforms the other algorithms; see Table 1. For highly repetitive strings, it is even faster than the precomputation of the LCP-array. Algorithm 2 requires $13n$ bytes (the suffix array can be discarded after the Φ-array is computed), whereas the others would take at least $17n$ bytes for a LZ-factorization (including the $4n$ bytes for the PrevOcc-array). The space consumption can further be reduced by streaming techniques (so that e.g. our algorithm takes only $9n$ bytes), but any LZ-factorization algorithm based on the arrays LPS and PrevOcc needs at least $8n$ bytes of memory.

The experiments also showed the expected space-time tradeoff. That is, our Algorithm 4 uses much less space than the above-mentioned algorithms (including Algorithm 2), but it is slower than these. For a fair comparison, we thus

Table 1. Running times in seconds of (from left to right): the precomputation of the LCP-array, the five algorithms from [6] (including LCP-array computation but without PrevOcc-array computation), and our new fast LZ-factorization algorithm

test case	LCP	LPF_simple	LPF_next_prev	LPF_sorting	LPF_online	LPF_optimal	LZ_OG
chr19.dna4	19.40	46.40	25.80	31.30	26.30	26.30	19.10
chr22.dna4	8.90	24.60	14.80	14.60	12.30	12.20	10.00
E.coli	0.80	2.00	1.20	1.40	1.20	1.20	0.90
bible	0.60	1.60	0.90	1.10	0.90	0.90	0.50
howto	8.10	20.60	11.80	13.40	11.60	11.50	6.70
fib_s14930352	2.20	6.90	3.40	3.80	3.30	3.30	0.80
fib_s9227465	1.40	4.20	2.30	2.40	2.10	2.10	0.50
fss10	1.70	5.60	2.80	3.00	2.70	2.70	0.70
fss9	0.40	1.10	0.60	0.70	0.60	0.60	0.20
p16Mb	3.50	8.70	5.00	5.90	4.90	4.90	3.90
p32Mb	8.50	20.60	11.60	13.90	11.60	11.60	9.70
rndA21_8Mb	1.60	4.00	2.40	2.80	2.30	2.30	1.90
rndA2_8Mb	1.50	3.90	2.20	2.60	2.20	2.20	1.50

Table 2. Comparison of the LZ-factorization algorithm CPS2 [4] with our Algorithm 4, called LZ_bwd (space usage is measured after the construction of the BWT). To show the effect of using a compressed suffix array, we varied the sample density of suffix array values. The algorithm LZ_bwd$x$ uses every x-th suffix array entry. That is, LZ_bwd1 uses a plain suffix array, whereas LZ_bwd32 uses only every 32-th suffix array entry. Non-sampled values are recovered by applying the LF-mapping iteratively.

test case	space in bytes per symbol							time in seconds						
	CPS2	LZ_bwd1	LZ_bwd2	LZ_bwd4	LZ_bwd8	LZ_bwd16	LZ_bwd32	CPS2	LZ_bwd1	LZ_bwd2	LZ_bwd4	LZ_bwd8	LZ_bwd16	LZ_bwd32
chr19.dna4	6.0	4.7	2.7	1.7	1.2	1.0	0.8	161.9	75.2	76.5	78.9	83.3	92.1	110.5
chr22.dna4	6.0	4.7	2.7	1.7	1.2	1.0	0.8	78.8	39.0	39.5	40.3	43.0	48.3	58.5
E.coli	6.0	4.7	2.7	1.7	1.2	1.0	0.8	7.6	3.9	4.0	4.3	4.5	5.0	6.2
bible	6.0	5.1	3.1	2.1	1.6	1.3	1.2	3.9	4.6	4.8	5.2	5.6	6.5	8.5
howto	6.0	5.1	3.1	2.1	1.6	1.4	1.2	54.2	60.1	61.7	65.2	72.1	84.9	110.1
fib_s14930352	6.0	4.6	2.6	1.6	1.1	0.8	0.7	2.1	14.6	15.0	14.7	14.7	14.6	14.7
fib_s9227465	6.0	4.6	2.6	1.6	1.1	0.8	0.7	1.3	8.9	9.1	8.9	8.9	9.1	8.9
fss10	6.0	4.6	2.6	1.6	1.1	0.8	0.7	1.8	11.5	11.6	11.6	11.6	11.5	11.6
fss9	6.0	4.6	2.6	1.6	1.1	0.9	0.7	0.4	2.0	2.0	2.0	2.0	2.0	2.0
p16Mb	6.0	5.0	3.0	2.0	1.5	1.3	1.2	33.7	24.8	26.3	28.5	33.8	43.7	63.9
p32Mb	6.0	5.0	3.0	2.0	1.5	1.3	1.1	73.7	52.5	54.5	59.7	69.3	88.1	126.7
rndA21_8Mb	6.0	5.1	3.1	2.1	1.6	1.3	1.2	17.6	13.0	13.9	15.6	19.0	26.2	40.6
rndA2_8Mb	6.0	4.6	2.6	1.6	1.1	0.8	0.7	11.7	6.8	6.8	6.9	7.2	7.5	8.4

compared it with CPS2 [4], one of the most space-efficient LZ-factorization algorithm that is currently available. (As a matter of fact, the online algorithm from [18] uses even less space than CPS2, but it is not competitive in terms of time; see the experimental results in [18,2].) Our results can be found in Table 2. Note that the construction time of the indexes is not included. Moreover, we stress that the space usage of our algorithm is measured after the construction of the BWT, the compressed suffix array, and the balanced parentheses sequence. This was done to show the effect of using compressed suffix arrays. However, the memory peak during the construction has to be taken into account as well. In our implementation, we first construct the suffix array and then derive the above-mentioned data structures from it. Currently, all known (in-memory) suffix array construction algorithms require at least $5n$ bytes; see [20,16]. Consequently, the memory peak of Algorithm 4 is $5n$ bytes. It goes without saying that our algorithm will benefit from further improvements in suffix array construction.

Algorithm 4 always uses less space than CPS2. It is faster than CPS2 for small alphabets ($|\Sigma| \leq 20$) except for some highly repetitive strings (Fibonacci strings and run rich strings). In these exceptional cases, CPS2 can avoid nearly all costly range minimum queries and therefore beats our algorithm. For larger alphabets (e.g. $|\Sigma| = 256$), Algorithm 4 is slightly slower than CPS2. This behavior can be attributed to the $\log|\Sigma|$ factor in its worst-case time complexity $\mathcal{O}(n \log |\Sigma|)$. However, our new algorithm has an advantage over CPS2. If space is extremely tight, we can use a compressed suffix array instead of the plain suffix array. Of course, if there are many factors in the LZ-factorization, this will slow down the computation (usual space-time tradeoff). But highly repetitive texts have few factors and thus the use of a compressed suffix array has little influence on the factorization time.

Future work: We would like to point out that the ideas described in this paper may help to improve the other algorithms.

Acknowledgment. We thank German Tischler for sharing his implementation of the algorithms from [6] with us, and Simon J. Puglisi for his implementation of the CPS2 algorithm (`http://goanna.cs.rmit.edu.au/~sjp/lz6n.tar.gz`). Special thanks go to an anonymous reviewer, who brought [15,18] to our attention.

References

1. Abouelhoda, M.I., Kurtz, S., Ohlebusch, E.: Replacing suffix trees with enhanced suffix arrays. Journal of Discrete Algorithms 2, 53–86 (2004)
2. Al-Hafeedh, A., Crochemore, M., Ilie, L., Kopylov, J., Smyth, W.F., Tischler, G., Yusufu, M.: A comparison of index-based Lempel-Ziv LZ77 factorization algorithms (2011) (submitted)
3. Burrows, M., Wheeler, D.J.: A block-sorting lossless data compression algorithm. Research Report 124, Digital Systems Research Center (1994)
4. Chen, G., Puglisi, S.J., Smyth, W.F.: Lempel-Ziv factorization using less time & space. Mathematics in Computer Science 1(4), 605–623 (2008)

5. Crochemore, M., Ilie, L.: Computing longest previous factor in linear time and applications. Information Processing Letters 106(2), 75–80 (2008)
6. Crochemore, M., Ilie, L., Iliopoulos, C.S., Kubica, M., Rytter, W., Waleń, T.: LPF computation revisited. In: Fiala, J., Kratochvíl, J., Miller, M. (eds.) IWOCA 2009. LNCS, vol. 5874, pp. 158–169. Springer, Heidelberg (2009)
7. Crochemore, M., Ilie, L., Smyth, W.F.: A simple algorithm for computing the Lempel-Ziv factorization. In: Proc. 18th Data Compression Conference, pp. 482–488. IEEE Computer Society, Los Alamitos (2008)
8. Ferragina, P., Manzini, G.: Opportunistic data structures with applications. In: Proc. IEEE Symposium on Foundations of Computer Science, pp. 390–398 (2000)
9. Gog, S., Fischer, J.: Advantages of shared data structures for sequences of balanced parentheses. In: Proc. 20th Data Compression Conference, pp. 406–415. IEEE Computer Society, Los Alamitos (2010)
10. Grossi, R., Gupta, A., Vitter, J.S.: High-order entropy-compressed text indexes. In: Proc. 14th Annual ACM-SIAM Symposium on Discrete Algorithms, pp. 841–850 (2003)
11. Gusfield, D., Stoye, J.: Linear time algorithms for finding and representing all the tandem repeats in a string. Journal of Computer and System Sciences 69(4), 525–546 (2004)
12. Kärkkäinen, J., Manzini, G., Puglisi, S.J.: Permuted longest-common-prefix array. In: Kucherov, G., Ukkonen, E. (eds.) CPM 2009 Lille. LNCS, vol. 5577, pp. 181–192. Springer, Heidelberg (2009)
13. Kasai, T., Lee, G.H., Arimura, H., Arikawa, S., Park, K.: Linear-time longest-common-prefix computation in suffix arrays and its applications. In: Amir, A., Landau, G.M. (eds.) CPM 2001. LNCS, vol. 2089, pp. 181–192. Springer, Heidelberg (2001)
14. Kolpakov, R., Kucherov, G.: Finding maximal repetitions in a word in linear time. In: Proc. 40th Annual Symposium on Foundations of Computer Science, pp. 596–604. IEEE Computer Society, Los Alamitos (1999)
15. Kreft, S., Navarro, G.: LZ77-like compression with fast random access. In: Proc. 20th Data Compression Conference, pp. 239–248. IEEE Computer Society, Los Alamitos (2010)
16. Nong, G., Zhang, S., Chan, W.H.: Linear suffix array construction by almost pure induced-sorting. In: Proc. Data Compression Conference, pp. 193–202. IEEE Computer Society, Los Alamitos (2009)
17. Ohlebusch, E., Fischer, J., Gog, S.: CST++. In: Chavez, E., Lonardi, S. (eds.) SPIRE 2010. LNCS, vol. 6393, pp. 322–333. Springer, Heidelberg (2010)
18. Okanohara, D., Sadakane, K.: An online algorithm for finding the longest previous factors. In: Halperin, D., Mehlhorn, K. (eds.) ESA 2008. LNCS, vol. 5193, pp. 696–707. Springer, Heidelberg (2008)
19. Pokrzywa, R., Polanski, A.: BWtrs: A tool for searching for tandem repeats in DNA sequences based on the Burrows-Wheeler transform. Genomics 96(5), 316–321 (2010)
20. Puglisi, S.J., Smyth, W.F., Turpin, A.: A taxonomy of suffix array construction algorithms. ACM Computing Surveys 39(2), 1–31 (2007)
21. Rodeh, M., Pratt, V.R., Even, S.: A linear time algorithm for data compression via string matching. Journal of the ACM 28, 16–24 (1981)

Succincter Text Indexing with Wildcards

Chris Thachuk

Department of Computer Science, University of British Columbia, Vancouver, Canada
cthachuk@cs.ubc.ca

Abstract. We study the problem of indexing text with wildcard positions, motivated by the challenge of aligning sequencing data to large genomes that contain millions of single nucleotide polymorphisms (SNPs) —positions known to differ between individuals. SNPs modeled as wildcards can lead to more informed and biologically relevant alignments. We improve the space complexity of previous approaches by giving a succinct index requiring $(2 + o(1))n \log \sigma + O(n) + O(d \log n) + O(k \log k)$ bits for a text of length n over an alphabet of size σ containing d groups of k wildcards. The new index is particularly favourable for larger alphabets and comparable for smaller alphabets, such as DNA. A key to the space reduction is a result we give showing how any compressed suffix array can be supplemented with auxiliary data structures occupying $O(n) + O(d \log \frac{n}{d})$ bits to also support efficient dictionary matching queries. We present a new query algorithm for our wildcard index that greatly reduces the query working space to $O(dm+m \log n)$ bits, where m is the length of the query. We note that compared to previous results this reduces the working space by two orders of magnitude when aligning short read data to the Human genome.

1   Introduction

The study of strings, their properties, and associated algorithms has played a key role in advancing our understanding of problems in areas such as compression, text mining, information retrieval, and pattern matching, amongst numerous others. A most basic and widely studied question in stringology asks: given a string T (the text) how many occurrences and in what positions does it contain a string P (the pattern) as a substring? It is well known that this problem can be solved in time proportional to the lengths of both strings [12]. However, it is often the case that we wish to repeat this question for many different pattern strings and a fixed text T of length n over an alphabet of size σ. The idea is to create a full-text index for T so that repeated queries can be answered in time proportional to the length of P alone. It was first shown by Weiner [22] in 1973 that the suffix tree data structure could be built in linear time for exactly this purpose. The ensuing years have seen the versatility of the suffix tree as it has been demonstrated to solve numerous other related problems.

While suffix trees use $O(n)$ words of space in theory, this does not translate to a space efficient data structure in practice. For this reason, Manber and Myers [15] proposed the suffix array data structure (see Figure 1). Though a great practical improvement over suffix trees, the $\Omega(n \log n)$ bit space requirement is often prohibitive for larger texts. Building in part on the pioneering work of Jacobson [11] into succinct data structures,

R. Giancarlo and G. Manzini (Eds.): CPM 2011, LNCS 6661, pp. 27–40, 2011.
© Springer-Verlag Berlin Heidelberg 2011

two seminal papers helped usher in the study of so-called succinct full-text indexes. Grossi and Vitter [8] proposed a compressed suffix array that occupies $O(n \log \sigma)$ bits; the same space required to represent the original string T. Soon after, Ferragina and Manzini [5] proposed the FM-index, a type of compressed suffix array that can be inferred from the Burrows-Wheeler transform of the text and some auxiliary structures, leading to a space occupancy proportional to $nH_k(T)$ bits, where $H_k(T)$ denotes the kth order empirical entropy of T. These and subsequent results have made it possible to answer efficiently the substring question on texts as large, or larger, than the Human genome.

We are interested in designing a succinct index to answer a generalized version of the substring question where the text T contains k wildcard positions that can match any character of a pattern. Our motivation arises in the context of aligning short read data, produced by second generation sequencing technology. Typically short reads are aligned against a so-called reference genome; however, the quantity of positions known to differ between individuals due to single nucleotide polymorphisms (SNPs) numbers in the millions [7]. Modeling SNPs as wildcards would yield more informed, and by extension, more accurate alignment of short reads.

Cole, Gottlieb & Lewenstein [4] were among the first to study the problem of indexing text sequences containing wildcards and proposed an index using $O(n \log^k n)$ words of space capable of answering queries in $O(m + \log^k n \log \log n + occ)$ time, where occ denotes the number of matching positions. This result was later improved by Lam $et\ al.$, [13] resulting in space usage of only $O(n)$ words and a query time no longer exponential in k. A key idea in their work was to build a type of dictionary of the text segments of $T = T_1 \phi^{k_1} T_2 \phi^{k_2} \ldots \phi^{k_d} T_{d+1}$ where each text segment T_i contains no wildcards and ϕ^{k_i} denotes the ith $wildcard\ group$ of size $k_i \geq 1$, for $1 \leq i \leq d \leq k$. The query time includes the term $\gamma = \sum_{i,j} \mathsf{prefix}(P[i..|P|], T_j)$ where $\mathsf{prefix}(P[i..|P|], T_j) = 1$ if T_j is a prefix of $P[i..|P|]$ and 0 otherwise. The authors also give a more detailed bound on γ based on prefix complexity.

Despite this improvement, $O(n)$ words of space is prohibitive for texts as large as the Human genome. Support for dictionary matching of text segments was also crucial in the approach of Tam $et\ al.$, [21] who proposed the first, and to our knowledge only, succinct index. They designed a dictionary structure using $(2 + o(1))n \log \sigma$ bits, based on a compressed suffix array, which therefore occupies most of the space required by their overall index. Very recently, Belazzougui [1] proposed a succincter dictionary based on the Aho-Corasick automaton having optimal query time. The compressed space occupancy was further improved by a slight modification given by Hon $et\ al.$, [10]. While these results are impressive, the wildcard matching problem benefits from an index that can report the text segments contained in P (dictionary problem), as well as the text segments which are prefixed by P and also fully contain P (substring problem). To draw a distinction, we will refer to this latter type as a $full\text{-}text\ dictionary$. In our first main contribution we show how a full-text dictionary can be built on top of any compressed suffix array using an additional $O(n) + O(d \log \frac{n}{d})$ bits of space and, in turn, how it can be used to provide a succincter index for texts containing wildcards. Our dictionary does not require any modification of the original string T and can therefore be used in applications benefiting from bidirectional search—a technique where both a forward and reverse index of T are cooperatively searched—which has been shown to significantly speed up the alignment of sequencing data [14].

In our view, the main challenge that must be overcome for successful wildcard matching is a reduction of the query working space. The fastest solution of Tam *et al.*, [21], matches our query time, if modified to use the same orthogonal range query structure we use, but requires a query working space of $O(n \log d + m \log n)$ bits. Acknowledging that the first term is impractical for large texts, they give a slower solution that reduces the working space to be proportional to the index itself. This makes the solution feasible, but constraining considering the fact that p parallel queries necessarily increase the working space by a factor of p. In our second main contribution we give an algorithm that reduces the query working complexity significantly to $O(dm + m \log n)$ bits. In particular, this is an improvement for any query length $m < \frac{n}{d} \log \sigma$. For our motivating problem, alignment of short reads (32-64 bases) to the Human genome (3 billion bases with 1-2 million SNPs), this reduces the working space by two orders of magnitude from gigabytes to tens of megabytes. Our result for indexing text with wildcards is summarized and compared with existing results in Table 1. Note that some proofs have been omitted from our results due to space constraints.

Table 1. A comparison of text indexes supporting wildcard characters. k, d, $\hat{d}$ is the # of wildcards, wildcard groups, and distinct wildcard group lengths, respectively; occ_1, occ_2, occ is the # of occurrences containing no wildcard group, 1 wildcard group, and overall, respectively; $\gamma = \sum_{i,j} \mathsf{prefix}(P[i..|P|], T_j)$, † = our result

Index Space		Query Time	Query Working Space	
$O(n \log^k n)$	words	$O(m + \log^k n \log \log n + occ)$	-	[4]
$O(n)$	words	$O(m \log n + \gamma + occ)$	$O(n)$	words [13]
$(3 + o(1)) n \log \sigma$ $+ O(d \log n)$ bits		$O\left(\begin{array}{l} m \left(\log \sigma + \min\left(m, \hat{d}\right) \log d \right) \\ + occ_1 \log n + occ_2 \log d + \gamma \end{array} \right)$	$O(n \log d + m \log n)$	bits [21]
		same as above with working space reduced by increasing query time		
$(3 + o(1)) n \log \sigma$ $+ O(d \log n)$ bits		$O\left(\begin{array}{l} m \left(\log \sigma + \min\left(m, \hat{d}\right) \log d \right) \\ + occ_1 \log n + occ_2 \log d + \gamma \log_\sigma d \end{array} \right)$	$O(n \log \sigma + m \log n)$	bits [21]
$(2 + o(1)) n \log \sigma$ $+ O(n) + O(d \log n)$ $+ O(k \log k)$ bits		$O\left(\begin{array}{l} m \left(\log \sigma + \min\left(m, \hat{d}\right) \frac{\log k}{\log \log k} \right) \\ + occ_1 \log n + occ_2 \frac{\log k}{\log \log k} + \gamma \end{array} \right)$	$O(dm + m \log n)$	bits †

2 Preliminaries

Let $T[1, n]$ be a string over a finite alphabet Σ of size σ. We denote its jth character by $T[j]$ and a substring from the ith to the jth position by $T[i..j]$. We assume that an end-of-text sentinel character $\$ \notin \Sigma$ has been appended to T ($T[n] = \$$) and $\$$ is lexicographically smaller than any character in Σ. For any substring X we use $|X|$ to denote its length and $\overline{X}$ to denote its reverse sequence. The suffix array SA of T is a permutation of the integers $[1, n]$ giving the increasing lexicographical order of the suffixes of T. Conceptually SA can be thought of as a matrix of all suffixes of T that have been sorted lexicographically and where $\mathsf{SA}[i] = j$ means that the ith lexicographically smallest suffix of T begins at position j.

A string X has a suffix array (SA) range $[a, b]$ with respect to SA if $a - 1$ $(n - b)$ suffixes of T are lexicographically smaller (larger) than X. If $a > b$ the range is said to be empty and X does not exist as a substring of T; otherwise, X occurs as a prefix of the $b - a + 1$ suffixes of T denoted by its range. The SA range for X can be found in a compressed suffix array by backward search using the LF-mapping which relates SA to T^{BWT}, the Burrows-Wheeler transform of T. T^{BWT} is also a string of length n where $T^{\text{BWT}}[i] = T[\text{SA}[i] - 1]$, if $\text{SA}[i] \neq 1$, and $T^{\text{BWT}}[i] = \$$ otherwise. See Figure 1 for an example. For details of backward search, the LF-mapping, existing implementations, and related topics we refer the reader to the excellent review by Navarro and Mäkinen [17]. In this work, we assume the availability of a compressed suffix array— such as the wavelet tree implementation of Grossi *et al.*, [9] —meeting the following space and time requirements, of which there are many (*cf.* [17]).

Lemma 1. *A compressed suffix array* SA *for T can be represented in $(1 + o(1))n \log \sigma$ bits of space, such that the suffix array range of every suffix of a string X can be computed in $O(|X| \log \sigma)$ time, and each match of X in T can be reported in an additional $O(\log n)$ time.*

In our dictionary construction, we also make use of the following well known data structures.

Lemma 2 (Raman et al., [20]). *A bit vector* B *of length n containing d 1 bits can be represented in $d \log \frac{n}{d} + O(d + n \frac{\log \log n}{\log n})$ bits to support the operations $\text{rank}_1(\text{B}, i)$ giving the number of 1 bits appearing in $\text{B}[1..i]$ and $\text{select}_1(\text{B}, i)$ giving the position of the ith 1 in B in $O(1)$ time.*

Lemma 3 (Grossi & Vitter [8]). *An array* L *of d integers where $\sum_{i=1}^{d} \text{L}[i] = n$ can be represented in $d(\lceil \lg(n/d) \rceil + 2 + o(1))$ bits to support $O(1)$ time access to any element.*

Lemma 4 (Munro & Raman [16]). *A sequence* BP *of d balanced parentheses can be represented in $(2 + o(1))d$ bits of space to support the following operations in $O(1)$ time:* $\text{rank}_{(}(\text{BP}, i)$, $\text{select}_{(}(\text{BP}, i)$, *and similarly for right parentheses, as well as:*

- $\text{findclose}(\text{BP}, l)$ *($\text{findopen}(\text{BP}, r)$): a index of matching right (left) parenthesis for left (right) parenthesis at position l (r)*
- $\text{enclose}(\text{BP}, i)$: *indexes (l, r) of closest matching pair to enclose $(i, \text{findclose}(\text{BP}, i))$ if such a pair exists and is undefined otherwise*

The matching statistics for a string X with respect to SA is an array ms of tuples such that $ms[i] = (q, [a, b])$ states that the longest prefix of $X[i..|X|]$ that matches anywhere in T has length q and suffix array range $[a, b]$. Very recently Ohlebusch *et al.*, [19] showed matching statistics can be efficiently computed with backward search if SA is *enhanced* with auxiliary data structures using $O(n)$ bits to represent so-called longest common prefix (lcp) intervals (*cf.* [19]). We leverage this result in the design of our succinct full-text dictionary and its search algorithm.

Lemma 5 (Ohlebusch et al., [19]). *The matching statistics of a pattern X with respect to text T over an alphabet of size σ can be computed in $O(|X| \log \sigma)$ time given a compressed enhanced suffix array of T.*

Finally, our wildcard matching algorithm makes use of an orthogonal range query data structure.

Lemma 6 (Bose *et al.*, [2]). *A set N of points from universe $M = [1..k] \times [1..k]$, where $k = |N|$, can be represented in $(1 + o(1))k \log k$ bits to support orthogonal range reporting in $O(occ \frac{\log k}{\log \log k})$ time, where occ is the size of the output.*

3 A Succinct Full-Text Dictionary

In the dictionary problem we are required to index a set of d text segments[1] $\mathcal{D} = \{T_1, T_2, \ldots, T_d\}$ so that we can match efficiently, in any input string P, all occurrences of text segments belonging to $\mathcal{D}$. We present a succinct *full-text dictionary* index that is also capable of efficiently identifying all text segments that contain P as a prefix, or more generally as a substring. We demonstrate the use of this additional functionality in our solution for wildcard matching.

3.1 A Compressed Suffix Array Representation of Text Segments

Let $T = \phi T_1 \phi T_2 \phi T_3 \phi \ldots \phi T_d \$$ be the concatenation of all d text segments, each prefixed by the character ϕ, followed by the traditional end-of-text sentinel $\$$, having total length n. Note that n is necessarily larger than the total number of characters in the dictionary. We define ϕ to be lexicographically smaller than any $c \in \Sigma$ and $\$$ to be lexicographically smaller than ϕ. We first build SA, the compressed suffix array for T. Consider any text segment $T_j \in \mathcal{D}$. There will be a contiguous range $[c, d]$ of suffixes in SA that are prefixed by the string T_j. Lemma 7 summarizes how we can use the SA range of T_j and its length to determine if it is a prefix of a given text P (and vice versa). Note that verifying the length condition is necessary in the case where P is a proper prefix of T_j and they share a common SA range.

Lemma 7. *Let SA be the compressed suffix array for T and let $[a, b]$ and $[c, d]$ be the non-empty suffix array ranges in SA for a string P and a text segment T_j respectively. Then T_j is a prefix of P if and only if $c \leq a \leq b \leq d$ and $|P| \geq |T_j|$. Similarly, P is a prefix of T_j if and only if $a \leq c \leq d \leq b$.*

3.2 Storing Text Segment Lengths

For Lemma 7 to apply, we must know both the SA range of a given text segment and also its length. By Lemma 3 we can store the lengths of all d text segments in a compressed integer array L using $d(\lceil \log(n/d) \rceil + 2 + o(1))$ bits ensuring constant time access. We store the lengths in L relative to the lexicographical order of text segments.

3.3 The Text Segment Interval Tree

The SA range of one text segment T_i will enclose the SA range of another T_j if T_i is a prefix of T_j. For instance, in the example of Figure 1 the text segment aca has SA

---

[1] To remain consistent with the section that follows, we refer to dictionary entries (patterns) as text segments.

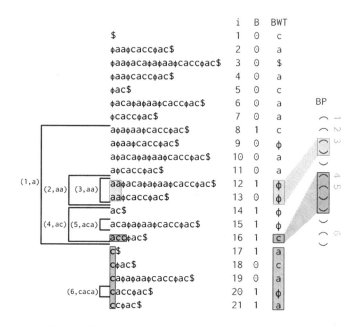

Fig. 1. A succinct full-text dictionary for the set of text segments {aa, aca, a, aa, cacc, ac}. Shown are the sorted suffixes of the string $T = \phi aa\phi aca\phi a\phi aa\phi cacc\phi ac\$$ representing the text segments. Text segment intervals are demarcated on the left. Three different queries (shaded intervals) are shown with their corresponding smallest enclosing text segment interval (if any).

range $[15, 15]$ and is enclosed by the SA range of the text segment ac ($[14, 16]$) and by the text segment a ($[8, 16]$). In general, it is also possible that many text segments begin at the same position, provided that they are different occurrences of the same string (*e.g.*, aa). This is by design since each text segment is followed by a character not found in Σ (either ϕ or $\$$). However, our construction requires us to distinguish between different occurrences of the same text segment string and we therefore introduce the concept of *text segment intervals*. When $t > 1$ text segments in the dictionary share a common SA range, we say that the text segment interval of occurrence a encloses the text segment interval of occurrence b, $1 \le a \ne b \le t$, if the suffix of T beginning with occurrence a is lexicographically smaller than the suffix beginning with occurrence b. In this way we are able to define a total order on all d text segment intervals based on their relative lexicographical order in SA. We assign *lex_ids*, a unique identifier for each text segment, based on this lexicographical order. Consider again the example in Figure 1. The text segment aa occurs as a prefix of $T[2..n]$ and $T[11..n]$. Since the suffix $T[2..n]$ is lexicographically smaller than $T[10..n]$, we say that the occurrence prefixing $T[2..n]$ encloses the other. Consequently, the text segment prefixing $T[2..n]$ ($T[11..n]$) is assigned *lex_id* 2 (3). We will refer to text segments or text segment intervals interchangeably.

In general the text segment intervals form a set of nested, non-crossing intervals (an interval tree) and can be represented by a sequence BP of d balanced parentheses; one pair for each text segment (see Figure 1). Conceptually, if we can identify the text segment interval having the largest *lex_id* that is a prefix of P, referred to as the *smallest enclosing text segment interval* of P, then we can immediately conclude that P is also prefixed by all intervals which enclose it.

Lemma 8. *Given the index pair* (l, r) *in* BP *corresponding to the smallest enclosing text segment interval for a string* P *the occ number of text segments that are prefixes of* P *can be counted in* $O(1)$ *time and reported in an additional* $O(occ)$ *time.*

3.4   Finding the Smallest Enclosing Text Segment Interval

We now describe how the smallest enclosing text segment interval can be determined given any non-empty SA range $[a, b]$ in SA for P. We wish to determine the pair (l, r) of indexes for the left and right parentheses in BP corresponding to this interval (or an undefined index range if P is not prefixed by any text segment). Unfortunately, we cannot directly infer where text segment intervals begin and end based on T^{BWT} alone. Therefore, we make use of a bit vector B of length n and set $B[k] = 1$ if and only if one or more text segment intervals begin at position k, or end at position $k - 1$. For the range $[a, b]$, end cases occur when $B[k] = 0$, for all k, where $a < k \leq n$ (all text segment intervals end before position a) or when $B[k] = 0$, for all k, where $1 \leq k \leq a$ (all text segment intervals begin after position a). Suppose otherwise and let $c = \arg\max_{1 \leq j \leq a}\{B[j] = 1\}$ and $d = \arg\min_{a < j \leq n}\{B[j] = 1\}$. Note that position c marks the largest position (up to a) when one or more text segment intervals begin or end (at $c - 1$). Our algorithm considers two main cases: either $B[c]$ marks the beginning of one or more intervals, or it only marks the ends of intervals.

Lemma 9. *Given two positions* c *and* d *of* B, *where* $c < d$, $B[c] = B[d] = 1$ *and* $B[k] = 0$, *for all* k, *where* $c < k < d$, *then* $B[c]$ *marks the beginning of* t *text segment intervals if and only if* $T^{\text{BWT}}[c..d - 1]$ *contains* t *occurrences of the character* ϕ.

Using Lemma 9 we are able to distinguish between the two main cases. If $B[c]$ marks the beginning of one or more text segment intervals, then T_j — the text segment interval with the largest *lex_id* beginning at position c — is the smallest enclosing text segment interval, provided $|T_j| \leq |P|$ (by condition of Lemma 7). If $|T_j| \leq |P|$, we can determine the largest *lex_id* beginning at position c by simply counting the occurrences of the character ϕ prior to position d in T^{BWT}. For example, consider the SA range $[12, 13]$ in Figure 1. We have $c = 12$ and $d = 14$. We determine the *lex_id* to be 3 by counting the number of ϕ up to position $d = 14$ in T^{BWT}. Conveniently and by construction, the *lex_id* corresponds to the rank of the left parenthesis denoting T_j in BP. It is worth noting that when $|T_j| > |P|$ special care is required to find the smallest enclosing text segment interval in worst case constant time. Details are given in the proof of Lemma 10, but the idea is to find the enclosing interval (if any) of the text segment interval having the *smallest lex_id* beginning at position c.

On the other hand, if $B[c]$ only marks the end of one or more text segment intervals (*e.g.*, SA range $[16, 16]$ in Figure 1) we can instead identify the right index for $T_{j'}$—the last text segment interval to end at position $c - 1$. The smallest enclosing text segment interval, if any, is therefore the one enclosing $T_{j'}$. Unfortunately, in this case we cannot infer how many intervals close prior to position c directly from T^{BWT}. For this reason,

---

Algorithm 1. Find smallest enclosing text segment interval

Input: a specifies the beginning of the non-empty suffix array interval for string P

Output: l, r where l (r) is the index of the left (right) parenthesis in BP corresponding to the smallest enclosing text segment interval of P if it exists, and an undefined interval otherwise

```
 1: c ← select₁(B, rank₁(B, a))
 2: d ← select₁(B, rank₁(B, a) + 1)
 3: if c or d is undefined then                    // handle end cases
 4:     return an undefined interval
 5: lex_id ← rank_φ(T^BWT, d − 1)
 6: if lex_id > rank_φ(T^BWT, c) then              // B[c] marks beginning of t.s. interval(s)
 7:     if L[lex_id] > |P| then
 8:         lex_id ← rank_φ(T^BWT, c − 1) + 1
 9:         l ← select_((BP, lex_id)
10:         l, r ← enclose(BP, l)
11:     else
12:         l ← select_((BP, lex_id)
13:         r ← findclose(BP, l)
14: else                                           // B[c] marks end of t.s. interval(s)
15:     r ← select_)(BP, R[rank₁(B, c)])
16:     l ← findopen(BP, r)
17:     l, r ← enclose(BP, l)
18: return l, r
```

---

we will employ another compressed integer array R to record the count of intervals that close prior to position k, for all $B[k] = 1$. We determine the appropriate index for R by simply counting the number of 1's up to position c in B. The corresponding entry in R gives us the rank of the right parenthesis for the last interval to close prior to position c, from which we can find the enclosing interval (if any). The entire procedure, including end cases, is summarized in Algorithm 1 and shown to be correct in Lemma 10.

Lemma 10. *Let* SA *be the compressed suffix array for* T *and let* $[a, b]$ *be the non-empty suffix array range in* SA *for a string* P. *In* $O(1)$ *time, Algorithm 1 either correctly identifies the indexes in* BP *corresponding to the smallest enclosing text segment interval of* P *if one exists, or it returns an undefined interval when it does not.*

3.5 The Overall Dictionary and Its Full-Text Capabilities

We have shown how all text segments occurring as a prefix of a string P having a non-empty SA range in SA can be reported efficiently. By enhancing SA with lcp-interval information using $O(n)$ bits, we can find the matching statistics for P in order to repeat the previous procedure for $1 \leq i \leq |P|$ (see Lemma 5). Importantly for our results on wildcard matching, we note that with a very minor modification, this same construction works when text segments are separated by more than one ϕ character and also when the first text segment is not preceded by a ϕ character. Note that the text segment interval tree can be built in a similar manner as an lcp-interval tree. Details are left for the full version. We have the following result.

Theorem 1. *Given a set of d text segments over an alphabet of size σ we can construct a succinct full-text dictionary, based on an enhanced compressed suffix array, using at most $(1 + o(1))n \log \sigma + O(n) + O(d \log \frac{n}{d})$ bits where n is the length of T, the text representation of the dictionary including ϕ characters, such that the occ_1 text segments contained in a string P can be counted in $O(|P| \log \sigma)$ time and reported in an additional $O(occ_1)$ time. Furthermore, all occ_2 text segments prefixed by P can be reported in $O(|P| \log \sigma + occ_2)$ time, and all occ_3 locations in T where P occurs as a substring can be reported in $O(|P| \log \sigma + occ_3 \log n)$ time.*

4 Matching Wildcards in Succinct Texts

Let T be a string over an alphabet $\Sigma \cup \{\phi\}$ of size σ where $\phi \notin \Sigma$ and $T[i] = \phi$ if and only if position i is a wildcard position in T. In particular, we denote the structure of the input string as $T = T_1 \phi^{k_1} T_2 \phi^{k_2} \ldots \phi^{k_d} T_{d+1}$ where each text segment T_i contains no wildcards and ϕ^{k_i} denotes the ith *wildcard group* of size $k_i \geq 1$, for $1 \leq i \leq d$. Our goal is to create an index for the purpose of identifying all the locations in T that exactly match any query pattern P, modulo wildcard positions. Similar to previous approaches [13,21], we classify the match into one of three cases: P contains no wildcard group (Type 1), P contains exactly one wildcard group (Type 2), and P contains more than one wildcard group (Type 3). Our solution for Type 2 matching is largely inspired by previous approaches [13,21], but differs in the details. Our algorithm for Type 3 matching is novel and can result in significantly reduced working space.

4.1 Overall Design of the Index

We first build the succinct full-text dictionary of Section 3. By design, the dictionary reports the match of a text segment T_j based on its lexicographical order (its *lex_id*) relative to other text segments; however, in the wildcard problem we are required to report the match based on T_j's position in T. Therefore, we store a permutation Π mapping the *lex_id*s of text segments to their relative position order in T. For instance, if T_j has *lex_id* k, then $\Pi[k] = j$. We find it convenient to store the following information for each text segment, in auxiliary arrays, indexed by this relative position order: length, SA range in SA (referenced as RSA), beginning position in T, and the size of the preceding wildcard group. Note that array L of the dictionary construction can be adapted to store lengths in this relative order with the use of Π. We also construct a compressed suffix array $\overline{SA}$ for $\overline{T}$, the reverse of T, and store the SA range of each $\overline{T_j}$ with respect to $\overline{SA}$ (referenced as $\overline{RSA}$). Note that $\overline{SA}$ does not need to support location reporting. We use simple arrays to store SA ranges resulting in $O(d \log n)$ bits combined space usage to store auxiliary information supporting constant time access. To support Type 2 matching we employ a range query data structure occupying $(1 + o(1))k \log k$ bits (see next section).

Lemma 11. *Given a text T of length n containing d groups of k wildcards the combined space required of the above indexes is $(2 + o(1))n \log \sigma + O(n) + O(d \log n) + O(k \log k)$ bits.*

All three matching types make use of the matching statistics of P with respect to SA. Types 2 and 3 matching also make use of the SA ranges of $\overline{P}$ with respect to $\overline{\text{SA}}$. Both can be computed in $O(m \log \sigma)$ time (by Lemmas 1 and 5) and require $O(m \log n)$ bits to store. We incorporate these times and working spaces into the results for each type. Type 1 matching is handled by the application of Lemma 1.

4.2   Type 2 Matching

A Type 2 match occurs when the alignment of P to T contains exactly (a portion of) one wildcard group. Specifically, we seek a pair of neighbouring text segments T_j and T_{j+1}, separated by a wildcard group of size k_j, where $P[i..|P|]$ aligns to the first $|P| - i + 1$ characters of T_{j+1} — referred to as the *suffix match* (of P) — and $P[1..i - 1 - k_j]$ aligns to the last $i - 1 - k_j$ characters of T_j — referred to as the *prefix match*. Let α_j (ω_j) be the the first (last) ϕ character of the jth wildcard group in T. End cases occur when the match begins or ends in $T[\alpha'_j..\omega'_j]$, where α'_j (ω'_j) is the position of α_j (ω_j) in T. For now, suppose this is not the case. For a fixed suffix $P[i..|P|]$ and wildcard group length k_j our strategy will be to (i) find all potential suffix matches, (ii) record the *lex_id* of the candidate text segments, (iii) find all potential prefix matches, and (iv) determine which candidate prefix matches are compatible with a *lex_id* recorded in step (ii).

$$\cdots \underline{\quad\quad} T_j \underline{\quad\quad} \dashv \phi \cdots \phi \vdash \underline{\quad\quad} T_{j+1} \underline{\quad\quad} \cdots$$
$$\quad\quad\quad\quad\quad\quad\quad \uparrow \quad\quad \uparrow$$
$$\quad\quad\quad\quad\quad\quad\quad \alpha_j \quad\quad \omega_j$$

Lemma 12. *Given a non-empty SA range $[a, b]$ in* SA *for a string X, the lex_ids (based on their lexicographical order) of text segments in T that contain X as a prefix will form a contiguous (possibly empty) range $[id_1, id_2]$ that can be reported in $O(1)$ time.*

By Lemma 12, we can identify the range $[id_1, id_2]$ of *lex_id*s corresponding to text segments that $P[i..|P|]$ is a prefix of in constant time using its stored SA range with respect to SA, completing steps (i)-(ii). Determining a range $[id_3, id_4]$ of *lex_id*s corresponding to text segments that $P[1..i - k_j - 1]$ is a suffix of is equivalent to determining all $\overline{T_t}$ that contain $\overline{P[1..i - k_j - 1]}$ as a prefix. Again, using a stored SA range with respect to $\overline{\text{SA}}$ this can be determined in constant time, completing step (iii). Now consider that the *lex_id* with respect to SA of a text segment T_{j+1} is relative to the rank of ω_j in T^{BWT}, the character which precedes it. Similarly, the relative rank of α_j in $\overline{T^{\text{BWT}}}$ determines the *lex_id* of $\overline{T_j}$, but in this case relative to $\overline{T}$. We make use of a permutation H to relate these *lex_id*s (α and ω values). Specifically, we set $\text{H}[\alpha_j] = \omega_j$, for $1 \leq j \leq k$. Therefore, we need to determine the entries in $\text{H}[id_3..id_4]$ that have a value in the range $[id_1, id_2]$. This is an orthogonal range query and by Lemma 6, H can be represented in $(1 + o(1))k \log k$ bits to report all *occ* matches in $O(occ \frac{\log k}{\log \log k})$ time. Once a *lex_id* ω_j has been verified, a match position can be reported in $O(1)$ time as the location of T_{j+1} with respect to T is known in addition to the length of the prefix match. This completes step (iv).

In general, we can repeat the above procedure for every combination of suffix length and wildcard group length bound by m. However, as pointed out by Tam *et al.*, [21] the number of distinct wildcard group sizes $\hat{d}$ is often a small constant, particularly in genomic sequences. We therefore only consider at most $\hat{d}$ lengths, provided they are not larger than m.

Now, consider the case when $P[i..|P|]$ aligns to a prefix of a wildcard group. To contain $P[i..|P|]$ as a prefix, the wildcard group must have a length $l \geq |P| - i + 1$. Let a be the first entry in SA denoting a suffix of T prefixed by at least $l - 1$ ϕ characters and let b be the last entry prefixed by any ϕ character. Then, similar to Lemma 12, $T^{\text{BWT}}[a..b]$ will contain a range $[id_1, id_2]$ giving ranks of ϕ characters in that interval. Some sub-sequence of $[id_1, id_2]$ will correspond to ω wildcards that begin groups having length l or longer. Therefore, Type 2 matches can be determined by reporting entries in $\mathsf{H}[id_3..id_4]$ having a value in $[id_1, id_2]$, where $[id_3, id_4]$ is defined as before. The case when a prefix of P aligns as a suffix of a wildcard group can be handled similarly. Note that the SA ranges of the at most m wildcard group lengths we are interested in can be determined in $O(m \log \sigma)$ time and stored in $O(m \log n)$ bits.

Lemma 13. *All Type 2 matches can be reported using $O(m \log n)$ bits of working space in $O(m(\log \sigma + \min(m, \hat{d}) \frac{\log k}{\log \log k}) + occ_2 \frac{\log k}{\log \log k})$ time.*

4.3 Type 3 Matching

Type 3 matches contain at least (portions of) two wildcard groups and therefore must fully contain at least one text segment. The general idea in previous approaches and in this paper is to consider this case as an extension of the dictionary matching problem: text segments contained within P are candidate positions, but we must verify if they can be extended to a full match of P. However, we execute this idea in an altogether novel manner that greatly reduces the working space over existing approaches. The complete details of our approach are given in Algorithm 2. We now highlight the main idea and give the intuition behind the correctness.

First, suppose that text segment T_j matches P starting at position i. Consider the conditions that must be satisfied to confirm that this match can be extended to a complete match of P in T. We must verify that (i) $P[1..i-1]$ can be matched to the text preceding T_j in T — referred to as the *prefix condition* — and (ii) $P[i + |T_j|..|P|]$ can be matched to the text following T_j in T — referred to as the *suffix condition*. If both conditions are verified, we can report that P matches T at position $x_j - i + 1$, where x_j is the start position of T_j in T.

$$\cdots \phi \vdash\!\!-\; T_{j-1} \;-\!\!\dashv \phi \cdots \phi \vdash\!\!-\; T_j \;-\!\!\dashv \phi \cdots \phi \vdash\!\!-\; T_{j+1} \;-\!\!\dashv \phi \cdots$$
$$\quad\;\; \uparrow \qquad\qquad\qquad\quad \uparrow \qquad\qquad\qquad \uparrow$$
$$\quad\; x_{j-1} \qquad\qquad\qquad x_j \qquad\qquad\quad x_{j+1}$$

For working space, we make use of an array W containing $d+1$ entries (one for each text segment) of m bits, with all entries set to zero using the constant time initialization technique [3]. During the course of the algorithm the ith bit of $\mathsf{W}[j]$ is set to 1 if the prefix condition is true for $P[1..i-1]$ with respect to T_j. There are exactly m stages of the algorithm ($i = 1, \ldots, m$) corresponding to the suffixes of P. In a given stage i we consider each text segment T_j found to be a prefix of the ith suffix of P. To verify

---

Algorithm 2. Report Type 3 matches

Input: a string P of length m, its matching statistics w.r.t. SA, SA ranges for all suffixes of $\overline{P}$
w.r.t. $\overline{\mathsf{SA}}$

Output: positions in T matching P, modulo wildcard positions

1: **for** $i = 1$ **to** m **do**
2:     let $(q, [a, b])$ be the matching statistics for $P[i..m]$
3:     use Algorithm 1 to find indexes (l, r) in BP denoting smallest enclosing text segment
       interval for SA range $[a, b]$
4:     **while** (l, r) is a defined interval in BP **do**
5:        $lex\_id \leftarrow \mathtt{rank}_{(}(\mathsf{BP}, l)$
6:        $j \leftarrow \Pi[lex\_id]$
7:        $[a_p, b_p], [a_s, b_s] \leftarrow$ SA range of $\overline{P[1..i-1-k_{j-1}]}$ w.r.t $\overline{\mathsf{SA}}$, SA range of $P[i + l_j + k_j..m]$ w.r.t SA
8:        $[c_s, d_s], [c_p, d_p] \leftarrow \mathsf{RSA}[j-1], \overline{\mathsf{RSA}}[j+1]$
9:        **if** $i \leq l_{j-1} + k_{j-1}$ **then**   // Case 1: P does not contain T_{j-1}
10:           **if** $k_{j-1} \geq i - 1$ **or** $[a_p, b_p]$ encloses $[c_p, d_p]$ **then**
11:              **if** $m - i + 1 < l_j + k_j + l_{j+1} - 1$ **then**
12:                 **if** $m - i \leq l_j + k_j$ **or** $[a_s, b_s]$ encloses $[c_s, d_s]$ **then**
13:                    **print** match at position $x_j - i + 1$
14:              **else**
15:                 set $(i + l_j + k_j)^{\mathrm{th}}$ bit of $\mathsf{W}[j+1]$ to 1
16:           **else**   // Case 2: P must contain T_{j-1}
17:              **if** i^{th} bit of $\mathsf{W}[j]$ is set to 1 **then**
18:                 **if** $m - i + 1 < l_j + k_j + l_{j+1} - 1$ **then**
19:                    **if** $m - i \leq l_j + k_j$ **or** $[a_s, b_s]$ encloses $[c_s, d_s]$ **then**
20:                       **print** match at position $x_j - i + 1$
21:                 **else**
22:                    set $(i + l_j + k_j)^{\mathrm{th}}$ bit of $\mathsf{W}[j+1]$ to 1
23:        $(l, r) \leftarrow \mathtt{enclose}(\mathsf{BP}, l)$

Notation: x_j, l_j, k_j denotes the position, length and wildcard group length (which follows) the
text segment T_j

---

the prefix and suffix conditions for T_j we first consider (line 9 of Algorithm 2): will $P[1..i-1]$ need to fully contain the previous text segment T_{j-1} in order to match in T? This breaks our algorithm into the two main cases. If not (Case 1), we check the prefix condition by checking whether $P[1..i-1]$ is compatible with the wildcard group to its left and the suffix of T_{j-1} to which it must align (line 10). If the prefix condition is satisfied, we consider (line 11): will $P[i + |T_j|..m]$ need to fully contain the next text segment T_{j+1} in order to match in T? If not (Case 1a), we check whether the suffix condition is satisfied by checking that $P[i + |T_j|..m]$ is compatible with the wildcard group to its right and the prefix of T_{j+1} to which it must align (line 12). If indeed the suffix condition is satisfied, we output a match (line 13). If yes (Case 1b), we set the $(i + l_j + k_j)^{\mathrm{th}}$ bit of entry $\mathsf{W}[j+1]$ to 1, to indicate that a prefix condition holds for $P[1..i + l_j + k_j - 1]$ with respect to T_{j+1} (line 15). The key idea here is that we only attempt to verify the suffix condition when T_j would be the last text segment to occur in P (i.e., Case 1a) and if not (Case 1b), we record information in W stating that we currently have a partial match, but for it to remain viable, T_{j+1} should be a prefix of $P[i + l_j + k_j..m]$. Case 2 occurs when P must contain the previous text segment T_{j-1} to satisfy the prefix condition (lines 16–22). Since stages of the algorithm proceed with increasing values of i, then the prefix condition would have been previously checked and, if satisfied, the ith bit of $W[j]$ would be set to 1. The remaining questions are answered as before: the suffix condition is verified if possible, and otherwise successful partial matches are again recorded in W.

Lemma 14. *All Type 3 matches can be reported in* $O(m \log \sigma + \gamma)$ *time using* $O(dm + m \log n)$ *bits of working space.*

Combining the results for the 3 types of matching we arrive at our main result.

Theorem 2. *Given a text* T *of length* n *containing* d *groups of* k *wildcards all matches of a pattern* P *of length* m *can be reported using* $O(dm + m \log n)$ *bits of working space in* $O(m(\log \sigma + \min(m, \hat{d}) \frac{\log k}{\log \log k}) + occ_1 \log n + occ_2 \frac{\log k}{\log \log k} + \gamma)$ *time with an index occupying* $(2 + o(1))n \log \sigma + O(n) + O(d \log n) + O(k \log k)$ *bits of space.*

5 Conclusions

We have presented a new succinct index for texts containing wildcard characters and also proposed a new query algorithm that can have a substantially reduced working space for short query patterns when compared to existing solutions. Ignoring lower order terms, our index requires $(2 + o(1))n \log \sigma + O(n)$ bits in comparison to that of Tam *et al.*, [21]—the only other succinct index for this problem—which requires $(3 + o(1))n \log \sigma$ bits. For alphabets such as proteins ($\sigma = 20$) or larger this can result in a substantially smaller index. However, for small alphabets such as DNA ($\sigma = 4$), the $O(n)$ term becomes quite significant. This term arises from the need to store auxiliary data structures for determining lcp parent intervals when computing matching statistics of a query string. Using a solution by Fischer *et al.*, [6] we can store the necessary lcp information using at most $2n + o(n)$ bits. This would yield an index of roughly the same size as Tam *et al.*,'s; however, it incurs a sublogarithmic slowdown (in n) when computing parent intervals. Ohlebusch and Gog [18] proposed a solution that computes parent intervals in constant time (for $\sigma = O(1)$) and has been demonstrated to use between $3n$–$5n$ bits in practice [19]. This approach would ensure no slowdown in query time at the expense of a larger index compared to that of Tam *et al.*, for the DNA alphabet. In either case, both our index and Tam *et al.*,'s store a compressed suffix array for both the text and its reverse. An interesting open question is whether we can eliminate the suffix array of the reverse text. Doing so would lead to a substantial space reduction, regardless of alphabet size.

Acknowledgments. The author would like to thank Anne Condon for helpful discussions, detailed feedback and suggestions as well as the anonymous reviewers for their constructive suggestions to improve the presentation of this manuscript.

References

1. Belazzougui, D.: Succinct dictionary matching with no slowdown. In: Symposium on Combinatorial Pattern Matching, pp. 88–100 (2010)
2. Bose, P., He, M., Maheshwari, A., Morin, P.: Succinct orthogonal range search structures on a grid with applications to text indexing. Algorithms and Data Structures, 98–109 (2009)
3. Briggs, P., Torczon, L.: An efficient representation for sparse sets. ACM Lett. Program. Lang. Syst. 2, 59–69 (1993)
4. Cole, R., Gottlieb, L.A., Lewenstein, M.: Dictionary matching and indexing with errors and don't cares. In: ACM Symposium on Theory of Computing, pp. 91–100 (2004)
5. Ferragina, P., Manzini, G.: Opportunistic data structures with applications. In: Symposium on Foundations of Computer Science, pp. 390–398 (2002)
6. Fischer, J., Makinen, V., Navarro, G.: Faster entropy-bounded compressed suffix trees. Theoretical Computer Science 410(51), 5354–5364 (2009)
7. Frazer, K., Ballinger, D., Cox, D., Hinds, D., Stuve, L., Gibbs, R., et al.: A second generation human haplotype map of over 3.1 million SNPs. Nature 449(7164), 851–861 (2007)
8. Grossi, R., Vitter, J.: Compressed suffix arrays and suffix trees with applications to text indexing and string matching. In: ACM Symposium on Theory of Computing, pp. 397–406 (2000)
9. Grossi, R., Gupta, A., Vitter, J.S.: High-order entropy-compressed text indexes. In: ACM-SIAM Symposium on Discrete Algorithms, pp. 841–850 (2003)
10. Hon, W., Ku, T., Shah, R., Thankachan, S., Vitter, J.: Faster Compressed Dictionary Matching. In: Symposium on String Processing and Information Retrieval, pp. 191–200 (2010)
11. Jacobson, G.: Succinct static data structures. Ph.D. thesis, Carnegie Mellon University (1989)
12. Knuth, D., Morris Jr., J., Pratt, V.: Fast pattern matching in strings. SIAM J. on Computing 6, 323 (1977)
13. Lam, T.W., Sung, W.K., Tam, S.L., Yiu, S.M.: Space efficient indexes for string matching with don't cares. In: Conference on Algorithms and Computation, pp. 846–857 (2007)
14. Lam, T., Li, R., Tam, A., Wong, S., Wu, E., Yiu, S.: High throughput short read alignment via bi-directional BWT. In: IEEE Conference on Bioinformatics and Biomedicine, pp. 31–36 (2009)
15. Manber, U., Myers, G.: Suffix arrays: a new method for on-line string searches. In: ACM-SIAM Symposium on Discrete Algorithms, pp. 319–327 (1990)
16. Munro, J., Raman, V.: Succinct representation of balanced parentheses and static trees. SIAM J. on Computing 31(3), 762–776 (2002)
17. Navarro, G., Mäkinen, V.: Compressed full-text indexes. ACM Computing Surveys 39(1), 2 (2007)
18. Ohlebusch, E., Gog, S.: A compressed enhanced suffix array supporting fast string matching. In: Symposium on String Processing and Information Retrieval, pp. 51–62 (2009)
19. Ohlebusch, E., Gog, S., Kügel, A.: Computing matching statistics and maximal exact matches on compressed full-text indexes. In: Symposium on String Processing and Information Retrieval, pp. 347–358 (2010)
20. Raman, R., Raman, V., Rao, S.: Succinct indexable dictionaries with applications to encoding k-ary trees and multisets. In: ACM-SIAM Symposium on Discrete Algorithms, pp. 233–242 (2002)
21. Tam, A., Wu, E., Lam, T.W., Yiu, S.M.: Succinct text indexing with wildcards. In: Symposium on String Processing and Information Retrieval, pp. 39–50 (2009)
22. Weiner, P.: Linear pattern matching algorithms. In: Symposium on Switching and Automata Theory, pp. 1–11 (1973)

Self-indexing Based on LZ77*

Sebastian Kreft and Gonzalo Navarro

Dept. of Computer Science, University of Chile, Santiago, Chile
{skreft,gnavarro}@dcc.uchile.cl

Abstract. We introduce the first self-index based on the Lempel-Ziv 1977 compression format (LZ77). It is particularly competitive for highly repetitive text collections such as sequence databases of genomes of related species, software repositories, versioned document collections, and temporal text databases. Such collections are extremely compressible but classical self-indexes fail to capture that source of compressibility. Our self-index takes in practice a few times the space of the text compressed with LZ77 (as little as 2.5 times), extracts 1–2 million characters of the text per second, and finds patterns at a rate of 10–50 microseconds per occurrence. It is smaller (up to one half) than the best current self-index for repetitive collections, and faster in many cases.

1 Introduction and Related Work

Self-indexes [26] are data structures that represent a text collection in compressed form, in such a way that not only random access to the text is supported, but also indexed pattern matching. Invented in the past decade, they have been enormously successful to drastically reduce the space burden posed by general text indexes such as suffix trees or arrays. Their compression effectiveness is usually analyzed under the k-th order entropy model [21]: $H_k(T)$ is the k-th order entropy of text T, a lower bound to the bits-per-symbol compression achievable by any statistical compressor that models symbol probabilities as a function of the k symbols preceding it in the text. There exist self-indexes able to represent a text $T_{1,n}$ over alphabet $[1, \sigma]$, within $nH_k(T) + o(n \log \sigma)$ bits of space for any $k \le \alpha \log_\sigma n$ and constant $\alpha < 1$ [10,7].

This k-th order entropy model is adequate for many practical text collections. However, it is not a realistic lower bound model for a kind of collections that we call *highly repetitive*. This is formed by sets of strings that are mostly near-copies of each other. For example, versioned document collections store all the history of modifications of the documents. Most versions consist of minor edits on a previous version. Good examples are the Wikipedia database and the Internet archive. Another example are software repositories, which store all the versioning history of software pieces. Again, except for major releases, most versions are

---

* Partially funded by Millennium Institute for Cell Dynamics and Biotechnology (ICDB), Grant ICM P05-001-F, Mideplan, Chile and, the first author, by Conicyt's Master Scholarship.

R. Giancarlo and G. Manzini (Eds.): CPM 2011, LNCS 6661, pp. 41–54, 2011.

minor edits of previous ones. In this case the versioning has a tree structure more than a linear sequence of versions. Yet another example comes from bioinformatics. Given the sharply decreasing sequencing costs, large sequence databases of individuals of the same or closely related species are appearing. The genomes of two humans, for example, share 99.9% to 99.99% of their sequence. No clear structure such as a versioning tree is apparent in the general case.

If one concatenates two identical texts, the statistical structure of the concatenation is almost the same as that of the pieces, and thus the k-th order entropy does not change. As a consequence, some indexes that are exactly tailored to the k-th order entropy model [10,7] are insensitive to the repetitiveness of the text. Mäkinen et al. [32,20] found that even the self-indexes that can compress beyond the k-th order entropy model [31,25] failed to capture much of the repetitiveness of such text collections.

Note that we are not aiming simply at *representing* the text collections to offer *extraction* of individual documents. This is relatively simple as it is a matter of encoding the edits with respect to some close sampled version; more sophisticated techniques have been however proposed for this goal [17,18,16]. Our aim is more ambitious: self-indexing the collection means providing not only access but indexed searching, just as if the text was available in plain form. Other restricted goals such as compressing the inverted index (but not the text) on natural-language text collections [12] or indexing text q-grams and thus fixing the pattern length in advance [5] have been pursued as well.

Mäkinen et al. [32,20] demonstrated that repetitiveness in the text collections translates into *runs* of equal letters in its Burrows-Wheeler transform [4] or runs of successive values in the Ψ function [11]. Based on this property they engineered variants of FM-indexes [7] and Compressed Suffix Arrays (CSAs) [31] that take advantage of repetitiveness. Their best structure, the Run-Length CSA (RLCSA) still stands as the best self-index for repetitive collections, despite of some preliminary attempts of self-indexing based on grammar compression [5].

Still, Mäkinen et al. showed that their new self-indexes were very far (by a factor of 10) from the space achievable by a compressor based on the Lempel-Ziv 1977 format (LZ77) [33]. They showed that the runs model is intrinsically inferior to the LZ77 model to capture repetitions. The LZ77 compressor is particularly able to capture repetitiveness, as it parses the text into consecutive maximal *phrases* so that each phrase appears earlier in the text. A self-index based on LZ77 was advocated as a very promising alternative approach to the problem.

Designing a self-index based on LZ77 is challenging. Even accessing LZ77-compressed text at random is a difficult problem, which we partially solved [16] with the design of a variant called LZ-End, which compresses only slightly less and gives some time guarantees for the access time. There exists an early theoretical proposal for LZ77-based indexing by Kärkkäinen and Ukkonen [14,13], but it requires to have the text in plain form and has never been implemented. Although it guarantees an index whose size is of the same order of the LZ77 compressed text, the constant factors are too large to be practical. Nevertheless, that was the first general compressed index in the literature and is the predecessor of all the Lempel-Ziv indexes that followed [25,6,30]. These indexes have used variants of the LZ78 compression format [34], which is more tractable but still too weak to capture high repetitiveness [32].

In this paper we face the challenge of designing the first self-index based on LZ77 compression. Our self-index can be seen as a modern variant of Kärkkäinen and Ukkonen's LZ77 index, which solves the problem of not having the text at hand and also makes use of recent compressed data structures. This is not trivial at all, and involves designing new solutions to some subproblems where the original solution [14] was too space-consuming. Some of the solutions might have independent interest.

The bounds obtained by our index are summarized in the following theorem.

Theorem 1. *Let $T_{1,n}$ be a text over alphabet $[1, \sigma]$, parsed into n' phrases by the LZ77 or LZ-End parsing. Then there exists an index occupying $2n' \log n + n' \log n' + 5n' \log \sigma + O(n') + o(n)$ bits, and able to locate the occ occurrences of a pattern $p_{1,m}$ in T in time $O(m^2 h + (m + occ) \log n')$, where h is the height of the parsing (see Def. 3). Extracting any ℓ symbols from T takes time $O(\ell h)$ on LZ77 and $O(\ell + h)$ on LZ-End. The space term $o(n)$ can be removed at the price of multiplying time complexities by $O(\log \frac{n}{n'})$.*

As the output of the Lempel-Ziv compressor has $n'(2 \log n + \log \sigma)$ bits, it follows that the index is asymptotically at most twice the size of the compressed text (for $\log \sigma = o(\log n)$; 3 times otherwise).

In comparison, the time complexity of RLCSA is $O(m \log n + occ \log^{1+\epsilon} n)$, that is, it depends less sharply on m but takes more time per occurrence reported.

We implemented our self-index over LZ77 and LZ-End parsings, and compared it with the state of the art on a number of real-life repetitive collections consisting of Wikipedia versions, versions of public software, periodic publications, and DNA sequence collections. We have left a public repository with those repetitive collections in http://pizzachili.dcc.uchile.cl/repcorpus.html, so that standardized comparisons are possible. Our implementations and that of the RLCSA are also available in there.

Our experiments show that in practice the smallest-space variant of our index takes 2.5–4.0 times the space of a LZ77-based encoding, it can extract 1–2 million characters per second, and locate each occurrence of a pattern of length 10 in 10–50 microseconds. Compared to the state of the art (RLCSA), our self-index always takes less space, less than a half on our DNA and Wikipedia corpus. Searching for short patterns is faster than on the RLCSA. On longer patterns our index offers competitive space/time trade-offs.

2   Direct Access to LZ-Compressed Texts

Let us first recall the classical LZ77 parsing [33], as well as the recent LZ-End parsing [16]. This involves defining what is a phrase and its source, and the number n' of phrases.

Definition 1 ([33]). *The LZ77 parsing of text $T_{1,n}$ is a sequence $Z[1, n']$ of phrases such that $T = Z[1]Z[2]\ldots Z[n']$, built as follows. Assume we have already processed $T_{1,i-1}$ producing the sequence $Z[1, p-1]$. Then, we find the longest prefix $T_{i,i'-1}$ of $T_{i,n}$ which occurs in $T_{1,i-1}$, set $Z[p] = T_{i,i'}$ and continue with $i = i' + 1$. The occurrence in $T_{1,i-1}$ of prefix $T_{i,i'-1}$ is called the* source *of the phrase $Z[p]$.*

Definition 2 ([16]). *The LZ-End parsing of text $T_{1,n}$ is a sequence $Z[1, n']$ of phrases such that $T = Z[1]Z[2]\ldots Z[n']$, built as follows. Assume we have already processed $T_{1,i-1}$ producing the sequence $Z[1, p-1]$. Then, we find the longest prefix $T_{i,i'-1}$ of $T_{i,n}$ that is a suffix of $Z[1]\ldots Z[q]$ for some $q < p$, set $Z[p] = T_{i,i'}$ and continue with $i = i' + 1$.*

We will store Z in a particular way that enables efficient extraction of any text substring $T_{s,e}$. This is more complicated than in our previous proposal [16] because these structures will be integrated into the self-index later. First, the last characters of the phrases, $T_{i'}$ of $Z[p] = T_{i,i'}$, are stored in a string $L_{1,n'}$. Second, we set up a bitmap $B_{1,n}$ that will mark with a 1 the ending positions of the phrases in $T_{1,n}$ (or, alternatively, the positions where the successive symbols of L lie in T). Third, we store a bitmap $S_{1,n+n'}$ that describes the structure of the sources in T, as follows. We traverse T left to right, from T_1 to T_n. At step i, if there are k sources starting at position T_i, we append $1^k 0$ to S (k may be zero). Empty sources (i.e., $i = i'$ in $Z[p] = T_{i,i'}$) are assumed to lie just before T_1 and appended at the beginning of S, followed by a 0. So the 0s in S correspond to text positions, and the 1s correspond to the successive sources, where we assume that those that start at the same point are sorted by shortest length first. Finally, we store a permutation $P[1, n']$ that maps targets to sources, that is, $P[i] = j$ means that the source of the ith phrase starts at the position corresponding to the jth 1 in S. Fig. 1(a) gives an example.

The bitmaps $B_{1,n}$ and $S_{1,n+n'}$ are sparse, as they have only n' bits set. They are stored using a compressed representation [29] so that each takes $n' \log \frac{n}{n'} + O(n') + o(n)$ bits, and rank/select queries require constant time: $rank_b(B, i)$ is the number of occurrences of bit b in $B_{1,i}$, and $select_b(B, j)$ is the position in B of the jth occurrence of bit b (similarly for S). The $o(n)$ term, the only one that does not depend linearly on n', can disappear at the cost of increasing the time for $rank$ to $O(\log \frac{n}{n'})$ [27]. Finally, permutations are stored using a representation [23] that computes $P[i]$ in constant time and $P^{-1}[j]$ in time $O(l)$, using $(1 + 1/l)n' \log n' + O(n')$ bits of space. We use parameter $l = \log n'$. Thus our total space is $n' \log n' + 2n' \log \frac{n}{n'} + n' \log \sigma + O(n') + o(n)$ bits.

To extract $T_{s,e}$ we proceed as follows. We compute $s' = rank_1(B, s-1) + 1$ and $e' = rank_1(B, e)$ to determine that we must extract characters from phrases s' to e'. For all phrases except possibly e' (where $T_{s,e}$ could end before its last position) we have their last characters in $L[s', e']$. For all the other symbols, we must go to the source of each phrase of length more than one and recursively extract its text: to extract the rest of phrase $s' \leq k \leq e'$, we compute its length as $l = select_1(B, k) - select_1(B, k-1)$ (except for $k = e'$, where the length is

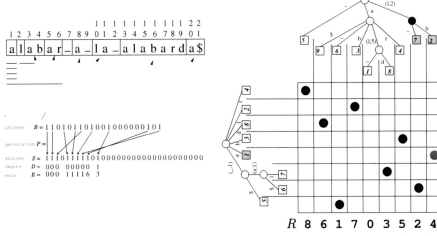

(a) The LZ77 parsing of the string 'alabar_a_la_alabarda$', showing the sources of each phrase on top. On the bottom, bitmap B marks the ends of phrases, the bitmap S marks the starting positions of sources, and the permutation P connects phrases to sources. We also show array D of depths and (virtual) array E of ending source positions (these arrays are exclusive).

(b) Top: The sparse suffix trie. The black node is the one we arrive at when searching for 'la', and the gray leaves of its subtree represent the phrases that start with 'la'. Left: The reverse trie for the string. The gray leaf is the node at which we stop searching for 'a'. Bottom: The range structure for the string. The gray dot marks the only primary occurrence of the pattern 'ala' (it is the only dot in the range defined by the gray nodes).

Fig. 1. Our self-index structure over the example text $T = $ 'alabar_a_la_alabarda$\$$' and part of the process of searching for $p = $ 'ala'

$l = e - select_1(B, k-1))$ and its starting position as $t = rank_0(S, select_1(S, P[k])) = select_1(S, P[k]) - P[k]$. Thus to obtain the rest of the characters of phrase k we recursively extract $T_{t,t+l-1}$

On LZ-End this method takes time $O(e - s + 1)$ if e coincides with the end of a phrase [16]. In general, a worst-case analysis [16] yields extraction time $O(e - s + h)$ for LZ-End and $O(h(e - s + 1))$ for LZ77, where h is a measure of how nested the parsing is.

Definition 3. *Let* $T = Z[1]Z[2]\ldots Z[n']$ *be a LZ-parsing of* $T_{1,n}$. *Then the* height *of the parsing is defined as* $h = \max_{1 \leq i \leq n} C[i]$, *where* C *is defined as follows. Let* $Z[i] = T_{a,b}$ *be a phrase whose source is* $T_{c,d}$. *Then* $C[b] = 1$ *and* $C[k] = C[(k - a) + c] + 1$ *for* $a \leq k < b$.

That is, h measures how many times a character is transitively copied in Z. While in the worst case h can be as large as n', it is usually a small value. It is limited by the longest length of a phrase [15], thus on a text coming from a

Markovian source it is $O(\log_\sigma n)$. On our repetitive collection corpus h is between 22 and 259 for LZ-End, and between 22 and 1003 for LZ77. Its average values, on the other hand, are 5–25 on LZ-End and 5–176 on LZ77.

Implementation considerations. As bitmaps B and S are very sparse in highly repetitive collections, we opted for δ-encoding the distances between the consecutive 1s, and adding a sampling where we store the absolute values and position in the δ-codes of every sth bit, where s is the sampling rate. So *select* consists in going to the previous sample and decoding at most s δ-codes, whereas *rank* requires a previous binary search over the samples.

3 Pattern Searches

Assume we have a text T of length n, which is partitioned into n' phrases using a LZ77-like compressor. Let $p_{1,m}$ be a search pattern. We call *primary occurrences* of p those overlapping more than one phrase or ending at a phrase boundary; and *secondary occurrences* the others. For example, in Fig. 1(a), the occurrence of 'lab' starting at position 2 is primary as it spans two phrases. The second occurrence, starting at position 14, is secondary.

We will find first the primary occurrences, and those will be used to recursively find the secondary ones (which, in turn, will be used to find further secondary occurrences).

3.1 Primary Occurrences

Each primary occurrence can be split as $p = p_{1,i}\,p_{i+1,m}$, where the left side $p_{1,i}$ is a nonempty suffix of a phrase and the (possibly empty) right side $p_{i+1,m}$ is the concatenation of zero or more consecutive phrases plus a prefix of the next phrase. To find primary occurrences we partition the pattern into two in every possible way. Then, we search for the left part in the suffixes of the phrases and for the right part in the prefixes of the suffixes of T starting at phrase boundaries. Then, we find which pairs of left and right occurrences are concatenated, thus representing actual primary occurrences of p.

Finding the Right Part of the Pattern. To find the right side $p_{i+1,m}$ of the pattern we use a suffix trie that indexes all the suffixes of T starting at the beginning of a phrase. In the leaves of the trie we store the identifiers of the phrases where the corresponding suffixes start. Conceptually, the identifiers form an array id that stores the phrase identifiers in lexicographic order of their suffixes. As we see later, we do not need to store id explicitly.

We represent the suffix trie as a Patricia tree [22], encoded using a succinct representation for labeled trees called *dfuds* [2]. As the trie has at most $2n'$ nodes, the succinct representation requires at most $2n' \log \sigma + O(n')$ bits. It supports a large number of operations in constant time, such as going to a child labeled c, going to the leftmost and rightmost descendant leaf, etc. To search for $p_{i+1,m}$ we descend through the tree using the next character of the pattern, skip as many characters as the skip value of the child indicates, and repeat the process until determining that $p_{i+1,m}$ is not in the set or reaching a node or an edge, whose leftmost and rightmost subtree leaves define the interval in array id whose suffixes start with $p_{i+1,m}$. Fig. 1(b) shows on top this trie, shading the range [8,9] of leaves found when searching for $p_{i+1,m} = \text{'la'}$.

Recall that, in a Patricia tree, after searching for the positions we need to check if they are actually a match, as some characters are not checked because of the skips. Instead of doing the check at this point, we defer it for later, when we connect both searches.

We do not explicitly store the skips, as they can be computed from the trie and the text. Given a node in the trie corresponding to a string of length l, we go to the leftmost and rightmost leaves and extract the corresponding suffixes from their $(l+1)$th symbols. The number s of symbols they share from that position is the skip. This takes $O(sh)$ time for LZ77 and LZ-End, since the extraction is from left to right and we have to extract one character at a time until they differ. Thus, the total time for extracting the skips as we descend is $O(mh)$.

Finding the Left Part of the Pattern. We have another Patricia trie that indexes all the reversed phrases, stored in the same way as the suffix trie. To find the left part of the pattern in the text we search for $(p_{1,i})^{rev}$ in this trie. The array that stores the leaves of the trie is called rev_id and is stored explicitly. The total space is at most $n' \log n' + 2n' \log \sigma + O(n')$ bits. Fig. 1(b) shows this trie on the left, with the result of searching for a left part $p_{1,i} = \text{'a'}$.

Connecting Both Searches. Actual occurrences of p are those formed by a phrase $rev\_id[j] = k-1$ and the following one $id[i] = k$, so that j and i belong to the lexicographical intervals found with the tries. To find those we use a $n' \times n'$ range structure that connects the consecutive phrases in both trees. If $id[i] = k$ and $rev\_id[j] = k-1$, the structure holds a point in (i,j).

The range structure is represented compactly using a wavelet tree [10,19], which requires $n' \log n' + O(n' \log \log n')$ bits. This can be reduced to $n' \log n' + O(n')$ [28]. The wavelet tree stores the sequence $R[1, n']$ so that $R[i] = j$ if (i,j) is a point (note there is only one j per i value). In $O(\log n')$ time it can compute $R[i]$, as well as find all the occ points in a given orthogonal range in time $O((occ+1)\log n')$. With such an orthogonal range search for the intervals of leaves found in both trie searches, the wavelet tree gives us all the primary occurrences. It also computes any $id[i] = rev\_id[R[i]] + 1$ in $O(\log n')$ time, thus we do not need to store id.

Fig. 1(b) gives an example, showing sequence R at the bottom. It also shows how we find the only primary occurrence of $p = \text{'ala'}$ by partitioning it into 'a' and 'la'.

At this stage we also verify that the answers returned by the searches in the Patricia trees are valid. It is sufficient to extract the text of one of the occurrences reported and compare it to p, to determine either that all or none of the answers are valid, by the Patricia tree properties.

Note that the structures presented up to now are sufficient to determine whether the pattern exists in the text or not, since p cannot appear if it does not have primary occurrences. If we have to report the occ occurrences, instead, we use bitmap B: An occurrence with partition $p_{1,i}$ and $p_{i+1,m}$ found at $rev\_id[j] = k$ is to be reported at text position $select_1(B,k) - i + 1$.

Overall, the data structures introduced in this section add up to $2n'\log n' + 4n'\log\sigma + O(n')$ bits. The occ primary occurrences are found in time $O(m^2h + m\log n' + occ\log n')$.

Implementation Considerations. As the average value for the skips is usually very low and computing them from the text phrases is slow in practice, we actually store the skips using *Directly Addressable Codes* [3]. These allow storing variable-length codes while retaining fast direct access. In this case arrays id and rev_id are only accessed for reporting the occurrences.

We use a practical *dfuds* implementation [1] that binary searches for the child labeled c, as the theoretical one [2] uses perfect hashing.

Instead of storing the tries we can do a binary search over the id or rev_id arrays. This alternative modifies the complexity of searching for a prefix/suffix of p to $O(mh\log n')$ for LZ77 or $O((m+h)\log n')$ for LZ-End. Independently, we could store explicitly array id, instead of accessing it through the wavelet tree. Although this alternative increases the space usage of the index and does not improve the complexity, it gives an interesting trade-off in practice.

3.2   Secondary Occurrences

Secondary occurrences are found from the primary occurrences and, recursively, from other previously discovered secondary occurrences. The idea is to locate all sources covering the occurrence and then finding their corresponding phrases. Each copy found is reported and recursively analyzed for sources containing it.

For each occurrence found $T_{i,i+m-1}$, we find the position pos of the 0 corresponding to its starting position in bitmap S, $pos = select_0(S,i)$. Then we consider all the 1s to the left of pos, looking for sources that start before the occurrence. For each such $S[j] = 1$, $j \leq pos$, the source starts in T at $t = rank_0(S,j)$ and is the sth source, for $s = rank_1(S,j)$. Its corresponding phrase is $f = P^{-1}[s]$, which starts at text position $c = select(B, f-1)+1$. Now we compute the length of the source, which is the length of its phrase minus one, $l = select_1(B,f) - select_1(B,f-1) - 1$. Finally, if $T_{t,t+l-1}$ covers the occurrence $T_{i,i+m-1}$, then this occurrence has been copied to $T_{c+i-t,c+i-t+m-1}$, where we report a secondary occurrence and recursively find sources covering it. The time per occurrence reported is dominated by that of computing P^{-1}, $O(\log n')$.

Consider the only primary occurrence of pattern 'la' starting at position 2 in our example text. We find the third 0 in the bitmap of sources at position 12. Then we consider all 1s starting from position 11 to the left. The 1 at position 11 maps to a phrase of length 2 that covers the occurrence, hence we report an occurrence at position 10. The second 1 maps to a phrase of length 6 that also covers the occurrence, thus we report another occurrence at position 15. The third 1 maps to a phrase of length 1, hence it does not cover the occurrence and we do not report it. We proceed recursively for the occurrences found at positions 10 and 15.

Unfortunately, stopping looking for 1s to the left in S as soon as we find the first source not covering the occurrence works only when no source contains another. We present now a general solution that requires just $2n' + o(n')$ extra bits and reports the occ secondary occurrences in time $O(occ \log n')$.[1]

Consider a (virtual) array $E[1, n']$ where $E[s]$ is the text position where the sth source ends. Then an occurrence $T_{i,i+m-1}$ is covered by source s if $s \le e = rank_1(S, pos)$ (i.e., s starts at or before i in T) and $E[s] \ge i + m - 1$ (i.e., s ends at or after $i + m - 1$ in T). Then we must report all values $E[1, e] \ge i + m - 1$. Fig. 1(a) shows E on our running example.

A *Range Maximum Query (RMQ)* data structure can be built on $E[1, n']$ so that it (*i*) occupies $2n' + o(n')$ bits of space; (*ii*) answers in constant time queries $\mathrm{RMQ}(i, j) = \arg\max_{i \le k \le j} E[k]$; (*iii*) it does *not* access E for querying [8]. We build such a data structure on E. The array E itself is not represented; any desired value can be computed as $E[s] = t + l - 1$, using the nomenclature given three paragraphs above, in time $O(\log n')$ as it involves computing $P^{-1}[s]$.

Thus $k = \mathrm{RMQ}(1, e)$ gives us the rightmost-ending source among those starting at or before i. If $E[k] < i + m - 1$ then no source in $[1, e]$ covers the occurrence. Else, we report the copied occurrence within phrase $P^{-1}[k]$ (and process it recursively), and now consider the intervals $E[1, k - 1]$ and $E[k + 1, e]$, which are in turn recursively processed with RMQs until no source covering the occurrence is found. This algorithm was already described by Muthukrishnan [24], who showed that it takes $2\,occ$ computations of RMQ to report occ occurrences. Each step takes us $O(\log n')$ time due to the need to compute the $E[k]$ values.

In practice: prevLess data structure. The best implemented RMQ-based solution requires in practice around $3n'$ bits and a constant but significant number of complex operations [8,9]. We present now an alternative development that, although offering worse worst-case complexities, in practice requires 2.88–$4.08n'$ bits and is faster (it takes 1–3 microseconds in total per secondary occurrence, whereas just one RMQ computation takes more than 1.5 microseconds, still ignoring the time to compute $E[k]$ values). It has, moreover, independent interest.

In early attempts to solve the problem of reporting secondary occurrences, Kärkkäinen [13] introduced the concept of *levels*. We use it in a different way.

Definition 4. *Source* $s_1 = [l_1, r_1]$ *is said to* cover *source* $s_2 = [l_2, r_2]$ *if* $l_1 < l_2$ *and* $r_1 > r_2$. *Let* cover(s) *be the set of sources covering a source* s. *Then the* depth *of source* s *is defined as* $depth(s) = 0$ *if* $cover(s) = \emptyset$, *and* $depth(s) = 1 + \max_{s' \in cover(s)} depth(s')$ *otherwise. We define* $depth(\varepsilon) = 0$. *Finally, we call* δ *the maximum depth in the parsing.*

In our example, the four sources 'a' and the source 'alabar' have depth zero, as all of them start at the same position. Source 'la' has depth 1, as it is contained by source 'alabar'.

---

[1] Thanks to the anonymous reviewer that suggested it.

We traverse S leftwards from pos. When we find a source not covering the occurrence, we look for its depth d and then consider to the left only sources with depth $d' < d$, as those at depth $\geq d$ are guaranteed not to contain the occurrence. This works because sources to the left with the same depth d will not end after the current source, and deeper sources to the left will be contained in those of depth d. Thus for our traversal we need to solve a subproblem we call $prevLess(D, s, d)$: Let $D[1, n']$ be the array of depths of the sources; given a position s and a depth d, we need to find the largest $s' < s$ such that $D[s'] < d$.

We represent D using a wavelet tree [10]. This time we need to explain its internal structure. The wavelet tree is a balanced tree where each node represents a range of the alphabet $[0, \delta]$. The root represents the whole range and each leaf an individual alphabet member. Each internal node has two children that split its alphabet range by half. Hence the tree has height $\lceil \log(1 + \delta) \rceil$. At the root node, the tree stores a bitmap aligned to D, where a 0 at position i means that $D[i]$ is a symbol belonging to the range of the left child, and 1 that it belongs to the right child. Recursively, each internal node stores a bitmap that refers to the subsequence of D formed by the symbols in its range. All the bitmaps are preprocessed for rank/select queries, needed for navigating the tree. The total space is $n' \log \delta + O(n')$.

We solve $prevLess(D, s, d)$ as follows. We descend on the wavelet tree towards the leaf that represents $d - 1$. If $d - 1$ is to the left of the current node, then no interesting values can be stored in the right child. So we recursively continue in the left subtree, at position $s' = rank_0(V, s)$, where V is the bitmap of the current node. Otherwise we descend to the right child, and the new position is $s' = rank_1(V, s)$. In this case, however, the answer could be at the left child. Any value stored at the left child is $< d$, so we are interested in the rightmost before position s. Hence $v_0 = select_0(V, rank_0(V, s - 1))$ is the last relevant position with a value from the left subtree. We find, recursively, the best answer v_1 from the right subtree, and return $\max(v_0, v_1)$. When the recursion ends at a leaf we return with answer -1. The running time is $O(\log \delta)$.

Using this operation we proceed as follows. We keep track of the smallest depth d that cannot cover an occurrence; initially $d = \delta + 1$. We start considering source s. Whenever s covers the occurrence, we report it, else we set $d = D[s]$. In both cases we then move to $s' = prevLess(D, s, d)$.

In the worst case the first source is at depth δ and then we traverse level by level, finding in each level that the previous source does not contain the occurrence. Therefore the overall time is $O(occ(\log n' + \delta \log \delta))$ to find occ secondary occurrences. This worst case is, however, rather unlikely. Moreover, in practice δ is small: it is also limited by the maximum phrase length, and in our test collections it is at most 46 and on average 1–4.

4   Experimental Evaluation

From the testbed in `http://pizzachili.dcc.uchile.cl/repcorpus.html` we have chosen four real collections representative of distinct applications: Cere

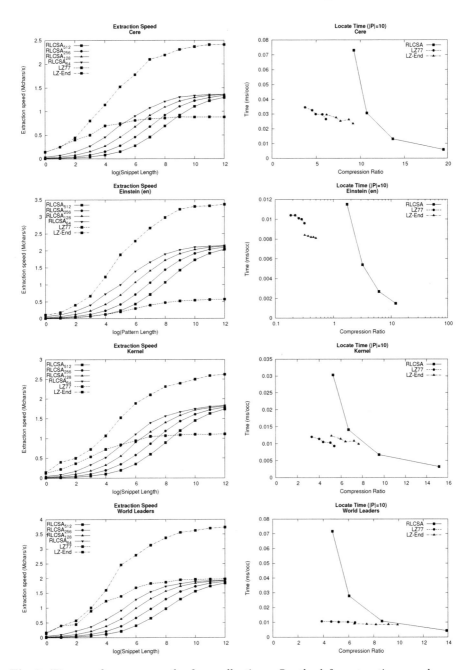

Fig. 2. Time performance on the four collections. On the left, extraction speed as a function of the extracted snippet size (higher is better). On the right, time per located occurrence for $m = 10$ as a function of the space used by the index, in percentage of text size (lower and leftwards is better). On the right the points for RLCSA refer to different sampling rates; for LZ77 and LZ-End the points refer to the 5 variants (LZ$_5$ is leftmost, LZ$_1$ is rightmost).

Table 1. Space statistics of our texts, giving the size when each symbol is represented with one, and compression achieved as a percentage of such representation: first public-domain compressors, then self-indexes

Collection	Cere	Einstein	Kernel	Leaders
Size	440MB	446MB	247MB	45 MB
p7zip	1.14%	0.07%	0.81%	1.29%
repair	1.86%	0.10%	1.13%	1.78%
bzip2	2.50%	5.38%	21.86%	7.11%
ppmdi	24.09%	1.61%	18.62%	3.56%
RLCSA	7.60%	0.23%	3.78%	3.32%
RLCSA_{512}	8.57%	1.20%	4.71%	4.20%
LZ77_5	3.74%	0.18%	3.31%	3.85%
LZ77_1	5.94%	0.30%	5.26%	6.27%
LZ-End_5	6.16%	0.32%	5.12%	6.44%
LZ-End_1	8.96%	0.48%	7.50%	9.63%

(37 DNA sequences of Saccharomyces Cerevisiae), Einstein (the version of the Wikipedia article on Albert Eintein up to Jan 12, 2010), Kernel (the 36 versions 1.0.x and 1.1.x of the Linux Kernel), and Leaders (pdf files of the CIA World Leaders report, from Jan 2003 to Dec 2009, converted with pdftotext).

We have studied 5 variants of our indexes, from most to least space consuming: (1) with suffix and reverse trie; (2) binary search on explicit id array and reverse trie; (3) suffix trie and binary search on rev_id; (4) binary search on explicit id array and on rev_id; (5) binary search on implicit id and on rev_id. In addition we test parsings LZ77 and LZ-End, so for example LZ-End_3 means variant (3) on parsing LZ-End.

Table 1 gives statistics about the texts, with the compression ratios achieved with a good Lempel-Ziv compressor (p7zip, www.7-zip.org), grammar compressor (repair, www.cbrc.jp/~rwan/en/restore.html), Burrows-Wheeler compressor (bzip2, www.bzip.org), and statistical high-order compressor (ppmdi, pizzachili.dcc.uchile.cl/utils/ppmdi.tar.gz). Lempel-Ziv and grammar-based compressors capture repetitiveness, while the Burrows-Wheeler one captures only some due to the runs, and the statistical one is blind to repetitiveness. We also give the space required by the RLCSA alone (which can count how many times a pattern appears in T but cannot locate the occurrences nor extract text at random), and RLCSA using a sampling of 512 (the minimum space that gives reasonable times for locating and extraction). Finally we show the most and least space consuming of our variants over both parsings.

Our least-space variants take 2.5–4.0 times the space of p7zip, the best LZ77 compressor we know of and the best-performing in our dataset. They are also always smaller than RLCSA_{512} (up to 6.6 times less) and even competitive with the crippled self-index RLCSA-with-no-sampling. The case of Einstein is particularly illustrative. As it is extremely compressible, it makes obvious how the RLCSA achieves much compression in terms of the runs of Ψ, yet it is unable to compress the sampling despite many theoretical efforts [20]. Thus even a sparse sampling has a very large relative weight when the text is so repetitive. The data our index needs for locating and extracting, instead, is proportional to the compressed text size.

Fig. 2 shows times for extracting snippets and for locating random patterns of length 10. We test RLCSA with various sampling rates (smaller rate requires more space). It can be seen that our LZ-End-based index extracts text faster than the RLCSA, while for LZ77 the results are mixed. For locating, our indexes operate within much less space than the RLCSA, and are simultaneously faster in several cases. See the extended version [15] for more results.

5 Conclusions

We have presented the first self-index based on LZ77 compression, showing it is particularly effective on highly repetitive text collections, which arise in several applications. The new indexes improve upon the state of the art in most aspects and solve an interesting standing challenge. Our solutions to some subproblems, such as that of *prevLess*, may be of independent interest.

Our construction needs 6–8 times the original text size and indexes 0.2–2.0 MB/sec. While this is usual in self-indexes and better than the RLCSA, it would be desirable to build it within compressed space. Another important challenge is to be able to restrict the search to a range of document numbers, that is, within a particular version, time frame, or version subtree. Finally, dynamizing the index, so that at least new text can be added, would be desirable.

References

1. Arroyuelo, D., Cánovas, R., Navarro, G., Sadakane, K.: Succinct trees in practice. In: ALENEX, pp. 84–97 (2010)
2. Benoit, D., Demaine, E., Munro, I., Raman, R., Raman, V., Rao, S.: Representing trees of higher degree. Algorithmica 43(4), 275–292 (2005)
3. Brisaboa, N.R., Ladra, S., Navarro, G.: Directly addressable variable-length codes. In: Karlgren, J., Tarhio, J., Hyyrö, H. (eds.) SPIRE 2009. LNCS, vol. 5721, pp. 122–130. Springer, Heidelberg (2009)
4. Burrows, M., Wheeler, D.: A block sorting lossless data compression algorithm. TRep. 124, DEC (1994)
5. Claude, F., Fariña, A., Martínez-Prieto, M., Navarro, G.: Compressed q-gram indexing for highly repetitive biological sequences. In: BIBE, pp. 86–91 (2010)
6. Ferragina, P., Manzini, G.: Indexing compressed text. J. ACM 52(4), 552–581 (2005)
7. Ferragina, P., Manzini, G., Mäkinen, V., Navarro, G.: Compressed representations of sequences and full-text indexes. ACM Trans. Alg. 3(2), article 20 (2007)
8. Fischer, J.: Optimal succinctness for range minimum queries. In: López-Ortiz, A. (ed.) LATIN 2010. LNCS, vol. 6034, pp. 158–169. Springer, Heidelberg (2010)
9. Gog, S., Fischer, J.: Advantages of shared data structures for sequences of balanced parentheses. In: DCC, pp. 406–415 (2010)
10. Grossi, R., Gupta, A., Vitter, J.: High-order entropy-compressed text indexes. In: SODA, pp. 841–850 (2003)

11. Grossi, R., Vitter, J.: Compressed suffix arrays and suffix trees with applications to text indexing and string matching. In: STOC, pp. 397–406 (2000)
12. He, J., Zeng, J., Suel, T.: Improved index compression techniques for versioned document collections. In: CIKM, pp. 1239–1248 (2010)
13. Kärkkäinen, J.: Repetition-Based Text Indexes. Ph.D. thesis, Univ. Helsinki, Finland (1999)
14. Kärkkäinen, J., Ukkonen, E.: Lempel-Ziv parsing and sublinear-size index structures for string matching. In: WSP, pp. 141–155 (1996)
15. Kreft, S.: Self-Index based on LZ77. MSc thesis, Univ. of Chile (2010), http://www.dcc.uchile.cl/gnavarro/algoritmos/tesisKreft.pdf
16. Kreft, S., Navarro, G.: LZ77-like compression with fast random access. In: DCC, pp. 239–248 (2010)
17. Kuruppu, S., Beresford-Smith, B., Conway, T., Zobel, J.: Repetition-based compression of large DNA datasets. In: RECOMB (2009), poster
18. Kuruppu, S., Puglisi, S.J., Zobel, J.: Relative lempel-ziv compression of genomes for large-scale storage and retrieval. In: Chavez, E., Lonardi, S. (eds.) SPIRE 2010. LNCS, vol. 6393, pp. 201–206. Springer, Heidelberg (2010)
19. Mäkinen, V., Navarro, G.: Rank and select revisited and extended. Theo.Comp.Sci. 387(3), 332–347 (2007)
20. Mäkinen, V., Navarro, G., Sirén, J., Välimäki, N.: Storage and retrieval of highly repetitive sequence collections. J. Comp. Biol. 17(3), 281–308 (2010)
21. Manzini, G.: An analysis of the Burrows-Wheeler transform. J. ACM 48(3), 407–430 (2001)
22. Morrison, D.: PATRICIA-Practical algorithm to retrieve information coded in alphanumeric. J. ACM 15(4), 514–534 (1968)
23. Munro, I., Raman, R., Raman, V., Rao, S.: Succinct representations of permutations. In: ICALP, pp. 345–356 (2003)
24. Muthukrishnan, S.: Efficient algorithms for document retrieval problems. In: SODA, pp. 657–666 (2002)
25. Navarro, G.: Indexing text using the Ziv-Lempel trie. J. Discr. Alg. 2(1), 87–114 (2004)
26. Navarro, G., Mäkinen, V.: Compressed full-text indexes. ACM Comp. Surv. 39(1), article 2 (2007)
27. Okanohara, D., Sadakane, K.: Practical entropy-compressed rank/select dictionary. In: ALENEX (2007)
28. Pătraşcu, M.: Succincter. In: FOCS, pp. 305–313 (2008)
29. Raman, R., Raman, V., Rao, S.: Succinct indexable dictionaries with applications to encoding k-ary trees and multisets. In: SODA, pp. 233–242 (2002)
30. Russo, L., Oliveira, A.: A compressed self-index using a Ziv-Lempel dictionary. Inf. Retr. 5(3), 501–513 (2008)
31. Sadakane, K.: New text indexing functionalities of the compressed suffix arrays. J. Alg. 48(2), 294–313 (2003)
32. Sirén, J., Välimäki, N., Mäkinen, V., Navarro, G.: Run-length compressed indexes are superior for highly repetitive sequence collections. In: Amir, A., Turpin, A., Moffat, A. (eds.) SPIRE 2008. LNCS, vol. 5280, pp. 164–175. Springer, Heidelberg (2008)
33. Ziv, J., Lempel, A.: A universal algorithm for sequential data compression. IEEE Trans. Inf. Theo. 23(3), 337–343 (1977)
34. Ziv, J., Lempel, A.: Compression of individual sequences via variable-rate coding. IEEE Trans. Inf. Theo. 24(5), 530–536 (1978)

Filling Scaffolds with Gene Repetitions: Maximizing the Number of Adjacencies

Haitao Jiang[1,2], Farong Zhong[3], and Binhai Zhu[1]

[1] Department of Computer Science, Montana State University,
Bozeman, MT 59717-3880, USA
{htjiang,bhz}@cs.montana.edu
[2] School of Computer Science and Technology, Shandong University, Jinan, China
[3] College of Computing, Zhejiang Normal University, Jinhua, Zhejiang, China
zfr@zjnu.cn

Abstract. In genome sequencing there is a trend not to complete the sequence of the whole genomes. Motivated by this Muñoz et al. recently studied the (one-sided) problem of filling an incomplete multichromosomal genome (or *scaffold*) H with respect to a complete target genome C such that the resulting genomic (or double-cut-and-join, DCJ for short) distance between H' and C is minimized, where H' is the corresponding filled scaffold. Jiang et al. recently extended this result to both the breakpoint distance and the DCJ distance and to the (two-sided) case when even C has some missing genes, and solved all these problems in polynomial time. However, when H and C contain duplicated genes, the corresponding breakpoint distance problem becomes NP-complete and there has been no efficient approximation or FPT algorithms for it. In this paper, we mainly consider the one-sided problem of filling scaffolds with gene repetitions so as to maximize the number of adjacencies between the two resulting sequences; namely, given an incomplete genome I and a complete genome G, both with gene repetitions, fill in the missing genes to obtain I' such that the number of adjacencies between I' and G is maximized. We prove that this problem is also NP-complete and present an efficient 1.33-approximation for the problem. The hardness result also holds for the two-sided problem for which a trivial factor-2 approximation exists. We also present FPT algorithms for some special cases of this problem.

1 Introduction

The recent genome sequencing technology makes it possible to sequence more organisms for genomic analysis. (Throughout this paper, we focus on unichromosomal genomes, each is represented as a sequence of genes. On the other hand, a multichromosomal genome is represented as sequences of genes.) In practice, the cost of finishing genome sequencing has not decreased at the same rate compared with the cost of random sequencing [2]. Consequently, many genomes released are not completely finished. Applying these incomplete genomes (scaffolds) for genomic analysis is far from ideal, as they could easily introduce unnecessary errors.

Hence, a natural combinatorial problem is to fill the missing genes into scaffolds. As one must find a biologically meaningful way of filling scaffolds, it makes sense to make use of some complete genomes (from some close species). Muñoz et al. [17] recently

R. Giancarlo and G. Manzini (Eds.): CPM 2011, LNCS 6661, pp. 55–64, 2011.

carried out this idea to fill an incomplete multichromosomal scaffold H to have H', such that the DCJ distance [19, 21] between H' and a given (complete) genome C is minimized. The DCJ (double-cut-and-join) distance is also called *rearrangement* distance, which is the minimum number of allowed rearrangement operations transforming one genome into the other. This was called the *one-sided* scaffold filling problem. It was shown that the one-sided scaffold filling problem (under the DCJ distance) can be solved in polynomial time [17].

Subsequently, Jiang *et al.* considered scaffold filling under the breakpoint distance and showed that even the two-sided problem (i.e., C is also incomplete) is polynomially solvable [13]. With a similar idea they also solved the two-sided scaffolding filling problem under the DCJ distance in polynomial time. However, when H and C both have gene repetitions, they proved that even the one-sided problem under the breakpoint distance is NP-complete [13].

In this paper, we extend our previous research on filling scaffolds with gene repetitions. There are two interesting points regarding this work. First, when there are gene repetitions, some practical way is missing to define the genomic (say breakpoint) distance between two genomes G_1, G_2. (The breakpoint distance for permutations was defined in [20].) The *exemplar genomic distance* between G_1 and G_2 (loosely speaking, the 'true' distance when the redundant genes are deleted), while biologically interesting, is too hard to compute [18]. (In fact, unless P=NP, there does not exist any polynomial time approximation for such a distance even when each gene is allowed to repeat three times [3, 5] or even two times [1, 15].) The 'breakpoint' distance between G_1 and G_2 used in [13] is very much the minimum common string partition between G_1 and G_2, for which no efficient FPT algorithm is known [6–8, 12, 14, 16]. Here we will use a different similarity measure, namely, the number of (common) adjacencies between G_1 and G_2, which has been used before in genomic analysis [4]. Not surprisingly, we proved that even the corresponding one-sided problem using this similarity measure is NP-complete, implying the two-sided problem to be NP-complete as well. Second, we design a factor-1.33 approximation for the one-sided problem — filling scaffolds with gene repetitions to maximize the number of adjacencies, improving upon a trivial factor-2 approximation for the general (two-sided) case. For genomic problems with gene repetitions, as far as we know, this is the first one which admits such a small approximation factor.

This paper is organized as follows. In Section 2, we give necessary definitions. In Section 3, we show the NP-completeness proof for the one-sided scaffold filling problem when gene duplications are allowed, using the number of adjacencies as the similarity measure. In Section 4, we present a factor-1.33 approximation for this problem. In Section 5, we discuss FPT algorithms for this problem. In Section 6, we conclude the paper.

2   Preliminaries

We first present some necessary definitions.

Given alphabet Σ, a string P is called a *permutation* if each element in Σ appears exactly once in P. We also use $c(P) = \Sigma$ to denote the multiset of elements in permutation P. G is a sequence on Σ if its elements form a multiset of Σ; e.g., $G = abcbacd$

with $\Sigma = \{a, b, c, d\}$ and $c(G) = \{a, a, b, b, c, c, d\}$. From now on, we focus on sequences with gene (letter) repetitions. An (unsigned) unichromosomal genome is just a sequence over Σ.

A *scaffold* (with gene/letter repetitions) is an incomplete sequence, i.e., with some missing elements. We use $+$ to denote scaffold filling, e.g., for a sequence A and an element multiset X, if A^* is a resulting sequence after filling all the elements in X into A, then $A^* = A + X$. Given two sequences A and B of length n, if $c(A) = c(B)$, then A and B are *related*. Given two related sequences A and B, two consecutive elements a_i and a_{i+1} in A form an *adjacency* if they are also consecutive in B (i.e., as $a_i a_{i+1}$ or $a_{i+1} a_i$), otherwise they form a *breakpoint*. The number of breakpoints in A, which is equal to that of B, is the breakpoint distance between A and B, denoted as $bd(A, B)$. Note that our breakpoint definition and the corresponding results all work when the letters (or genes) are possibly *signed*.

Given two related permutations P and Q of length n, let the number of breakpoints in P and Q be $b(P, Q)$ and let the number of adjacencies in P and Q be $a(P, Q)$, then we have

$$a(P, Q) + b(P, Q) = n - 1,$$

moreover, when $a(P, Q) = n - 1$ then we have $P = Q$ or P is the reversal of Q. However, when we are given two sequences G_1, G_2, over the same multiset of letters (i.e., possibly with duplications of some or all letters), then the latter relation is not true any more. Example. Let $G_1 = bcidabeb, G_2 = bebcidab$, then the number of adjacencies between G_1 and G_2 is 7 (which is the maximum), but G_1 is not equal to G_2.

The (*two-sided*) scaffold filling problem is defined as follows.

Scaffold Filling to Maximize the Number of (String) Adjacencies (SF-MNSA)
Input: two incomplete sequences G_1 and G_2 and two multisets of elements X and Y, where $X = c(G_2) - c(G_1)$ and $Y = c(G_1) - c(G_2)$.
Question: maximize the number of adjacencies $a(G_1 + X, G_2 + Y)$.

Note that if G_1 and G_2 were related (i.e., $c(G_2) = c(G_2)$), then we would have $X = Y = \emptyset$. In the above definition, when either X or Y is empty, we have the *one-sided* scaffold filling problem, which is formally defined as follows.

One-sided SF-MNSA
Input: an incomplete sequence I and a complete sequence G, with $X = c(G) - c(I) \neq \emptyset$ and $Y = c(I) - c(G) = \emptyset$.
Question: maximize the number of adjacencies $a(I + X, G)$.

In the next section we show that the One-sided SF-MNSA problem is NP-complete which implies that SF-MNSA is also NP-complete.

3   Hardness of SF-MNSA

In this section, we prove that SF-MNSA is NP-complete; in fact, even the One-sided SF-MNSA problem is NP-complete.

It is easy to see that One-sided SF-MNSA is in NP. We try to make a reduction from the NP-complete Exact Cover by 3-Sets (X3C) [11]. Recall that the input for X3C is a set of 3-sets $S = \{S_1, S_2, ..., S_m\}$. Each set S_i contains exactly 3 elements from a base set $X = \{x_1, x_2, ..., x_n\}$, where $n = 3q$ for some integer q. The problem is to decide whether there are q 3-sets in S which cover each element in X exactly once.

The main idea of our proof is to first show that a special case of X3C, X3C-1, is NP-complete. X3C-1 has the property that each 3-set S_u can share at most one element with any other 3-set S_v. Then we reduce X3C-1 to One-sided SF-MNSA.

Lemma 1. *X3C-1 is NP-complete.*

Proof. Again, it is easy to show that X3C-1 is in NP, so we focus on reducing X3C to X3C-1. Given an X3C instance with $3q$ elements and m 3-sets, we construct an instance of X3C-1 with $3q + 6m$ elements and $5m$ 3-sets. Assume that we are given a 3-set $S_u = \{x_i, x_j, x_k\}$, we construct five 3-sets as follows: $S'_u = \{ \{x_i, y_{u,1}, y_{u,2}\}, \{x_j, y_{u,3}, y_{u,4}\}, \{x_k, y_{u,5}, y_{u,6}\}, \{y_{u,1}, y_{u,3}, y_{u,5}\}, \{y_{u,2}, y_{u,4}, y_{u,6}\} \}$. Now in the new set of 3-sets $S' = \cup_u S'_u$ for all u, we have $5m$ 3-sets. Obviously, any pair of 3-sets in S' share at most one element. If $S_u = \{x_i, x_j, x_k\}$ is selected for a solution for S then we select $\{x_i, y_{u,1}, y_{u,2}\}$, $\{x_j, y_{u,3}, y_{u,4}\}$, and $\{x_k, y_{u,5}, y_{u,6}\}$ as a part of solution for S'. If $S_u = \{x_i, x_j, x_k\}$ is not selected for a solution for S then we select $\{y_{u,1}, y_{u,3}, y_{u,5}\}$ and $\{y_{u,2}, y_{u,4}, y_{u,6}\}$.

If the input X3C instance has a solution, one can easily use the selected q 3-sets (from S) to obtain $q + 2m$ 3-sets to cover the $3q + 6m$ elements contained in S'. Now assume that S' has an exact cover of size $q + 2m$. First, the base set contains $3q + 6m$ elements. By the construction of S'_u, if a solution other than $\{\{x_i, y_{u,1}, y_{u,2}\}, \{x_j, y_{u,3}, y_{u,4}\}, \{x_k, y_{u,5}, y_{u,6}\}\}$ or $\{\{y_{u,1}, y_{u,3}, y_{u,5}\}$ and $\{y_{u,2}, y_{u,4}, y_{u,6}\}\}$ is selected to cover elements $y_{u,j}, j = 1..6$, then either we have to cover some elements more than once or we fail to cover all of the $3q + 6m$ elements. Then, clearly the $q + 2m$ 3-sets selected for S' implies an exact cover of size q for S.

Therefore, we can conclude that the given X3C instance has a solution of size q iff the constructed X3C-1 instance has a solution of size $q + 2m$. It is easy to see that this reduction takes polynomial time. So the lemma is proven.   □

Theorem 1. *SF-MNSA is NP-complete.*

Proof. Following Lemma 1, we reduce X3C-1 to One-sided SF-MNSA. Let the instance of X3C-1 be $S = \{S_1, S_2, ..., S_m\}$. Each set S_i contains exactly 3 elements from a base set $Y = \{y_1, y_2, ..., y_n\}$, where $n = 3q$ for some integer q and $|S_u \cap S_v| \leq 1$ for all u, v. The problem is to decide whether there are q 3-sets in S which cover each element in Y exactly once.

Without loss of generality, we assume that both m and $3m - 3q$ are even, and the elements in each 3-set $S_u = \{y_{u1}, y_{u2}, y_{u3}\}$ are ordered by the indices of the elements, i.e., $u1 \leq u2 \leq u3$. Let

$$T_u = g_u f_u y_{u1} y_{u2} y_{u3} f'_u g'_u$$

and

$$M_u = g'_u f_u f'_u g_u.$$

Let y_i appear in all 3-sets of S for a total of n_i times (so $\sum_i n_i = 3m$). For each i, we map each of the $n_i - 1$ copies of y_i as a new letter z_j, we hence have a total of $3m - 3q$

new z_j letters. (Note that z_j could be empty — if the corresponding y_i appears exactly once.) Let

$$V_i = p_i z_i p_i'.$$

We construct two sequences G, I as follows.

$$G = r_1 r_2 \cdots r_{3m-3q-1} T_1 T_2 \cdots T_{m-1} T_m V_1 V_2 \cdots V_{3m-3q-1} V_{3m-3q}.$$

$$I = z_1 r_1 z_2 r_2 \cdots r_{3m-3q-1} z_{3m-3q} p_1 p_2 \cdots p_{3m-3q-1} p_{3m-3q} p_1' p_2' \cdots p_{3m-3q-1}' p_{3m-3q}'$$
$$M_1 M_3 \cdots M_{m-1} M_2 M_4 \cdots M_m.$$

Note that in the incomplete sequence I, each y_i is missed exactly once. We claim that there is a solution for the X3C-1 instance iff the maximum number of adjacencies generated by inserting these missing y_i's back into I is $4q$.

'$\rightarrow$' It is easy to see that if X3C-1 has a solution then we can obtain $4q$ adjacencies between G and I'. To obtain $4q$ of adjacencies between G and I', we simply insert a triple of elements (y_i, y_j, y_k) such that they are inserted back between f_u and f_u' in M_u, with $S_u = \{y_i, y_j, y_k\}$ being in the X3C-1 solution.

'$\leftarrow$' Suppose that one can insert the missing y_i's back into I to obtain I' such that there are $4q$ adjacencies between G and I', first notice that due to the construction of G and I, y_i's have to be inserted in M_j's, as inserting three missing y_i's anywhere before M_1 can only generate at most 3 adjacencies. Second, due to that any pair of 3-sets have at most one common element, we cannot insert two y_i's in some M_j and one y_i in some M_k, as that can only generate at most 3 adjacencies. Then, to insert three missing y_i's (say (y_i, y_j, y_k), ordered by their indices) between f_l and f_l' in M_l to generate 4 adjacencies, we must have $T_l = g_l f_l y_i y_j y_k f_l' g_l'$, or, in other words, $S_l = \{y_i, y_j, y_k\}$. This results in a solution for X3C-1.

It is easy to see that this construction takes polynomial time. As One-sided SF-MNSA is a special case of SF-MNFA, the theorem is proven.   □

In the next section, we present some approximation algorithms for SF-MNSA.

4   Approximation Algorithms for SF-MNSA

We first show some properties for SF-MNSA, which implies an easy 2-approximation for SF-MNSA. Then we try to improve the approximation factor to 1.33 for the One-Sided SF-MNSA problem.

4.1   A 2-Approximation Algorithm for SF-MNSA

Lemma 2. *Suppose that k is the optimal solution value for SF-MNSA, then we need to insert at least $\lceil k/2 \rceil$ and at most k genes into G_1 and G_2 when $c(G_1) \cap c(G_2) \neq \emptyset$.*

Proof. This lemma follows from the fact that when a missing gene x is inserted into G_1 or G_2, at least one adjacency and at most two adjacencies can be formed.   □

From the above lemma, it is easily seen that the optimal solutions are obtained in a way, (1) either a sequence of m missing genes $z_1, z_2, ..., z_m$ are inserted to obtain $m + 1$ adjacencies (say $y_1 z_1 z_2 \cdots z_m y_2$), (2) or a sequence of l missing genes $x_1, x_2, ..., x_l$ are inserted to form l adjacencies (say $x_1x_2\cdots x_ly$). Note that $m, l \geq 1$. We call the corresponding missing genes (and the inserted substrings) type-1 and type-2 respectively.

The above lemma also implies that one can easily obtain a factor-2 approximation — just make all missing genes type-2 and insert them into G_1 and G_2. We next show how to improve this factor to 1.33 for the One-sided SF-MNSA problem.

4.2   A 1.33-Approximation Algorithm for One-Sided SF-MNSA

Recall that in the One-sided SF-MNSA problem, we have input I and G, and we need to insert the missing genes into I. Let k_1 be the number of missing genes inserted. (By Lemma 2, $\lceil k/2 \rceil \leq k_1 \leq k$.) Let b_i be the number of new adjacencies obtained in an optimal solution, subtract by the total length of inserted substrings (of type-1 missing genes) of length i. (Example, $I = abcd$, $G = acbacd$, and we need to insert two type-1 substrings of length one: a and c. Four new adjacencies are formed, so $b_1 = 4 - 2 = 2$.) Then we have,

Lemma 3. *Suppose that $OPT = k$ is the optimal solution value for One-sided SF-MNSA, then*

$$OPT = k_1 + b_1 + b_2 + \cdots + b_q \leq \frac{4}{3}(k_1 + b_1/2 + b_2/4).$$

Proof. By definition, $OPT = k_1 + b_1 + b_2 + \cdots + b_q$. As each inserted substring counted in b_j has j genes for $j \geq 3$ and following Lemma 2, $b_3 + b_4 + \cdots b_q \leq \frac{1}{3}(k_1 - b_1 - 2b_2)$. Hence,

$$OPT \leq \frac{4}{3}(k_1 + b_1/2 + b_2/4). \qquad \square$$

From Lemma 3, it is easy to see that our algorithm hinges on approximating $b_1/2 + b_2/4$, which uses a greedy idea. The algorithm goes as follows. At Step 1, for each missing gene a to be inserted into I, we use a greedy method to scan from left to right to find a 2-substring cd in I such that the insertion of a results in a 3-substring cad which contains two adjacencies $\langle ca \rangle$ and $\langle ad \rangle$. At Step 2, for each pair of missing genes x, y to be inserted into I, we scan in I, again left to right, to find whether there is a 2-substring wz such that x, y can be inserted into it to obtain three adjacencies, i.e., either $\langle wxyz \rangle$ or $\langle wyxz \rangle$. We insert all such pairs of missing genes, in a greedy fashion, into I. At Step 3, we insert the remaining missing genes in an arbitrary fashion into I, provided that each inserted missing gene generates one adjacency.

We have the following lemma regarding this greedy algorithm.

Lemma 4. *Let b_1', b_2' be the number of new adjacencies subtracted by the total number of missing genes inserted at Step 1 and Step 2 of our greedy algorithm respectively. Then $b_1' + b_2' \geq \frac{b_1}{2} + \frac{b_2}{4}$.*

Proof. Let k_1', k_2' be the number of missing genes inserted at Step 1 and Step 2 respectively. (So $b_1' = k_1'$ and $b_2' = k_2'/2$.) First, note that each of the k_1' inserted missing genes can destroy at most two type-1 adjacencies and at most two type-1 2-substrings in an optimal solution. Let b_{11}' be the number of missing genes inserted at Step 1 which destroy exactly two type-1 adjacencies in some optimal solution. Let b'_{12} be the number of missing genes inserted at Step 1 which destroy one type-1 adjacency and one type-1 2-substring in some optimal solution. Let b'_{13} be the number of missing genes inserted at Step 1 which destroy exactly two type-1 2-substrings in some optimal solution. Obviously,

$$k'_1 = b'_1 = b'_{11} + b'_{12} + b'_{13}.$$

We show an example for a, one of the b'_{13} inserted missing genes that destroy two type-1 2-substrings in OPT (i.e., counted into b_2). Let $G = \cdots \alpha a \beta \cdots \gamma ab\delta \cdots \alpha uv \beta \cdots$ and let $I = \alpha \cdots \alpha \beta \cdots \gamma \delta \cdots \beta \cdots a \cdots$. We need to insert a, b, u, v into I. Due to the greedy fashion of the algorithm, a is inserted between α, β in I (destroying the possibility of inserting uv at the same location). On the other hand, due to the insertion of a (instead of ab), ab cannot be inserted in between γ and δ.

Next, we need to show that at Step 2, each of the inserted type-1 2-substrings can destroy at most three type-1 2-substrings in some optimal solution. This is again easy to see. Suppose that we need to insert xy into I to obtain three adjacencies $\langle \alpha x y \beta \rangle$. Due to the greedy fashion something else (like uv) are inserted between α, β instead. Then xw and yz could be destroying two other locations for the optimal insertion of type-2 2-substrings.

Now, putting all together,

$$b_1 \le 2b'_{11} + b'_{12},$$

and

$$b_2 \le 3b'_2 + b'_{12} + 2b'_{13}.$$

Then

$$\frac{b_1}{2} + \frac{b_2}{4} \le \frac{2b'_{11} + b'_{12}}{2} + \frac{3b'_2 + b'_{12} + 2b'_{13}}{4}$$
$$= (b'_{11} + \frac{3b'_{12}}{4} + \frac{b'_{13}}{2}) + \frac{3b'_2}{4}$$
$$\le b'_1 + b'_2 \qquad \qquad \Box$$

From the above lemma, we have the following theorem.

Theorem 2. *There is a greedy algorithm which approximates One-sided SF-MNSA with a factor of 1.33.*

Proof. Following the greedy algorithm and Lemma 4, we have the approximation solution value APP, which satisfies the following inequalities.

$$APP = k_1 + b'_1 + b'_2$$
$$\ge k_1 + \frac{b_1}{2} + \frac{b_2}{4}$$
$$\ge \frac{3}{4} OPT$$
$$= OPT/1.33$$

So the theorem is proven. $\Box$

5   Exact Algorithms for Some Variants of One-Sided SF-MNSA

In the previous section, we design a factor-2 approximation for SF-MNSA and a factor-1.33 approximation for One-sided SF-MNSA. In practice, it might be interesting to solve SF-MNSA with exact algorithms.

An FPT (Fixed-Parameter Tractable) algorithm for a decision problem Π with input parameters $k_1, \cdots, k_i$ is an algorithm which solves the problem in $O(f(k_1, \cdots, k_i)n^c)$ $= O^*(f(k_1, \cdots, k_i))$ time, where f is any function only on $k_1, \cdots, k_i$, n is the input size and c is some fixed constant not related to $k_1, \cdots, k_i$. For convenience we also say that Π is in FPT. More details on FPT algorithms can be found in [9, 10].

Unfortunately, we do not know how to design an FPT algorithm only parameterized on k (i.e., the maximum number of adjacencies), even for the One-sided SF-MNSA problem. This makes an interesting open question. In the following, we present two simple FPT algorithms for two practical variants of One-sided SF-MNSA. While not trying to claim too much credits for these simple algorithms, we hope that these algorithms could help shed light on answering the open question. On the other hand, these two variants are closely related to computational genomics, hence are meaningful practically.

5.1   FPT Algorithm for One-Sided d-SF-MNSA

Firstly, in *d-SF-MNSA* each gene in G_1 or G_2 only appears at most d times. So for One-sided d-SF-MNSA each gene in I, G appears at most d times. Then, let us assume that we need to insert an x into I. Certainly, we hope to insert x into I to obtain at least one adjacency xy (or yx). Now let us look at G, as x appears in G at most d times, x has at most $2d$ neighbors (at least one of them should be y). As we have no information on what the right y should be, we have to try over all such (at most) $2d$ possibilities in I for each inserted x. Since we might have to insert a total of k letters, the running time is

$$O^*(((2d) \times (2d))^k) = O^*((2d)^{2k}).$$

Theorem 3. *One-sided d-SF-MNSA can be solved in $O^*((2d)^{2k})$ time.*

5.2   FPT Algorithm for One-Sided SF-MNSA^c

In *SF-MNSA^c*, each gene is selected from a set of c letters Σ. For One-sided SF-MNSA^c, the algorithm is similar to that of One-sided d-SF-MNSA, with a simple twist. Assume that one needs to insert a missing x into I to obtain some adjacency xy (or yx), one just needs to find the neighbors of x in G. It is easy to see that if we have two substrings xy's in G then we can match either one of them with the intended adjacency xy (introduced due to the insertion of x in I). As Σ contains only c letters, we could have c choices for the intended adjacencies xy (resp. yx). When we insert x into I, we also have c possibilities to obtain xy. Therefore, we need to try over all such (at most) c^2 possibilities for each inserted x. By Lemma 2, we need to insert at most k genes into I, so the running time of the algorithm is

$$O^*(c^{2k}).$$

Theorem 4. *One-sided SF-MNSA^c can be solved in $O^*(c^{2k})$ time.*

6   Concluding Remarks

In this paper, we extend the scaffold filling problem when gene repetitions are allowed. We use the number of adjacencies between two genomes as the similarity measure. It is not surprising that this SF-MNSA problem is NP-complete. A very interesting open problem is whether one can improve the factor-2 approximation for SF-MNSA. Another open problem is whether an FPT algorithm only parameterized on k exists for the (One-sided) SF-MNSA problem. Finally, in reality when inserting missing genes we cannot insert them inside some important contigs (i.e., we cannot insert missing genes anywhere as we want), but we do not know how to handle this problem. Some algorithms in [13] have the same problem and more study is needed along this line.

Acknowledgments

This research is partially supported by NSF of China under grant 60928006 and by the Opening Fund of Top Key Discipline of Computer Software and Theory in Zhejiang Provincial Colleges at Zhejiang Normal University. We also thank anonymous reviewers for several useful comments.

References

1. Blin, G., Fertin, G., Sikora, F., Vialette, S.: The exemplar breakpoint distance for non-trivial genomes cannot be approximated. In: Das, S., Uehara, R. (eds.) WALCOM 2009. LNCS, vol. 5431, pp. 357–368. Springer, Heidelberg (2009)
2. Chain, P., Grafham, D., Fulton, R., Fitzgerald, M., Hostetler, J., Muzny, D., Ali, J., et al.: Genome project standards in a new era of sequencing. Science 326, 236–237 (2009)
3. Chen, Z., Fu, B., Fowler, R., Zhu, B.: On the inapproximability of the exemplar conserved interval distance problem of genomes. J. Combinatorial Optimization 15(2), 201–221 (2008)
4. Chen, Z., Fu, B., Xu, J., Yang, B., Zhao, Z., Zhu, B.: Non-breaking similarity of genomes with gene repetitions. In: Ma, B., Zhang, K. (eds.) CPM 2007. LNCS, vol. 4580, pp. 119–130. Springer, Heidelberg (2007)
5. Chen, Z., Fu, B., Zhu, B.: The approximability of the exemplar breakpoint distance problem. In: Cheng, S.-W., Poon, C.K. (eds.) AAIM 2006. LNCS, vol. 4041, pp. 291–302. Springer, Heidelberg (2006)
6. Chen, X., Zheng, J., Fu, Z., Nan, P., Zhong, Y., Lonardi, S., Jiang, T.: Computing the assignment of orthologous genes via genome rearrangement. In: Proc. of the 3rd Asia-Pacific Bioinformatics Conf. (APBC 2005), pp. 363–378 (2005)
7. Chrobak, M., Kolman, P., Sgall, J.: The greedy algorithm for the minimum common string partition problem. In: Jansen, K., Khanna, S., Rolim, J.D.P., Ron, D. (eds.) RANDOM 2004 and APPROX 2004. LNCS, vol. 3122, pp. 84–95. Springer, Heidelberg (2004)
8. Damaschke, P.: Minimum Common String Partition Parameterized. In: Crandall, K.A., Lagergren, J. (eds.) WABI 2008. LNCS (LNBI), vol. 5251, pp. 87–98. Springer, Heidelberg (2008)
9. Downey, R., Fellows, M.: Parameterized Complexity. Springer, Heidelberg (1999)
10. Flum, J., Grohe, M.: Parameterized Complexity Theory. Springer, Heidelberg (2006)
11. Garey, M.R., Johnson, D.S.: Computers and Intractability: A Guide to the Theory of NP-Completeness. W.H. Freeman, New York (1979)

12. Goldstein, A., Kolman, P., Zheng, J.: Minimum common string partition problem: Hardness and approximations. In: Fleischer, R., Trippen, G. (eds.) ISAAC 2004. LNCS, vol. 3341, pp. 484–495. Springer, Heidelberg (2004)

13. Jiang, H., Zheng, C., Sankoff, D., Zhu, B.: Scaffold filling under the breakpoint distance. In: Tannier, E. (ed.) RECOMB-CG 2010. LNCS, vol. 6398, pp. 83–92. Springer, Heidelberg (2010)

14. Jiang, H., Zhu, B., Zhu, D., Zhu, H.: Minimum common string partition revisited. In: Lee, D.-T., Chen, D.Z., Ying, S. (eds.) FAW 2010. LNCS, vol. 6213, pp. 45–52. Springer, Heidelberg (2010)

15. Jiang, M.: The zero exemplar distance problem. In: Tannier, E. (ed.) RECOMB-CG 2010. LNCS(LNBI), vol. 6398, pp. 74–82. Springer, Heidelberg (2010)

16. Kaplan, H., Shafrir, N.: The greedy algorithm for edit distance with moves. Inf. Process. Lett. 97(1), 23–27 (2006)

17. Muñoz, A., Zheng, C., Zhu, Q., Albert, V., Rounsley, S., Sankoff, D.: Scaffold filling, contig fusion and gene order comparison. BMC Bioinformatics 11, 304 (2010)

18. Sankoff, D.: Genome rearrangement with gene families. Bioinformatics 16(11), 909–917 (1999)

19. Tesler, G.: Efficient algorithms for multichromosomal genome rearrangements. J. Computer and System Sciences 65, 587–609 (2002)

20. Watterson, G., Ewens, W., Hall, T., Morgan, A.: The chromosome inversion problem. J. Theoretical Biology 99, 1–7 (1982)

21. Yancopoulos, S., Attie, O., Friedberg, R.: Efficient sorting of genomic permutations by translocation, inversion and block interchange. Bioinformatics 21, 3340–3346 (2005)

String Comparison and Lyndon-Like Factorization Using V-Order in Linear Time[*]

David E. Daykin[1], Jacqueline W. Daykin[2,4], and W.F. Smyth[3,4]

[1] Department of Mathematics, University of Reading, UK
[2] Department of Computer Science
Royal Holloway & King's College, University of London, UK
J.Daykin@cs.rhul.ac.uk, jackie.daykin@kcl.ac.uk
[3] Algorithms Research Group,
Department of Computing & Software
McMaster University, Hamilton, Ontario, Canada
smyth@mcmaster.ca
[4] Centre for Stringology & Applications
Digital Ecosystems & Business Intelligence Institute
Curtin University, Perth WA 6845, Australia

Abstract. In this paper we extend previous work on Unique Maximal Factorization Families (UMFFs) and a total (but non-lexicographic) ordering of strings called V-order. We describe linear-time algorithms for string comparison and Lyndon factorization based on V-order. We propose extensions of these algorithms to other forms of order.

1 Introduction

This paper presents algorithms for a generalization of Lyndon words known as circ-UMFFs; that is, families of strings which permit the unique maximal factorization of any given string [7,8,10]. For over half a century, both the combinatorics and algorithmics of Lyndon words have been studied extensively [3,14,17,20]. Lyndon words have been applied in tackling a surprisingly wide range of problems: the Burrows-Wheeler Transform and data compression [2,15], musicology [1], bioinformatics [13], cryptanalysis [18], string combinatorics [16,20], and free Lie algebras [19]. We therefore study UMFFs as generalizations of Lyndon words in order to extend the range of applications. UMFFs can be characterized by the following simple lemma:

Lemma 1 ([7]). *A set $\mathcal{W}$ of strings over an alphabet Σ with $\Sigma \subseteq \mathcal{W}$ is an UMFF if and only if $xy, yz \in \mathcal{W}$ and y nonempty imply $xyz \in \mathcal{W}$.*

Definition 1 ([8,10]). *Let x denote a string over an alphabet Σ.*

* *If for some positive integer k, $x = w_1 w_2 \cdots w_k$, where every (nonempty) factor w_i, $1 \le i \le k$, is an element of some set $\mathcal{W}$, then $w_1 w_2 \cdots w_k$ is said to be a **factorization of x over** $\mathcal{W}$, written $F_{\mathcal{W}}(x)$.*

---

[*] The work of the third author was supported in part by the Natural Sciences & Engineering Research Council of Canada.

R. Giancarlo and G. Manzini (Eds.): CPM 2011, LNCS 6661, pp. 65–76, 2011.

* $\mathcal{W}$ is a **factorization family** (FF) if for every nonempty $\boldsymbol{x}$, there exists a factorization $F_{\mathcal{W}}(\boldsymbol{x})$.
* An FF $\mathcal{W}$ is a **unique maximal factorization family** (UMFF) if, in the factorization of every $\boldsymbol{x}$, no factor $\boldsymbol{w_i} \in \mathcal{W}$ can be extended to left or to right so that the extension $\boldsymbol{w_i'} \in \mathcal{W}$. That is, $\mathcal{W}$ satisfies Lemma 1, and thus each factor in a factorization is **maximal**.
* An UMFF $\mathcal{W}$ is a **circ-UMFF** if it contains exactly one rotation of every primitive $\boldsymbol{x}$.

As shown in [7,8], for a circ-UMFF $\mathcal{W}$, the factors $\boldsymbol{w_i}$ of every $\boldsymbol{x}$ satisfy an **UMFF-order** $>_{\mathcal{W}}$, which is a total order over all the elements of $\mathcal{W}$, such that

$$\boldsymbol{w_1} \geq_{\mathcal{W}} \boldsymbol{w_2} \geq_{\mathcal{W}} \cdots \geq_{\mathcal{W}} \boldsymbol{w_k}. \tag{1}$$

Thus, by virtue of its defining properties, a circ-UMFF satisfies, in terms of UMFF-order, an analogue or generalization of the Lyndon factorization theorem [3], which guarantees that the Lyndon words $\mathcal{L}$, in terms of lexicographic order (**lexorder**), provide a unique maximal factorization (1) of every string $\boldsymbol{x}$. In other words, $\mathcal{L}$ is an *example* of a circ-UMFF $\mathcal{W}$, raising the possibility that other orderings of the strings $\boldsymbol{x} \in \Sigma^+$ may yield alternative circ-UMFFs distinct from $\mathcal{L}$ (as in fact described in [7]).

In particular, we consider here the V-word circ-UMFF first introduced in [7]. Whereas Lyndon words are defined using lexorder, the V-word circ-UMFF is defined using V-order [4,6] (see below). Hence the analogues of Lyndon words are known as V-words, which are ordered according to a simple extension of lexorder called **lexextension**, a combination of lexorder and V-order.

The Lyndon factorization for both RAM and PRAM algorithms has linear complexity [5,12,14,20]. We match this linear complexity for sequential string factorization over the V-word circ-UMFF. Our novel algorithm requires that substrings $\boldsymbol{u}, \boldsymbol{v}$ are compared, not in lexorder, but in V-order. We also achieve string comparison using V-order in linear time, using a method based on the current longest common suffix of the substrings $\boldsymbol{u}, \boldsymbol{v}$ and implemented using a simple doubly-linked list.

Hence we have distinct methods for string factorization. Once a string has been decomposed over a circ-UMFF, say the Lyndon words, the factors are then maximal and thus final. Given an alternative circ-UMFF, say V-words, the Lyndon factors can be re-factored allowing for deeper burrowing into the string, for example yielding a subsequence of V-word prefixes, suffixes or substrings of all the Lyndon factors. Thus, in a natural way, algorithms for additional circ-UMFFs may lead to computing further subsequences of interest.

Designing these algorithms also suggested new combinatorial results for V-order and V-words. We compare and contrast the ordering of words in lexorder with that of V-order. We show, for instance, that any subsequence of a string precedes the string in V-order, but of course this is not necessarily true in lexorder. While similar to the co-Lyndon circ-UMFF [10], the V-word alphabet is in reverse order to that of the defining V-order.

V-order is defined using a star tree of letter deletions as follows. Let Σ be a totally ordered alphabet, and let $\boldsymbol{u} = u_1 u_2 ... u_n$ be a string over Σ. Define $h \in \{1, ..., n\}$ by $h = 1$ if $u_1 \leq u_2 \ ... \ \leq u_n$; otherwise, by the unique value such that $u_{h-1} > u_h \leq u_{h+1} \leq u_{h+2} \leq \ ... \ \leq u_n$. Let $\boldsymbol{u}^* = u_1 u_2 ... u_{h-1} u_{h+1} ... u_n$, where the star * indicates deletion of the "V" letter u_h. Write $\boldsymbol{u}^{s*}$ for $(...(\boldsymbol{u}^*)^* ...)^*$ with $s \geq 0$ stars[1]. Let $\mathrm{L} = max\{u_1, u_2, ..., u_n\}$ and let $k = k_L$ be the number of occurrences of L in $\boldsymbol{u}$. Then the sequence $\boldsymbol{u}, \boldsymbol{u}^*, \boldsymbol{u}^{2*}, ...$ ends $\mathrm{L}^k, ..., \mathrm{L}^2, \mathrm{L}^1, \mathrm{L}^0 = \varepsilon$. In the **star tree** each string $\boldsymbol{u}$ over Σ labels a vertex, and there is a directed edge from $\boldsymbol{u}$ to $\boldsymbol{u}^*$, with ε as the root.

Definition 2. *We define the V-order $\prec$ between distinct strings $\boldsymbol{u}, \boldsymbol{v}$. Firstly $\boldsymbol{u} \succ \boldsymbol{v}$ if $\boldsymbol{v}$ is in the path $\boldsymbol{u}, \boldsymbol{u}^*, \boldsymbol{u}^{2*}, ..., \varepsilon$. If $\boldsymbol{u}, \boldsymbol{v}$ are not in a path, there are smallest s, t with $\boldsymbol{u}^{(s+1)*} = \boldsymbol{v}^{(t+1)*}$. Put $\boldsymbol{c} = \boldsymbol{u}^{s*}$ and $\boldsymbol{d} = \boldsymbol{v}^{t*}$, then $\boldsymbol{c} \neq \boldsymbol{d}$ but length $\boldsymbol{c} =$ length $\boldsymbol{d} = m$ say. Let j be the greatest i in $1 \leq i \leq m$ such that $\boldsymbol{c}[i] \neq \boldsymbol{d}[i]$. If $\boldsymbol{c}[j] < \boldsymbol{d}[j]$ in Σ then $\boldsymbol{u} \prec \boldsymbol{v}$. Clearly $\prec$ is a total order.*

Surprisingly, although V-order is not immediately intuitive, it gives rise to V-words which have analogous structure to Lyndon words, as shown in the following example.

Example 1. Let $\Sigma = \{a < b < c < d\}$, and the string $\boldsymbol{s} = cbaabdcba$. Then the Lyndon factorization of $\boldsymbol{s}$ is $c \geq_{\mathcal{L}} b \geq_{\mathcal{L}} aabdcb \geq_{\mathcal{L}} a$, while the V-word factorization is $cbaab \geq_V dcba$.

2   String Comparison

In this section we discuss the comparison in V-order of two finite strings that, to avoid trivialities, we suppose to be not equal. First we describe a new algorithm to do V-order comparisons of strings in worst-case time linear in their lengths, thus asymptotically the same as comparisons in lexorder. We then go on to prove new combinatorial results related to the V-order comparison of strings that suggest the possibility of an alternative linear-time V-order comparison algorithm. We suppose throughout that $\boldsymbol{x}$ is defined on a finite ordered alphabet Σ of size $\sigma = |\Sigma|$. If a string $\boldsymbol{x}$ is less than a string $\boldsymbol{y}$ in lexorder, we write $\boldsymbol{x} < \boldsymbol{y}$ and $\boldsymbol{y} > \boldsymbol{x}$; if $\boldsymbol{x}$ is less than $\boldsymbol{y}$ in V-order, we write $\boldsymbol{x} \prec \boldsymbol{y}$ and $\boldsymbol{y} \succ \boldsymbol{x}$.

   In order to simplify the description of our algorithm, we suppose without loss of generality that the letters $\lambda \in \Sigma$ are drawn from the first σ natural numbers. We represent strings $\boldsymbol{x} = \boldsymbol{x}[1..n]$ using added sentinel positions $\boldsymbol{x}[0]$ and $\boldsymbol{x}[n+1]$ for processing convenience. At each position $i \in \{1, ..., n\}$ we define a triple $(\lambda, \mathit{left}, \mathit{right})$, where λ is the letter at position i, $\mathit{left} = i-1$, and $\mathit{right} = i+1$; for $i = 0$, $(\lambda, \mathit{left}, \mathit{right}) = (\sigma+1, -1, 1)$ and for $i = n+1$, $(\lambda, \mathit{left}, \mathit{right}) = (\sigma+1, n, n+2)$. The elements of the triple at position i are denoted $\boldsymbol{x}[i].\lambda$, $\boldsymbol{x}[i].\mathit{left}$, and $\boldsymbol{x}[i].\mathit{right}$. Thus we represent $\boldsymbol{x}$ as a linked list to facilitate the deletion of positions in $\boldsymbol{x}$ required to locate parent nodes in the star tree used to define V-order. Note that

---

[1] Note that this star operator, as defined in [6] and [7], is distinct from the Kleene star operator.

the linked list operations (for the left- and right-pointers) consist of increment, decrement and read, which can be implemented in constant time.

We suppose that two unequal strings x_1 and x_2 are given, both defined on Σ, of lengths $n_1 > 0$ and $n_2 \geq n_1$ respectively. We describe a $\Theta(n_2)$-time algorithm to solve the following problem:

(P1) Determine whether $x_1 \prec x_2$ or $x_1 \succ x_2$.

> — *Left-extend the suffix of nondecreasing letters, then delete.*
> — *Position i is next right of the deleted position δ.*
> **function** $delete(x, i)$: x, i, δ
> $prev \leftarrow x[i].left$
> **while** $x[i].\lambda \geq x[prev].\lambda$ **do**
> $i \leftarrow prev;\ prev \leftarrow x[i].left$
> $\delta \leftarrow i;\ i \leftarrow x[\delta].right$
> $x[i].left \leftarrow prev;\ x[prev].right \leftarrow i$

> — *Perform n_2-n_1 successive deletions.*
> **function** $reduce(x_2, n_2, n_1)$: x_2, i_2
> $i_2 \leftarrow n_2+1$
> **for** $j \leftarrow 1$ **to** n_2-n_1 **do**
> $(x_2, i_2, \delta) \leftarrow delete(x_2, i_2)$

Fig. 1. Reduction of x_2 to n_1 letters

Recall that in order to compare two strings in V-order, it is necessary first to identify a common ancestor v (perhaps the empty string ε) in the star tree, and then to compare the children of v. The common ancestor is found by performing iterated deletes in the given strings according to the following rule:

> Delete the letter at the rightmost position $i \in \{1, \ldots, n\}$ of x such that $x[i] \leq x[i+1]$ and $x[i] < x[i-1]$. (Note that the sentinel positions 0 and $n+1$ ensure the correct application of this rule.)

Since no comparison can be made until x_2 is reduced to the same length as x_1, we begin by performing n_2-n_1 deletions from x_2, as shown in Figure 1. Then we continue by performing single deletions from *both* x_1 and x_2, checking at each step to determine whether or not the current deletion has made the reduced x_1 equal to the reduced x_2. To facilitate this checking, we introduce the idea of the ***longest matching suffix*** (LMS): the longest suffix of the reduced x_1 that matches a suffix of the reduced x_2. Of course LMS is a subsequence (possibly empty) of both x_1 and x_2; it is useful to identify the leftmost position in LMS as LMS-ℓ, and the position immediately to the left of LMS-ℓ as LMS-$\ell\ell$. Clearly, when LMS(x_1, x_2) equals the length of the reduced strings, the common ancestor of x_1 and x_2 in the star tree has been found (ε when LMS is empty). We make the following observations:

— *Recompute the positions ℓ_1, ℓ_2 that define LMS-ℓ.*
function LMS($\boldsymbol{x_1}, \ell_1, \boldsymbol{x_2}, \ell_2$) : ℓ_1, ℓ_2
$i_1 \leftarrow \boldsymbol{x_1}[\ell_1].left;\;\; i_2 \leftarrow \boldsymbol{x_2}[\ell_2].left$
while $\boldsymbol{x_1}[i_1].\lambda = \boldsymbol{x_2}[i_2].\lambda$ **and** $i_1 > 0$ **do**
    $i_1 \leftarrow \boldsymbol{x_1}[i_1].left;\;\; i_2 \leftarrow \boldsymbol{x_2}[i_2].left$
$\ell_1 \leftarrow \boldsymbol{x_1}[i_1].right;\;\; \ell_2 \leftarrow \boldsymbol{x_2}[i_2].right$

— *Determine whether or not $\boldsymbol{x_1} \prec \boldsymbol{x_2}$.*
function $prec(\boldsymbol{x_1}, n_1, \boldsymbol{x_2}, n_2)$: *boolean*
$(\boldsymbol{x_2}, i_2) \leftarrow reduce(\boldsymbol{x_2}, n_2, n_1)$
$(\ell_1, \ell_2) \leftarrow$ LMS($\boldsymbol{x_1}, n_1+1, \boldsymbol{x_2}, n_2+1$)
    — *Since $\boldsymbol{x_1} \neq \boldsymbol{x_2}$, $\boldsymbol{x_1}[\ell_1].left = 0$ implies that $\boldsymbol{x_1}$ lies*
    — *on the upward path from $\boldsymbol{x_2}$ in the star tree.*
if $\boldsymbol{x_1}[\ell_1].left = 0$ **then return** TRUE
$i_1 \leftarrow n_1+1$ — i_2 *already set by* $reduce(\boldsymbol{x_2}, n_2, n_1)$
repeat
    $(\boldsymbol{x_1}, i_1, \delta_1) \leftarrow delete(\boldsymbol{x_1}, i_1);\;\; (\boldsymbol{x_2}, i_2, \delta_2) \leftarrow delete(\boldsymbol{x_2}, i_2)$
        — *Remark 2: possibly LMS-ℓ was deleted in at least*
        — *one of $\boldsymbol{x_1}, \boldsymbol{x_2}$; if so, shift LMS-$\ell$ right in both $\boldsymbol{x_1}, \boldsymbol{x_2}$.*
    if $\delta_1 = \ell_1$ **or** $\delta_2 = \ell_2$ **then**
        $\ell_1 \leftarrow \boldsymbol{x_1}[\ell_1].right;\;\; \ell_2 \leftarrow \boldsymbol{x_2}[\ell_2].right$
        — *Remark 3: possibly LMS-$\ell\ell$ was deleted in one*
        — *of $\boldsymbol{x_1}, \boldsymbol{x_2}$, while LMS-$\ell$ was not deleted in either.*
    elsif $\ell_1 = i_1$ **or** $\ell_2 = i_2$ **then**
        $(\ell_1, \ell_2) \leftarrow$ LMS($\boldsymbol{x_1}, \ell_1, \boldsymbol{x_2}, \ell_2$)
until $\boldsymbol{x_1}[\ell_1].left = 0$ — ($\boldsymbol{x_1}[\ell_1].left = 0 \Leftrightarrow \boldsymbol{x_2}[\ell_2].left = 0$)
    — *V-order is determined by the lexorder of the last deleted letters.*
if $\boldsymbol{x_1}[\delta_1].\lambda < \boldsymbol{x_2}[\delta_2].\lambda$ **then**
    return TRUE
else
    return FALSE

Fig. 2. Match the reduced $\boldsymbol{x_1}$ and $\boldsymbol{x_2}$ to compare $\boldsymbol{x_1} : \boldsymbol{x_2}$

Remark 1. A position (not LMS-ℓ) of $\boldsymbol{x_1}$ (respectively, $\boldsymbol{x_2}$) is deleted within LMS($\boldsymbol{x_1}, \boldsymbol{x_2}$) if and only if the corresponding position (also not LMS-ℓ) within $\boldsymbol{x_2}$ (respectively, $\boldsymbol{x_1}$) is also deleted. In this case LMS cannot be extended to the left, but is reduced in length by one.

Remark 2. LMS cannot be extended to the left if *either* LMS-ℓ is deleted in either $\boldsymbol{x_1}$ or $\boldsymbol{x_2}$ *or* positions left of LMS-$\ell\ell$ are deleted in both $\boldsymbol{x_1}$ and $\boldsymbol{x_2}$. In the latter case, LMS is unchanged; in the former it is reduced in length by one.

Remark 3. The LMS may possibly (but not necessarily) be extended to the left only if *either* LMS-$\ell\ell$ is deleted in both $\boldsymbol{x_1}$ and $\boldsymbol{x_2}$ *or* LMS-$\ell\ell$ is deleted in $\boldsymbol{x_1}$ (respectively, $\boldsymbol{x_2}$) and a position left of LMS-$\ell\ell$ is deleted in $\boldsymbol{x_2}$ (respectively, $\boldsymbol{x_1}$).

These are the essential facts exploited in the function *prec* (Figure 2) that returns TRUE or FALSE depending on whether or not $\boldsymbol{x_1} \prec \boldsymbol{x_2}$.

Lemma 2. *Procedure* prec *executes in time and space linear in* $|x_2|$.

Proof. We consider the efficiency of function *prec*. In terms of space, *prec* requires only three storage locations for each position in x_1 and x_2, namely the triple $(\lambda, \textit{left}, \textit{right})$, in addition to constant space for variables and program storage; thus its space requirement is $\Theta(n_1+n_2) = \Theta(n_2)$. To estimate time usage, first consider the function *delete*. This function may left-extend the current suffix of nondecreasing letters, but the deleted position is always at the left end of the suffix. It follows that over all the executions of *delete* (at most n_1+n_2 of them), each position in x_1 and x_2 is visited at most twice, with constant-time processing corresponding to each visit. These visits comprise either the scanning of letters from right to left, or, while re-scanning from left to right, a deletion. (Any deletions within LMS are achieved by the function *delete* that updates the linked list in constant time.) Similarly, the calls to function LMS (at most n_1 altogether) go from right to left without backtracking, and so each position is visited at most once, each with constant-time processing. We conclude that the time requirement for *prec* is $\Theta(n_1+n_2)$; that is, linear in the lengths of the compared strings.   □

In order to describe our combinatorial results, we let L denote the largest letter of Σ that actually appears in a string x, and suppose that L occurs in x with frequency $k > 0$. The **V-form** of x is then defined as

$$V_k(x) = x = x_0 L x_1 L \cdots x_{k-1} L x_k, \tag{2}$$

for possibly empty $x_i, i = 0, 1, \ldots, k$. Observe that if L_1 is the second largest letter in x, then V-form can be recursively (and independently) applied to each x_i that includes L_1 with frequency $k_1 > 0$:

$$V_{k_1}(x_i) = x_{i,0} L_1 x_{i,1} \cdots x_{i,k_1-1} L_1 x_{i,k_1}.$$

Denote by $A^T(x)$ the set of ancestors of x in the star tree, and by $A_j(x)$ the jth ancestor, $0 \le j \le |x|$ (the zeroth ancestor is x itself, the $|x|^{\text{th}}$ ancestor is the empty string ε).

Lemma 3. *If* $V_k(x) = x_0 L x_1 L \cdots x_{k-1} L x_k$, *then*

$$A_{t_j}(x) = x_0^T L x_1^T \cdots x_{k-1}^T L x_k^T,$$

where $j \in 0..k$, $t_j = |x_j| + |x_{j+1}| + \cdots + |x_k|$, *and*

$$x_h^T = x_h \text{ for } 0 \le h < j,$$
$$= \varepsilon \text{ for } j \le h \le k.$$

Thus $A_{t_0}(x) = L^k$, *while for* $j \ge 1$, $A_{t_j}(x)$ *has suffix* L^{k-j+1} *and* $A_{t_{j-1}}(x) = A_{|x_j|-1}(A_{t_j}(x))$.

The next lemma is an extension of result (C) in [7, Section 4], stated but not explicitly proved. It provides the interesting insight that if v is any proper subsequence of x, then $v \prec x$.

Lemma 4. *Given a string x of length n, let $v = x[i_1]x[i_2] \cdots x[i_r]$ for $1 \le i_1 < i_2 < \cdots < i_r$, $0 \le r < n$. Then $v \prec x$.*

Proof. Write $V_k(x)$ as in (2) and observe that if v includes j occurrences of L, $0 \le j < k$, then v must have ancestor L^j in the star tree but no ancestor L^k. Thus in this case $v \prec x$.

Suppose then that $j = k$ and write $V_k(v) = v_0 L v_1 L \cdots v_{k-1} L v_k$, also in V-form (2). Since $|v| < |x|$, it follows that there is some least integer $k' \in 0..k$ such that $|v_{k'}| < |x_{k'}|$, hence that v has ancestor

$$A_{s_{k'}}(v) = x_0^T L x_1^T L \cdots x_{k-1}^T L x_k^T, \tag{3}$$

where $s_{k'} = |v_{k'}| + |v_{k'+1}| + \cdots + |v_k|$, and

$$x_h^T = x_h \text{ for } 0 \le h < k',$$
$$= \varepsilon \text{ for } k' \le h \le k.$$

Then $A_{s_{k'}}(v) = z L^{k-k'}$ for some string z that either is empty or has suffix L. On the other hand, we know from Lemma 3 that x has ancestor

$$A_{t_{k'}}(x) = x_0^T L x_1^T L \cdots x_{k-1}^T L x_k^T, \tag{4}$$

where $t_{k'} = |x_{k'}| + |x_{k'+1}| + \cdots + |x_k| > s_{k'}$. Comparing (3) and (4), we see that $A_{s_{k'}}(v) = A_{t_{k'}}(x)$ is a common ancestor of v and x, whereas $\overline{v} = A_{s_{k'} - |v_{k'}|} = z v_{k'} L^{k-k'}$ and $\overline{x} = A_{t_{k'} - |x_{k'}|} = z x_{k'} L^{k-k'}$ are distinct. It follows that $v \prec x$ (with $|v| < |x|$ and largest letter L) if and only if $v_{k'} \prec x_{k'}$ (with $|v_{k'}| < |x_{k'}|$ and largest letter $L_1 < L$), and so to prove the lemma we need only consider the proper substrings $v_{k'}$ and $x_{k'}$. We can therefore continue recursively, at each stage replacing v/x-strings by proper v/x-substrings in which every letter is less than the largest letter occurring in the preceding stage. Since x is finite, and since the v-substring is always shorter than the x-substring, it must at some stage be true that the occurrences of the current largest letter in the v-substring are fewer than its occurrences in the x-substring. Hence $v \prec x$, as required. $\square$

We conclude this section by stating a lemma from [7] which we apply in the design of the V-word factorization algorithm VF.

Lemma 5. *Given distinct strings v and x with corresponding V-forms*

$$v = v_0 L_v v_1 L_v \cdots v_{j-1} L_v v_j, \quad x = x_0 L_x x_1 L_x \cdots x_{k-1} L_x x_k,$$

let $h \in 0..\max(j, k)$ denote the least integer such that $v_h \ne x_h$. Then $v \prec x$ (respectively, $v \succ x$) if and only if one of the following conditions holds:
(a) $L_v < L_x$ (respectively, $L_v > L_x$);
(b) $L_v = L_x$ and $j < k$ (respectively, $j > k$);
(c) $L_v = L_x$ and $j = k$ and $v_h \prec x_h$ (respectively, $v_h \succ x_h$).

3   Lyndon-Like Factorization

If a string $x = uv$, then vu is said to be a *rotation* (cyclic shift) of x, specifically the $|u|^{th}$ rotation $R_{|u|}(x)$, where $|u| \in \{0, \ldots, |x|\}$. Note that $R_0(x) = R_{|x|}(x)$. Then a string x of length n is a **V-word** iff $x \prec R_i(x)$ for every $i \in \{1, \ldots, n-1\}$; that is, iff x is the unique minimum in V-order over all rotations of x. Clearly no V-word can be a repetition. For given x, not a repetition, we write VW(x) to denote the rotation of x that is its V-word. Notice that the V-form (2) of a V-word x must, by the properties of V-order, begin with the largest letter L in x. Thus in (2) $x_0 = \varepsilon$, and VW(x) must take the form

$$\text{L}x_1 \cdots \text{L}x_{k-1}\text{L}x_k. \tag{5}$$

Accordingly the k rotations of (5) (those commencing $\text{L}x_1, \text{L}x_2, \ldots, \text{L}x_k$) are the only ones that need to be considered in order to determine VW(x). This means that we seek the minimum rotation of $X = x_1x_2 \cdots x_k$, say $Z = z_1z_2 \cdots z_k$, where the individual components x_i and z_i, $1 \le i \le k$, are compared in V-order, while X and Z are compared in the lexorder of the components. (For this comparison we need to interpret $x_i = \varepsilon$ as a least string in V-order.) As noted in the introduction, this hybrid order is called the **lexextension** of V-order $\prec$, here denoted by $<_V$. More formally:

Definition 3 ([7]). *Let $u_1, u_2, \ldots, u_m$ and $v_1, v_2, \ldots, v_n$ be given strings over Σ, and let $u = u_1u_2\ldots u_m$ and $v = v_1v_2\ldots v_n$. Then the lexicographic extension $\prec_{LEX}$ order is defined as:*
$u \prec_{LEX} v$ if either u is a proper prefix of v, that is $u_i = v_i$ for $1 \le i \le m < n$, or $u_i = v_i$ for $1 \le i < m$ and $u_{i+1} \prec v_{i+1}$.

In other words $u_1u_2\ldots u_m \prec v_1v_2\ldots v_n$ in the lexicographic order of strings, using not $<$ but $\prec$.

Hence, as shown in [7]:

Lemma 6. *(a) The string (5) is a V-word iff $x_1x_2 \cdots x_k$ is a Lyndon word under lexextension; (b) The V-words form a circ-UMFF $\mathcal{V}$.*

Example 2. If $\Sigma = \{1 < 2 < 3 < 4\}$, then the *V*-ordering of all rotations of the string 13142 is $42131 \prec 14213 \prec 31421 \prec 13142 \prec 21314$. Hence 13142 is not a *V*-word, while 42131 is a *V*-word.

Example 3. Consider the string $s = 5554213151421353142162131461 3142$. From Definition 3 and Example 2, the factorization of s into V-words is
5554213151421353 1421 $\ge_V$ 621314 $\ge_V$ 613142.

Thus, denoting by VF(x) the factorization of x into V-words, we have

(P2) Given a nonempty string x on a finite ordered alphabet Σ, compute VF(x).

In the remainder of the paper we consider the computational problem (P2). In order to compute $\text{VF}(\boldsymbol{x})$, we need to identify the unique subset of V-words $\boldsymbol{w}_j$, $j = 1, 2, ..., J$, such that $\boldsymbol{x} = \boldsymbol{w_1 w_2 \cdots w_J}$ and $\boldsymbol{w}_j \geq_V \boldsymbol{w}_{j+1}$ for every $j \in \{1, \ldots, J-1\}$. The following lemmas make clear that the order $<_V$ is not always the same as V-order $\prec$:

Lemma 7 ([7]). *Given letters* $\lambda, \mu \in \Sigma$, $\lambda \prec \mu \iff \mu >_V \lambda$.

Lemma 8. *For any strings* $\boldsymbol{u}$ *and* $\boldsymbol{v} \prec \boldsymbol{w}$ *defined on* Σ, *all with maximum letter less than* L, $\text{L}\boldsymbol{uv} <_V \text{L}\boldsymbol{uw}$.

Proof. By Lemma 6 both $\text{L}\boldsymbol{uv}$ and $\text{L}\boldsymbol{uw}$ are V-words. The result is immediate for $\boldsymbol{u} = \varepsilon$, and for $\boldsymbol{v} = \varepsilon$ it follows from the corresponding result for prefixes of general circ-UMFFs [8]. It suffices therefore to show that $\boldsymbol{uv} \prec \boldsymbol{uw}$ for nonempty $\boldsymbol{u}$ and $\boldsymbol{v}$. The proof is by induction. Suppose first that $\boldsymbol{u}$ is a single letter $\lambda < \text{L}$, and consider the sequence of deletions in the star tree that yields a common ancestor $\boldsymbol{z}$ of $\lambda \boldsymbol{v}$ and $\lambda \boldsymbol{w}$. Three outcomes are possible: $\boldsymbol{z}$ has prefix $\mu > \lambda$, $\boldsymbol{z} = \varepsilon$, and $\boldsymbol{z}$ has prefix λ. In the first of these cases, the prefix λ has been deleted on both upward paths (from $\lambda \boldsymbol{v}$ and from $\lambda \boldsymbol{w}$) in the star tree, and so therefore at some point λ was the strictly least letter remaining in each of the reduced strings. Thus apart from the deletion of λ, the deletions from $\lambda \boldsymbol{v}$ and $\lambda \boldsymbol{w}$ are the same as they would have been from $\boldsymbol{v}$ and $\boldsymbol{w}$, and so $\lambda \boldsymbol{v} \prec \lambda \boldsymbol{w}$. Consider the second case. Since $\boldsymbol{z} = \varepsilon$, the previous deletions were from strings consisting of single letters, α, β say, where distinct α, β are maximal letters in $\lambda \boldsymbol{v}$, $\lambda \boldsymbol{w}$ respectively. Suppose that one of α, β is equal to λ, for otherwise the previous argument applies. If $\beta = \lambda$, then $\alpha > \lambda$ contradicting $\boldsymbol{v} \prec \boldsymbol{w}$. If $\alpha = \lambda$, then $\beta \geq \lambda$, and due to distinctness, $\alpha < \beta$ as required. In the third case, $\boldsymbol{z} = \lambda \boldsymbol{z}^*$, where $\boldsymbol{z}^*$ is the root for $\boldsymbol{v}$ and $\boldsymbol{w}$, and therefore at the deciding positions on the two paths below the root, the same letters determine $\lambda \boldsymbol{v} \prec \lambda \boldsymbol{w}$ that also determine $\boldsymbol{v} \prec \boldsymbol{w}$. Thus the lemma holds for a single letter λ.

Suppose now that the lemma holds for every prefix $\boldsymbol{u}$ of length $\ell \geq 1$, and consider a prefix $\overline{\boldsymbol{u}}$ of length $\ell+1$. Here there are two cases to consider: either the common ancestor $\boldsymbol{z}$ of $\overline{\boldsymbol{u}}\boldsymbol{v}$ and $\overline{\boldsymbol{u}}\boldsymbol{w}$ has prefix $\overline{\boldsymbol{u}}$, or not. If so, then $\overline{\boldsymbol{u}}$ has played no role in the decision, and $\boldsymbol{v} \prec \boldsymbol{w} \implies \overline{\boldsymbol{u}}\boldsymbol{v} \prec \overline{\boldsymbol{u}}\boldsymbol{w}$. If not, there exist ancestors $\boldsymbol{u}'\boldsymbol{v}^*$ and $\boldsymbol{u}''\boldsymbol{w}^*$ of $\overline{\boldsymbol{u}}\boldsymbol{v}$ and $\overline{\boldsymbol{u}}\boldsymbol{w}$, respectively, such that $|\boldsymbol{u}'| = |\boldsymbol{u}''| = \ell$. In fact, since the first letter deleted from $\overline{\boldsymbol{u}}$ must be the same in both cases, we see moreover that $\boldsymbol{u}' = \boldsymbol{u}''$, and we can conclude therefore from the inductive assumption that $\boldsymbol{u}'\boldsymbol{v} \prec \boldsymbol{u}'\boldsymbol{w}$, hence that $\boldsymbol{uv} \prec \boldsymbol{uw}$, as required. $\square$

Figure 3 gives pseudocode for a linear-time algorithm VF to solve (P2), based on Duval's Lyndon factorization algorithm [14] as presented in [20]. Algorithm VF essentially considers two adjacent substrings of $\boldsymbol{x}[1..n]$: a lefthand one $\boldsymbol{u}$ (that may be a repetition repeated *rep* times, and whose positions are tracked by i) of length $rep \times \ell_1$ with a prefix of k_1 L's; and a righthand one $\boldsymbol{v}$ (tracked by j) of length ℓ_2 with a prefix of k_2 L's. At each step of the algorithm j is

procedure VF(x)

   — *Given the input string $x\$$, output in ascending sequence*
     the rightmost positions of every V-word in VF(x)

$h \leftarrow 0$ — *the total length of VF already output*

while $h < n$ **do**   — *continue as long as total* VF $< n$

    RESET

    while $x[j] \leq$ L **do**

      if $x[i] = x[j] =$ L **then**

        — *Extend the L-prefix of either u or v.*

        if $j - i = k_1$ **then** $k_1 \leftarrow k_1 + 1$; $rep \leftarrow rep + 1$; $\ell_1 \leftarrow \ell_1 + 1$

        else

            if $i - h \leq k_1$ **then** $k_2 \leftarrow k_2 + 1$

            $\ell_2 \leftarrow \ell_2 + 1$; $i \leftarrow i + 1$

      elsif $x[j] =$ L **then**

        — *The copy v of u truncated by a mismatched L:*
          output u^{rep} and restart $u \leftarrow v$.

        output (h, rep, ℓ_1); RESET; $j \leftarrow j - 1$

      elsif $x[i] =$ L **then**

        — *u contains a mismatched L: if $k_2 = 0$, continue u*
          by incrementing ℓ_1; otherwise, concatenate $u \leftarrow u^{rep}v$.

        if $j = k_1 + 1$ **then** $\ell_1 \leftarrow k_1 + 1$

        else $\ell_1 \leftarrow (rep \times \ell_1) + 1$

        $rep \leftarrow 1$; $i \leftarrow h + 1$

        if $k_2 > 0$ **then** $\ell_1 \leftarrow \ell_1 + \ell_2$; $k_2, \ell_2 \leftarrow 0$

      elsif $x[i] = x[j]$ **then**

        — *Corresponding non-L positions in u & v are equal.*

        $\ell_2 \leftarrow \ell_2 + 1$

        if $\ell_2 = \ell_1$ **then** $rep \leftarrow rep + 1$; $k_2, \ell_2 \leftarrow 0$; $i \leftarrow h + 1$

        else $i \leftarrow i + 1$

      else

        — *Unequal non-L positions: compare u & extended v.*

        $j' \leftarrow j$; **while** $x[j+1] <$ L **do** $j \leftarrow j + 1$

            — *If $u \succ v$, output u^{rep}; otherwise, concatenate $u \leftarrow u^{rep}v$.*

        if not $prec(x[h+1..h+\ell_1], \ell_1, x[j' - \ell_2..j], j - j' + \ell_2 + 1)$ **then**

            output (h, rep, ℓ_1)

        $\ell_1 \leftarrow j - h$; $rep \leftarrow 1$; $k_2, \ell_2 \leftarrow 0$; $i \leftarrow h + 1$

      $j \leftarrow j + 1$

    output (h, rep, ℓ_1); **output** $(h, 1, \ell_2)$

Fig. 3. Compute VF(x)

incremented by one, and then tests are performed to determine whether as a result u^{rep} and v should be concatenated into a single V-word, or whether each of rep occurrences of u (and perhaps also v, if L has increased) should be output as the current V-word(s), or whether no decision can currently be taken. The variable h gives the rightmost position of the last V-word output (thus one less than the starting position of u). VF invokes three other routines: the function $prec$ of Section 2, an initialization procedure RESET:

$$i \leftarrow h+1; \ j \leftarrow h+2; \ \mathrm{L} \leftarrow \boldsymbol{x}[i]$$
$$k_1, \ell_1, rep \leftarrow 1; \ k_2, \ell_2 \leftarrow 0$$

and a procedure **output** (h, rep, ℓ) that for $\ell > 0$ outputs the rightmost positions h of V-word prefixes of $\boldsymbol{u}^{rep}\boldsymbol{v}$ (and updates the global variable h). For processing convenience, the input string to procedure VF is actually $\boldsymbol{x}\$$, where $\$$ is a sentinel letter greater than any letter of Σ.

Lemma 9. *Procedure* $\mathrm{VF}(\boldsymbol{x})$ *executes correctly in time and space linear in* $|\boldsymbol{x}|$.

Proof. Analogous to Lyndon factorization, V-word factorization maintains current candidates for factors (corresponding to the current maximum letter L) in the form $\boldsymbol{u}^{rep}\boldsymbol{v}$, subject to the following rules:

* $\boldsymbol{u} = \mathrm{L}^{k_1}\boldsymbol{u}^*$, where $\boldsymbol{u}^*$ contains no letter $\lambda > \mathrm{L}$ and no substring $\mathrm{L}^k, k \geq k_1$;
* $\boldsymbol{v} = \mathrm{L}^{k_2}, 0 \leq k_2 < k_1$, or $\boldsymbol{v} = \mathrm{L}^{k_1}\boldsymbol{v}^*$, where $\boldsymbol{v}^*$ is a proper prefix of $\boldsymbol{u}^*$;
* if it becomes true that $\boldsymbol{v}^* = \boldsymbol{u}^*$, set $rep \leftarrow rep+1$, $\boldsymbol{v} \leftarrow \varepsilon$.

Only in the case that distinct letters $\lambda_1 < \mathrm{L}$ and $\lambda_2 < \mathrm{L}$ are found at corresponding positions i in $\boldsymbol{u}$ and j in $\boldsymbol{v}$, respectively, does it become necessary to call *prec* to determine whether or not $\boldsymbol{u} \prec \boldsymbol{v}$ — with $\boldsymbol{v}$ first extended to the right until some letter $\lambda \geq \mathrm{L}$ (possibly the sentinel $\$$) is found. We remark that this case can occur only if $k_2 = k_1$, with the current $\ell_2 < \ell_1$.

To establish the linearity of VF's execution time, note first that all the outputs taken together require at most $O(n)$ time. Also, the restarts that occur when $\boldsymbol{x}[j] = \mathrm{L} \neq \boldsymbol{x}[i]$ can total at most $O(n)$ time. Otherwise, all operations performed at each increment of j require only constant time except for the call to *prec*. Observe that the total number of positions tested in *prec* over all calls is at most the sum of all the terms $j-j'+\ell_1+\ell_2+1$. If a call to *prec* in procedure VF yields $\boldsymbol{u} \succ \boldsymbol{v}$, then an output occurs, and $\boldsymbol{u} \leftarrow \boldsymbol{v}$; if not, however, then $\boldsymbol{u} \leftarrow \boldsymbol{u}\boldsymbol{v}$. In the latter case, $\boldsymbol{u}$ necessarily increases in size, so that the next call to *prec*, if it occurs, requires that correspondingly more positions need to be tested. We see that repeated consecutive invocations of *prec*, without output, must yield a worst case for total time, since each call must include all the positions already tested in previous calls. Note however that the result $\boldsymbol{u} \prec \boldsymbol{v}$ can occur only if the number of L's in $\boldsymbol{v}$ is at least equal to the number of L's in $\boldsymbol{u}$, subject to the constraint noted above that $k_2 = k_1$; thus the lengths of both $\boldsymbol{u}$ and $\boldsymbol{v}$ must approximately double at each step, as in the example

$$\boldsymbol{x} = \mathrm{L1L2 \ L1L3 \ L1L2L1L4} \cdots,$$

where we first find $\mathrm{L1} \prec \mathrm{L2}$, then $\mathrm{L1L2} \prec \mathrm{L1L3}$, $\mathrm{L1L2L1L3} \prec \mathrm{L1L2L1L4}$, and so on. This example in fact must constitute a worst case for the number of positions tested in *prec*; assuming without loss of generality that $|\boldsymbol{x}| = 2^m$, we find that the total number of positions tested will be

$$2^m + 2^{m-1} + \cdots + 2 = 2(|\boldsymbol{x}|-1). \qquad \square$$

References

1. Chemillier, M.: Periodic musical sequences and Lyndon words. In: Soft Computing - A Fusion of Foundations, Methodologies and Applications, vol. 8-9, pp. 611–616. Springer, Heidelberg (2004)
2. Crochemore, M., Désarménien, J., Perrin, D.: A note on the Burrows-Wheeler transformation. In: Scientific Commons (2005), http://en.scientificcommons.org/16732444
3. Chen, K.T., Fox, R.H., Lyndon, R.C.: Free differential calculus, IV — the quotient groups of the lower central series. Ann. Math. 68, 81–95 (1958)
4. Daykin, D.E.: Ordered ranked posets, representations of integers and inequalities from extremal poset problems. In: Rival, I. (ed.) Graphs and order, Proceedings of a Conference in Banff (1984). NATO Advanced Sciences Institutes Series C: Mathematical and Physical Sciences, vol. 147, pp. 395–412. Reidel, Dordrecht-Boston (1985)
5. Daykin, D.E.: Algorithms for the Lyndon unique maximal factorization. J. Combin. Math. Combin. Comput. (to appear)
6. Danh, T.-N., Daykin, D.E.: The structure of V-order for integer vectors. In: Hilton, A.J.W. (ed.) Congr. Numer., vol. 113, pp. 43–53 (1996)
7. Daykin, D.E., Daykin, J.W.: Lyndon-like and V-order factorizations of strings. J. Discrete Algorithms 1–3/4, 357–365 (2003)
8. Daykin, D.E., Daykin, J.W.: Properties and construction of unique maximal factorization families for strings. Internat. J. Found. Comput. Sci. 19–4, 1073–1084 (2008)
9. Daykin, D.E., Daykin, J.W., Iliopoulos, C.S., Smyth, W.F.: Generic algorithms for factoring strings (in preparation)
10. Daykin, D.E., Daykin, J.W., (Bill) Smyth, W.F.: Combinatorics of Unique Maximal Factorization Families (UMFFs). Fund. Inform. 97–3, Special StringMasters Issue Janicki, R., Puglisi, S.J., Rahman, M.S. (eds.), pp. 295–309 (2009)
11. Daykin, D.E., Daykin, J.W., Smyth, W.F.: Sequential and Parallel Algorithms for Lyndon Factorization using V -Order (in preparation)
12. Daykin, J.W., Iliopoulos, C.S., Smyth, W.F.: Parallel RAM algorithms for factorizing words. Theoret. Comput. Sci. 127, 53–67 (1994)
13. Delgrange, O., Rivals, E.: STAR: an algorithm to Search for Tandem Approximate Repeats. Bioinformatics 20–16, 2812–2820 (2004)
14. Duval, J.P.: Factorizing words over an ordered alphabet. J. Algorithms 4, 363–381 (1983)
15. Gil, J., Scott, D.A.: A bijective string sorting transform (submitted)
16. Iliopoulos, C.S., Smyth, W.F.: Optimal algorithms for computing the canonical form of a circular string. Theoret. Comput. Sci. 92–1, 87–105 (1992)
17. Lothaire, M.: Combinatorics on Words. Addison-Wesley, Reading (1983); 2nd edn. Cambridge University Press, Cambridge (1997).
18. Perret, L.: A chosen ciphertext attack on a public key cryptosystem based on Lyndon words. In: Proc. International Workshop on Coding and Cryptography, pp. 235–244 (2005)
19. Reutenauer, C.: Free Lie algebras, London Math. Soc. Monographs New Ser., vol. 7, p. 288. Oxford University Press, Oxford (1993)
20. Smyth, B.: Computing patterns in strings. p. 423. Pearson, Addison-Wesley (2003)

A d-Step Approach for Distinct Squares in Strings*

Antoine Deza, Frantisek Franek, and Mei Jiang

Advanced Optimization Laboratory
Department of Computing and Software
McMaster University, Hamilton, Ontario, Canada L8S 4K1
{deza,franek,jiangm5}@mcmaster.ca

Abstract. We present an approach to the problem of maximum number of distinct squares in a string which underlines the importance of considering as key variables both the length n and $n - d$ where d is the size of the alphabet. We conjecture that a string of length n and containing d distinct symbols has no more than $n - d$ distinct squares, show the critical role played by strings satisfying $n = 2d$, and present some properties satisfied by strings of length bounded by a constant times the size of the alphabet.

Keywords: string, distinct squares, primitively rooted distinct squares, d-step approach.

1 Introduction

The problem of the number of distinct squares when the types of the squares in a string are counted rather than the occurrences, was first introduced by Fraenkel and Simpson [3] showing that the number of distinct squares in a string of length n is bounded from above by $2n$ and giving a lower bound of $n - o(n)$ asymptomatically approaching n from below for primitively rooted squares. Let us remark that a primitively rooted square is a square whose generator is primitive, i.e. not a repetition. Later, Ilie [4] provided a simpler proof of the main lemma of [3] and slightly improved the upper bound to $2n - \Theta(\log n)$ in [5]. It is believed, that the number of distinct squares is bounded by the length of the string.

In this paper we investigate the problem of primitively rooted distinct squares in relationship to the alphabet of the string. Let us denote by $\sigma_d(n)$ the maximum number of primitively rooted distinct squares over all strings of length n containing exactly d distinct symbols. We conjecture that $\sigma_d(n) \leq n - d$, and point to possible avenues for investigating the conjecture.

* Supported in part by grants from the Natural Sciences and Engineering Research Council of Canada for the first 2 authors, and by the Canada Research Chair Programme and Mathematics of Information Technology and Complex Systems grants for the first author, and by Queen Elizabeth II Graduate Scholarship in Science and Technology for the third author.

R. Giancarlo and G. Manzini (Eds.): CPM 2011, LNCS 6661, pp. 77–89, 2011.

$$n - d$$

	1	2	3	4	5	6	7	8	9	10	11
1	**1**	**1**	1	1	1	1	1	1	1	1	·
2	1	**2**	**2**	3	3	4	5	6	7	7	·
3	1	2	**3**	**3**	4	4	5	6	7	8	·
4	1	2	3	**4**	**4**	5	5	6	7	8	·
5	1	2	3	4	**5**	**5**	6	6	7	8	·
d 6	1	2	3	4	5	**6**	**6**	7	7	8	·
7	1	2	3	4	5	6	**7**	**7**	8	8	·
8	1	2	3	4	5	6	7	**8**	**8**	9	·
9	1	2	3	4	5	6	7	8	**9**	**9**	·
10	1	2	3	4	5	6	7	8	9	**10**	·
11	·	·	·	·	·	·	·	·	·	·	·

Fig. 1. $(d, n - d)$ table: entries computed for $\sigma_d(n)$ with $1 \le d \le 10$ and $1 \le n - d \le 10$

Similarly as in [2], which was dealing with the maximum number of runs in a string with respect to the string's alphabet, we present some elementary structures of the entries for $\sigma_d(n)$ presented in a so-called $(d, n - d)$ table whose rows are indexed by d and columns are indexed by $n - d$, and point to ways of applying reductions to the problem of distinct squares. A fragment of the table for $d \le 10$ and $n - d \le 10$ is shown in Fig. 1.

Several regularities can be observed in the fragment of the $(d, n - d)$ table: first observe that $\sigma_d(n) \le n - d$ is satisfied by all known entries. There are several other regularities that can be observed in the table; some are proven analytically in section 2, some are shown to be equivalent with the conjectured upper bound for $\sigma_d(n)$, some are shown to lead to a slightly stronger upper bound – see section 3. In section 4 we investigate the structure of relatively short square-maximal strings on the main diagonal. In section 5, we discuss possible ways to investigate the conjectured upper bound using the methods and insight presented in section 4.

First we introduce the notation used in this paper. $S_d(n)$ denotes the set of strings of length n with exactly d distinct symbols; $s(x)$ denotes the number of primitively rooted distinct squares in a string x; $\sigma_d(n) = \max\{ s(x) \mid x \in S_d(n) \}$. $\mathcal{A}(x)$ denotes the alphabet set of a string x; a **singleton** of x refers to a symbol in a string x that occurs exactly once, a **pair** refers to a symbol that occurs exactly twice, a **triple** refers to a symbol that occurs exactly three times, and in general an **k-tuple** (k times).

2   Some Basic Properties of the $(d, n - d)$ Table

The following auxiliary lemma will be used later to investigate the structure of square-maximal strings.

Lemma 1. *Let x be a square-maximal string of length n with exactly d symbols, and let every symbol of x occur at most 3 times. Then every pair in x must be adjacent.*

Proof. Let $x \in S_d(n)$ be square-maximal. Let us assume that x has a non-adjacent pair of C's. Case (i): if the pair does not occur in any square, then we can create a string y by moving the C's to the end. This will not destroy any square of x, but we gain a new square CC, which contradicts the square-maximality of x. Case (ii): if the pair occurs in at least one square, let us move the two C's to the end of the string. For every square $uCvuCv$ of x destroyed by the removal of the C's, we gain a new square $uvuv$: if $uvuv$ already existed in some other part of x, every symbol of uv would have to occur in x at least 4 times, which is not possible. Thus every destroyed square $uCvuCv$ is replaced by a new square $uvuv$, in addition we gain a new square CC. This contradicts the square-maximality of x. □

The next proposition summarizes basic properties of the $(d, n-d)$ table.

Proposition 1. *For any $2 \le d \le n$:*

(a) $\sigma_d(n) \le \sigma_d(n+1)$, *i.e. the values are non-decreasing when moving left-to-right along a row.*

(b) $\sigma_d(n) \le \sigma_{d+1}(n+1)$, *i.e. the values are non-decreasing when moving top-to-bottom along a column.*

(c) $\sigma_d(n) < \sigma_{d+1}(n+2)$, *i.e. the values are strictly increasing when moving left-to-right and top-to-bottom along descending diagonals.*

(d) $\sigma_d(2d) = \sigma_d(n) = \sigma_{d+1}(n+1)$ *for $n \le 2d$, i.e. the values under and on the main diagonal along a column are constant.*

(e) $\sigma_d(n) \ge n-d$ *for $n \le 2d$, i.e. the values under and on the main diagonal are at least as big as conjectured; $\sigma_d(2d+1) \ge d$ and $\sigma_d(2d+2) \ge d+1$.*

(f) $\sigma_d(2d) - \sigma_{d-1}(2d-1) \le 1$, *i.e. the difference between the value on the main diagonal and the value immediately above it is no more than 1.*

Proof. (a) Let $x \in S_d(n)$ be square-maximal. Let y be x appended with a symbol $a \in \mathcal{A}(x)$. Then $y \in S_d(n+1)$, and $\sigma_d(n+1) \ge s(y) \ge s(x) = \sigma_d(n)$.

(b) Let $x \in S_d(n)$ be square-maximal. Let y be x appended with a symbol $a \notin \mathcal{A}(x)$. Then $y \in S_{d+1}(n+1)$, and $\sigma_{d+1}(n+1) \ge s(y) = s(x) = \sigma_d(n)$.

(c) Let $x \in S_d(n)$ be square-maximal, let $a \notin \mathcal{A}(x)$. Define a new string y as x concatenated with aa. Then $y \in S_{d+1}(n+2)$, and $\sigma_{d+1}(n+2) \ge s(y) = s(x) + 1 > s(x) = \sigma_d(n)$.

(d) Let $n \le 2d$ and let $x \in S_{d+1}(n+1)$ be square-maximal. Since $2(d+1) \ge n+2 > n+1$, x has a singleton. Let y be x with the singleton removed. Then $y \in S_d(n)$ and $s(y) \ge s(x)$ as no square can be destroyed while some squares can be created. Thus, $\sigma_d(n) \ge s(y) \ge s(x) = \sigma_{d+1}(n+1)$. By (b), $\sigma_d(n) \le \sigma_{d+1}(n+1)$, so $\sigma_d(n) = \sigma_{d+1}(n+1)$ for $n \le 2d$.

(e) Let $n \leq 2d$ and consider the string $x = aabbcc\ldots$ consisting of $n - d$ adjacent pairs. Then $x \in S_{n-d}(2n - 2d)$ and $s(x) = n - d$. By (d), $\sigma_d(n) = \sigma_{n-d}(2n - 2d) \geq s(x) = n - d$. Let consider the strings $y = aaabbcc\ldots$ consisting of $d - 1$ adjacent pairs except for the first 3 entries being aaa, and $z = aababaccdd\ldots$ consisting of $d-2$ adjacent pairs except for the first 6 entries being $aababa$. We have $\sigma_d(2d + 1) \geq s(y) = d$ and $\sigma_d(2d + 2) \geq s(z) = d + 1$.

(f) Let $x \in S_d(2d)$ be square-maximal. Case (i): if x has a singleton, let y be x with the singleton removed, then $y \in S_{d-1}(2d - 1)$ and $s(y) \geq s(x)$. It follows that $\sigma_d(2d) = s(x) \leq s(y) \leq \sigma_{d-1}(2d - 1)$, and since $\sigma_d(2d) \geq \sigma_{d-1}(2d - 1)$ by (b), therefore we get $\sigma_d(2d) = \sigma_{d-1}(2d - 1)$. Case (ii): if x does not have a singleton, then x consists of pairs, and by Lemma 1, x consists of adjacent pairs, and thus $\sigma_d(2d) = s(x) = d$. Consider the string $z = aaabbcc\ldots$ consisting of $d-2$ adjacent pairs except for the first 3 entries being aaa. We have $\sigma_{d-1}(2d - 1) \geq s(z) = d - 1 = \sigma_d(2d) - 1$, i.e., $\sigma_d(2d) - \sigma_{d-1}(2d - 1) \leq 1$.   □

3   Main Results

This sections contains several propositions that are equivalent with the conjectured upper bound for $\sigma_d(n)$. We also present conditions that lead to a slightly stronger upper bound in Theorems 3 and 4. It can be observed in the $(d, n - d)$ table, that the known values on the main diagonal are identities, i.e. $\sigma_d(2d) = d$ – which is equivalent to $\sigma_d(2d) \leq d$ by Proposition 1(e). The next theorem shows that, indeed, this observation is equivalent with the conjectured bound. In essence, the theorem shows that if the upper bound is violated, then there must be a violation on the main diagonal.

Theorem 1. *The conjectured upper bound $\sigma_d(n) \leq n - d$ holding true for all strings is equivalent with the statement: $\sigma_d(2d) \leq d$ for every $d \geq 2$.*

Proof. Let $n \geq d \geq 2$, $\sigma_d(n) \leq n - d$ clearly implies that $\sigma_d(2d) \leq d$; that is, by Proposition 1(e), $\sigma_d(2d) = d$. To prove the other direction, we consider case (i) $2d > n$: by Proposition 1(d) we have $\sigma_d(n) = \sigma_{n-d}(2n - 2d) \leq n - d$, and case (ii) $n > 2d$: by Proposition 1(b) we have $\sigma_d(n) \leq \sigma_{n-d}(2n - 2d) \leq n - d$.   □

Another observation of the $(d, n - d)$ table given in Figure 1 is that the value on the main diagonal and the value of its right neighbour are identical. Theorem 2 shows that the inequality is equivalent with the conjectured upper bound, while the equality gives rise to a slightly stronger upper bound given in Theorem 4.

Theorem 2. *The conjectured upper bound $\sigma_d(n) \leq n - d$ holding true for all strings is equivalent with the statement: $\sigma_d(2d + 1) - \sigma_d(2d) \leq 1$ for every $d \geq 2$.*

Proof. The statement follows from the conjectured upper bound is clear. Let us, thus prove the opposite direction. We shall prove by contradiction that

$\sigma_d(2d) \leq d$ for $d \geq 2$. Let $d \geq 2$ be the least such that $\sigma_d(2d) > d$. From the computed values of the $(d, n - d)$ table it follows that $d > 10$. Let $x \in S_d(2d)$ be square-maximal. If x does not have a singleton, then $n = 2d$ and x consists of pairs, and thus by Lemma 1, x consists of adjacent pairs and $\sigma_d(2d) = d$, a contradiction. Thus, x must have a singleton. Let y be x with the a singleton removed. Then $y \in S_{d-1}(2d - 1)$ and $s(y) \geq s(x)$. Thus, $\sigma_{d-1}(2d - 1) \geq s(y) \geq s(x) = \sigma_d(2d)$. Moreover, $\sigma_{d-1}(2d-1) \leq \sigma_{d-1}(2d-2)+1 \leq d-1+1 = d$. Thus, $d \geq \sigma_{d-1}(2d - 1) = \sigma_d(2d) > d$, a contradiction. Therefore, $\sigma_d(2d) \leq d$ for every $d \geq 2$ and the conjectured upper bound follows by applying Theorem 1.   □

Another observation of the $(d, n - d)$ table given in Figure 1 is that not only $\sigma_d(2d)$ is bounded by d, but also it is true for $\sigma_d(2d + 1)$. Theorem 3 shows that this property implies a slightly stronger upper bound.

Theorem 3. *If $\sigma_d(2d + 1) \leq d$ for every $d \geq 2$, then $\sigma_d(n) \leq n - d - 1$ for $n > 2d \geq 4$ and $\sigma_d(n) = n - d$ for $n \leq 2d$.*

Proof. We have $d \leq \sigma_d(2d) \leq \sigma_d(2d+1) \leq d$ and so $\sigma_d(2d) = \sigma_d(2d+1) = d$. It implies that $\sigma_d(n) = n - d$ for $n \leq 2d$. For $n > 2d$ we have, by Proposition 1(b), $\sigma_d(n) \leq \sigma_{n-d-1}(2n - 2d - 1) \leq n - d - 1$.   □

Theorem 4. *If $\sigma_d(2d) = \sigma_d(2d+1)$ for every $d \geq 2$, then $\sigma_d(n) \leq n - d - 1$ for $n > 2d \geq 4$ and $\sigma_d(n) = n - d$ for $n \leq 2d$.*

Proof. The results follow from Theorem 3 and the fact that $\sigma_d(2d) = \sigma_d(2d + 1) = d$ for every $d \geq 2$. To show that $\sigma_d(2d) = \sigma_d(2d + 1) = d$ for every $d \geq 2$, let us argue by contradiction. Let d be the smallest such that $\sigma_d(2d) = \sigma_d(2d+1) > d$. From the values in the $(d, n - d)$ table calculated so far, we know that $d > 10$. Thus $d - 1 = \sigma_{d-1}(2d - 2) = \sigma_{d-1}(2d - 1)$. However, by Proposition 1(f), $\sigma_{d-1}(2d - 1) + 1 \geq \sigma_d(2d)$. It follows that $d - 1 \geq \sigma_d(2d) - 1$. i.e. $d \geq \sigma_d(2d)$, a contradiction.   □

4   Structure of Relatively Short Square-Maximal Strings

In this section we investigate square-maximal strings that are short relative to the size of their alphabets. The main goal of this investigation is to either find a counterexample on the main diagonal if there is one, or to show that there are no counterexamples on the main diagonal, as this would prove the conjectured upper bound for all strings. We show that a square-maximal string from the main diagonal either complies with the conjectured upper bound or has to have many singletons based on the facts that such string (a) cannot contain pairs, see Lemma 4, and (b) if it contains a triple, it is must be a very special triple, implying the existence of a symbol occurring at least 6 times, see Lemma 8. We hope that it might be possible to show that counterexamples on the main diagonal do not exist by showing that their structure would be impossible. We discuss this in Conclusion.

Lemma 2 shows the structure of the square-maximal strings on the main diagonal if they are in compliance with the conjectured upper bond and they are identical with the value of its right neighbour.

Lemma 2. *If $\sigma_d(2d) = \sigma_d(2d + 1)$ for every $d \geq 2$, then for any $d \geq 2$, $x \in S_d(2d)$ square-maximal, x is up to relabeling of the alphabet, unique and equal to $x = (aabbcc\ldots)$.*

Proof. If x contains only pairs, by Lemma 1 all these pairs have to be adjacent. If x did not consist only of pairs, then it would have to have a singleton. Let y be a string obtained from x by removing a singleton. $y \in S_{d-1}(2d-1)$ and $s(y) \geq s(x)$. Thus $d - 1 = \sigma_{d-1}(2d - 2) = \sigma_{d-1}(2d - 1) \geq s(y) \geq s(x) = \sigma_d(2d) = d$ which is contradiction. Therefore x contains only pairs and is up to relabeling, unique and equal to $x = (aabbcc\ldots)$. □

Auxiliary Lemma 3 will be used to estimate the number of squares that span from one part of a string to the other part and relies on the result of Fraenkel and Simpson [3].

Lemma 3. *Consider non-empty strings w, u, and v. The number of distinct squares of the string wuv that start in w and end in v is at most $|w| + |v|$ where $|w|$, respectively $|v|$, denotes the length of w, respectively v.*

Proof. We discuss two cases: Case (i) $|w| \leq |v|$: we count the rightmost occurrences of squares. By Fraenkel-Simpson [3], there are at most two such squares starting at the same position. Thus, there are at most $2|w|$ squares that start in w, and $2|w| \leq |w| + |v|$. Case (ii) $|w| > |v|$: let $\overline{x}$ denote the reversal of the string x. By the previous argument, there are at most $2|\overline{v}|$ squares of the string $\overline{wuv} = \overline{v}\,\overline{u}\,\overline{w}$ starting in $\overline{v}$. It follows that there are at most $2|v|$ squares of wuv that end in v and $2|v| < |w| + |v|$. □

Lemma 4 shows that the square-maximal strings in first unknown position on the main diagonal either comply with the conjectured upper bound or cannot contain a pair.

Lemma 4. *Let $\sigma_{d'}(2d') \leq d'$ where $d' < d$. Let $x \in S_d(2d)$ be square-maximal. Then either $s(x) = \sigma_d(2d) = d$ or x does not contain a pair.*

Proof. Let assume that $s(x) = \sigma_d(2d) > d$ and x contains a pair of C's at positions i_0 and i_1, so $x[i_0] = x[i_1] = C$. If the pair occurs in at most 1 square, then we can replace the first C with a new symbol $\hat{C} \notin \mathcal{A}(x)$. Let y be x with $x[i_0]$ replaced by $\hat{C}$. Then $y \in S_{d+1}(2d)$ and $\sigma_{d+1}(2d) \geq s(y) = s(x) - 1 = \sigma_d(2d) - 1$. Since $2d - (d + 1) < d$, we get $2d - (d + 1) \geq \sigma_{d+1}(2d) \geq \sigma_d(2d) - 1$, i.e. $d - 1 \geq s(x) - 1$, and so $d \geq s(x)$, a contradiction. Therefore, the pair must occur in at least two squares, in fact in a non-trivial run $x = \cdots uvCwuvCwu \cdots$, where $|u| \geq 1$. Let us form a new string y by removing all the symbols between the C's: $y = \cdots uvCCwu \cdots$. By doing this, we may have destroyed $|u| + 1$ squares – $uvCwuvCw$ and its $|u|$ rotations. The type of any square of u is preserved, as y has u as a substring. The same is true for w, v, wu, and uv. Thus, we may have destroyed the squares of wuv that start in w and end in v. By Lemma 3, we may have destroyed at most $|w| + |v|$ squares. So, altogether, we may have destroyed at most $|w| + |u| + |v| + 1$ squares, but we created a new one: CC. Thus $s(y) \geq s(x) - (|w| + |u| + |v|)$. Clearly, $\mathcal{A}(y) = \mathcal{A}(x)$, and so $y \in S_d(2d - k)$ where $k = |w| + |u| + |v|$. By the assumption of this lemma as $2d - k - d = d - k < d$, we have $d - k \geq \sigma_d(2d - k) \geq s(y) \geq s(x) - k$, and thus $d \geq s(x)$, a contradiction. $\square$

Lemmas 5 and 6 use the same scenario investigating the square-maximal strings in the first unknown position on the main diagonal and showing that they either comply with the conjectured upper bound or may contain only very specific triples.

Lemma 5. *Let $\sigma_{d'}(2d') \leq d'$ where $d' < d$. Let $x \in S_d(2d)$ be square-maximal. Then either $s(x) = \sigma_d(2d) = d$ or if x contains a triple, then the triple has to occur in two distinct runs.*

Proof. Let assume that $s(x) = \sigma_d(2d) > d$. Let $x[i_0] = x[i_1] = x[i_2] = C$ be a triple in x. We first show all three symbols occur in some runs. Assume that $x[i_0]$ does not occur in any run. Let $\hat{C}$ be a symbol $\notin \mathcal{A}(x)$. Let y be x with $x[i_0]$ replaced by $\hat{C}$. Then $y \in S_{d+1}(2d)$ and $\sigma_{d+1}(2d) \geq s(y) = s(x) = \sigma_d(2d)$. Since $2d - (d + 1) < d$, we get $2d - (d + 1) \geq \sigma_{d+1}(2d) \geq \sigma_d(2d)$, i.e. $d - 1 \geq \sigma_d(2d)$, a contradiction. For $x[i_2]$ not occurring in any run, the proof is the same. If $x[i_1]$ does not occur in any run, then none of the elements of the triple occur in any run. Then we can remove $x[i_1]$ forming a string $y \in S_d(2d - 1)$ such that $d - 1 \geq \sigma_d(2d - 1) \geq s(y) \geq s(x) = \sigma_d(2d)$, a contradiction. We then show the three symbols cannot occur in the same run. Assume they do occur in the run $uvCwuvCwuvCwu$. We can proceed as in the proof of Lemma 4 and remove wuv between the first and second C. $\square$

Lemma 6. *Let $\sigma_{d'}(2d') \leq d'$ where $d' < d$. Let $x \in S_d(2d)$ be square-maximal. Then either $s(x) = \sigma_d(2d) = d$, or if x has a triple $x[i_0] = x[i_1] = x[i_2] = C$ occurring in two distinct runs $u_1v_1x[i_0]w_1u_1v_1x[i_1]w_1u_1 = u_1v_1Cw_1u_1v_1Cw_1u_1$ and $u_2v_2x[i_1]w_2u_2v_2x[i_2]w_2u_2 = u_2v_2Cw_2u_2v_2Cw_2u_2$, then $|u_1| \geq 1$ and $|u_2| \geq 1$ and either u_2v_2 is not a suffix of u_1v_1 or w_1u_1 is not a prefix of w_2u_2.*

Proof. Let us assume that $s(x) = \sigma_d(2d) > d$. If $|u_1| = 0$, then $x[i_0]$ occurs in a single square $v_1Cw_1v_1Cw_1$. Let $\hat{C}$ be a symbol $\notin \mathcal{A}(x)$ and let y be x with $x[i_0]$ replaced by $\hat{C}$. Then $y \in S_{d+1}(2d)$ and $\sigma_{d+1}(2d) \geq s(y) = s(x) - 1 = \sigma_d(2d) - 1$. Since $2d - (d + 1) < d$, we get $2d - (d + 1) \geq \sigma_{d+1}(2d) \geq \sigma_d(2d) - 1$, i.e. $d - 1 \geq \sigma_d(2d) - 1$, and so $d \geq \sigma_d(2d)$, a contradiction. It follows that $|u_1| \geq 1$. For $|u_2| = 0$, the proof is the same. Thus, $|u_1| \geq 1$ and $|u_2| \geq 1$. Let us assume that both u_2v_2 is a suffix of u_1v_1 and w_1u_1 a prefix of w_2u_2. Let us form a new string y from x by removing $w_1u_1v_1$ between $x[i_0]$ and $x[i_1]$ and removing $w_2u_2v_2$ between $x[i_1]$ and $x[i_2]$, that is $y = x[1..i_0]x[i_1]x[i_2..2d] = x[1..i_0 - 1]CCCx[i_2 + 1..2d]$. It follows that $y \in S_d(2d - k)$ where $k = |w_1| + |u_1| + |v_1| + |w_2| + |u_2| + |v_2|$. How many squares we might have destroyed? We might have destroyed $|u_1| + 1$ squares of

$u_1 v_1 C w_1 u_1 v_1 C w_1 u_1$ and $|u_2| + 1$ squares of $u_2 v_2 C w_2 u_2 v_2 C w_2 u_2$. From $w_1 u_1 v_1$, $u_1 v_1$ has been preserved, $w_1 u_1$ is a prefix of $w_2 u_2$ that was preserved, so the only squares we might have destroyed are the ones starting in w_1 and ending in v_1, and by Lemma 3 there are at most $|w_1| + |v_1|$ of them. Similarly for $w_2 u_2 v_2$. Thus we might have destroyed at most $|w_1| + |u_1| + |v_1| + |w_2| + |u_2| + |v_2| + 2 = k + 2$ squares, and we gained one (CCC). It follows that $s(y) \geq s(x) - k - 1$. Replace the first C in y by a new symbol $\hat{C} \notin \mathcal{A}(x)$ to form a string z. Then $z \in S_{d+1}(2d - k)$ and $s(z) = s(y)$. Thus $\sigma_{d+1}(2d - k) \geq s(z) = s(y) \geq s(x) - k - 1 = \sigma_d(2d) - k - 1$. Since $2d - k - d - 1 = 2d - |w_1| - |u_1| - |v_1| - |w_2| - |u_2| - |v_2| - d - 1 < d$, we have $2d - k - d - 1 \geq \sigma_{d+1}(2d - k) \geq s(x) - k - 1$, so $2d - k - d - 1 \geq s(x) - k - 1$ and so $d \geq s(x)$, a contradiction. It follows that either $u_2 v_2$ is not a suffix of $u_1 v_1$, in which case $u_1 v_1$ is a suffix of $u_2 v_2$, or $w_1 u_1$ is not a prefix of $w_2 u_2$, in which case $w_2 u_2$ is a prefix of $w_1 u_1$. □

Lemma 7 shows that the square-maximal strings cannot contain parallel k-tuples. A k-tuple of C's occurring at positions $\{i_1, \cdots i_k\}$ and a k-tuple of D's occurring at positions $\{j_1, \cdots j_k\}$ are *parallel* if $i_1 < j_1 < i_2 < j_2 < \cdots < i_k < j_k$.

Lemma 7. *Let $x \in S_d(2d)$ be square-maximal. Then x cannot contain two parallel k-tuples for any $k \geq 2$.*

Proof. Let us assume that x contains two parallel k-tuples of C's and D's. Let us move all D's to the end of the string x, forming a new string $y \in S_d(2d)$. Any primitively rooted square that contains m of the D's must also contain at least m of the C's. If we remove the D's from the square, we create a new square. Since it contains the C's and since the original square was primitively rooted, the new square also must be primitively rooted. For illustration: $[uCvDw][uCvDw]$ will become $[uCvw][uCvw]$. Moving the D's to the end creates a new square DD and so $s(y) > s(x)$, a contradiction with the square-maximality of x. □

Lemma 8 utilizes the previous lemmas and shows that any square-maximal string in the first unknown position on the main diagonal either complies with the conjectured upper bound, or if if it contains a triple, it must be a very specific one giving rise to a symbol that must occur at least 6 times. Thus, each triple occurring must be balanced by an existence of a unique set of 5 occurrences of a certain symbol. Though the symbol may not be unique to a particular triple, the set of occurrences are mutually disjoint. Thus, every triple with its assigned set of 5 occurrences is balanced by an existence of at least 4 singletons unique to the triple and its assigned set.

Lemma 8. *Let $\sigma_{d'}(2d') \leq d'$ where $d' < d$. Let $x \in S_d(2d)$ be square-maximal. Then either $s(x) = \sigma_d(2d) = d$ or x has at least $\lceil \frac{2d}{3} \rceil$ singletons.*

Proof. Let us assume that $s(x) = \sigma_d(2d) > d$. From Lemma 4 it follows, that x does not have any pair. From Lemmas 5 and 6, any triple $x[i_0] = x[i_1] = x[i_2] = C$ of x must be special, i.e. it must satisfy

1. $x[i_0]$ and $x[i_1]$ occur in a run $r_1 = u_1v_1Cw_1u_1v_1Cw_1u_1$, where $|u_1| \geq 1$,
2. $x[i_1]$ and $x[i_2]$ occur in a run $r_2 = u_2v_2Cw_2u_2v_2Cw_2u_2$, where $|u_2| \geq 1$, and where $i_1 - i_0 \neq i_2 - i_1$ as otherwise the two runs would merge into a single one,
3. either u_1v_1 is a proper suffix of u_2v_2, or w_2u_2 is a proper prefix of w_1u_1.

Let us discuss the case when u_1v_1 is a proper suffix of u_2v_2; the case of w_2u_2 being a proper prefix of w_1u_1 is the same just argued from the opposite direction. Let the run $r_1 = u_1v_1Cw_1u_1v_1Cw_1u_1$ start at position t of x. Consider $a = x[t]$. If there is no occurrence of a in $x[t+1..i_0-1]$, then we can replace all occurrences of a in $x[1..i_0-1]$ with a new symbol, forming a string y, while destroying a single square $u_1v_1Cw_1u_1v_1Cw_1$ of x. Thus $y \in S_{d+1}(2d)$, $2d-d-1 \geq \sigma_{d+1}(2d) \geq s(y) = s(x)-1 = \sigma_d(2d)-1$, so $d \geq \sigma_d(2d)$, a contradiction. Thus a occurs at least twice in $x[t..i_0-1] = u_1v_1$. Since u_1v_1 is a suffix of u_2v_2, a occurs at least 4 more times – twice in each occurrence of u_2v_2. Thus, $x[t]$ occurs in x at least six times, the last occurrence before the last C. We assign to the triple the sequence of positions of the 5 first occurrences of a after the position t and denote it by $As(C) = \langle j_0, j_1, j_2, j_3, j_4 \rangle$, where $t < j_0 < j_1 < j_2 < j_3 < j_4 < i_2$ and $j_0 < i_0$ and t is the start of the run r_1 and $x[t] = x[j_0] = x[j_1] = x[j_2] = x[j_3] = x[j_4]$. Of course, if the short appendix used was w_2u_2, then $As(C) = \langle j_0, j_1, j_2, j_3, j_4 \rangle$, where $i_0 < j_4 < j_3 < j_2 < j_1 < j_0 < t$ and $i_2 < j_0$ and t is the end of the run r_2 and $x[t] = x[j_0] = x[j_1] = x[j_2] = x[j_3] = x[j_4]$. Below, we will show that such assignments are mutually disjoint, i.e. if C's and D's are different triples, then $As(C) \cap As(D) = \emptyset$.

Now we can estimate the number of singletons in x. Let m_0 be the number of triples in x. Let m_1 be the number of multiply occurring symbols that are not assigned to triples – since there are no pairs, it follows that such symbols occur at least 4 times. Finally, let m_2 be the number of singletons in x. The following 2 inequalities must hold: $2d \geq 8m_0 + 4m_1 + m_2$ and $d \leq 2m_0 + m_1 + m_2$ which clearly yields $3m_2 \geq 2d$ and so $m_2 \geq \lceil \frac{2d}{3} \rceil$.

A proof of the claim that the assignments are mutually disjoint: Let $As(C) = \langle j_0, j_1, j_2, j_3, j_4 \rangle$ and let $As(D) = \langle k_0, k_1, k_2, k_3, k_4 \rangle$. If $x[j_0] \neq x[k_0]$, then $As(C) \cap As(D) = \emptyset$. Bellow, we discuss the case when $x[j_0] = x[k_0] = a$.

In Lemma 6 it is shown that a triple of C's can exist in x only if it occurs in two distinct non-trivial runs $u_1v_1Cw_1u_1v_1Cw_1u_1$ and $u_2v_2Cw_2u_2v_2Cw_2u_2$. We refer to u_1v_1 and w_2u_2 as the appendices, and we say that u_1v_1 is a short appendix if u_1v_1 is a proper suffix of u_2v_2, similarly we say that w_2u_2 is a short appendix if it is a proper prefix of w_1u_1. Thus, Lemma 6 also stipulates that at least one of the appendices must be short.

Let us consider two different triples, one of C's and one of D's and let us assume that the first C precedes the first D. We must discuss all the possible configurations of the two triples. For better readability, we will denote by C_1 the first occurrence of C, by C_2 the second occurrence of C etc. Similarly for D's.

The C's occur in two non-trivial runs $r_1 = u_1v_1C_1w_1u_1v_1C_2w_1u_1$ and $r_2 = u_2v_2C_2w_2u_2v_2C_3w_2u_2$, while the D's occur in two non-trivial runs $r_3 = u_3v_3D_1w_3u_3v_3D_2w_3u_3$ and $r_4 = u_4v_4D_2w_4u_4v_4D_3w_4u_4$.

1. C_3 occurs before D_1, i.e. the triples do not interleave (schematically $C_1\,C_2\,C_3$ $D_1\,D_2\,D_3$).

   (a) First we consider the case when the appendix determining $As(C)$ and the appendix determining $As(D)$ are on the opposite sides.

   Thus, the short appendix determining $As(C)$ is on the left and the short appendix determining $As(D)$ is on the right. Then we are guarantied the following pattern of occurrences of a in x (for the C's, the a's are shown in bold, for the D's, the a's are shown underscored):
   $x = \cdots a\,a\;C_1\;a\,a\;C_2\;a\,a\;C_3\;D_1\;\underline{a}\,\underline{a}\;D_2\;\underline{a}\,\underline{a}\;D_3\;\underline{a}\,\underline{a}\cdots$, so $x[j_4]$ occurs before C_3, while the $x[k_4]$ occurs after D_1. Therefore $j_4 < k_4$ and so $As(C) \cap As(D) = \emptyset$.

   (b) Next we consider the case when the appendix determining $As(C)$ and the appendix determining $As(D)$ are facing each other.

   Thus, for the C's we are using the right appendix, for the D's the left appendix. Then we are guarantied the following pattern of occurrences of a in x (for the C's, the a's are shown in bold, for the D's, the a's are shown underscored):
   $x = \cdots C_1\;a\,a\;C_2\;a\,a\;C_3\;\underline{a}\,\underline{a}\;D_1\;\underline{a}\,\underline{a}\;D_2\;\underline{a}\,\underline{a}\;D_3\cdots$, and thus $x[j_0]$ occurs at or to the left of a (shown in bold), while $x[k_0]$ occurs at or to the right of $\underline{a}$ (shown underscored). It is possible that two a's between C_3 and D_1 are the same. However, since we do not take the first occurrence of a for the assignments, $As(C) \cap As(D) = \emptyset$.

   (c) Here we consider the case when the appendix determining $As(C)$ and the appendix determining $As(D)$ are on the same side.

   Without loss of generality, we can assume that both appendices used are on the left. Then we are guarantied the following pattern of occurrences of a in x (for the C's, the a's are shown in bold, for the D's, the a's are shown underscored):
   $x = \cdots a\,a\;C_1\;a\,a\;C_2\;a\,a\;C_3\;\underline{a}\,\underline{a}\;D_1\;\underline{a}\,\underline{a}\;D_2\;\underline{a}\,\underline{a}\;D_3\cdots$. Why cannot the first two $\underline{a}$'s be the same as the last two a's? If it were the case, then C would be in the appendix for the D's, i.e. a part of the run r_3 and hence repeat later. So, again $As(C) \cap As(D) = \emptyset$.

2. Case $x = \cdots C_1\;D_1\;D_2\;C_2\cdots$ is not possible.
   If either D_1 or D_2 occurred in $u_1 v_1$, then there would be a D preceding C_1. Thus both D_1 and D_2 occur in w_1, but then D occurs at least 4 times, a contradiction.

3. Case $x = \cdots C_1\;D_1\;C_2\;D_2\;D_3\;C_3\cdots$ is not possible.
   As in the previous case, D_1 must occur in w_1 and D_2 together with D_3 must occur in v_2, hence D must occur at least 4 times, a contradiction.

4. Case $x = \cdots C_1\;D_1\;C_2\;D_2\;C_3\;D_3\cdots$.
   This case is not possible by Lemma 7 as the triples of C's and D's are parallel.

5. Case $x = \cdots C_1\;C_2\;D_1\;C_3\;D_2\;D_3\cdots$.
   We denote by $w_2^{(1)}$ the first occurrence of w_2 in x, by $w_2^{(2)}$ the second occurrence of w_2 in x, etc.

If D_1 occurred in $(u_2v_2)^{(2)}$, there would be a D preceding C_2. Hence D_1 must occur in $w_2^{(1)}$ and hence D_2 occurs in $w_2^{(2)}$. Since the distance between C_2 and C_3 is the period of r_2, and the distance between D_1 and D_2 is the period of r_3, and the distances are equal, it follows that $r_2 = r_3 = u_2v_2C_2w'_2D_1w''_2u_2v_2C_3w'_2D_2w''_2u_2$ (note that $u_3 = u_2$ and $v_3 = v_2Cw'_2$ and $w_3 = w''_2$.)

Schematically:

$r_1:$ $u_1v_1C_1w_1u_1v_1C_2w_1u_1$

$r_2 = r_3:$ $u_2v_2C_2w'_2D_1w''_2u_2v_2C_3w'_2D_2w''_2u_2$

$r_4:$ $u_4v_4D_2w_4u_4v_4D_3w_4u_4$

Now consider the two runs r_1 and r_2. Since D_1 cannot occur in $(w_1u_1)^{(2)}$, it follows that the w_1u_1 is a prefix of w'_2 and hence of $w'_2D_1w''_2u_2$, and so the appendix $w'_2D_2w''_2u_2$ is long and by Lemma 6, u_1v_1 must be a short appendix and is used to determine $As(C)$.

Now consider the two runs r_3 and r_4. Since C_3 cannot occur in $(u_4v_4)^{(1)}$, u_4v_4 is a suffix of w'_2 and hence of $u_2v_2C_3w'_2$, and so the appendix $u_2v_2C_2w'_2$ is long. By Lemma 6, w_4u_4 must be a short appendix and is used to determine $As(D)$.

(a) Let a occur twice in u_1 and in twice in u_4 (*the dots indicate the occurrences*).

$r_1:$ $\overset{..}{u_1}v_1C_1w_1\overset{..}{u_1}v_1C_2w_1\overset{..}{u_1}$

$r_2 = r_3:$ $u_2v_2C_2w'_2D_1w''_2u_2v_2C_3w'_2D_2w''_2u_2$

$r_4:$ $u_4v_4D_2w_4\underset{..}{u_4}v_4D_3w_4\underset{..}{u_4}$

Then a occurs twice in each occurrence of u_1 and hence $x[j_4]$ occurs in or before $u_1^{(3)}$. Similarly, a occurs twice in each occurrence of u_4 and hence $x[k_4]$ occurs in or after $u_4^{(1)}$. Thus $As(C) \cap As(D) = \emptyset$.

(b) Let a occur only once in u_1 and twice in u_4.

$r_1:$ $\overset{.}{u_1}\overset{.}{v_1}C_1w_1\overset{.}{u_1}\overset{.}{v_1}C_2w_1\overset{.}{u_1}$

$r_2 = r_3:$ $u_2v_2C_2w'_2D_1w''_2\overset{.}{u_2}\overset{.}{v_2}C_3w'_2D_2w''_2u_2$

$r_4:$ $u_4v_4D_2w_4\underset{..}{u_4}v_4D_3w_4\underset{..}{u_4}$

Then a must occur in v_1. Since u_1v_1 is a suffix of u_2v_2 and since w_1u_1 is a prefix of w'_2, we have 7 occurrences of a from the left and 6 occurrences of a from the right, so again $As(C) \cap As(D) = \emptyset$.

(c) Let a occur twice in u_1 and only once in u_4.
This is symmetric to the previous case, we will have 6 occurrences of a from the left, and 7 occurrences of a from the right.

(d) Let a occur in u_1 only once and in u_4 also only once.

$$r_1: \qquad \overset{\cdot}{u_1} \overset{\cdot}{v_1} C_1 w_1 \overset{\cdot}{u_1} \overset{\cdot}{v_1} C_2 w_1 \overset{\cdot}{u_1}$$

$$r_2 = r_3: \qquad \overset{\cdot}{u_2} \overset{\cdot}{v_2} C_2 w'_2 D_1 w''_2 \overset{\cdot}{u_2} \overset{\cdot}{v_2} C_3 w'_2 D_2 w''_2 u_2$$

$$r_4: \qquad \overset{\cdot}{u_4} \overset{\cdot}{v_4} D_2 w_4 \overset{\cdot}{u_4} \overset{\cdot}{v_4} D_3 w_4 \overset{\cdot}{u_4}$$

From the left there are 8 occurrences of a: a must occur in v_1 and since $u_1 v_1$ is a suffix of $u_2 v_2$, it must occur twice in $(u_2 v_2)^{(2)}$, and since u_1 is a substring of w'_2, a must occur in all occurrences of w'_2. Similarly, there are 8 occurrences of a from the right. Even though it is possible the the last four occurrences from the left and the last four occurrences from the right are the same, the first 6 occurrences from the left and 6 occurrences from the right are disjoint, and so $As(C) \cap As(D) = \emptyset$. □

Theorem 5 stresses the fact that the first position on the main diagonal violating the conjectured upper bound implies an existence of a counterexample higher up. Similarly as Theorems 1 and 2, this is yet another reformulation of the conjectured upper bound.

Theorem 5. *The conjectured upper bound $\sigma_d(n) \leq n - d$ holding true for all strings is equivalent with the statement: $\sigma_d(4d) \leq 3d$ for every $d \geq 2$.*

Proof. The statement clearly follows from the conjectured upper bound. We shall prove the opposite direction by contradiction. Let us assume that the conjectured upper bound does not hold. By Theorem 1, it follows that there is a counterexample x on the main diagonal, i.e. a square-maximal $x \in S_d(2d)$ with $s(x) = \sigma_d(2d) > d$. Let us consider the first column d of the table in which the counterexample occurs, from the table as computed so far, we know that $d > 10$. By Lemma 8, x has at least $\lceil \frac{2d}{3} \rceil$ singletons. If we remove $\lceil \frac{2d}{3} \rceil$ singletons from x, we get a string $y \in S_{d'}(n')$ such that $s(y) \geq s(x) > d$ where $d' = d - \lceil \frac{2d}{3} \rceil$ and $n' = 2d - \lceil \frac{2d}{3} \rceil$. Moreover, $4d' = 4(d - \lceil \frac{2d}{3} \rceil) = 4d - 4 \cdot \lceil \frac{2d}{3} \rceil = 4d - 2d - \lceil \frac{2d}{3} \rceil = 2d - \lceil \frac{2d}{3} \rceil = n'$, thus $n' = 4d'$. So we have $\sigma_{d'}(4d') \geq s(y) \geq s(x) = \sigma_d(2d) > d$ and since $3d' = 4d' - d' = n' - d' = (2d - \lceil \frac{2d}{3} \rceil) - (d - \lceil \frac{2d}{3} \rceil) = d$, $\sigma_{d'}(4d') > 3d'$. Thus, we have a counterexample from $S_{d'}(4d')$. □

5   Conclusions

The methods used in section 4 illustrate two possible approaches to investigate the conjectured upper bound for all strings. One is to show that the first counterexample on the main diagonal cannot have a pair, a triple, a quadruple, ... or an k-tuple, i.e. it cannot exist. This approach is represented by Lemma 4. The other approach is to show that if the first counterexample on the main diagonal contains a k-tuple, then it must contain a symbol with a frequency $> k$. This also leads to the conclusion that a counterexample cannot exist. This approach is represented by the proof of Lemma 8. Thus, Lemmas 4 and 8 illustrate the usefulness of investigating the more orderly world of the strings on the main diagonal.

Let us just remark that our approach was inspired by a similar $(d, n - d)$ table used for investigating the Hirsch bound for the diameter of bounded polytopes. The associated Hirsch $(d, n - d)$ table exhibits similar regularities as the $(d, n - d)$ table considered in this paper. The Conjecture of Hirsch was recently disproved by Santos [7] exhibiting a violation on the main diagonal with $d = 43$ which was further improved to $d = 20$, see [6]. Similarly, we hope that the structure of square-maximal strings is richer for $n = 2d$ and therefore this could be the focus of investigation for tackling the conjectured upper bound. For instance, while for known values there is only essentially a single square-maximal string on the main diagonal and it has a well-described structure, the further up from the diagonal, the more irregular and unpredictable the set of square-maximal strings and their structures are.

An analogue of Theorem 5 for the maximal number of runs given in [1] shows that the conjectured upper bound of $n - d$ for the number of runs holding true for all strings equivalent with the upper bound of $8d$ for strings in $S_d(9d)$ for every $d \geq 2$.

References

1. Baker, A., Deza, A., Franek, F.: On the structure of relatively short run-maximal strings, AdvOL Technical Report 2011/02, Department of Computing and Software, McMaster University, Hamilton, Ontario, Canada
2. Deza, A., Franek, F.: A *d*-step analogue for runs on strings, AdvOL Technical Report 2010/02, Department of Computing and Software, McMaster University, Hamilton, Ontario, Canada
3. Fraenkel, A.S., Simpson, J.: How Many Squares Can a String Contain? Journal of Combinatorial Theory Series A 82(1), 112–120 (1998)
4. Ilie, L.: A simple proof that a word of length n has at most 2n distinct squares. Journal of Combinatorial Theory Series A 112(1), 163–164 (2005)
5. Ilie, L.: A note on the number of squares in a word. Theoretical Computer Science 380(3), 373–376 (2007)
6. Matschke, B., Santos, F., Weibel, C.: The width of 5-prismatoids and smaller non-Hirsch polytopes (2011), http://www.cs.dartmouth.edu/~weibel/hirsch.php
7. Santos, F.: A counterexample to the Hirsch conjecture, arXiv:1006.2814v1 (2010)

Tractability Results for the Consecutive-Ones Property with Multiplicity

Cedric Chauve[1], Ján Maňuch[1,2], Murray Patterson[2], and Roland Wittler[1,3]

[1] Department of Mathematics, Simon Fraser University, Burnaby, BC, Canada
[2] Department of Computer Science, UBC, Vancouver, BC, Canada
[3] Technische Fakultät, Universität Bielefeld, Bielefeld, Germany
cchauve@sfu.ca, {jmanuch,murrayp}@cs.ubc.ca,
roland@cebitec.uni-bielefeld.de

Abstract. A binary matrix has the Consecutive-Ones Property (C1P) if its columns can be ordered in such a way that all 1's in each row are consecutive. We consider here a variant of the C1P where columns can appear multiple times in the ordering. Although the general problem of deciding the C1P with multiplicity is NP-complete, we present here a case of interest in comparative genomics that is tractable.

1 Introduction

A binary matrix M has the *Consecutive-Ones Property* (C1P) if there exists a permutation of its columns such that all 1's in each row are consecutive. Deciding if a matrix has the C1P can be done in linear-time and space [3,5,11,9,10]. This problem has been considered in genomics, for problems such as physical mapping [2,7] or ancestral genome reconstruction [1,4,8].

Recently, Wittler and Stoye in [12], motivated by handling duplicated genes in reconstructing ancestral gene clusters, introduced a generalized problem: Given several sets of genes and a maximum multiplicity for each gene, decide whether there exists a sequence of genes which meets the multiplicity constraint for each gene and in which each set of genes occurs consecutively. This can be phrased in terms of a binary matrix, where a column corresponds to a gene and a set of genes is represented by a row containing a 1 for each gene in the set in the respective column and 0's in all other columns. Now each column c of the matrix is given a multiplicity threshold $\boldsymbol{m}(c)$: M satisfies the *mC1P* (for C1P with multiplicity) if there is a sequence S of columns of M, in which at most $\boldsymbol{m}(c)$ occurrences of column c can appear, and for each row r of M, the columns containing 1 in r appear consecutively somewhere in S. The sequence S corresponds then to a valid gene order. Deciding if a binary matrix M with multiplicity satisfies the mC1P is tractable if every row of M contains at most two entries 1 (which corresponds in gene clusters models to gene adjacencies) [12], but the problem is NP-complete if M contains rows with at most three entries 1 [13]. The mC1P can also be related to gene proximity analysis with duplicated genes [6].

In this work, we present a tractability result for a restricted mC1P decision problem. After some technical preliminaries (Section 2), we give in Section 3 a

R. Giancarlo and G. Manzini (Eds.): CPM 2011, LNCS 6661, pp. 90–103, 2011.

tractability result for a family of matrices where every row of M has (i) at most one entry 1 in columns with multiplicity greater than one, or (ii) exactly two entries 1 in columns with multiplicity greater than one and no other entries. This result is motivated by handling telomeres in ancestral gene order reconstruction (described in Appendix A). Our proofs rely on two classical concepts: PQ-trees and Eulerian cycles in graphs. We conclude by discussing future work.

2   Preliminaries

Let M be a binary matrix, with m rows $\mathcal{R} = \{r_1, \ldots, r_m\}$, n columns $\mathcal{C} = \{c_1, \ldots, c_n\}$ and ℓ entries 1. We represent a row r of M as a subset of $\mathcal{C}$, defined as the set of c_i such that $M[r, c_i] = 1$. A *multiplicity vector* $\boldsymbol{m}$ for M is a sequence of positive integers $[\boldsymbol{m}(c_1), \ldots, \boldsymbol{m}(c_n)]$: $\boldsymbol{m}(c_i)$ is called the multiplicity of column c_i. A column c with multiplicity $\boldsymbol{m}(c) > 1$ is called a *multicolumn* and a row r containing a multicolumn (i.e., $M[r, c] = 1$ for some column c with $\boldsymbol{m}(c) > 1$) is called a *multirow*. A multirow that does not contain any other multirow is called *minimal*. We say a binary matrix M with multiplicity vector $\boldsymbol{m}$ has *matched multirows* if, for every multirow $r \subseteq \mathcal{C}$ that contains at least two entries 1 in non-multicolumns, there exists a row $\hat{r}$ which is a copy of r where all entries in multicolumns have been discarded (i.e., switched from 1 to 0). We denote by $\hat{M}$ the binary matrix obtained from M by discarding all multicolumns. In this work, we assume that all matrices we deal with have matched multirows unless otherwise stated. Figure 1 illustrates the above definitions.

$$
\begin{array}{c|ccccccc}
M & 1 & 2 & 3 & 4 & 5 & a & b \\
\hline
r_1 & 1 & 1 & 0 & 0 & 0 & 1 & 1 \\
\hat{r}_1 & 1 & 1 & 0 & 0 & 0 & 0 & 0 \\
r_2 & 1 & 1 & 1 & 0 & 0 & 0 & 0 \\
r_3 & 0 & 0 & 1 & 1 & 1 & 0 & 1 \\
\hat{r}_3 & 0 & 0 & 1 & 1 & 1 & 0 & 0 \\
r_4 & 0 & 0 & 0 & 1 & 1 & 0 & 1 \\
\hat{r}_4 & 0 & 0 & 0 & 1 & 1 & 0 & 0 \\
r_5 & 1 & 0 & 0 & 1 & 1 & 0 & 0 \\
\end{array}
\qquad
\begin{array}{c|ccccc}
\hat{M} & 1 & 2 & 3 & 4 & 5 \\
\hline
r_1 & 1 & 1 & 0 & 0 & 0 \\
r_2 & 1 & 1 & 1 & 0 & 0 \\
r_3 & 0 & 0 & 1 & 1 & 1 \\
r_4 & 0 & 0 & 0 & 1 & 1 \\
r_5 & 1 & 0 & 0 & 1 & 1 \\
\end{array}
$$

Fig. 1. Left: Binary matrix M, with matched multirows. Let $\boldsymbol{m}(1) = \cdots = \boldsymbol{m}(5) = 1$ and $\boldsymbol{m}(a) = \boldsymbol{m}(b) = 2$: a and b are multicolumns and r_1, r_3 and r_4 are multirows. Row r_3 is not minimal, because it contains r_4. Right: The corresponding matrix $\hat{M}$. Since in $\hat{M}$, by definition $\hat{r}_i = r_i$ for all multirows r_i, the matched multirows are discarded.

Definition 1. *Matrix M has the* Consecutive-Ones Property with multiplicity *(mC1P) for multiplicity vector $\boldsymbol{m}$ if there exists a sequence $S = s_1 \ldots s_p$ on the alphabet $\mathcal{C}$ such that it meets*

(1) the consecutivity requirement: *for each row r of M there are two integers j, k, with $j < k$ such that $r = \{s_j, s_{j+1}, \ldots, s_k\}$ (the columns in r are consecutive in S), and*

(2) the multiplicity requirement: *each c_i appears at most $\boldsymbol{m}(c_i)$ times.*

The sequence S is then called an mC1P-*ordering of M.*

Given row $\{1, 2, 3, 4\}$, an example of a sequence that satisfies condition (1) of the above definition is the sequence 5142435, since $\{1, 2, 3, 4\} = \{1, 4, 2, 4, 3\}$.

The mC1P generalizes the classical Consecutive-Ones Property (C1P), where $m(c_i) = 1$ for every c_i. Lemma 1 below, whose proof is straightforward, relates both problems.

Lemma 1. *Every mC1P-ordering of M with multiplicity vector m contains a C1P-ordering of $\hat{M}$ as a subsequence. As a consequence, if a binary matrix M has the mC1P, then $\hat{M}$ has the C1P.*

This lemma suggests that, to decide if M has the mC1P for a given multiplicity vector m, we can first check if $\hat{M}$ has the C1P, and then extend a C1P-ordering of $\hat{M}$ into an mC1P-ordering of M by adding copies of multicolumns. Note that the matrix $\hat{M}$ in Figure 1 does not have C1P, and hence, M does not have mC1P. However, if we omit column r_5, then 12345 is a C1P ordering of $\hat{M}$, which can be extended to the following mC1P-ordering of M: $ab12345b$. To account for the fact that there can be an exponential number of C1P-orderings of $\hat{M}$, we use PQ-trees, a linear size structure that can describe all C1P-orderings of $\hat{M}$, defined below. For a more complete treatment of PQ-trees, we refer the reader to [3,9].

Definition 2. *A PQ-tree on $\mathcal{C}$ is a rooted ordered tree with leaves labeled by $\mathcal{C}$ and two kinds of internal nodes, P-nodes and Q-nodes. Each P-node has at least two children and each Q-node has at least three children.*

The frontier $F(T)$ *of a PQ-tree T is the sequence of $\mathcal{C}$ obtained by reading the labels of its leaves from left to right. The* frontier *of a node N in T is the frontier of the subtree rooted at N. Let $\{F(N)\}$ be the set of elements appearing in the sequence $F(N)$.*

Two PQ-trees are equivalent if one can be obtained from the other by applying a sequence of the following transformation rules: (RP) arbitrarily permute the children of a P-node; (RQ) reverse the order of the children of a Q-node.

Theorem 1. [3] *If a binary matrix M has the C1P, there exists a unique equivalence class PQ_M of PQ-trees with the property that there is a one-to-one correspondence between the C1P-orderings of M and the frontiers of the PQ-trees of PQ_M, and a PQ-tree belonging to PQ_M can be constructed in linear time.*

Each PQ-tree in the equivalence class PQ_M satisfies the following properties (that are implicitly given in [3,9]) which we will use in this paper.

Property 1. Let M be a binary matrix that has C1P with rows $\mathcal{R}$ and T a PQ-tree in the equivalence class PQ_M. Then

1. for every row $r \in \mathcal{R}$, there is a node N in T such that either $\{F(N)\} = r$, if N is a P-node, or r is consecutive in $F(N)$, if N is a Q-node;
2. for every node N different from the root of T, there is a row $r \in \mathcal{R}$ such that $\{F(N)\} \subseteq r$; and
3. for every Q-node N, and every two consecutive children N_1 and N_2 of N, there is a row $r \in \mathcal{R}$ such that $\{F(N_1)\} \cup \{F(N_2)\} \subseteq r$.

Finally, we recall briefly the technique used to prove that matrices with two entries 1 per row (usually called matrices of *degree* 2) form a class of tractable instances for deciding the mC1P as we will use it to prove our main result. Such matrices can be naturally represented as a collection of adjacency constraints $\mathcal{A} = \{\{a_i, b_i\}\}_{i=1}^{m}$ on the set $\mathcal{C}$, where $a_i \neq b_i$ and the collection is a set (no duplicate elements). Collection $\mathcal{A}$ is *consistent* with respect to $\boldsymbol{m}$ if there is a sequence S on $\mathcal{C}$ such that each adjacency is consecutive in S. We will refer to this sequence as a *consistency sequence* of $\mathcal{A}$ and $\boldsymbol{m}$. Note that an mC1P-ordering of M is a consistency sequence of the corresponding collection $\mathcal{A}$ and $\boldsymbol{m}$, and vice versa, and hence, M has the mC1P for $\boldsymbol{m}$ if and only if $\mathcal{A}$ is consistent with respect to $\boldsymbol{m}$. Given a collection of adjacencies $\mathcal{A}$, we define the graph $G_{\mathcal{A}}$ with vertex set $\mathcal{C}$ and edges given by adjacencies.

Theorem 2. [12] *A collection of adjacencies $\mathcal{A}$ is consistent with respect to a multiplicity vector $\boldsymbol{m}$ if and only if for all $c_i \in \mathcal{C}$, $\mathrm{degree}_{G_{\mathcal{A}}}(c_i) \leq 2\boldsymbol{m}(c_i)$ and for each connected component $B \subseteq \mathcal{C}$ of $G_{\mathcal{A}}$, for at least one $c_i \in B$, $\mathrm{degree}_{G_{\mathcal{A}}}(c_i) < 2\boldsymbol{m}(c_i)$.*

The above theorem relies on the fact that the graph $G_{\mathcal{A}}$ satisfying the above conditions can be extended to a multigraph on $\mathcal{C} \cup \{c_0\}$ that has an Eulerian cycle. It can be easily seen that the proof presented in [12] applies to generalized adjacencies, where we allow $a_i = b_i$ and the collection to be a multiset, and we require that each adjacency in $\mathcal{A}$ appears in S in a unique position. Note that $G_{\mathcal{A}}$ is now a multigraph with self-loops. We have the following corollary.

Corollary 1. *A collection of generalized adjacencies $\mathcal{A}$ is consistent with respect to a multiplicity vector $\boldsymbol{m}$ if and only if for all $c_i \in \mathcal{C}$, $\mathrm{degree}_{G_{\mathcal{A}}}(c_i) \leq 2\boldsymbol{m}(c_i)$ and for each connected component $B \subseteq \mathcal{C}$ of $G_{\mathcal{A}}$, for at least one $c_i \in B$, $\mathrm{degree}_{G_{\mathcal{A}}}(c_i) < 2\boldsymbol{m}(c_i)$.*

3   A Tractable Case of the mC1P Decision Problem

Our main result is that deciding the mC1P is tractable for a large family of matrices with constraints on the maximum number of entries 1 in multicolumns a row can have. The motivation for studying this particular family of matrices arises from incorporating information on telomeres in ancestral gene order reconstruction (Appendix A).

Theorem 3. *Let M be a binary matrix and $\boldsymbol{m}$ a multiplicity vector such that (1) M has matched multirows, and (2) each row contains either (i) at most one entry 1 in multicolumns, or (ii) two entries 1 in multicolumns and no other entries. Deciding if M has the mC1P for $\boldsymbol{m}$ can be done in polynomial time and space.*

We split the proof into two parts. In Section 3.1, we consider the case (2i) where M with multiplicity vector $\boldsymbol{m}$ contains a single multicolumn, and we show that deciding if M has the mC1P for $\boldsymbol{m}$ can be done efficiently using PQ-trees.

Then, in Section 3.2, we show how to handle the general case using Corollary 1 which relies on Eulerian cycles. Finally, in Section 3.3, we give an algorithm for building a PQ-tree which describes all sequences that satisfy the consecutivity requirement (condition (1) of Definition 1).

3.1 The Case of a Single Multicolumn

We assume that the multiplicity vector m defines only one multicolumn denoted by c'. According to Lemma 1, M satisfies the mC1P only if $\hat{M}$ has the C1P, which can be checked in linear time (Theorem 1). Assume that $\hat{M}$ has the C1P and let T be a PQ-tree from the equivalence class $PQ_{\hat{M}}$. We then aim at finding a PQ-tree from $PQ_{\hat{M}}$ (by applying operations (RP) and (RQ) on T) whose frontier can be extended to a valid mC1P-ordering by inserting copies of c'. We say that inserting a copy of c' into $F(T)$ *breaks* a row r of $\hat{M}$ if r is not consecutive in the resulting sequence. An example is given in Figure 2.

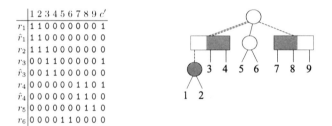

Fig. 2. Left: Binary matrix M, with matched multirows. Let $m(c') = 2$. Right: PQ-tree belonging to the equivalence class $PQ_{\hat{M}}$. P-nodes are represented by circular nodes and Q-nodes by rectangular nodes. An example of a valid mC1P-ordering is $c'\,1\,2\,3\,4\,c'\,7\,8\,9\,5\,6$ which is obtained by taking the equivalent PQ-tree with frontier $1\,2\,3\,4\,7\,8\,9\,5\,6$ and inserting two copies of c' into the corresponding positions. Notice that inserting c' between 2 and 3 would break row r_2.
Illustration of Algorithm 1. LCA($\hat{r}_1$) and the respective segments of LCA($\hat{r}_{3,4}$) are highlighted in gray and the respective paths are depicted by dashed lines. The upper left edge is contained in two paths. Here, $K_1 = 1$ and $K_2 = 1$, thus $K = 2 \leq m(c') = 2$.

Recall that rows are subsets of $\mathcal{C}$. As M has matched multirows, all rows in $\hat{M}$ are also rows in M. Since the consecutivity of the 1's in each row of $\hat{M}$ in the frontier $F(T)$ has to be maintained when inserting copies of c', no c' can be inserted into a position where it breaks any row of $\hat{M}$. Lemma 2 below is a consequence of this observation.

Lemma 2. *Let M be a binary matrix with matched multirows, and m be a multiplicity vector defining exactly one multicolumn c'. Assume that M has the mC1P, and let T be a PQ-tree from $PQ_{\hat{M}}$ and T' an extension of T whose frontier $F(T')$ is an mC1P-ordering of M.*

1. *If the root of T is a P-node, then, for each child node N of the root, c' can only appear as the first or last element of the frontier $F(N)$ in T'.*
2. *If the root of T is a Q-node, the copies of c' in T' can only appear as the first and/or last element of the frontier $F(T')$.*

Proof. It follows by Property 1.2 that for every child N of the root of T, any pair of consecutive leaves in $F(N)$ belongs to a row of $\hat{M}$, and hence, inserting c' between these leaves breaks this row.

In addition, if the root of T is a Q-node, then by Property 1.3, for any two consecutive children N_1 and N_2 of the root, there is a row of $\hat{M}$ that contains elements of $F(N_1)$ and of $F(N_2)$. This prevents the insertion of c' into root between N_1 and N_2 as this would break such a row. Hence c' can appear only at the extremities of $F(T')$. $\square$

Lemma 2 rules out many positions in $F(T)$ where to insert copies of c': indeed, copies of c' can only be inserted at extremities of the subsequences of $F(T)$ formed by children of the root (and only at the extremities of $F(T)$, if the root is a Q-node). On the other hand, each multirow specifies a position where a copy of c' must be inserted. These two constraints give rise to a polynomial algorithm which we describe in the following.

Algorithm 1 starts with a PQ-tree for $\hat{M}$ and works in two stages. First (Step 3), based on Lemma 2, it checks if there is a way to permute nodes in the subtrees rooted at each child of the root such that for each multirow $r = \hat{r} \cup \{c'\}$, rows in $\hat{r}$ appear as a prefix or a suffix of the frontier of some child. To satisfy the consecutivity requirement for each multirow r it is enough to add copies of c' to $F(T)$ before or after the frontier of the child of the root containing $\hat{r}$. To satisfy the multiplicity requirement, we need to permute the children of the root and possibly reverse the order of the frontier of some children. The basic idea is that we can save one copy of c' if a child requiring a copy of c' on the right is followed by a child requiring a copy of c' on the left. Whether enough copies of c' can be saved to satisfy the multiplicity requirement is checked in Steps 4–5.

Let $r = \hat{r} \cup \{c'\}$ be a multirow. By Property 1.1, there is in T either a P-node that contains exactly the columns in $\hat{r}$ in its subtree, or a Q-node with a segment of two or more consecutive children which together contain exactly the columns in $\hat{r}$ in their subtrees. This node is the least common ancestor in T of the columns in $\hat{r}$, and hence, will be denoted by $\text{LCA}(\hat{r})$.

Now to argue that Algorithm 1 is correct. If condition 3.c.i applies, r would require the insertion of a copy of c' within $F(U)$ in any PQ-tree of $PQ_{\hat{M}}$, which contradicts Lemma 2.

The paths indicate positions where copies of c' have to be added to the frontier so that the consecutivity requirement is satisfied. Following Lemma 2, we have to verify whether we can transform T such that all paths lie on the outside of the subtree of a child of the root of T. If conditions 3.c.ii–3.c.iv apply, there are two or more competing multirows, and we cannot transform T such that all of the corresponding paths lie on the outside of the subtree of a child of the root of T. Paths that are sub-paths of one another are excluded by not considering any

---

Algorithm 1. Deciding the mC1P for a matrix M with matched multirows and a multiplicity vector $\boldsymbol{m}$ defining a single multicolumn c'.

---

1. Check if $\hat{M}$ has the C1P.
2. If not, return false, else let T be a PQ-tree from $PQ_{\hat{M}}$.
3. For each minimal multirow $r = \hat{r} \cup \{c'\}$ in M do
   a. Locate $N := \text{LCA}(\hat{r})$.
   b. Let P_r be the path from N to the root of T.
   c. For each edge $e = \{U, V\}$ in P_r, where U is the parent of V do
      i. If U is a Q-node and V is neither its first nor its last child, return false;
      ii. If the root of T is a Q-node and e also belongs to the path $P_{r'}$ defined by another minimal multirow r', return false;
      iii. If U is not the root of T and e also belongs to the path defined by another minimal multirow, return false;
      iv. If U is the root of T and e also belongs to the paths defined by at least two other minimal multirows, return false.
4. If the root of T is a Q-node, return true.
5. If the root of T is a P-node:
   a. Let K_1 and K_2 be the number of children of the root of T belonging to exactly one or two paths defined by minimal multirows, respectively.
   b. $K := \left\lceil \frac{K_1}{2} \right\rceil + K_2 + \begin{cases} 1 \text{ if } K_1 = 0 \text{ and } K_2 > 0, \\ 0 \text{ otherwise.} \end{cases}$
   c. Return $K \leq \boldsymbol{m}(c')$

---

multirow $r = \hat{r} \cup \{c'\}$ which contains another multirow $r' = \hat{r}' \cup \{c'\}$ (line 3). These rows do not need to be considered at this stage, because in any ordering with c' adjacent to the elements in $\hat{r}'$, since $\hat{r}' \subseteq \hat{r}$, c' is also adjacent to the elements in $\hat{r}$. If the root of T is a P-node, we have to consider the children of the root node separately: We could insert a copy of c' on both sides of a frontier of a child of the root, i.e., at most two paths can join above such a child node. In levels below the root, only one path can be moved to the border of the subtree, i.e., no two edges can join.

If conditions 3.c.i–iv do not apply for a multirow r, there is a way to transform T (with rules (RP) and (RQ)) in the nodes on the path P_r (excluding the root) so that the frontier of $N = \text{LCA}(\hat{r})$ appears as a prefix or suffix of the frontier of N', where N' is a child of the root lying on the path P_r. Next, we will show that all these transformations can be performed simultaneously without any conflict. Obviously, the conflicts could only occur if the paths P_r share vertices other than root. Condition 3.c.iv guarantees that there are never three or more minimal multirows in the same subtree rooted at a child N' of the root. Condition 3.c.iii guarantees that if there are two minimal multirows in the same subtree rooted at a child N' of the root, their paths must meet only in N', and hence, one can appear as a prefix and one as a suffix of the frontier of N'. However, if the root is a Q-node, by Lemma 2, column c' can be attached only on one side of the frontier of N', and hence, only one minimal multirow can appear in the subtree rooted at N', which is checked in condition 3.c.ii.

Hence, if Step 3 succeeds for all rows, there is a PQ-tree in $PQ_{\hat{M}}$ from which we can obtain a sequence of the columns fulfilling the consecutivity requirement of M by inserting copies of c' into its frontier at positions indicated by the paths of multirows. Steps 4–5 check if the multiplicity constraint imposed by $\boldsymbol{m}$ can be satisfied. Note, that if the root of T is a Q-node (Step 4), then the multiplicity constraint is satisfied since $\boldsymbol{m}(c') \geq 2$.

In Step 5, we count the number of copies of c' required to satisfy all multirows. The position where to insert these copies are given by the paths. Since the root of T is a P-node, we can rearrange the children of the root such that one copy of c' would coincide with two paths (from neighboring children). For instance, we can greedily pair nodes with one path each, using $\lceil K_1/2 \rceil$ copies and then include nodes with two paths (one path on each side) in-between, requiring one further copy each, K_2 in total. If $K_1 = 0$ and $K_2 > 0$, chaining the two-path nodes results in $K_2 + 1$ copies of c'. It is easy to see that this joining process is optimal.

If the number of required copies of c' does not exceed the given maximum multiplicity $\boldsymbol{m}(c')$, the given matrix M with multiplicity vector $\boldsymbol{m}$ has the mC1P. Finally, to complete the proof of the correctness of the algorithm, we only need to notice that the result of Algorithm 1 does not depend on the choice of the PQ-tree T of $PQ_{\hat{M}}$, as the LCAs and paths are invariant under the transformation rules (RQ) and (RP).

The analysis of the time and space complexity of Algorithm 1 is as follows. First, Steps 1 and 2 can be completed in $O(m + n + \ell)$ time and space using the algorithm described in [9]; note that T can then be encoded in $O(n)$ space. Next, Step 3 is composed of at most m iterations, each of them requiring time $O(n)$, the maximum length of a path from N to the root of T, as each path is obviously processed in time linear in its length. This gives an $O(mn)$ time complexity for Step 3. For similar reasons, Step 4 can be achieved in time $O(mn)$, which gives an overall worst-case time complexity of $O(mn)$. This completes the proof of the case of a single multicolumn in Theorem 3.

3.2   Completing the Proof of Theorem 3

Proof (Proof of Theorem 3). Given matrix M with multiplicity vector $\boldsymbol{m}$ and having matched multirows, let $\mathcal{C}'$ be its set of multicolumns. A multirow containing multicolumn $c' \in \mathcal{C}'$, will be called a c'-*multirow*. Algorithm 2 works in the same two stages as Algorithm 1. However, the second stage is more complex. It requires building the collection of generalized adjacencies $\mathcal{A}$ on set $\mathcal{C}' \cup \{c_0\}$ by replacing each child of the root of the PQ-tree T for $\hat{M}$ by an adjacency and then applying Corollary 1.

Correctness of Step 1 follows from the correctness of the first stage of Algorithm 1. If Step 1 succeeds, we can assume that the root of T is a P-node (the case when the root is a Q-node is handled in Step 1), and hence, it is enough to satisfy the multiplicity requirement by permuting the children of the root and possibly reversing the order of the frontiers of some children. Let π be this order of children of the root. In Step 2, the algorithm constructs the multiset of

---

Algorithm 2. Deciding the mC1P for a matrix M with matched multirows and a multiplicity vector $\boldsymbol{m}$.

1. Run the first 4 steps of Algorithm 1, where c' is any element of $\mathcal{C}'$.
2. Construct a multiset of generalized adjacencies $\mathcal{A}$ on set $\mathcal{C}' \cup \{c_0\}$ as follows. For every child N of the root of T do
   a. If N belongs to exactly one path defined by multirows, say by a c'-multirow, add adjacency $\{c', c_0\}$ to $\mathcal{A}$;
   b. If N belongs to two paths defined by multirows, say by a c'-multirow and a d'-multirow (c' and d' may be equal), add adjacency $\{c', d'\}$ to $\mathcal{A}$.
3. Report if $\mathcal{A}$ is consistent with respect to $\boldsymbol{m}$ (use Corollary 1).

---

generalized adjacencies $\mathcal{A}$ whose consistency sequence (produced in Step 3) describes the way to do this as follows. Children that belong to zero paths defined by multirows will not introduce any adjacency constraints and can be placed at the end of π in any order and orientation. For any other child of the root, we have a unique position in the consistency sequence, hence we can order and orient these children based on these positions. Next, we insert copies of multicolumns as follows. For each subsequence $c_1 c_2 c_3$ of the consistency sequence, where adjacency $\{c_1, c_2\}$ corresponds to child N_1 and $\{c_2, c_3\}$ to N_2, if $c_2 \neq c_0$, we insert a copy of c_2 between the frontiers of N_1 and N_2 in $F(T)$. Hence, the number of copies of a multicolumn $c' \in \mathcal{C}'$ is equal to the number of its occurrences in the consistency sequence. Therefore, the frontier $F(T)$ with all required copies of multicolumns inserted satisfies the multiplicity requirement given by $\boldsymbol{m}$. It is easy to see that if there is an mC1P ordering of M, then we can extract from it an ordering of the children of the root which gives this consistency sequence.

The analysis of the time complexity is as follows. The first stage of the algorithm is a subroutine of Algorithm 1, and hence, has a time and space complexity of order $O(mn)$. Since the number of children of the root of T that belong to at least one path defined by multirows is at most m, the number of adjacencies in $\mathcal{A}$ is at most m, and hence, building $\mathcal{A}$ takes time $O(m)$. Finally, checking the degree conditions (applying Corollary 1) takes time $O(n)$. Hence, the total time and space complexity of the algorithm is $O(mn)$.

Finally, Algorithm 2 can also be easily extended to the case when the matrix also contains rows of degree 2 containing two multicolumns, as follows. First, we run Steps 1 and 2 where we ignore multirows containing two multicolumns. Then, we add to $\mathcal{A}$ also an adjacency for every such multirow. Finally, we run Step 3 of the algorithm on this new collection $\mathcal{A}$. It is easy to see that the time complexity of this new algorithm is still $O(mn)$. Hence, the theorem holds. □

3.3 Building a PQ-Tree Which Describes All Sequences That Satisfy the Consecutivity Requirement

Here, we describe how a given PQ-tree $T \in PQ_M$ can be augmented to a PQ-tree T' which represents the set of all sequences S, up to "pumping" occurrences

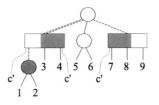

Fig. 3. Augmented PQ-tree T' for the matrix given in Figure 2. (In fact, to get an augmented PQ-tree from the original PQ-tree shown in Figure 2, no modifications are necessary other than attaching leaf nodes labeled c' at appropriate locations.) Only the trees in the equivalence class of T' where the left side of the right Q-node is placed adjacent to the left Q-node have shortened frontiers that meet the multiplicity requirement $(m(c') = 2)$, for example, $c'\,1\,2\,3\,4\,c'\,7\,8\,9\,5\,6$.

of multicolumns, that satisfy the consecutivity requirement (condition (1) of Definition 1) in that the frontier of any tree in the equivalence class of T' is such a sequence S. However, not all frontiers meet the multiplicity requirement (condition (2) of Definition 1). For some trees in the equivalence class of T', the respective frontier contains pairs of adjacent occurrences of a multicolumn c', each of which can be replaced by one occurrence of c' without breaking any row of M (violating the consecutivity requirement). This reduces the number of used copies of the multicolumns. Only such shortened frontiers which meet the multiplicity requirement are valid mC1P orderings, and, in fact, the set of such shortented frontiers is exactly the set of valid mC1P orderings of M. Figure 3 shows an example.

To construct an augmented PQ-tree T', we process the original tree T in a bottom-up fashion along the paths P_r defined in Algorithm 1, starting with the LCAs. We replace an LCA by a new Q-node which has a copy of its corresponding

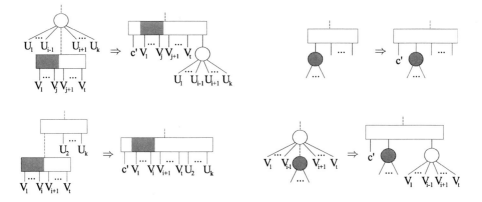

Fig. 4. Transformation rules for the LCAs to construct an augmented PQ-tree. An LCA and its parent node are replaced by the nodes shown on the right. The LCA (or the segment of an LCA, respectively) are highlighted in gray.

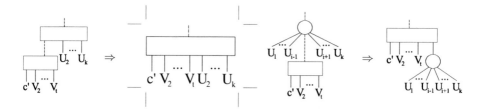

Fig. 5. Transformation rules for bottom-up iteration to construct an augmented PQ-tree. A newly created Q-node and its parent node are replaced by the nodes shown on the right.

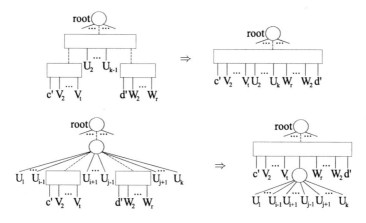

Fig. 6. Special transformation rules for bottom-up iteration to construct an augmented PQ-tree. A newly created Q-node two levels below the root node and its parent node are replaced by the nodes shown on the right.

multicolum c' as its first child and further children, depending on whether the LCA itself and its parent are P or Q-nodes. These intuitive transformation rules are detailed in Figure 4.

Then, any parent node of a newly obtained Q-node is refined to a new Q-node, moving up the copy of c', as shown in Figure 5.

This process is iterated until we reach the root node. Since a node that is a child of the root can be contained in two paths, separate (but similar) rules are required, illustrated in Figure 6.

Further specific rules which apply if an LCA is a child of the root of T or if the root node is a Q-node are straightforward. In some cases, after generating the tree as described above, simplifications can be carried out, such as replacing a P-node with a single child by a direct edge or substituting a Q-node with two children by a P-node.

Analogously to Algorithm 1, that only checks if a matrix has the mC1P, the above construction of an augmented PQ-tree T' can be carried out in $O(mn)$ time.

4   Conclusion

In the present work, we extend the domain of tractable instances of deciding the C1P with multiplicity for binary matrices. Our approach relies on previously used techniques to decide the C1P and simpler instances of the mC1P, and answers a natural problem in reconstructing ancestral gene orders. Several questions remain open. Naturally, one can ask to relax the condition that M has matched multirows, which is crucial in our proofs. It seems however that the problem becomes hard in this case, and some less rigid constraints on M would then have to be introduced to recover tractability. Also it is open to exhibit an extension of the notion of the PQ-tree that could encode all mC1P-orderings of a binary matrix that satisfies this property. Even for the case of a matrix with matched multirows, our techniques lead to a data structure which only captures the consecutivity requirement but not the multiplicity requirement. From an algorithmic complexity point of view, our algorithm has an $O(mn)$ time complexity, and it remains open to see if this case can be solved in $O(m + n + \ell)$ time.

Acknowledgments. We would like to thank Éric Tannier for suggesting the idea of using the mC1P for dealing with telomeres in ancestral genome reconstruction. Part of this research was funded by an NSERC Discovery Grant to C.C. and J.M.

References

1. Adam, Z., Turmel, M., Lemieux, C., Sankoff, D.: Common intervals and symmetric difference in a model-free phylogenomics, with an application to streptophyte evolution. J. Comput. Biol. 14, 436–445 (2007)
2. Alizadeh, F., Karp, R., Weisser, D., Zweig, G.: Physical mapping of chromosomes using unique probes. J. Comput. Biol. 2, 159–184 (1995)
3. Booth, K.S., Lueker, G.S.: Testing for the consecutive ones property, interval graphs, and graph planarity. J. Comput. Syst. Sci. 13, 335–379 (1976)
4. Chauve, C., Tannier, E.: A methodological framework for the reconstruction of contiguous regions of ancestral genomes and its application to mammalian genome. PLoS Comput. Biol. 4, paper e1000234 (2008)
5. Habib, M., McConnell, R.M., Paul, C., Viennot, L.: Lex-BFS and partition refinement, with applications to transitive orientation, interval graph recognition and consecutive ones testing. Theoret. Comput. Sci. 234, 59–84 (2000)
6. Landau, G.M., Parida, L., Weimann, O.: Gene proximity analysis across whole genomes via PQ trees. J. Comput. Biol. 12, 1289–1306 (2005)
7. Lu, W.-F., Hsu, W.-L.: A test for the Consecutive Ones Property on noisy data – application to physical mapping and sequence assembly. J. Comput. Biol. 10, 709–735 (2003)
8. Ma, J., Zhang, L., Suh, B.B., Raney, B.J., Burhans, R.C., Kent, W.J., Blanchette, M., Haussler, D., Miller, W.: Reconstructing contiguous regions of an ancestral genome. Genome Res. 16, 1557–1565 (2006)
9. McConnell, R.M.: A certifying algorithm for the consecutive-ones property. In: SODA 2004, pp. 761–770. ACM, New York (2004)

10. Meidanis, J., Porto, O., Telles, G.P.: On the consecutive ones property. Discrete Appl. Math. 88, 325–354 (1998)
11. Hsu, W.-L.: A simple test for the Consecutive Ones Property. J. Algorithms 43, 1–16 (2002)
12. Wittler, R., Stoye, J.: Consistency of sequence-based gene clusters. In: Tannier, E. (ed.) RECOMB-CG 2010. LNCS, vol. 6398, pp. 252–263. Springer, Heidelberg (2010)
13. Wittler, R., Maňuch, J., Patterson, M., Stoye, J.: Consistency of sequence-based gene clusters (unpublished manuscript)

Appendix A   Ancestral Gene Orders and Telomeres

In the context of ancestral genome reconstruction as approached in [4], $\mathcal{C}$ is an alphabet of genomic markers that are believed to appear uniquely in the extinct ancestral genome. An *ancestral synteny* is a set of markers that are believed to have been consecutive along a chromosome of the ancestor. A set of ancestral syntenies can then be represented by a binary matrix M: columns represent markers and the 1 entries of a given row define an ancestral synteny. If all ancestral syntenies are true positives (i.e., represent sets of markers that were consecutive in the ancestor), then M has the C1P. Otherwise, some ancestral syntenies are false positives, and a general approach to handle such conflicts is to remove from M some rows (optimizing some criterion) such that the resulting matrix M' has the C1P. Each subtree rooted at a child of the root of the resulting PQ-tree represents a set of markers that are believed to have been contiguous in the ancestral genome (with partial information regarding the order of the markers along this segment), called a CAR (Contiguous Ancestral Region) following [8].

A CAR is an ancestral chromosomal segment, but it is not guaranteed to be a complete ancestral chromosome. In fact, it is common that the number of CARs obtained is larger than the expected number of ancestral chromosomes. This raises the following natural question: which CARs are believed to form complete ancestral chromosomes, or more generally, to contain an extremity of an ancestral chromosome (an ancestral telomere)? Indeed, a CAR with two ancestral telomeres is in fact a complete ancestral chromosome. Moreover, when CARs are grouped into syntenic sets, that is, sets of CARs that are believed to belong to the same ancestral chromosome, each such syntenic set of CARs can contain only two ancestral telomeres.

We address this question as follows. A column c' with multiplicity (bounded, for example, by twice the maximum expected number of ancestral chromosomes, or more generally with infinite multiplicity) can then be used to represent telomeres, that is, virtual extremities of ancestral chromosomes. Then any ancestral synteny that contains putatively a marker that is an extremity of an ancestral chromosome (for example because the ancestral synteny is telomeric in two extant descendants of the considered ancestor) can be represented by two rows in M: a row representing the ancestral synteny, plus a copy of this row with an additional entry 1 in column c' (hence M has matched multirows). This structure, as seen earlier, ensures that if M has the mC1P, then the occurrences of c are located at the extremities of the CARs. Otherwise (M does not have the mC1P), some rows can be discarded to result in a matrix M' that has the mC1P, with the same property. The assumption that M has matched multirows is fundamental to leave open the possibility for any ancestral synteny to be at the extremity of a CAR or to be embedded inside a CAR.

Considering several columns with multiplicity can be used to model more precise knowledge about possible ancestral telomeres, provided that the fundamental assumption that the matrix M has matched multirows is maintained, and that any ancestral synteny (i.e., row) contains at most one putative ancestral telomere, which are the assumptions of our main result.

Forest Alignment with Affine Gaps and Anchors

Stefanie Schirmer and Robert Giegerich

Practical Computer Science, Bielefeld University, Bielefeld, Germany
{sschirme,robert}@techfak.uni-bielefeld.de

Abstract. We present two enhancements to Jiang's tree alignment algorithm, motivated by experience with its use for RNA structure alignment. One enhancement is the introduction of an affine gap model, which can be accommodated with a runtime increase by a constant factor. The second enhancement is a speed-up of the alignment algorithm when certain nodes in the trees are pre-aligned by a so-called anchoring. Both enhancements are included in a new implementation of the tool *RNAforester*. We also argue that tree alignment should be parameterized by a user-described set of edit operations, generalizing over the traditional, atomic edit operations.

Keywords: RNA structure alignment, forest alignment, affine gap costs, anchored alignment.

1 Introduction and Motivation

Classical Tree Alignment and its use in RNA Structure Comparison. Tree alignment methods have a wide variety of applications when it comes to comparing objects that are represented as ordered labeled trees. For example, text documents or physical objects composed recursively from smaller constituents have a natural tree representation. In bioinformatics, RNA secondary structure is conveniently expressed as a tree, incorporating the relationships of adjacency and embedding between structural components. Tree alignment [10] is the generalization of sequence alignment from sequences to trees. It has been implemented in the tool *RNAforester* [7,8], which is widely distributed with the Vienna RNA package [9] and has performed well in large scale studies such as [14], in spite of high computational cost of the tree alignment algorithm compared to simpler, competing methods.

A particular virtue of tree alignments, and an advantage compared to alternative methods such as tree edit distance based on node mappings [20], is that they can be used to infer a structural alignment of the leaf sequences of the trees, i.e. the underlying RNA sequences in the RNA application. These derived sequence alignments have been shown, in another large study, to be helpful in determining structural conservation with other sequence based methods [3].

However, two shortcomings of the present method have become apparent: (1) The tree alignment sometimes leads to a rather scattered sequence alignment, using a large number of small gaps. A good alignment may be available, which

R. Giancarlo and G. Manzini (Eds.): CPM 2011, LNCS 6661, pp. 104–117, 2011.

may be more plausible by using fewer but larger gaps, but the algorithm is not aware of this criterion. (2) The high computational cost of the algorithm is an obstacle for its use in many cases, and this is particularly annoying when we need to compare structures in the aforementioned search of structure conservation. In this situation, structures are pre-selected and are only compared when they are known to have a similar overall shape. It should be possible to make use of such knowledge for improved efficiency, without compromising alignment quality.

Contributions of this Article. The main contributions of this work are threefold:

(1) We generalize the tree alignment algorithm to accommodate an *affine gap model*: Gaps are scored with a (large) gap opening penalty, and a moderate penalty that grows linearly with the size of the gap. We show that this can be achieved with a constant runtime factor ≈ 7 compared to the previous model. (2) We define the notion of an *anchoring* as a partial bijection between two trees, and construct alignments consistent with these anchorings, with a speed-up depending on the number of anchors. (3) In applying these improvements to RNA structure comparison, we observe that the classical tree alignment model – based on the operations of matching, deletion and insertion of tree nodes – is too atomic in a real-world scenario such as ours. We propose - as a new research problem – a model of tree alignment based on a set of general tree rewrite rules.

These ideas and further variants of tree alignment algorithms, such as a local tree similarity and multiple tree alignment, have been implemented and evaluated in the first author's PhD thesis [16], but space does not permit to describe these variants in the present paper.

Structure of this Article. The next section gives a short review of related work, recalls the definitions and gives a graphical explanation of the original tree alignment algorithm. Thereafter, we describe our new techniques of affine gap modeling and of anchored tree alignment. These two sections are actually independent, as both contributions are orthogonal to each other. We then discuss asymptotic efficiency and report from the evaluation of the bioinformatics tool. Finally, we discuss the cases where the present tree alignment model is insufficient, and propose a generalization.

2   State of the Art

Tree Edit Distance. The most widely studied model of tree matching and comparison is the *tree edit distance* model [17,20], which is based on partial bijective mappings between two trees, where the mappings preserve ancestorship, and also sibling order in the case of ordered trees. A mapping is scored by summing individual scores based on the label pairs of the mapped nodes. Unmapped nodes are considered deleted or inserted, and do not contribute to the score. The mapping of maximal score defines the edit distance between the two trees. Under a unit cost model, the tree edit distance leads to the largest common subtree of two trees. Ideas similar to our contribution to the tree alignment distance have been developed for the tree edit distance [18,19].

The tree edit distance does not produce tree alignments, although some authors use "tree alignment" as a synonym for "mapping". The unmapped nodes are not brought into any particular, tree-like arrangement, and there is no ovious way to transform the maximal common subtree into a common super-tree, given an arbitrary scoring function. Our definition of tree alignment corresponds to a topological embedding in the terminology of [15].

In the domain of RNA structure analysis, the leaves of the trees carry the RNA sequence information, and the inner nodes represent structure. In this context, it is often desired to derive a sequence alignment from a structure (tree) alignment, and hence, the following approach is more adequate.

Tree Alignment Distance. The tree alignment algorithm by Jiang et al. in [10] is the foundation for the method described in this work. Formal definitions will be given below. Tree alignment seeks a tree in which both trees can be embedded homeomorphically. Run with unit cost scoring, it leads to the smallest common super-tree of two input trees. Insertions and deletions are explicitly embedded in the super-tree, allowing a richer set of scoring schemes. We will make use of this property when introducing composite gaps with affine gap scoring.

Each tree alignment indicates a mapping in the sense of the tree edit distance model, but not vice versa. Hence, the search space of the tree alignment distance is smaller than for the tree edit distance. The algorithm for ordered trees presented in [10] has time complexity $\mathcal{O}(|T_1| \cdot |T_2| \cdot (deg(T_1) + deg(T_2))^2)$, where $|T_i|$ is the number of nodes in tree T_i and $deg(T_i)$ is the degree of tree T_i, so the algorithm is faster than all known ones for the tree edit distance, if the degrees are smaller than the depths of the trees.

In [7], Hoechsmann et al. extend the tree alignment algorithm to compute local forest alignments of forests F_1 and F_2 with a time complexity of $\mathcal{O}(|F_1| \cdot |F_2| \cdot deg(F_1) \cdot deg(F_2) \cdot (deg(F_1) + deg(F_2)))$. The algorithm uses a dense two-dimensional dynamic programming table, and considerably reduces the space requirements compared to previous versions of the algorithm with sparse, four-dimensional tables.

Based on the forest alignment model for the alignment of two trees, a multiple alignment model is developed by Hoechsmann et al. in [8]. This is done with the profile alignment method, which can be transferred from strings to trees and forests. A forest profile representation for RNA secondary structure alignments is presented together with an algorithm to compute the profile alignment, implemented in the tool *RNAforester*.

Seeded mappings. An interesting cross-breed of mapping and alignment is the method of *seeded tree alignment* [11], which, in spite of its name, computes mappings rather than alignments. These mappings can be constrained by seed mappings (a set of node pairs required to map onto each other) which preserve the lowest common ancestor relationship. Such preservation, where enforced, selects a specific common super-tree structure and makes the mappings compatible with, while still more abstract than, tree alignments. On the other hand, it requires the lowest common ancestor nodes of two seed nodes to be mapped onto each other, which is more restrictive than requiring the existence of a compatible super-tree. Hence, this approach is incomparable to our use of anchors explained below, which does not require preservation of lowest common ancestors.

Arc Annotated Sequences. With a particular focus on the RNA structure problem, the alignment problem has been reformulated in terms of arc annotated sequences. See [2] for a review. Recent contributions present alignment algorithms that include pseudo-knotted structures [12,1], which cannot be represented as trees. Neither of these approaches considers affine gaps or anchorings.

3   Forest Alignment Algorithm

RNA Molecules as Trees and Forests. An RNA sequence or primary structure is represented as a string on the alphabet $\{A, C, G, U\}$. For secondary structures, we have to take sequence and basepairings into account.

We represent a secondary structure as a rooted ordered forest, where the sequence is at the leaves. This forest is defined on the above alphabet. Additionally, we introduce *P-nodes*, labeled with P, to represent the base pair bond. Such a node always has two or more child nodes, because the outmost nodes are leaves representing the paired bases. In between, there may be an arbitrary number of P-nodes and bases, according to the nested substructure that is enclosed by this basepair. Thus, our representation is a forest on the alphabet $\mathcal{A} = \{A, C, G, U, P\}$. See Fig. 1 for examples. The forest alignment algorithm to be presented is not restricted to RNA structure comparison, but can be used to compare arbitrary trees and forests.

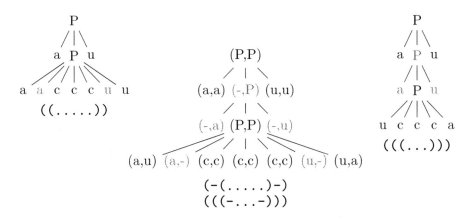

Fig. 1. An example for the forest alignment model. Left and right, there are two hairpin structures and their forest representations. As closed structures, they are forests of length one. In the middle, there is one possible alignment forest. Under each forest, we indicate the encoded structure or structural alignment in dot-bracket notation.

Tree and Forest Alignment. The notion of *tree alignment* is best introduced by the analogy to sequence alignment. An alignment of two *sequences* of characters can be seen as a *sequence* of character pairs, now allowing for a gap character "−". These character pairs in turn are interpreted as the edit operations *Replace*, *Delete*, and *Insert*. Transferring this analogy to trees, an alignment of two node-labeled trees is a tree whose nodes carry label pairs.

Definition 1. *An* alignment tree *is a tree labeled with pairs from the alphabet* $\{\mathcal{A} \cup \{-\} \times \mathcal{A} \cup \{-\}\}/\{(-,-)\}$.
A tree alignment *of trees F and G is an alignment tree A, which can be transformed (1) to F by projecting the pair node labels to their left component and contracting the resulting tree to remove all gaps (nodes labeled "−")from it, and (2) to G in the same way after projecting the pair node labels to their right component.*

The tree alignment definition can be extended to the alignment of *forests* by seeing a forest as a tree with a fictitious root. With the use of tuples rather than pairs, the definition generalizes to multiple forest alignments. An example forest alignment of two RNA hairpin structures is shown in Fig. 1.

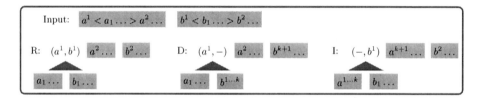

Fig. 2. Graphical explanation for the forest alignment algorithm. Here, $a^1 < a_1 \ldots >$ $a^2 \ldots$ denotes a forest whose first tree has root label a^1 and subtrees $a_1 \ldots$, and k splits a forest at all possible points. On top are the two forests that we want to align, underneath are three cases corresponding to the edit operations. We indicate one node of the alignment forest, and the resulting subforests who are to be aligned to become children (below) and sibling (right) alignment forest.

Figure 2 explains the structural recursion of the forest alignment algorithm. The algorithm's other ingredients are a dynamic programming table, indexed by sub-forests, and a scoring function $\sigma : \mathcal{A} \cup \{-\} \times \mathcal{A} \cup \{-\} \to \mathbb{R}$. This function defines the local score contribution of the edit operations, and the score $\sigma(A)$ of an alignment is the score sum of its edit operations.

The optimal similarity score of the forest alignment of forests F and G is the maximum score over all possible alignments

$$\sigma_{FA}(F, G) = max\{\sigma(A)|\ A \text{ is an alignment of } F \text{ and } G \ \}.$$

In a concrete application, it is sometimes meaningless to align certain node labels. In this case, one may choose a score of $-\infty$ to exclude this situation from optimal alignments, or else incorporate such restrictions into the recurrence. The latter way is more difficult technically, but preserves the capability to derive meaningful statistics about the search space, such as number of alignments and score average.

With RNA, we score five different cases of edit operations, where indel is either insertion or deletion: *single base match*, *single base replacement*, *single base indel*, *pair match*, and *pair indel*. If base indels, base matches and base replacements are scored 0, the alignment becomes independent of the concrete sequence content of the two RNA molecules. Such purely structural alignments are helpful in searching for conservation of structure in sequences with a large evolutionary distance. Should we present two sequences without any base pairs – represented by forests consisting of leaves only – to our algorithm, it reverts to straightforward sequence alignment, based on the scoring of base matches, replacements and indels.

4   Forest Alignment with Affine Gap Costs

The forest alignment algorithm inserts gaps in the input forests to construct an alignment forest.

Definition 2. *A* singleton gap *is a single node in a forest, which is labeled with the gap symbol "−".*

In the above forest alignment algorithm, each singleton gap adds a contribution to the score according to the cost function. That means, for adjacent gaps, the score function is linear in the number of singleton gaps. This procedure can lead to many small gaps, and a scattered alignment, which is not reasonable.

From a biological point of view, it is much more unlikely to open a new gap than to extend an existing one. We need to consider a series of adjacent gaps as one large unit.

We will use "singleton" for single nodes labeled "−" and reserve the term "gap" for larger units:

Definition 3. *A* gap in a tree *is a set of nodes labeled "−", which is maximal and connected under the union of the parent-child and direct-sibling relations.*
A gap in an alignment A *of F and G is a gap in either the left or right projection of A.*

Note that a gap in the tree alignment can appear as several gaps in the derived sequence alignment. For example, when several successive base pairs are deleted in F, this will be one gap in the tree alignment, but show as two gaps in the derived sequence alignment. This is exactly what we want.

For composite gaps, we suggest an *affine gap cost* model. In this model, we have high gap opening costs, whereas the costs for gap extension are low. The cost function can be written as $w(l) = w_{open} + (l - 1) \cdot w_{extend}$, where l is the number of singletons in the composite gap. If we build up the score in a structurally recursive fashion, we do not know gap size l in advance and have to compute the score in multiple steps. To be able to do so, we have to keep track in the recurrences whether we opened a gap before and are in *gap mode* already.

When aligning sequences rather than forests, this leads to three modes, for each of which the three edit operations may be computed: Starting in normal mode (no gap mode), we stay in this mode if we begin the alignment with a replacement. If we start the alignment with an insertion, we open a gap in the first sequence, and the first sequence enters gap mode. Similarly, if we begin the alignment with a deletion, we open a gap in the second sequence, and the second sequence enters gap mode. All three cases, the original (no gap) and the two additional cases (left sequence gap mode and right sequence gap mode) each contain our usual case distinction for the three edit operations. This is known as Gotoh's algorithm in bioinformatics [6].

In a forest alignment A of forests F and G, we do not only have to align the rest of the sequence of trees (*over*), but also the forest of children trees of the aligned nodes (*down*). This two dimensional recursion causes two directions in which we traverse the forest, and two types of gap modes. A deletion at the start of the alignment, for example, introduces a gap in the first tree of G. For the rest of the forests of F and G, we may say that the latter has now entered *(left) sibling gap mode*, because its left sibling already opened a gap and paid the opening costs. For the forests consisting of children trees of F and G, we may say that the latter has now entered *parent gap mode*, because its parent has already opened a gap. In this way, we can score gap openings and gap extensions differently.

The modes of F and G combine in seven different ways, which leads to an algorithm running seven "copies" of the original one. Figure 3 explains the case analysis, using the graphical conventions of Fig. 2.

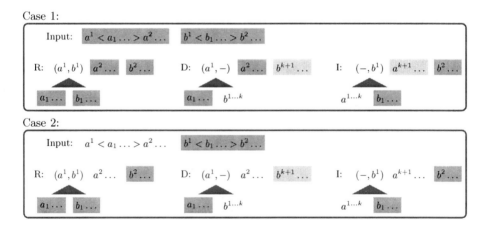

Fig. 3. Case distinction for the tree alignment algorithm with affine gap costs. No-gap-mode is blue , parent gap mode is yellow and sibling gap mode green . We have seven cases for each reasonable combination of gap modes. Compare also Fig. 2. Figure continued on next page.

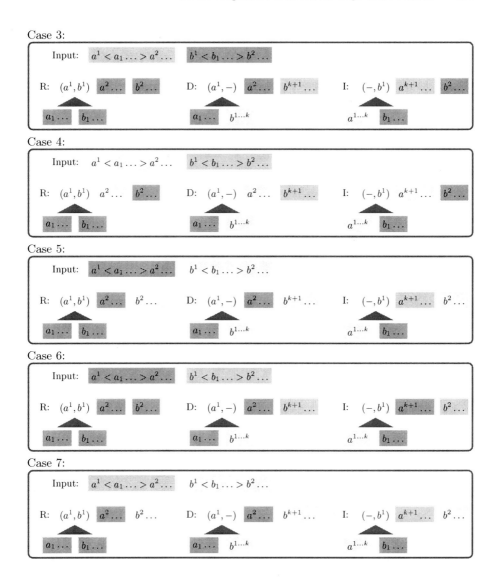

Fig. 4. Remaining cases of Fig. 3

From these recurrences, variants of the algorithm used with sequence alignment can be carried over. One may restrict the search space further by enforcing an insert-before-delete convention for adjacent indels. One may also allow for "oscillating" gaps, scoring a switch from delete to insert mode or vice versa as a gap extension rather than a new gap opening. We remark without proof that this merges cases $\{2, 4, 5, 7\}$ as well as $\{3, 6\}$, leaving us with 3 dynamic programming tables to compute rather than 7.

5   Speed-Up by Anchoring

Often, structures to be aligned come from the same RNA family with a conserved abstract shape [4]. Simply said, abstract shapes record the (tree-like)
arrangement of RNA helices, but abstract from their size and from unpaired
regions. A shape like "[[] [] []]" indicates a clover-leaf structure for sequences
of any length. This should be reflected in the resulting alignment, and may also
be exploited to speed up the algorithm.

   The idea is to use the shape "brackets" to determine anchor points in the
structures. The corresponding anchor nodes must be aligned with each other.
This constrains the alignment algorithm, and only substructures between the
anchor points have to be aligned by our usual algorithm. In contrast to the
approach of [11] discussed earlier, our anchors do not imply that lowest common ancestors of anchors are matched – they are still candidates for deletion or
insertion.

   It is not essential that the anchoring is derived from abstract shapes. It can
be provided, for example, by expert annotation – if it satisfies the following
definition.

Definition 4. *An* anchoring *is a partial mapping function between the nodes of
two forests, with the following constraints: 1) it is a bijection, 2) it preserves the
ancestor relation, 3) it preserves the sibling ordering relation.*

Figure 5 gives an example of an anchoring. By Definition 3 and 4, no gap can contain an anchor, which is why affine gaps and anchoring are orthogonal concepts
that work well in cooperation.

Definition 5. *The* anchored alignment *of forest F with n anchors a_i, and G
with n anchors $b_i, 0 \leq i \leq n$ is the best alignment of F and G, with nodes (a_i, b_i)
for $0 \leq i \leq n$.*

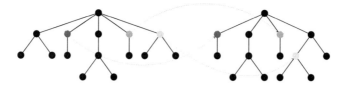

Fig. 5. Anchored alignment input trees with constraints

   In the anchoring variant of the algorithm, we align two forests F and G, such
that anchor a_i in F is aligned to b_i in G for all n anchors of the common shape of
F and G, $0 \leq i \leq n$. To prune unsuccessful alignments, we make sure that, after
each alignment step, we have the same anchors in both remaining substructures
of the down/over alignment to ensure correct parent-sibling relation. In fact – it

suffices to check for the same *number* of anchors, because the anchors must have a consistent tree-like arrangement in both forests due to Definition 4. If in the *match* case, the subtree rooted at the matched node of the first tree has α anchors, the number of anchors in the subtree rooted at the matched node of the second tree, α', must be the same. For the *insertion* or *deletion* case, the running index k is restricted, as in the substructures for the down alignment, the correct number of anchors has to be present. This, again, has to be the same number of anchors in the first and the second component of the alignment.

With this straightforward idea, a challenge had to be faced during implementation: The anchored variant of this algorithm is naturally described in a top down fashion. In the unanchored alignment, the recurrences have been translated into a bottom up computation method, as common in dynamic programming. To overcome this problem, we also implemented a top down variant of the previous algorithm. Based on this, the anchoring was incorporated.

6   Computational Complexity and Performance

For $n = |F| = |G|$ and $d = deg(F) = deg(G)$, the time complexity for alignment of F and G is $\mathcal{O}(n^2 d^2)$ [7,10]. This remains asymptotically the same with affine gap scoring, but a constant factor ≈ 7 is expected. In our current implementation, we measured an average runtime factor of 8.02 for alignments of folded sequences of ≈ 100 nucleotides in length, and 7.79 for those of ≈ 200 nucleotides in length. Allowing gaps to oscillate between F and G without opening penalty, as explained above, would merge cases and reduce the constant factor further.

Using $k - 1$ evenly placed anchors, such that both F and G are split into k parts of about equal size, n is divided by k in the complexity expression. The complexity is thus reduced from n^2 to $k(\frac{n}{k})^2$. In general, d may also be reduced, but with RNA structure trees, we tend to have $d \leq 30$ anyway. Efficiency with $k - 1$ anchors is then $\mathcal{O}(\frac{n^2}{k} d^2)$. Space is reduced in the same way as in [7].

Table 1. Speed-up factor gained by shape anchoring, for ten members of each of the above Rfam families, folded by *RNAcast* [13]

RNA Family	Common Shape	Anchors	Avg. Speed-up by Anchoring
5S rRNA (RF00001)	[[] []]	3	1.27
Spot 42 (RF00021)	[] [] []	3	3.30
Cobalamin (RF00174)	[[] [[] [[] []]]]	7	2.96
T-box (RF00230)	[[[] []] [[] []]]	7	3.23

As intended, affine gap scoring helps to improve structure alignments. We show excerpts from three alignments of two introns of Arabidopsis thaliana. See Figs. 6, 7, 8 and the explanation given there.

```
global optimal score: -9
Intron_7            GUCUGUUACACGCCGAGAUCGGACUCCGAGUGAUAUCCUC-GA-CGGAUCUGU--
Intron_8            GUCUGUUACACGAGAGAUCGGUCUCCCGGAUCGAGCCCUCGACGGAUCUGAUUCC

Intron_7            UCCGAUCUUGUGUUUCUCUGUUACUUGAU-UCGAUUACUCUGUUACUAUUCUCGU
Intron_8            GAUCUGUUUCUCUGU-UACUUGAUUCGAUUACUGUUACUAUGUU-C----UCUCU

Intron_7            UCUUUGUUACUACUACUACUACUA
Intron_8            -CG-U--U-CU--U--UG-U--UA


Intron_7            ((..((.(((...(((((((((...(((((((((.....-((-((((.....--
Intron_8            .........(((((((.((((....(((((((((.....(((((((.......

Intron_7            )))).)).........))))))))).)-))))...............))))).
Intron_8            ))))))))......-..)))))))))...))))..........-.----)))))

Intron_7            ....))).)).))...........
Intron_8            -))-)--.-..--.--..-.--..
```

Fig. 6. Example of a tree alignment, which is scattered due to the original, linear gap score model and the default score parameters of *RNAforester*. This alignment has 24 matched basepairs and 23 singleton gaps, which appear as 15 composite gaps. Scoring type: global similarity; Scoring parameters: pair match: 10; pair indel: -5; base match: 1; base replacement: 0; base indel: -10;

```
global optimal score: -125
Intron_7            GUCUGUUACACGCCGAGAUCGGACUCCGAGUGAUAUCCU-C-GACGGAUCUGU--
Intron_8            GUCUGUUACACGAGAGAUCGGUCUCCCGGAUCGAGCCCUCGACGGAUCUGAUUCC

Intron_7            UCCGAUCUUGUGUUUCUCUGUUACUUGAU-UCGAUUACUCUGUUACUAUUCUCGU
Intron_8            GAUC-UGUUUCUCUGUUACUUGAUUCGAUUACUGUUACUAUGUU-C----UCUCU

Intron_7            UCUUUGUUACUACUACUACUACUA
Intron_8            CG--U--U-CU-------UUGUUA


Intron_7            ((..((.(((...(((((((((...(((((((((....-.-(((((.....--
Intron_8            .........(((((((.((((....(((((((((.....(((((((.......

Intron_7            )))).)).........))))))))).)-))))...............))))).
Intron_8            ))))-))))........)))))))))...))))..........-.----)))))

Intron_7            ....))).)).))...........
Intron_8            ))--)--.-..-------......
```

Fig. 7. Choosing the affine gap score model and a higher gap opening penalty for pair and base indels, the gaps are contracted to longer composite gaps. This alignment has also 24 matched basepairs, but only 11 composite gaps. Scoring type: affine global similarity; Scoring parameters: pair match: 10; pair indel open: -20; pair indel: -5; base match: 1; base replacement: 0; base indel: -10; base indel open: -20;

```
global optimal score: -161
Intron_7                    GUCUGUU--ACACGCCGAGA-UCGGACUCCGAGUGAUAUCCU-C-GACGGAUCUG
Intron_8                    GUCUGUUACACG---AGAGAUCGGUCUCCCGGAUCGAGCCCUCGACGGAUCUGAU

Intron_7                    U--UCCGAUCUUGUGUUUCUCUGUUACUUGAU-UCGAUUACUCUGUUACUAUUCU
Intron_8                    UCCGAUC-UGUUUCUCUGUUACUUGAUUCGAUUACUGUUACUAUGUUC----UCU

Intron_7                    CGUUCUUUGUUACUACUACUACUACUA
Intron_8                    CU-----CGUU-CU-------UUGUUA

Intron_7                    ((..((.--(((...(((((-(((((...(((((((((....-.-(((((((....
Intron_8                    .........(((---(((((.(((((....(((((((((.....((((((((....

Intron_7                    .--)))).)).........))))))))))).)-))))...............)))
Intron_8                    ...))))-))))........))))))))))...))))...........----)))

Intron_7                    )).....))).)).))...........
Intron_8                    ))-----))).-..-------.....
```

Fig. 8. Choosing an even higher gap opening penalty for pair indels also contracts gaps between brackets/basepairs. This alignment now has 27 matched base pairs, and also 11 composite gaps, in positions different from those of the previous alignment. Scoring type: affine global similarity; Scoring parameters: pair match: 10; pair indel open: -30; pair indel: -5; base match: 1; base replacement: 0; base indel: -10; base indel open: -20;

7   Towards a Generalized Model of Forest Alignment

Semantically, a P-node and its two outmost children constitute a unit of RNA structure. The alignment model allows to break up this unit and recode it ambiguously in many different forms, because the pairing bases can be "removed" from their P-node by intervening gaps. See Fig. 9. Such semantic ambiguity [5] precludes the use of probabilistic scoring schemes, as the most likely alignment tree is not the most likely alignment.

Fig. 9. Three alignment trees of the trees F = a and G = P(b,c), where the middle and right "mean" the same. The right one is considered artefactual, as the b is removed from its P-node.

As a generalization, we want to be able to describe meaningful, non-atomic edit operations explicitly, depending on the meaning of the trees to be aligned. Then, a generalized tree alignment algorithm should be based on these edit operations.

The above case could be described by two edit rules

```
(1)  x <-> P(x,y)              (2)  z <-> P(x,y)
```

Rule 1 says that, in the example of Fig. 9, x binds to a in F and to b in G, and implies the leftmost alignment. Rule 2 says that a and b are not matched, and the implied alignment is the middle one, where b stays with its P-node, while a becomes a $(a, -)$ (or (a, c), which also makes sense). Explicit rules would also allow us to rule out alignment nodes such as (a, P), which are legal in the general tree edit model, but meaningless in our semantically richer application context. Our present implementation avoids this situation in an ad-hoc manner. As a general model, tree alignment parameterized by explicit matching rules appears a well-motivated challenge for future research in combinatorial pattern matching.

Acknowledgements

Thanks to I. Hofacker and P. Stadler for a first discussion of affine gap scoring, once upon a time at the Benasque RNA meeting. We also thank the anonymous reviewers for valuable hints.

References

1. Backofen, R., Landau, G.M., Möhl, M., Tsur, D., Weimann, O.: Fast RNA Structure Alignment for Crossing Input Structures. In: Proceedings of the 20th Annual Symposium on Combinatorial Pattern Matching, pp. 236–248 (2009)
2. Blin, G., Touzet, H.: How to compare arc-annotated sequences: The alignment hierarchy. In: Crestani, F., Ferragina, P., Sanderson, M. (eds.) SPIRE 2006. LNCS, vol. 4209, pp. 291–303. Springer, Heidelberg (2006)
3. Bremges, A., Schirmer, S., Giegerich, R.: Fine-tuning structural RNA alignments in the twilight zone. BMC Bioinformatics 11, 222 (2010)
4. Giegerich, R., Voss, B., Rehmsmeier, M.: Abstract shapes of RNA. Nucleic Acids Research 32(16), 4843–4851 (2004)
5. Giegerich, R., Höner zu Siederdissen, C.: Semantics and Ambiguity of Stochastic RNA Family Models. IEEE/ACM Transactions on Computational Biology and Bioinformatics 8(2), 499–516 (2011), DOI, http://doi.ieeecomputersociety.org/10.1109/TCBB.2010.12
6. Gotoh, O.: An improved algorithm for matching biological sequences. J. Mol. Biol. 162(3), 705–708 (1982)
7. Hoechsmann, M., Toeller, T., Giegerich, R., Kurtz, S.: Local similarity in RNA secondary structures. Proc. IEEE Comput. Soc. Bioinform. Conf. 2, 159–168 (2003)
8. Hoechsmann, M., Voss, B., Giegerich, R.: Pure multiple RNA secondary structure alignments: A progressive profile approach. IEEE/ACM Transactions on Computational Biology and Bioinformatics 1, 53–62 (2004)
9. Hofacker, I.L., Fontana, W., Stadler, P.F., Bonhoeffer, L.S., Tacker, M., Schuster, P.: Fast folding and comparison of RNA secondary structures. Monatshefte für Chemie / Chemical Monthly 125(2), 167–188 (1994)

10. Jiang, T., Wang, L., Zhang, K.: Alignment of trees – an alternative to tree edit. *Theor. Comput. Sci.*, 143 (1): 137–148 (1995)
11. Lozano, A., Pinter, R.Y., Rokhlenko, O., Valiente, G., Ziv-Ukelson, M.: Seeded Tree Alignment. IEEE/ACM Trans. Comput. Biol. Bioinformatics 5(4), 503–513 (2008)
12. Möhl, M., Will, S., Backofen, R.: Fixed Parameter Tractable Alignment of RNA Structures Including Arbitrary Pseudoknots. In: Proceedings of the 19th Annual Symposium on Combinatorial Pattern Matching, pp. 69–81 (2008)
13. Reeder, J., Giegerich, R.: Consensus Shapes: An Alternative to the Sankoff Algorithm for RNA Consensus Structure Prediction. Bioinformatics 21(17), 3516–3523 (2005)
14. Ritchie, W., Legendre, M., Gautheret, D.: RNA stem loops: to be or not to be cleaved by RNAse III. RNA 13(4), 457–462 (2007)
15. Rosselló, F., Valiente, G.: An algebraic view of the relation between largest common subtrees and smallest common supertrees. Theor. Comput. Sci. 362(1), 33–53 (2006)
16. Schirmer, S.: Comparing forests. PhD thesis, Faculty of Technology, Bielefeld University (to appear)
17. Tai, K.C.: The tree-to-tree correction problem. J. ACM 26, 422–433 (1979)
18. Touzet, H.: Tree edit distance with gaps. Inf. Process. Lett. 85(3), 123–129 (2003)
19. Touzet, H.: A linear tree edit distance algorithm for similar ordered trees. In: Apostolico, A., Crochemore, M., Park, K. (eds.) CPM 2005. LNCS, vol. 3537, pp. 334–345. Springer, Heidelberg (2005)
20. Zhang, K., Shasha, D.: Simple fast algorithms for the editing distance between trees and related problems. SIAM J. Comput. 18(6), 1245–1262 (1989)

Phylogenetic Footprinting and Consistent Sets of Local Aligments

Wolfgang Otto[1,2], Peter F. Stadler[4,1,2,5−8], and Sonja J. Prohaska[3,2]

[1] Max Planck Institute for Mathematics in the Sciences, Inselstraße 22,
D-04103 Leipzig, Germany
[2] Interdisciplinary Center for Bioinformatics, University of Leipzig,
Härtelstraße 16-18, D-04107 Leipzig, Germany
[3] Computational EvoDevo Group, Department of Computer Science,
University of Leipzig, Härtelstraße 16-18, D-04107 Leipzig, Germany
[4] Bioinformatics Group, Department of Computer Science,
University of Leipzig, Härtelstraße 16-18, D-04107 Leipzig, Germany
[5] Fraunhofer Institut für Zelltherapie und Immunologie,
Perlickstraße 1, D-04103 Leipzig, Germany
[6] Center for noncoding RNA in Technology and Health, University of Copenhagen,
Grønnegårdsvej 3, DK-1870 Frederiksberg C, Denmark
[7] Department of Theoretical Chemistry, University of Vienna,
Währingerstraße 17, A-1090 Wien, Austria
[8] Santa Fe Institute, 1399 Hyde Park Rd., Santa Fe, NM 87501, USA

Abstract. The problem of constructing alternative local multiple sequence alignments from a collection of local pairwise alignments arises naturally in phylogenetic footprinting, a technique used to identify regulatory elements by comparative sequence analysis. Based on a theoretical discussion of the problem we devise an efficient heuristic and introduce the software tool `tracker2` for this task. Tests on both biological and random data demonstrated the heuristic yields excellent results at very short runtimes.

Keywords: alignment consistency, phylogenetic footprinting, combinatorial optimization, `tracker2`.

1 Introduction

The discovery of functional sequence elements in genomic DNA data is an important research topic in bioinformatics [1]. Most individual binding motifs, in particular transcription factor binding sites (TFBS) are short and gapless. Their overrepresentation in the surrounding of co-regulated genes makes them detectable by motif discovery approaches such as `meme` [2] and `footprinter` [3]. Regulatory sequence elements are often (but not always) subject to stabilizing selection and hence evolve much more slowly than adjacent non-functional DNA. Such *phylogenetic footprints* are therefore detectable by comparative sequence analysis. This task is frequently referred to as *phylogenetic footprinting*. A large class of tools combines pattern search with the explicit analysis of conservation, see e.g. [4–7]. However, pattern discovery approaches usually fail when not only

R. Giancarlo and G. Manzini (Eds.): CPM 2011, LNCS 6661, pp. 118–131, 2011.
© Springer-Verlag Berlin Heidelberg 2011

small promoter-proximal regions but large intergenic regions are under investigation. Techniques based on global or local sequence alignments are successfully employed in such cases [8, 9].

As an alternative to the analysis of a single global alignment it has been suggested to start from local alignments between all pairs of sequences of interest [10]. This is appealing in particular when large stretches of orthologous sequences need to be analyzed. This approach leads, however, to contradictory signals of sequence similarity that require a sophisticated post-processing. Since regulatory modules need not be co-linear, the construction of alternative clusters of pairwise alignments has been suggested. Here we revisit the approach of [10]. We discuss its theoretical foundation, starting with ideas from [11] on the consistency of alignments, see also [12]. An efficient heuristic for the assembly of maximal local multiple sequence alignments from local pairwise alignments is implemented in the software tool `tracker2`.

2   Theory

Definitions and Basic Properties. Following [11, 13] we consider sequence alignments as vertex-labeled, undirected graphs. Each position of an input sequence corresponds to a vertex. The so-called alignment edges, that is, matches or mismatches between sequence positions, form the edges of the graph.

We formalize this picture as follows: Consider a set X of m sequences x_a, $1 \leq a \leq m$, with not necessarily equal lengths $|x_a|$ over a common alphabet Σ. The symbol x_{ai}, $1 \leq i \leq |x_a|$ refers to the i-th letter of the a-th sequence and the site space $\mathcal{S}(X)$ of X comprises all indices refering to letters in X, i.e., $\mathcal{S}(X) = \{(a, i) | 1 \leq i \leq |x_a|, \ 1 \leq a \leq m\}$.

Definition 1. *A multiple alignment of X is a undirected graph $\Gamma(\mathcal{S}(X))$ over the site space of X with vertex set $\mathbb{V} = \mathcal{S}(X)$, vertex labels $x : \mathbb{V} \to \Sigma$, $(a, i) \mapsto x_{ai}$, and an edge set A satisfying the following three conditions:*

1. *The connected components of $(\mathbb{V}, A)$ are complete graphs. These complete graphs correspond to the alignment columns.*
2. *If (a, i) and (a, j) are contained in the same connected component, then $i = j$. Thus, every alignment column contains at most one position from each sequence.*
3. *There is a partial order $\preceq$ on the set of connected components so that for any two components P and Q containing vertices $(a, i) \in P$ and $(a, j) \in Q$ the ordering $i \leq j$ along the sequence implies $P \preceq Q$. Alignment columns therefore never cross each other.*

We note that alignments can be stored and manipulated more efficiently in e.g. as a partially ordered sets of alignment columns, a point of view that we will adopt later in this contribution. The graph structure introduced here, however, appears more convenient for theoretical analysis, in particular when starting from collections of pairwise alignments whose union in general does not form an alignment.

Given a set X of m sequences, whose lengths are bounded by n, and a set of edges E over $\mathbb{V}$, it can be decided in $O(n^2 m^2)$ time whether $G = (\mathbb{V}, E)$ is an alignment. This is done by first inserting directed edges $((a, i), (a, j))$ for all nodes (a, i) and (a, j) with $i < j$, corresponding to the order of letters in the strings, in $O(n^2 m^2)$ time and finding the strongly connected components of G [14] in linear time to the number of edges in G which is in $O(nm^2)$. If these components are complete graphs and have at most one position from each sequence, which can be checked in $O(nm^2)$, also the partial order $\preceq$ is well defined for all pairs of components.

Lemma 1. *Let $\Gamma(\mathcal{S}(X))$ be an alignment of a set of strings X, and let $Y \subseteq \mathcal{S}(X)$ be a subset of the sitespace of X. Then $\Gamma(\mathcal{S}(X))[Y]$, i.e., the subgraph of $\Gamma(\mathcal{S}(X))$ induced by Y, is an alignment.*

Proof. By construction, the vertex set $\mathbb{V}' = Y$ is a subset of the vertex set $\mathbb{V} = \mathcal{S}(X)$. Furthermore, the label $x : \mathbb{V}' \to \Sigma$ satisfies $(a, i) \mapsto x_{ai}$. The subgraph of $\Gamma(\mathcal{S}(X))$ defined by Y thus equals the induced subgraph $\Gamma(\mathcal{S}(X)[Y]$. Every induced subgraph of a complete graph is again complete, thus property (i) is satisfied. The other two properties are satisfied for all subgraphs.

Note that Y not only represents the sitespace of single subsequences of a subset of X but of an arbitrary number of subsequences for each sequence in X and that this subsequences do not have to be consecutive.

For simplicity we write from now on $\Gamma(X)$ for $\Gamma(\mathcal{S}(X))$ and $\Gamma[Y]$ or, if necessary, $\Gamma(X)[Y]$ to denote the restriction of an alignment of X to the site space Y of a subset of subsequences of X.

In particular every pair of sequences x_a, x_b gives rise to a pairwise alignment as an induced subgraph $\Gamma(X)[\{x_a, x_b\}]$. Similarly, every individual subset of alignment edges of $\Gamma(X)$ can also be interpreted as an alignment.

Definition 2. *Let X be a set of sequences and let $\mathcal{C} = \{\Gamma(Y_k)\}$ be a collection of alignments with $Y_k \subseteq \mathcal{S}(X)$. We say that $\mathcal{C}$ is* consistent *if there is an alignment $\Gamma(X)$ so that $\Gamma(X)[Y_k] = \Gamma(Y_k)$, i.e., the given alignments $\Gamma(Y_k)$ are the restrictions of $\Gamma(X)$ to the subset of subsequences Y_k.*

Given a set X of sequences and a collection $\mathcal{C} = \{\Gamma(Y_k)\}$ of alignments we write $\bigcup\{\mathcal{C}\}$ to denote the union of the sets of all alignment edges in each of the $\Gamma(Y_k)$.

Lemma 2. *Let X be a collection of sequences and let $\mathbb{V}$ be the site space of X. A collection $\mathcal{C}$ of alignments on X is consistent if and only if the transitive closure of the graph $(\mathbb{V}, \bigcup\{\mathcal{C}\})$ is an alignment.*

Proof. An alignment graph is transitive since, by definition, it is a disjoint union of complete graphs. Consistency of $\mathcal{C}$, on the other hand, implies that $(\mathbb{V}, \bigcup\{\mathcal{C}\})$ is a subgraph of an alignment Γ. In particular the connected components of Γ contain at most one vertex from each sequence, and hence the connected components of $(\mathbb{V}, \bigcup\{\mathcal{C}\})$ also contain at most one vertex from each sequence. Now observe that the transitive closure of a graph equals the disjoint union of the transitive closures of its connected components. Thus, the transitive closure of $(\mathbb{V}, \bigcup\{\mathcal{C}\})$ is a transitive subgraph of Γ, and thus itself an alignment.

Finally, β denotes a weighting function that assigns each alignment Γ the weight $\beta(\Gamma)$. Note that β does not have to be additive in terms of the alignment edges. In fact, the weight of each input alignment $\Gamma(Y_k)$ can be assigned arbitrarily in our setting.

Combinatorial Optimization Problem. Based on Lemma 2, we can now formalize the combinatorial optimization problem.

Definition 3 (Maximal Consistent Alignment Subset Problem). *Given a set of strings X and a collection $\mathcal{C} = \{\Gamma(Y_k)\}$ of alignments of subsequences of the elements of X, the* Maximal Consistent Alignment Subset Problem *is to find a maximum sub-collection $\mathcal{C}' \subseteq \mathcal{C}$ that is consistent.*

Here, maximality can be defined either in terms of cardinality or in terms of the weights $\beta(\Gamma(Y_k))$. Note that as an alternative one might want to optimize the sum-of-pair score of the multiple alignment M formed by combining the alignment edges of the members of $\mathcal{C}'$.

In practice, this is of particular interest in two settings:

1. All $\Gamma(Y_k)$ are individual alignment edges. This version of our problem is the problem faced by consistency-based alignment procedures. For example, T-coffee [15] takes a "library" of alignment edges and then employs a heuristic approach to extract a collection of alignment edges consistent with a multiple alignment of maximal score. In practice, the pairwise alignment edges are often computed from pairwise alignments.
2. All $\Gamma(Y_k)$ are local pairwise sequence alignments. This has been a starting point for footprinting tool tracker [10].

Instead of the maximum consistent subset we are interested in the collection of all maximal consistent subsets of $\mathcal{C}$ in particular in the context of phylogenetic footprinting.

Clearly, consistency is hereditary, i.e., the consistent subsets of $\mathcal{C}$ form an independence system [16]. Collections of alignments do not form a matroid or greedoid, however. Distinct maximal consistent subsets of $\mathcal{C}$ therefore may have different cardinalities. The canonical greedy algorithm, furthermore, is not guaranteed to find an optimally scoring consistent subset of $\mathcal{C}$ [17]. In fact, our problem is NP-complete in general, because the multiple alignment problem is the special case with $\mathcal{C}$ being the collection of all possible alignment edges on X. Hardness results for multiple sequence alignment are proved in [18].

Concatenation of Alignments. The construction of the transitive closure outlined above has an alternative interpretation as a concatenation or transfer operation between pairwise alignments. Consider two pairwise alignments P and Q on the site space $\mathbb{V}$ with edge sets $E(P)$ and $E(Q)$. If P and Q have no sites in common, or if P and Q are inconsistent, we set $P \bullet Q = \emptyset$. If P and Q overlap in exactly one sequence, then $\{(x, i), (z, k)\}$ is an edge in the concatenated alignment $P \bullet Q$ if and only if there is a vertex $(y, j) \in \mathbb{V}$ such that $\{(x, i), (y, j)\}$ is an alignment edge in P and $\{(y, j), (z, k)\}$ is an alignment edge in Q, and $a \neq c$. The vertex

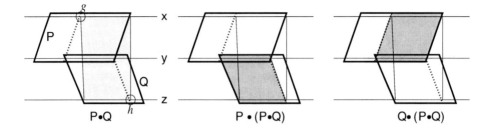

Fig. 1. Concatenation of pairwise alignments

set belonging to $P \bullet Q$ is the minimal subset of vertices form P and Q so that all edges in A are supported. It will be convenient, furthermore, to set $P \bullet Q = (\mathbb{V}, E(P) \cap E(Q))$ for any pair of consistent pairwise alignments of the same sequences, to ensure that the $\bullet$-operation is well-defined for any pair of pairwise alignments.

Note that the edge set $E(P \bullet Q)$ is the relational composition $E(P) \circ E(Q)$ if P and Q have exactly one sequence in common [11]: $P \bullet Q$ is the pairwise alignment of sequences a and c that is *implied* by the alignment edges of P and Q. By construction, furthermore, $\{P, Q, P \bullet Q\}$ is consistent provided $\{P, Q\}$ is consistent. Examples for the concatenation of pairwise alignments are shown in Fig. 1.

We observe that the $\bullet$ operation is commutative and idempotent by definition. It is not associative, however. If the alignments have both or no sequences in common, the repeated application of the operation will not produce any additional alignments. Otherwise, we observe that $P \bullet (P \bullet Q)$ is a non-trivial subset of alignment edges of Q: To see this just note that $\{(i, a), (j, b)\} \in P$ concatenated with $\{(i, a), (k, c)\} \in P \bullet Q$ yields the edge $\{(j, b), (k, c)\}$, which by construction of $P \bullet Q$ is contained already in Q. Analogously, $Q \bullet (P \bullet Q) \subseteq P$. Further concatenations do not lead to additional distinct alignments, see Fig. 1. For instance we have $(P \bullet Q) \bullet (P \bullet (P \bullet Q)) = Q \bullet (P \bullet Q)$.

For a collection of pairwise alignments $\mathcal{C}$ the transitive closure w.r.t. the $\bullet$ operation, $\mathfrak{T}(\mathcal{C})$ is well defined as the collection of all pairwise alignments that can be generated from $\mathcal{C}$ by repeated application of the $\bullet$ operator. By construction, the union of the alignment edges in all pairwise alignments of $\mathfrak{T}(\mathcal{C})$ equals the transitive closure of $(\mathbb{V}, \bigcup\{\mathcal{C}\})$. A set $\mathcal{C}$ of pairwise alignments is therefore consistent if and only if the set of pairwise alignments generated by $\bullet$ from $\mathcal{C}$ is consistent.

Differences of alignments are also well-defined in terms of their graphs. For example, $P \setminus (Q \bullet (P \bullet Q))$ specifies the part of P in Fig. 1 which cannot be extended to an alignment of all three sequences. In this way, it is possible to refer to specific parts of the transitive closure of any consistent collection of pairwise alignments.

Alignments as paired intervals. In [10] local alignments were treated as matches (or pairings) between two sequence intervals, disregarding the exact position

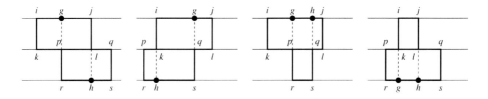

Fig. 2. Construction of additional boundaries when alignments are concatenated

of the individual alignment edges within the intervals. In the present formalism developed above this can be implemented by representing only the *delimiting edges*, i.e. the left- and rightmost edge w.r.t. to the alignment's position on the sequences it contains. More formally, a pairwise alignment $\Gamma(X) = (\mathbb{V}, A)$ over the two sequences $X = \{x, y\}$ is then specified by the two edges $([x, b_x], [y, b_y])$ and $([x, e_x], [y, e_y])$, where $b_x = \min\{i_x | ([x, i_x], [y, i_y]) \in A\}$, $e_x = \max\{i_x | ([x, i_x], [y, i_y]) \in A\}$, and b_y and e_y is defined analogously. It will be convenient in to following to specify an interval on sequence x as a triple $[x, b, e]$ where b and e are the begin an end coordinates. Pairwise alignments are then defined as unordered pairs of intervals $\Gamma(X) = \{[x, b_x, e_x], [y, b_y, e_y]\}$.

Clearly, the reduction of an whole alignment to its outer edges only, is an approximation. In order to construct the concatenation $P \bullet Q$ of two alignments P and Q, we would need also the inner alignment edges to determine the begin and end, denoted by g and h in the example of Fig. 1, of the new alignment $P \bullet Q$. Since the exact edges are not available in this interval approximation, we compute them using a linear interpolation scheme [10]. For instance, in the first case in Fig. 2 with $k \leq p \leq l \leq q$, we compute

$$g = i + (p - k)\frac{j - i}{l - k} \quad \text{and} \quad h = r + (l - p)\frac{s - r}{q - p} \tag{1}$$

Analogous equations are easily derived for the other three cases in Fig. 2.

In the interval approximation model it also makes sense to relax the consistency conditions for alignments: For example, we may want to require that two vertices of the form (a, i) and (a, j) in the same connected component satisfy $|i - j| \leq \varepsilon$. Analogously, for two edges with $\{(a, i), (b, j)\}$ and $\{(a, k), (b, l)\}$ with $i < k - \varepsilon$ we require $j < l + \varepsilon$. Consistency as defined above is recovered as the special case $\varepsilon = 0$.

3   Heuristic Algorithm

The structure of the collection $\mathcal{C}$ of alignments is an independence system. This suggests to explore greedy-like heuristics. We therefore construct a multiple alignment $\mathbf{M}$ iteratively by adding one pairwise alignment $P \in \mathcal{C}$ after the other so that the sum of the scores $\beta(P)$ of the incorporated pairwise alignments is maximized at the end.

Since it will be ensured by the construction procedure that $\mathbf{M}$ is an alignment, we can represent it as a partially ordered list of alignment columns, each of which corresponds to one connected components in Definition 1. Since we know that these are complete subgraphs, there is no need to store the alignment edges explicitly. The insertion of the pairwise alignments into the growing multiple alignment $\mathbf{M}$ and the subsequent consistency checks can therefore be performed very efficiently in the interval approximation. In this setting it is convenient to combine the columns in $\mathbf{M}$ that are consecutive and involve the same sequences to *thick columns*. In analogy to pairwise alignments, a thick column is represented as a set of intervals of the form $[z, b_z^c, e_z^c]$, one for each involved sequence z.

The quality of the final multiple alignment $\mathbf{M}$ depends crucially on the order in which the pairwise alignments from $\mathcal{C}$ are inserted. Intuitively, the optimal collection $\mathcal{C}'$ will contain in particular all those alignments that are "biologically correct". These are unknown in real life, of course. However, partial alignments that are supported by many other alignments are at least good candidates. We therefore adopt the idea, which proved successful in $\mathtt{T\text{-}coffee}$ [15], namely to introduce an extended score $\sigma(\Gamma)$ that consists of to the basis score $\beta(\Gamma)$ of an alignment Γ and "bonus contributions" that are added when Γ is well-supported by other alignments.

Extended Scores. A pairwise alignment $A \in \mathcal{C}$ is supported by $B \in \mathcal{C}$ if A and B align the same regions, i.e., if $A \cap B \neq \emptyset$. Similarly, B and C together support A if $A \cap (B \bullet C) \neq \emptyset$. Bonus scores are computed from all pairs and triples of alignments and in each case, the extended score of $\sigma(A)$ is increased by $\beta_A(X)$ which is the fraction of score of $X \in \{A \cap B, A \cap (B \bullet C)\}$ proportional to the relative size of overlapping region of X and A, i.e., $\beta_A(X) = |X|/|A| \times \beta(A)$. The extended score is then defined as

$$\sigma(A) = \beta(A) + \sum_B \beta_A(A \cap B) + \sum_{B,C} \beta_A(A \cap (B \bullet C)) \qquad (2)$$

We remark that there is no *a priori* theoretical reason for this particular form of the extended score.

Greedy Heuristic. We order $\mathcal{C}$ by the extended alignment scores σ and treat it as a queue. If alignments have the same extended score σ, we sort them by the basis score γ and if also γ is equal, we use the input order. In the beginning $\mathcal{C}' = \emptyset$ and the multiple alignment $\mathbf{M}$ is the graph with vertex $\mathbb{V}$ without any edges. In each insertion step, we remove the top-scoring alignment A from the queue $\mathcal{C}$, and check whether $\mathcal{C}' \cup \{A\}$ is consistent. If so, we add A to $\mathcal{C}'$, insert A into the graph $\mathbf{M}$, and compute its transitive closure, i.e., we update the thick columns of which $\mathbf{M}$ is composed.

In practice, the consistency test and the insertion of a candidate alignment $A = \{[x, b_x^A, e_x^A], [y, b_y^A, e_y^A]\}$ into $\mathbf{M}$ is performed simultaneous, investigating one thick column $c = \{[z, b_z^c, e_z^c]\}$ of $\mathbf{M}$ at a time. We assume that the columns c in $\mathbf{M}$ are ordered relative to $\preceq$, so that e.g. the left-most column c in $\mathbf{M}$ that intersects A can be determined efficiently.

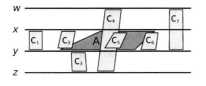

Fig. 3. Possible locations of columns relative to the entries in alignment A. Column c_1 is a prefix in both species S_2 and S_3, while c_7 is a suffix, c_4 is independent, and c_5 overlaps in both species. The remaining columns c_2 (prefix in S_2), c_3 (overlap in S_3) and c_6 (suffix in S_3) touch A in only one species.

There are four possibilities for the relative positioning of the sequence intervals $[x, b_x^A, e_x^A]$ and $[y, b_y^A, e_y^A]$ of the pairwise alignment A and the thick column c that need to be distinguished, see Fig. 3:

independence: c contains no entry $[i, b_i^c, e_i^c]$ with $i \in \{x, y\}$. In this case we have no information about the $\preceq$-order of A and c at sequence i.

overlap: c contains an entry $[i, b_i^c, e_i^c]$ and we have $[i, b_i^A, e_i^A] \cap [i, b_i^c, e_i^c] \neq \emptyset$ for one sequence $i \in \{x, y\}$. Thus c can be extended by the information of A about the other sequence $\{x, y\} \setminus \{i\}$.

prefix: c contains an entry $[i, b_i^c, e_i^c]$ and we have $e_i^c < b_i^A$. Here, c is in front of A for sequence i. This information is important to maintain the partial order $\preceq$ and we have to remember the column c as the closest prefix p_i of A for sequence i detected so far.

suffix: c contains an entry $[i, b_i^c, e_i^c]$ and we have $b_i^c > e_i^A$. In this case c is behind A for sequence i. If c is the first column following A on sequence i we remember, we have to remember, analogously, c as the closest suffix s_i of A at i.

An update of c is only necessary if the column overlaps with at least one sequence. Otherwise, only the prefix, suffix, or order information needs to be updated: If c is a prefix for one sequence and we already know a prefix c' from a earlier stage of the algorithm *and* there is a closest suffix column s for the other sequence of A, then A provides additional information about the order of s and c. In this case we need to update the partial order of the columns: we have to move c and columns c' with $c' \preceq c$ between s and c to the front of s. Also, we may detect that A cannot be inserted into **M** at this stage: if $s \preceq c$ or if c is prefix and suffix at the same time, the insertion of A would create an inconsistency in form of a crossing, see Fig. 4. In this case A is rejected.

A second special case arises if c is the first suffix to be encountered for both sequences or, if it is the first suffix for one sequence and there is a column $c' \preceq c$ that is the currently closest suffix for the other sequence. Then we can stop the column check since all subsequent columns must be also suffix columns or independent.

If A and c overlap we assume, without loss of generality, that the overlap is at sequence x and denote the overlapping interval by $[x, \bar{b}_x, \bar{e}_x]$. To update c and A we additional need to determine the corresponding overlap $[y, \bar{b}_y, \bar{e}_y]$

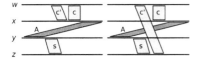

Fig. 4. Switching columns. In the first case we have $c \preceq A$ and $A \preceq s$. Thus we move c and all columns c' with $c' \preceq c$ infront of s. In the second case we also have $s \preceq c$ so that insertion of A would lead to a contradiction.

at sequence y of A and the corresponding overlap $[w, \bar{b}_w, \bar{e}_w]$ for all other entries $[w, b_w, e_w]$ in the thick colum c. Note that this amounts to computing $A \bullet \{[x, b_x, e_x], [w, b_w, e_w]\}$ and $A \bullet \{[y, b_y, e_y], [w, b_w, e_w]\}$ by means of the linear interpolation described in equation 1.

The update of A and the thick column itself then consists out of three steps: *First*, we separate the part in front of the overlap from the alignment or the column. This part is then inserted as a new column c' in front of c: If A starts before c the new column c' is the prefix $\{[x, b_x^A, \bar{b}_x - 1], [y, b_y^A, \bar{b}_y - 1]\}$ of A. In this case A is replaced by the remaining part $\{[x, \bar{b}_x, e_x^A], [y, \bar{b}_y, e_y^A]\}$. Otherwise if c starts before of A we insert the prefix part $c' = \{[w, b_w^c, \bar{b}_w - 1] : [w, b_w^c, e_w^c] \in c\}$ of the column c and replace c by the remainder $\{[w, \bar{b}_w, e_w^c] : [w, b_w^c, e_w^c] \in c\}$. *Second*, the overhanging suffix is separated from A or c. If A ends behind c we shorten A to $\{[x, \bar{b}_x, \bar{e}_x], [y, \bar{b}_y, \bar{e}_y]\}$ and save the remaining part $\{[x, \bar{e}_x + 1, e_x^A], [y, \bar{e}_y + 1, e_y^A]\}$ as A'; this rest must then be checked in the following steps against the following thick columns of **M**. If c ends behind A we append the prefix part $c' = \{[w, \bar{e}_w + 1, e_w^c] : [w, b_w^c, e_w^c] \in c\}$ as a new column behind c and set c to $\{[w, \bar{b}_w, \bar{e}_w] : [w, b_w, e_w] \in c\}$. *Third*, if A contains additional information about column c, i.e., if c has no entry for sequence y, we expand the thick column c by the new entry $[y, \bar{b}_y, \bar{e}_y]$. These three cases are represented graphically in Fig 5.

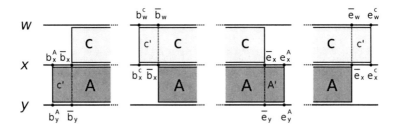

Fig. 5. Update of a column c by an alignment A. Given the overlap at sequence x beginning in $\bar{b}_x$ and ending in $\bar{e}_x$ and the corresponding overlaps at w and y determined by linear interpolation (labeled with a bar) together with the begin and end values of A and c, the first two figures represent the splitting of the alignment or the column in a prefix part c', that is inserted in front of c, and the remaining overlap part of A and c. The last two figures show the splitting at the overlap end where the suffix part A' of A is used for the update of the next column or the suffix part c' of c is inserted behind c.

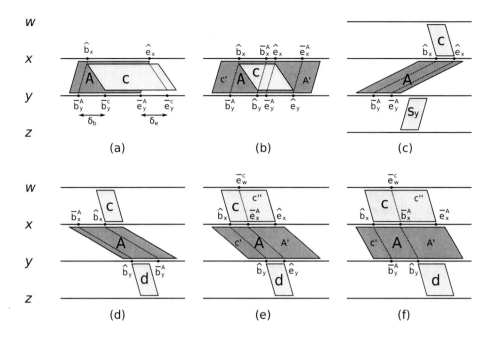

Fig. 6. Determination of the alignment overlap. The figures illustrate the special cases that affect the assumed overlap: (a) determination of the contradiction rate, (b) overlap in case of contradictions, (c) order information by former column, (d) order information by later column, (e) merge of columns and (f) overlap end correction. For more details see text.

It is important to note that the start and end positions of the overlap, $\bar{b}_x$ and $\bar{e}_x$, do not always coincide with the start and end positions of the intersection of A and c, which we denote by $\hat{b}_x$ and end $\hat{e}_x$. There are three special cases to consider:

1. The first complication arises when the thick column c already includes an interval $[y, b^c_y, e^c_y]$ for the second sequence y of A. In this case, there are two ways to compute the boundaries of the overlap, namely using A or using c to estimate the coordinates. We denote the two estimates by $\bar{b}^A_y$, $\bar{e}^A_y$ and $\bar{b}^c_y$ and $\bar{e}^c_y$, resp. If A is consistent with c then the two estimates are the same and correspond to the real intersection $\hat{b}_y$ and $\hat{e}_y$. Otherwise, A causes a contradiction with c and the maximum of the values $\delta_b = |\bar{b}^A_y - \bar{b}^c_y|$ and $\delta_e = |\bar{e}^A_y - \bar{e}^c_y|$ measures the size δ of this contradiction, see Fig. 6(a). If $\delta > \epsilon$ we regard A as incompatible and reject it. Otherwise, i.e., if $\delta \leq \epsilon$, the contradiction is small enough to ignore it and we update c. Since c already has an entry at sequence y that is created by an earlier alignment, i.e., an more trustworthy alignment with higher extended score, we do not update the overlapping part. Hence it remains to insert the prefix part c' of A in front of the overlap, and to insert the remaining part A' of A behind the overlap. See Fig. 6(b).

2. The second special case occurs when A overlaps with c at sequence x and is independent at sequence y but we already have found a closest suffix s_y at y in a previous column update. Then A provides additional information about the order of s_y and c and we have to move c and all columns c' with $c' \preceq c$ between s and to the front of s. If $s \prec c$ the insertion of A would create an inconsistency in form of a crossing and we therefore reject A. Otherwise we update c, as normal with $\bar{b}_x = \hat{b}_x$ and $\bar{e}_x = \hat{e}_x$. See also Fig. 6(c).

3. A overlaps with c at sequence x, A is independent of c for sequence y and we have not found a suffix for y so far. We then check the columns behind c for the first column $d \succ c$ that overlaps A in y. If no such column exists, c is updated as normal with $\bar{b}_x = \hat{b}_x$ and $\bar{e}_x = \hat{e}_x$. Otherwise, we have three subcases depending on the position of $\hat{b}_x$ and the position $\bar{b}_x^A$ of the intersection start $\hat{b}_y$ of A and d on sequence x.

(a) If $\bar{b}_x < \hat{b}_x$, see Fig. 6(d), A tells us that $d \preceq c$ and hence we have to move d and all columns d' with $d' \preceq d$ between c and d to the front of c. If we have, in addition, $c \preceq d$, then the insertion of A would create an inconsistency in form of a crossing and we reject A. Otherwise we stop the update of c and continue with the update of d.

(b) If $\bar{b}_x = \hat{b}_x$, see Fig. 6(e), then A connects the column c with d and we move d and all columns d' with $d' \preceq d$ between c and d to the front of c. In addition, we merge the columns c and d along the overlap. Beginning at $\bar{b}_x$, the overlap ends at $\bar{e}_x^A = \min\{\hat{e}_x, \bar{e}_x^A\}$, where $\bar{e}_x^A$ is the position of the end $\hat{e}_y$ of the intersection of A and d on sequence x.

(c) If $\bar{b}_x > \hat{b}_x$, see Fig. 6(f), A and c are updated as normal except that $\bar{b}_x^A$, i.e. the position $\hat{b}_y$ of the begin of the intersection of A and d on sequence x, is the first position that has to be merged with c. Therefore we set the end of the overlap to $\bar{e}_x^A = \min\{\hat{e}_x, \bar{b}_x^A - 1\}$. The prefix and suffix parts of A and c are separated and inserted as in the case of a normal update described above.

The steps outlined above reduce A to the the remaining part A'. In the followin step, we attempt to merge A' with the column following after c in $\mathbf{M}$. This process is interated either until A', and hence the complete alignment A, is rejected as incompatible, or until A' is empty, in which case the insertion is complete. If A' is not empty after we have checked and updated all columns in $\mathbf{M}$ we append A' as new column following the closest prefix of the current columns. In no closest prefix has yet been determined, we insert A' as first column in $\mathbf{M}$.

The procedure outlines above ensures that after the successful insertion of a single alignment A, all thick columns of $\mathbf{M}$ are still transitive closures, have at most on entry per sequence and obey the partial order $\preceq$. Hence $\mathbf{M}$ is still a multiple alignment, composed now of $\mathcal{C} \cup \{A\}$. If the insertion of A was not successful and $\mathbf{M}$ remains unchanged. In both cases the necessary conditions for the insertion of the next pairwise alignment are satisfied. After all alignments in the queue $\mathcal{C}$ has been processed, the algorithm returns the set $\mathcal{C}'$ of consistent alignments and the corresponding multiple alignment $\mathbf{M}$.

Alternative Solutions. The greedy procedure above can be repeated on the set $\mathcal{C} \setminus \mathcal{C}'$ of pairwise alignment that are inconsistent with the approximately optimal solution found in the first pass. After having extracted a maximal consistent set from $\mathcal{C} \setminus \mathcal{C}'$ we try to add additional alignments from $\mathcal{C}'$. This yields another maximal consistent subset of $\mathcal{C}$. The procedure is iterated by initially removing all alignments from $\mathcal{C}$ that have already been incorporated in a previous solution. We stop when every pairwise alignment is included in at least one consistent subset. In the worst case, therefore, we obtain $O(|\mathcal{C}|)$ solutions each comprising $O(|\mathcal{C}|)$ consistent alignments.

Complexity. The complexity depends on the number $m = |X|$ of sequences, on the upper bound n for the length of these sequences and on the number $c = |\mathcal{C}|$ of alignments in the collection $\mathcal{C}$. The calculation of the extended scores considers all pairs and and triple of alignments in $\mathcal{C}$ and requires $O(c^3)$ time in the interval approximation. The subsequent assembly depends on the update of columns by alignments in $\mathcal{C}$. Let m_i be the number of sequences represented in column i of the multiple alignment $\mathbf{M}$. The update of a single (thick) column takes a constant number of operations for each entry and hence is in $O(m_i)$ for the entire column. On the other hand we have $\sum_i m_i \leq mn$. Therefore the effort for inserting a complete (local) alignment $A \in \mathcal{C}$ is bounded by $O(nm)$. The greedy heuristic thus produces a solution in $O(c^3 + cnm)$ time and the determination of all alternative solutions requires $O(c^3 + c^2nm)$ time. In practice, however, the number of columns of $\mathbf{M}$ intersected by an alignment A is much smaller than the theoretical upper bound of $O(nm)$, and this number is further reduced by using thick columns. For datasets of practical interest we observe that the effort required to compute a set of maximal consistent alignments that cover all input alignments at least once is dominated by computation of the extended alignment scores. The amount of memory required in addition to storing the input collection $\mathcal{C}$ of pairwise alignments is determined by the size of the multiple alignment $\mathbf{M}$ and hence is in $O(nm)$.

Quality of the solutions. The greedy algorithm employed here cannot be guaranteed to produce optimal solutions. We have not attempted to investigate rigorously whether it can guarantee e.g. a constant-factor approximation. It appears, however, that this is not the case if the collection $\mathcal{C}$ of input alignments is arbitrary. On the other hand, we are mostly interested in the biologically relevant instances, in which $\mathcal{C}$ consists of locally optimal (or at least nearly optimal) alignments. In order to test how well the greedy heuristics works on such data sets we used about 30 multiple alignments of different classes of ncRNAs comprising of 5-11 sequences with average lengths about 120 bases and low pairwise similarity provided in BRaliBase [19] as source of homologous sequence sets. For each set we computed pairwise local blast-alignments. These instances are small enough to compute the optimal maximum compatible set by exhaustive enumeration, i.e., an exponential-time algorithm. We used only sets of pairwise alignments that contained at least four different maximal consistent subsets. In all cases, our heuristic returned the correct optimal solution. Interestingly, these solution also agree very well with the manually curated reference alignments.

In order to test the efficiency of our approach, we constructed 5000 random alignments with an average length of $50nt$ in 100 sequences with an average length of $200nt$. On a 2.66GHz Quad Core CPU tracker2 determined all maximal consistent subsets in the entire test set in only 27 seconds.

Availability. The source code of tracker2 is available at
http://www.bioinf.uni-leipzig.de/Software/tracker2/.

Acknowledgement

We thank Carlos E. Ferreira for stimulating discussions and two anonymous reviewers for their helpful comments on the manuscript.

References

1. Elnitski, L., Jin, V.X., Farnham, P.J., Jones, S.J.M.: Locating mammalian transcription factor binding sites: A survey of computational and experimental techniques. Genome Res. 16, 1455–1464 (2006)
2. Bailey, T.L., Williams, N., Misleh, C., Li, W.W.: MEME: discovering and analyzing DNA and protein sequence motifs. Nucleic Acids Res. 34, W369–W373 (2006)
3. Blanchette, M., Tompa, M.: Discovery of regulatory elements by a computational method for phylogenetic footprinting. Genome Res. 12, 739–748 (2002)
4. Liu, Y., Liu, X., Wei, L., Altman, R., Batzoglou, S.: Eukaryotic regulatory element conservation analysis and identification using comparative genomics. Genome Res. 14, 451–458 (2004)
5. Siddharthan, R., Siggia, E., van Nimwegen, E.: PhyloGibbs: A gibbs sampling motif finder that incorporates phylogeny. PLoS Comput. Biol. 1, e67 (2005)
6. van Nimwegen, E.: Finding regulatory elements and regulatory motifs: a general probabilistic framework. BMC Bioinformatics 8 (suppl. 6), S4 (2007)
7. Gordân, R., Narlikar, L., Hartemink, A.J.: Finding regulatory DNA motifs using alignment-free evolutionary conservation information. Nucleic Acids Res. 38, e90 (2010)
8. Margulies, E.H., Blanchette, M., Haussler, D., Green, E.D.: Identification and characterization of multi-species conserved sequences. Genome Res. 13, 2507–2518 (2003)
9. Zhang, Z., Gerstein, M.: Of mice and men: phylogenetic footprinting aids the discovery of regulatory elements. J. Biol. 2, 11 (2003)
10. Prohaska, S., Fried, C., Flamm, C., Wagner, G., Stadler, P.F.: Surveying phylogenetic footprints in large gene clusters: Applications to Hox cluster duplications. Mol. Phyl. Evol. 31, 581–604 (2004)
11. Morgenstern, B., Stoye, J., Dress, A.W.M.: Consistent equivalence relations: a set-theoretical framework for multiple sequence alignments. Technical report, University of Bielefeld, FSPM (1999)
12. Corel, E., Pitschi, F., Morgenstern, B.: A min-cut algorithm for the consistency problem in multiple sequence alignment. Bioinformatics 26, 1015–1021 (2010)
13. Morgenstern, B., Frech, K., Dress, A., Werner, T.: DIALIGN: finding local similarities by multiple sequence alignment. Bioinformatics 14(3), 290–294 (1998)

14. Tarjan, R.E.: Depth first search and linear graph algorithms. SIAM J. Computing 1, 146–160 (1972)
15. Notredame, C., Higgins, D.G., Heringa, J.: T-Coffee: A novel method for fast and accurate multiple sequence alignment. J. Mol. Biol. 302, 205–217 (2000)
16. Euler, R.: On a classification of independence systems. Math. Methods Operations Res. 27, 123–136 (1983)
17. Helman, P., Moret, B.M.E., Shapiro, H.D.: An exact characterization of greedy structures. SIAM J. Discrete Math. 6, 274–283 (1993)
18. Elias, I.: Settling the intractability of multiple alignment. J. Comp. Biol. 13, 1323–1339 (2006)
19. Wilm, A., Mainz, I., Steger, G.: An enhanced RNA alignment benchmark for sequence alignment programs. Algorithms Mol. Biol. 1, 19 (2006)

Unique Perfect Phylogeny Is NP-Hard

Michel Habib[1] and Juraj Stacho[2]

[1] LIAFA – CNRS and Université Paris Diderot – Paris VII,
Case 7014, 75205 Paris Cedex 13, France
`habib@liafa.jussieu.fr`
[2] Caesarea Rothschild Institute, University of Haifa
Mt. Carmel, 31905 Haifa, Israel
`stacho@cs.toronto.edu`

Abstract. We answer, in the affirmative, the following question proposed by Mike Steel as a \$100 challenge: *"Is the following problem NP-hard? Given a ternary[1] phylogenetic X-tree $\mathcal{T}$ and a collection $\mathcal{Q}$ of quartet subtrees on X, is $\mathcal{T}$ the only tree that displays $\mathcal{Q}$?"* [28, 29] As a particular consequence of this, we show that the unique chordal sandwich problem is also NP-hard.

1 Introduction

One of the major efforts in molecular biology has been the computation of phylogenetic trees, or *phylogenies*, which describe the evolution of a set of species from a common ancestor. A phylogenetic tree for a set of species is a tree in which the leaves represent the species from the set and the internal nodes represent the (hypothetical) ancestral species. One standard model for describing the species is in terms of *characters*, where a character is an equivalence relation on the species set, partitioning it into different *character states*. In this model, we also assign character states to the (hypothetical) ancestral species. The desired property is that for each state of each character, the set of nodes in the tree having that character state forms a connected subgraph. When a phylogeny has this property, we say it is *perfect*. The Perfect Phylogeny problem [18] then asks *for a given set of characters defining a species set, does there exist a perfect phylogeny?* Note that we allow that states of some characters are unknown for some species; we call such characters *partial*, otherwise we speak of *full* characters. This approach to constructing phylogenies has been studied since the 1960s [7, 23–25, 32] and was given a precise mathematical formulation in the 1970s [10–13]. In particular, Buneman [6] showed that the Perfect Phylogeny problem reduces to a specific graph-theoretic problem, the problem of finding a chordal completion of a graph that respects a prescribed colouring. In fact, the two problems are polynomially equivalent [21]. Thus, using this formulation, it has been proved that the Perfect Phylogeny problem is NP-hard in [3] and independently in [30]. These two results rely on the fact that the input may contain

---

[1] Some formulations of this question use the term "binary", as in "rooted binary tree".

R. Giancarlo and G. Manzini (Eds.): CPM 2011, LNCS 6661, pp. 132–146, 2011.
© Springer-Verlag Berlin Heidelberg 2011

partial characters. In fact, the characters in these constructions only have two states. If we insist on full characters, the situation is different as for any fixed number r of character states, the problem can be solved in time polynomial [1] in the size of the input (and exponential in r). In particular, for $r = 2$ (or $r = 3$), the solution exists if and only if it exists for every pair (or triple) of characters [13, 22]. Also, when the number of characters is k (even if there are partial characters), the complexity [26] is polynomial in the number of species (and exponential in k).

Another common formulation of this problem is the problem of a *consensus tree* [9, 17, 30], where a collection of subtrees with labelled leaves is given (for instance, the leaves correspond to species of a partial character). Here, we ask for a (phylogenetic) tree such that each of the input subtrees can be obtained by contracting edges of the tree (we say that the tree *displays* the subtree). The problem does not change [28] if we only allow particular input subtrees, the so-called *quartet trees*, which have exactly six vertices and four leaves. This follows from the fact that every ternary phylogenetic tree can be uniquely described by a collection of quartet trees [28]. However, a collection of quartet trees does not necessarily uniquely describe a ternary phylogenetic tree.

This leads to a natural question: *what is the complexity of deciding whether or not a collection of quartet trees uniquely describes a (ternary) phylogenetic tree?* This question was first posed in 1992 in [30], later conjectured to be NP-hard [28] and listed on M. Steel's personal webpage [29] where he offers $100 for the first proof of NP-hardness.

In this paper, we are the first to answer this question by showing that the problem is indeed NP-hard. That is, we prove the following theorem.

Theorem 1. *It is NP-hard to determine, given a ternary phylogenetic X-tree $\mathcal{T}$ and a collection $\mathcal{Q}$ of quartet subtrees on X, whether or not $\mathcal{T}$ is the only phylogenetic tree that displays $\mathcal{Q}$.*

(We note that an alternative proof of this theorem recently appeared on arxiv [4]. The proof uses different techniques and extends the hardness result of [30].)

In light of this, we note that there are special cases of the problem that are known to be solvable in polynomial time. For instance, this is so if the collection $\mathcal{Q}$ contains a subcollection $\mathcal{Q}'$ with the same set $\mathcal{L}$ of labels of leaves and with $|\mathcal{Q}'| = |\mathcal{L}| - 3$. However, finding such a subcollection is known to be NP-complete. For these and similar results, we refer the reader to [2].

We prove Theorem 1 by describing a polynomial-time reduction from the uniqueness problem for ONE-IN-THREE-3SAT, which is NP-hard by [20].[2]

Theorem 2. [20] *It is NP-hard to decide, given an instance I to ONE-IN-THREE-3SAT, and a truth assignment σ that satisfies I, whether or not σ is the unique satisfying truth assignment for I.*

---

[2] We extract this from [20] by encoding the problem as the relation $\{001, 010, 100\}$. We check that this relation is not: 0-valid, 1-valid, Horn, anti-Horn, affine, 2SAT, or complementive. Then the uniqueness of the satisfiability problem corresponding to this relation is $CoNP$-hard by [20] and thus NP-hard (assuming Turing reductions).

Our construction in the reduction is essentially a modification of the construction of [3] which proves NP-hardness of the Perfect Phylogeny problem. Recall that the construction of [3] produces instances $\mathcal{Q}$ that have a perfect phylogeny if and only if a particular boolean formula Φ is satisfiable. We immediately observed that these instances $\mathcal{Q}$ have, in addition, the property that Φ has a unique satisfying assignment if and only if there is a unique minimal restricted chordal completion of the partial partition intersection graph of $\mathcal{Q}$ (for definitions see Section 2). This is precisely one of the two necessary conditions for uniqueness of perfect phylogeny as proved by Semple and Steel in [27] (see Theorem 4). Thus by modifying the construction of [3] to also satisfy the other condition of uniqueness of [27], we obtained the construction that we present in this paper. Note that, however, unlike [3] which uses 3SAT, we had to use a different NP-hard problem in order for the construction to work correctly. Also, to prove that the construction is correct, we employ a variant of the characterization of [27] that uses the more general chordal sandwich problem [15] instead of the restricted chordal completion problem (see Theorem 7). In fact, by way of Theorems 5 and 6, we establish a direct connection between the problem of perfect phylogeny and the chordal sandwich problem, which apparently has not been yet observed. (Note that the connection to the (restricted) chordal completion problem of coloured graphs as mentioned above [6, 21] is a special case of this.)

Finally, as a corollary, we obtain the following result using [8].

Corollary 1. *The Unique chordal sandwich problem is NP-hard. Counting the number of minimal chordal sandwiches is $\#P$-complete.*

The paper is structured as follows. In Section 2, we introduce definitions and some preliminary work. In Sections 3 and 4, we present our hardness reduction, first informally and then formally. Then we sketch the proof of one of the main theorems (Theorem 8) in Section 5, and conclude with some open questions.

2   Preliminaries

We mostly follow the terminology of [27, 28] and graph-theoretical notions of [31].

Let X be a non-empty set. An X-*tree* is a pair (T, ϕ) where T is tree and $\phi : X \to V(T)$ is a mapping such that $\phi^{-1}(v) \neq \emptyset$ for all vertices $v \in V(T)$ of degree at most two. An X-tree (T, ϕ) is *ternary* if all internal vertices of T have degree three. Two X-trees (T_1, ϕ_1), (T_2, ϕ_2) are *isomorphic* if there exists an isomorphism $\psi : V(T_1) \to V(T_2)$ between T_1 and T_2 that satisfies $\phi_2 = \psi \circ \phi_1$.

An X-tree (T, ϕ) is a *phylogenetic X-tree* (or a *free X-free* in [27]) if ϕ is a bijection between X and the set of leaves of T. A *partial partition* of X is a partition of a non-empty subset of X into at least two sets. If A_1, A_2, ..., A_t are these sets, we call them *cells* of this partition, and denote the partition $A_1|A_2|\ldots|A_t$. If $t = 2$, we call the partition a *partial split*. A partial split $A_1|A_2$ is trivial if $|A_1| = 1$ or $|A_2| = 1$. A *quartet tree* is a ternary phylogenetic tree with a label set of size four, that is, a ternary tree T with 6 vertices, 4 leaves labelled

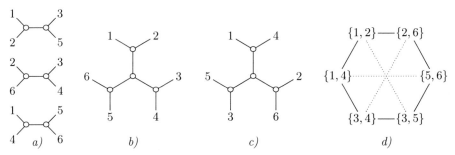

Fig. 1. *a)* quartet trees $\mathcal{Q}$, *b)* and *c)* two X-trees displaying $\mathcal{Q}$ and distinguished by $\mathcal{Q}$, *d)* the graph $\text{int}^*(\mathcal{Q})$; the dotted lines represent the edges of $\text{forb}(\mathcal{Q})$

a, b, c, d, and with only one non-trivial partial split $\{a, b\}|\{c, d\}$ that it displays. Note that such a tree is unambiguously defined by this partial split. Thus, in the subsequent text, we identify the quartet tree $\mathcal{T}$ with the partial split $\{a, b\}|\{c, d\}$, that is, we say that $\{a, b\}|\{c, d\}$ is both a quartet tree and a partial split.

Let $\mathcal{T} = (T, \phi)$ be an X-tree, and let $\pi = A_1|A_2|\ldots|A_t$ be a partial partition of X. Let $F \subseteq E(T)$ be a set of edges of T. We say that F *displays* π in $\mathcal{T}$ if for all distinct $i, j \in \{1 \ldots t\}$, the sets $\phi(A_i)$ and $\phi(A_j)$ are subsets of the vertex sets of different connected components of $T - F$. We say that $\mathcal{T}$ *displays* π if there is a set of edges that displays π in $\mathcal{T}$. Further, an edge e of T is *distinguished* by π if every set of edges that displays π in $\mathcal{T}$ contains e.

Let $\mathcal{Q}$ be a collection of partial partitions of X. An X-tree $\mathcal{T}$ *displays* $\mathcal{Q}$ if it displays every partial partition in $\mathcal{Q}$. An X-tree $\mathcal{T} = (T, \phi)$ is *distinguished* by $\mathcal{Q}$ if every internal edge of T is distinguished by some partial partition in $\mathcal{Q}$; we also say that $\mathcal{Q}$ *distinguishes* $\mathcal{T}$. The set $\mathcal{Q}$ *defines* $\mathcal{T}$ if $\mathcal{T}$ displays $\mathcal{Q}$, and all other X-trees that display $\mathcal{Q}$ are isomorphic to $\mathcal{T}$. Note that if $\mathcal{Q}$ defines $\mathcal{T}$, then $\mathcal{T}$ is necessarily a ternary phylogenetic X-tree, since otherwise "resolving" any vertex either of degree four or more, or with multiple labels results in a non-isomorphic X-tree that also displays $\mathcal{Q}$ (also, see Proposition 2.6 in [27]). See Figure 1 for an illustration of these concepts.

The *partial partition intersection graph* of $\mathcal{Q}$, denoted by $\text{int}(\mathcal{Q})$, is a graph whose vertex set is $\{(A, \pi) \mid \text{where } A \text{ is a cell of } \pi \in \mathcal{Q}\}$ and two vertices (A, π), (A', π') are adjacent just if the intersection of A and A' is non-empty.

A graph is *chordal* if it contains no induced cycle of length four or more. A *chordal completion* of a graph $G = (V, E)$ is a chordal graph $G' = (V, E')$ with $E \subseteq E'$. A *restricted chordal completion* of $\text{int}(\mathcal{Q})$ is a chordal completion G' of $\text{int}(\mathcal{Q})$ with the property that if A_1, A_2 are cells of $\pi \in \mathcal{Q}$, then (A_1, π) is not adjacent to (A_2, π) in G'. A restricted chordal completion G' of $\text{int}(\mathcal{Q})$ is *minimal* if no proper subgraph of G' is a restricted chordal completion of $\text{int}(\mathcal{Q})$.

The problem of perfect phylogeny is equivalent to the problem of determining the existence of an X-tree that display the given collection $\mathcal{Q}$ of partial partitions. In [6], it was given the following graph-theoretical characterization.

Theorem 3. [6, 28, 30] *Let $\mathcal{Q}$ be a set of partial partitions of a set X. Then there exists an X-tree that displays $\mathcal{Q}$ if and only if there exists a restricted chordal completion of $\operatorname{int}(\mathcal{Q})$.*

Of course, the X-tree in the above theorem might not be unique. For the problem of uniqueness, Semple and Steel [27, 28] describe necessary and sufficient conditions for when a collection of partial partitions defines an X-tree.

Theorem 4. [27] *Let $\mathcal{Q}$ be a collection of partial partitions of a set X. Let $\mathcal{T}$ be a ternary phylogenetic X-tree. Then $\mathcal{Q}$ defines $\mathcal{T}$ if and only if:*

(i) $\mathcal{T}$ displays $\mathcal{Q}$ and is distinguished by $\mathcal{Q}$, and
(ii) there is a unique minimal restricted chordal completion of $\operatorname{int}(\mathcal{Q})$.

In order to simplify our proof of Theorem 1, we now describe a variant of the above theorem that, instead, deals with the notion of chordal sandwich [15].

Let $G = (V, E)$ and $H = (V, F)$ be two graphs on the same set of vertices with $E \cap F = \emptyset$. A *chordal sandwich* of (G,H) is a chordal graph $G' = (V, E')$ with $E \subseteq E'$ and $E' \cap F = \emptyset$.[3] A chordal sandwich G' of (G,H) is *minimal* if no proper subgraph of G' is a chordal sandwich of (G,H).

The *cell intersection graph* of $\mathcal{Q}$, denoted by $\operatorname{int}^*(\mathcal{Q})$, is the graph whose vertex set is $\{A \mid \text{where } A \text{ is a cell of } \pi \in \mathcal{Q}\}$ and two vertices A, A' are adjacent just if the intersection of A and A' is non-empty. Let $\operatorname{forb}(\mathcal{Q})$ denote the graph whose vertex set is that of $\operatorname{int}^*(\mathcal{Q})$ in which there is an edge between A and A' just if A,A' are cells of some $\pi \in \mathcal{Q}$. See Figure 1d for an example.

The correspondence between the partial partition intersection graph and the cell intersection graph is captured by the following theorem.

Theorem 5. *Let $\mathcal{Q}$ be a collection of partial partitions of a set X. There is a one-to-one mapping between the minimal restricted chordal completions of $\operatorname{int}(\mathcal{Q})$ and the minimal chordal sandwiches of $(\operatorname{int}^*(\mathcal{Q}), \operatorname{forb}(\mathcal{Q}))$.*

The proof of this theorem follows easily from the following lemma.

Lemma 1. *Let G be a graph, and let G^+ be a graph obtained from G by substituting complete graphs[4] for the vertices of G. Then there is a one-to-one mapping between minimal chordal completions of G and G^+.*

This combined with Theorem 3 yields that there is a phylogenetic X-tree that displays $\mathcal{Q}$ if and only if there exists a chordal sandwich of $(\operatorname{int}^*(\mathcal{Q}), \operatorname{forb}(\mathcal{Q}))$. Conversely, we can express every instance to the chordal sandwich problem as a corresponding instance to the problem of perfect phylogeny as follows.

Theorem 6. *Let (G, H) be an instance to the chordal sandwich problem. Then there is a collection $\mathcal{Q}$ of partial splits such that there is a one-to-one mapping between the minimal chordal sandwiches of (G, H) and the minimal restricted chordal completions of $\operatorname{int}(\mathcal{Q})$. In particular, there exists a chordal sandwich for (G, H) if and only if there exists a phylogenetic tree that displays $\mathcal{Q}$.*

---

[3] We say that E are the *forced* edges and F are the *forbidden* edges.
[4] Replacing v by a complete graph K and adding edges $\{ux \mid x \in V(K) \wedge uv \in E(G)\}$.

Proof. (Sketch) Without loss of generality, we may assume that each connected component of G has at least three vertices. As usual, $G = (V, E)$ and $H = (V, F)$ where $E \cap F = \emptyset$. The collection $\mathcal{Q}$ satisfying the claim is defined as follows: for every edge $xy \in F$, we construct the partial split $D_x | D_y$, where D_x are the edges of E incident to x, and D_y are the edges of E incident to y.   □

As a corollary, we obtain the following desired characterization.

Theorem 7. *Let $\mathcal{Q}$ be a collection of partial partitions of a set X. Let $\mathcal{T}$ be a ternary phylogenetic X-tree. Then $\mathcal{Q}$ defines $\mathcal{T}$ if and only if:*

(i) $\mathcal{T}$ displays $\mathcal{Q}$ and is distinguished by $\mathcal{Q}$, and
(ii) there is a unique minimal chordal sandwich of $\left(\mathrm{int}^*(\mathcal{Q}), \mathrm{forb}(\mathcal{Q}) \right)$.

The main technical advantage of this theorem over Theorem 4 is that it is less restrictive; it allows us to construct instances with arbitrary sets of forbidden edges rather than just with forbidden edges between vertices of the same colour. This makes our proof of Theorem 1 much simpler and more manageable.

3   Construction

Consider an instance I to ONE-IN-THREE-3SAT. The instance I consists of n *variables* $v_1, \ldots, v_n$ and m *clauses* $\mathcal{C}_1, \ldots, \mathcal{C}_m$ each of which is a disjunction of exactly three *literals* (i.e., variables v_i or their negations $\overline{v_i}$).

To simplify the presentation, we shall denote literals by capital letters X, Y, etc., and indicate their negations by $\overline{X}$, $\overline{Y}$, etc. (For instance, if $X = v_i$ then $\overline{X} = \overline{v_i}$, and if $X = \overline{v_i}$ then $\overline{X} = v_i$.)

By standard arguments, we may assume that no variable appears twice in the same clause. First, we discuss how to find a collection $\mathcal{Q}_I$ of quartet trees arising from the instance I that satisfies the following theorem.

Theorem 8. *There is a one-to-one mapping between the satisfying assignments of the instance I and the minimal chordal sandwiches of $(\mathrm{int}^*(\mathcal{Q}_I), \mathrm{forb}(\mathcal{Q}_I))$.*

Before describing the collection $\mathcal{Q}_I$, let us briefly review the construction from [3] that proves NP-hardness of the Perfect Phylogeny problem. For convenience, we describe it in terms of the chordal sandwich problem whose input is a graph with (forced) edges and forbidden edges. In the construction from [3], one similarly considers a collection $\mathcal{C}_1, \ldots, \mathcal{C}_m$ of 3-literal clauses, and treats it as an instance of 3-SATISFIABILITY. From this instance, one constructs a graph where each variable v_i corresponds to two *shoulders* S_{v_i} and $S_{\overline{v_i}}$, and where each literal W in a clause $\mathcal{C}_j$ corresponds to a pair of *knees* K_W^j and $K_{\overline{W}}^j$. In addition, there are two special vertices *the head H* and *the foot F*. All shoulders are adjacent to the head while all knees are adjacent to the foot. Further, if v_i occurs in the clause $\mathcal{C}_j$ (positively or negatively), then the vertices H, S_{v_i}, $K_{v_i}^j$, F, $K_{\overline{v_i}}^j$, $S_{\overline{v_i}}$ form an induced 6-cycle (see Figure 2a). Also, if $\mathcal{C}_j = X \vee Y \vee Z$, then the vertices K_X^j,

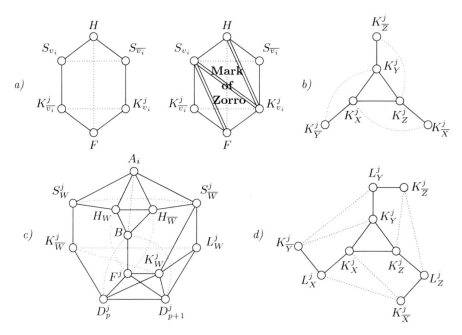

Fig. 2. *a)* and *b)* configurations from [3], *c)* and *d)* configurations from our construction (note that in *c)* the literal W is either v_i or $\overline{v_i}$, and is the p-th literal of the clause $\mathcal{C}_j$)

K_Y^j, K_Z^j induce a triangle with pendant edges $K_X^j K_{\overline{Y}}^j$, $K_Y^j K_{\overline{Z}}^j$, and $K_Z^j K_{\overline{X}}^j$ (we call this the *clause gadget*, see Figure 2b).

Finally, the edge between H and F is forbidden in the desired chordal sandwich, and so is the edge between S_{v_i} and $S_{\overline{v_i}}$, and between $K_{v_i}^j$ and $K_{\overline{v_i}}^j$ for all meaningful indices i and j (the dotted edges in Figure 2).

The main idea of this construction is that each of the 6-cycles allows only two possible chordal sandwiches: either the path $H, K_{v_i}^j, S_{v_i}, F$ is added, or the path $H, K_{\overline{v_i}}^j, S_{\overline{v_i}}, F$ is added (the authors of [3] call this path the "Mark of Zorro"). These two choices correspond to assigning v_i the value *true* or *false*, respectively, and the construction ensures that this choice is consistent over all clauses. This only produces satisfying assignments to 3-SATISFIABILITY, since we notice that no chordal sandwich adds a triangle on $K_{\overline{X}}^j, K_{\overline{Y}}^j, K_{\overline{Z}}^j$.

We can try to use this construction to prove Theorem 1. We immediately observe that the truth assignments satisfying the clauses $\mathcal{C}_1, \ldots, \mathcal{C}_m$ are in one-to-one correspondence with the minimal chordal sandwiches of the above graph G. This is a little technical to prove. To do this, one first observes that different assignments add a different mark of Zorro to at least one 6-cycle. For the converse, one needs to find out which edges are forced in the sandwich after the marks of Zorro are added according to a satisfying assignment. It turns out that these edges yield a chordal sandwich, and thus a minimal chordal sandwich.

From G, using Theorems 5 and 6, one can construct a collection $\mathcal{Q}$ of partial splits (phylogenetic trees) such that the satisfying assignments of the clauses

$\mathcal{C}_1, \ldots, \mathcal{C}_m$ are in one-to-one correspondence with the minimal chordal sand-wiches of $(\mathrm{int}^*(\mathcal{Q}), \mathrm{forb}(\mathcal{Q}))$. In particular, this collection $\mathcal{Q}$ satisfies the con-dition (ii) of Theorem 7 if and only if the clauses $\mathcal{C}_1, \ldots, \mathcal{C}_m$ have a unique satisfying assignment. Since this is NP-hard to determine [20], it would seem like we almost have a proof of Theorem 1. Unfortunately, we are missing a crucial piece which is the phylogenetic tree $\mathcal{T}$ satisfying the condition (i) of Theorem 7 for the collection $\mathcal{Q}$. A straightforward construction of such a tree based on [27] yields a phylogenetic tree that is distinguished by $\mathcal{Q}$, but whose internal nodes may have degree higher than three. If we try to fix this (by "resolving" the high-degree nodes in order to get a ternary tree), the resulting tree may no longer be distinguished by $\mathcal{Q}$. Moreover, the collection $\mathcal{Q}$ may not consist of quartet trees only. For all these reasons, we need to modify the construction of G.

First, we discuss how to modify G so that it corresponds to a collection of quartet trees. To do this, we must ensure that the neighbourhood of each vertex consists of two cliques (with possibly edges between them). We construct a new graph G_I by modifying G as follows. Instead of one head H, we now have, for each variable v_i, two *heads* H_{v_i}, $H_{\overline{v_i}}$, and an *auxiliary head* A_i. For a literal W in the clause $\mathcal{C}_j$, we now have two *shoulders* S_W^j and $S_{\overline{W}}^j$, and, as before, two *knees* K_W^j and $K_{\overline{W}}^j$, but also an additional *auxiliary knee* L_W^j. Further, for each clause $\mathcal{C}_j$, we have a *foot* F^j and three *auxiliary feet* D_1^j, D_2^j, and D_3^j. Finally, we have one additional vertex B known as *the backbone*. The resulting modifications to the 6-cycles and the clause gadgets can be seen in Figures 2c and 2d. (The forbidden edges are again indicated by dotted lines.) Note that, unlike in the case of G, this is not a complete description of G_I as we need to add some additional (forced) edges and forbidden edges not shown in these diagrams in order to make the reduction work. This is rather technical and we omit this for brevity.

From the construction, we conclude that, just like in G, the "6-cycles" of G_I (Figure 2c) admit only two possible kinds of sandwiches, and this is consistent over different clauses. However, unlike in G, the chordal sandwiches of G_I no longer correspond to satisfying assignments of 3-SATISFIABILITY but rather to satisfying assignments of ONE-IN-THREE-3-SAT. Fortunately, this problem is also NP-hard as is its uniqueness variant as previously discussed (see Theorem 2).

Now, from G_I, we construct a collection $\mathcal{Q}_I$ of quartet trees. To do this, we cannot just use Theorem 6 as before, since this may create partial partitions that do not correspond to quartet trees. Moreover, even if we use [28] to replace these partitions by an equivalent collection of quartet trees, this process may not preserve the number of solutions. We need a more careful construction.

We recall that the each vertex v of G_I belongs to two cliques that completely cover its neighbourhood; we assign greek letters to these two cliques (to distin-guish them from vertices), and associate them with v.

In particular, we use the following symbols: α_W, β_W^j, $\gamma_1^j, \gamma_2^j, \gamma_3^j$, λ^j, δ, μ where W is a literal and $j \in \{1 \ldots m\}$. They define specific cliques of G_I as follows. The letter α_W defines the clique of G_I consisting of all heads and shoulders of W. The letter β_W^j corresponds to the clique formed by the shoulder S_W^j and the knees $K_{\overline{W}}^j$, $L_{\overline{W}}^j$ (if exists). Further, λ^j yields a clique on F^j, D_1^j, D_2^j, D_3^j,

K^j_X, K^j_Y, K^j_Z where $\mathcal{C}_j = X \vee Y \vee Z$, while the clique for γ^j_p where $p \in \{1,2,3\}$ is formed by D^j_p, $K^j_{\overline{W}}$, L^j_U where W and U are the p-th and $(p-1)$-th (modulo 3) literals of $\mathcal{C}_j$. Finally, δ corresponds to the clique containing B and all heads H_W whereas μ correspond to the clique with B and all feet F^j.

From this, we construct the collection $\mathcal{Q}_I$ by considering every forbidden edge uv of G_I and by constructing a partial partition with two cells in which one cell is the set of cliques assigned to u and the other is the set of cliques assigned to v. Since we assign to each vertex of G_I exactly two cliques, this yields partitions corresponding to quartet trees. For instance, in Figure 2d, we have a forbidden edge $K^j_X K^j_{\overline{X}}$ where K^j_X is assigned cliques $\beta^j_{\overline{X}}$, λ^j and $K^j_{\overline{X}}$ is assigned β^j_X, γ^j_1. This yields a quartet tree $\{\beta^j_{\overline{X}}, \lambda^j\}|\{\beta^j_X, \gamma^j_1\}$. The complete definition of $\mathcal{Q}_I$ can be found in Section 4. Finally, since by construction every vertex of G_I is incident to at least one forbidden edge, we conclude that $G_I = \text{int}^*(\mathcal{Q}_I)$.

This completes the overview of the proof of Theorem 8. Its actual proof is quite technical and involved, but it is along the same lines as the uniqueness property we discussed for G, i.e., one describes the edges forced by an assignment and proves that this yields a chordal sandwich. (We sketch this in Section 5.)

To complete the result, we need to explain how to construct a phylogenetic tree corresponding to a satisfying assignment for $\mathcal{C}_1, \ldots, \mathcal{C}_m$ (as an instance of ONE-IN-THREE-3SAT) and show that it displays and is distinguished by the trees in $\mathcal{Q}_I$. Instead of giving a formal definition here, we discuss a small example. (The complete description is rather technical and is presented in Section 4.)

The example instance I^+ consists of four variables v_1, v_2, v_3, v_4 and three clauses $\mathcal{C}_1 = v_1 \vee v_2 \vee v_3$, $\mathcal{C}_2 = \overline{v_1} \vee v_2 \vee v_4$, and $\mathcal{C}_3 = v_3 \vee \overline{v_2} \vee \overline{v_4}$. The unique satisfying assignment assigns true to v_1, v_4 and false to v_2, v_3. The corresponding phylogenetic tree $\mathcal{T} = (T, \phi)$ is shown in Figure 3.

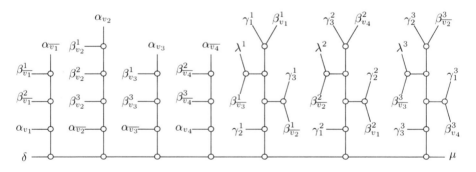

Fig. 3. The phylogenetic tree for the example instance I^+

For instance, one of the quartet trees in $\mathcal{Q}_{I^+}$ is $\pi = \{\alpha_{v_1}, \beta^1_{v_1}\}|\{\alpha_{\overline{v_1}}, \beta^1_{\overline{v_1}}\}$ representing the forbidden edge of G_{I^+} between $S^1_{v_1}$ and $S^1_{\overline{v_1}}$. It is easy to verify $\mathcal{T}$ displays π. Another example from $\mathcal{Q}_{I^+}$ is $\pi' = \{\beta^1_{\overline{v_1}}, \lambda^1\}|\{\beta^1_{v_1}, \gamma^1_1\}$ representing the forbidden edge $K^1_{v_1} K^1_{\overline{v_1}}$. Again, it is displayed by $\mathcal{T}$, but this time one internal edge of T is contained in every set of edges of T that displays π' in T; hence, this edge is distinguished by π'. This way we can verify all other quartet trees in $\mathcal{Q}_{I+}$ and conclude that they are displayed by T and they distinguish T.

Now, with the help of Theorem 7, this allows us to prove that given an instance I to ONE-IN-THREE-3SAT and a satisfying assignment φ for I, one can in polynomial time construct a phylogenetic tree T and a collection of quartet trees $\mathcal{Q}$ such that T is the unique tree defined by $\mathcal{Q}$ if and only if φ is the unique satisfying assignment for I. Combined with Theorem 2, this proves Theorem 1.

That concludes this section. In the next sections, we formally describe the above constructions and sketch some proofs. For full details of proofs, we invite the reader to see our arxiv version of this paper [19].

4   Formal Description

Let I be an instance to ONE-IN-THREE-3SAT consisting of variables $v_1, \ldots, v_n$ and clauses $C_1, \ldots, C_m$. A *truth assignment* σ assigns to each variable v_i a truth value *true* or *false*; we indicate this by writing $v_i = 1$ or $v_i = 0$, respectively, and extend this notation to literals. A truth assignment σ is a *satisfying assignment for I* if in each clause C_j exactly one the three literals evaluates to true.

For each $i \in \{1 \ldots n\}$, we let Δ_i denote all indices j such that v_i or $\overline{v_i}$ appears in the clause C_j. Let $\mathcal{X}_I$ be the set consisting of the elements:

a) $\alpha_{v_i}, \alpha_{\overline{v_i}}$ for each $i \in \{1 \ldots n\}$,

b) $\beta_{v_i}^j, \beta_{\overline{v_i}}^j$ for each $i \in \{1 \ldots n\}$ and each $j \in \Delta_i$,

c) $\gamma_1^j, \gamma_2^j, \gamma_3^j, \lambda^j$ for each $j \in \{1 \ldots m\}$, and

d) δ and μ.

Consider the following collection of 2-element subsets of $\mathcal{X}_I$:

a) $B = \{\mu, \delta\}$,   b) for each $i \in \{1, \ldots, n\}$:
$$H_{v_i} = \{\alpha_{v_i}, \delta\}, \ H_{\overline{v_i}} = \{\alpha_{\overline{v_i}}, \delta\}, \ A_i = \{\alpha_{v_i}, \alpha_{\overline{v_i}}\},$$
$$S_{v_i}^j = \{\alpha_{v_i}, \beta_{v_i}^j\}, \ S_{\overline{v_i}}^j = \{\alpha_{\overline{v_i}}, \beta_{\overline{v_i}}^j\} \text{ for all } j \in \Delta_i,$$

c) for each $j \in \{1 \ldots m\}$ where $C_j = X \vee Y \vee Z$:
$$K_{\overline{X}}^j = \{\beta_{\overline{X}}^j, \gamma_1^j\}, \ K_{\overline{Y}}^j = \{\beta_{\overline{Y}}^j, \gamma_2^j\}, \ K_{\overline{Z}}^j = \{\beta_{\overline{Z}}^j, \gamma_3^j\},$$
$$K_X^j = \{\beta_{\overline{X}}^j, \lambda^j\}, \ K_Y^j = \{\beta_{\overline{Y}}^j, \lambda^j\}, \ K_Z^j = \{\beta_{\overline{Z}}^j, \lambda^j\},$$
$$L_X^j = \{\beta_{\overline{X}}^j, \gamma_2^j\}, \ L_Y^j = \{\beta_{\overline{Y}}^j, \gamma_3^j\}, \ L_Z^j = \{\beta_{\overline{Z}}^j, \gamma_1^j\},$$
$$D_1^j = \{\gamma_1^j, \lambda^j\}, \ D_2^j = \{\gamma_2^j, \lambda^j\}, \ D_3^j = \{\gamma_3^j, \lambda^j\}, \ F^j = \{\lambda^j, \mu\}.$$

The collection $\mathcal{Q}_I$ of quartet trees is defined as follows:

$$\mathcal{Q}_I = \bigcup_{i \in \{1 \ldots n\}} \{A_i | B\} \cup \bigcup_{j \in \{1 \ldots m\}} \{D_1^j | B, D_2^j | B, D_3^j | B\}$$

$$\cup \bigcup_{\substack{i\in\{1...n\} \\ j,j'\in \Delta_i}} \left\{S^j_{v_i}|S^{j'}_{\overline{v_i}}\right\} \cup \bigcup_{\substack{i\in\{1...n\} \\ j,j'\in \Delta_i \text{ and } j<j'}} \left\{S^j_{v_i}|K^{j'}_{\overline{v_i}}, S^j_{v_i}|K^{j'}_{v_i}\right\} \cup \bigcup_{\substack{i\in\{1...n\} \\ j\in \Delta_i \text{ and } j<j'\leq m}} \left\{K^j_{\overline{v_i}}|F^{j'}, K^j_{v_i}|F^{j'}\right\}$$

$$\cup \bigcup_{\substack{1\leq i'<i\leq n \\ j\in\Delta_i}} \left\{H_{v_{i'}}|S^j_{v_i}, H_{\overline{v_{i'}}}|S^j_{v_i}, H_{v_{i'}}|S^j_{\overline{v_i}}, H_{\overline{v_{i'}}}|S^j_{\overline{v_i}}\right\} \cup \bigcup_{\substack{i\in\{1...n\} \\ j\in\{1...m\}}} \left\{H_{\overline{v_i}}|F^j, H_{v_i}|F^j\right\}$$

$$\cup \bigcup_{\substack{j\in\{1...m\} \\ \text{where } \mathcal{C}_j=X\vee Y\vee Z}} \left\{\begin{array}{l} K^j_{\overline{X}}|K^j_X, \ K^j_{\overline{Y}}|K^j_Y, \ K^j_{\overline{Z}}|K^j_Z, \ K^j_{\overline{X}}|L^j_X, \ K^j_{\overline{Y}}|L^j_Y, \ K^j_{\overline{Z}}|L^j_Z \\ S^j_Y|K^j_X, \ S^j_Z|K^j_Y, \ S^j_X|K^j_Z, \ S^j_Z|L^j_X, \ S^j_X|L^j_Y, \ S^j_Y|L^j_Z \end{array}\right\}$$

Let T_I be the tree defined as follows:

$$V(T_I) = \left\{y_0, y_1, y'_1, \ldots, y_n, y'_n\right\} \cup \left\{a_1, a'_1, \ldots, a_n, a'_n\right\} \cup \left\{u_0, u_1, \ldots, u_m\right\}$$
$$\cup \left\{x^j_1, x^j_2, x^j_3, x^j_4, x^j_5, x^j_6, b^j_1, b^j_2, b^j_3, g^j_1, g^j_2, g^j_3, \ell^j\right\}^m_{j=1} \cup \left\{c^j_i, z^j_i \mid j \in \Delta_i\right\}^n_{i=1}$$

$$E(T_I) = \left\{y_1y'_1, y_2y'_2, \ldots, y_ny'_n\right\} \cup \left\{a_1y'_1, a_2y'_2, \ldots a_ny'_n\right\} \cup \left\{c^j_i z^j_i \mid j \in \Delta_i\right\}^n_{i=1}$$
$$\cup \left\{y_0y_1, y_1y_2, y_2y_3, \ldots, y_{n-1}y_n\right\} \cup \left\{y_nu_1, u_1u_2, u_2u_3, \ldots, u_{m-1}u_m, u_mu_0\right\}$$
$$\cup \left\{u_jx^j_1, x^j_1x^j_2, x^j_2x^j_3, x^j_2x^j_4, x^j_4x^j_5, x^j_4x^j_6, b^j_1x^j_6, b^j_2x^j_3, b^j_3x^j_5, g^j_1x^j_6, g^j_2x^j_1, g^j_3x^j_3, \ell^j x^j_5\right\}^m_{j=1}$$
$$\cup \left\{a'_i z^{j_1}_i, z^{j_1}_i z^{j_2}_i, \ldots, z^{j_{t-1}}_i z^{j_t}_i, z^{j_t}_i y'_i \ \middle| \ j_1 < j_2 < \ldots < j_t \text{ are elements of } \Delta_i\right\}^n_{i=1}$$

Let σ be a satisfying assignment for the instance I, and let ϕ_σ be the mapping of $\mathcal{X}_I$ to $V(T_I)$ defined as follows:

a) for each $i \in \{1 \ldots n\}$:

   if $v_i = 1$, then $\phi_\sigma(\alpha_{v_i}) = a_i$, $\phi_\sigma(\alpha_{\overline{v_i}}) = a'_i$, $\phi_\sigma(\beta^j_{\overline{v_i}}) = c^j_i$ for all $j \in \Delta_i$,

   if $v_i = 0$, then $\phi_\sigma(\alpha_{\overline{v_i}}) = a_i$, $\phi_\sigma(\alpha_{v_i}) = a'_i$, $\phi_\sigma(\beta^j_{v_i}) = c^j_i$ for all $j \in \Delta_i$,

b) for each $j \in \{1 \ldots m\}$ where $\mathcal{C}_j = X \vee Y \vee Z$:

   if $X = 1$, then $\phi_\sigma(\beta^j_X) = b^j_1$, $\phi_\sigma(\beta^j_{\overline{Y}}) = b^j_2$, $\phi_\sigma(\beta^j_{\overline{Z}}) = b^j_3$,

   $\phi_\sigma(\gamma^j_1) = g^j_1$, $\phi_\sigma(\gamma^j_2) = g^j_2$, $\phi_\sigma(\gamma^j_3) = g^j_3$, $\phi_\sigma(\lambda^j) = \ell_j$,

   if $Y = 1$, then $\phi_\sigma(\beta^j_Y) = b^j_1$, $\phi_\sigma(\beta^j_{\overline{Z}}) = b^j_2$, $\phi_\sigma(\beta^j_{\overline{X}}) = b^j_3$,

   $\phi_\sigma(\gamma^j_2) = g^j_1$, $\phi_\sigma(\gamma^j_3) = g^j_2$, $\phi_\sigma(\gamma^j_1) = g^j_3$, $\phi_\sigma(\lambda^j) = \ell_j$,

   if $Z = 1$, then $\phi_\sigma(\beta^j_Z) = b^j_1$, $\phi_\sigma(\beta^j_{\overline{X}}) = b^j_2$, $\phi_\sigma(\beta^j_{\overline{Y}}) = b^j_3$,

   $\phi_\sigma(\gamma^j_3) = g^j_1$, $\phi_\sigma(\gamma^j_1) = g^j_2$, $\phi_\sigma(\gamma^j_2) = g^j_3$, $\phi_\sigma(\lambda^j) = \ell_j$,

c) $\phi_\sigma(\delta) = y_0$ and $\phi_\sigma(\mu) = u_0$.

Theorem 9. *If σ is a satisfying assignment for I, then $\mathcal{T}_\sigma = (T_I, \phi_\sigma)$ is a ternary phylogenetic $\mathcal{X}_I$-tree that displays $\mathcal{Q}_I$ and is distinguished by $\mathcal{Q}_I$.*

5   Proof of Theorem 8

To explain the proof, we need the following naming convention adopted from [3]. If W is a literal in the clause $\mathcal{C}_j$, we say that S^j_W is a *shoulder of the clause* $\mathcal{C}_j$

as well as a *shoulder of the literal* W. It is a *a true shoulder* if $W = 1$; otherwise, a *false shoulder*. Similarly, the vertex K^j_W and L^j_W (if exists) are *knees of the clause* $\mathcal{C}_j$ as well as *knees of the literal* W. A knee of W is a *true knee* if $W = 1$; otherwise, a *false knee*. The vertices A_i, D^j_p, H_W, F^j for all meaningful choices of indices are respectively called *A-vertices, D-vertices, H-vertices,* and *F-vertices*.

Let G_σ be the graph constructed from $\mathrm{int}^*(\mathcal{Q}_I)$ by performing the following.

(i) make B adjacent to all true knees and true shoulders.

Let G'_σ be constructed from G_σ by performing the following steps.

(ii) make {true knees, true shoulders} into a complete graph,
(iii) for all $i \in \{1 \ldots n\}$, make A_i adjacent to all true knees of the literals $v_i, \overline{v}_i$,
(iv) for all $1 \leq i' \leq i \leq n$, make H_{v_i}, $H_{\overline{v}_i}$ adjacent to all true knees and true shoulders of the literals $v_{i'}$, $\overline{v}_{i'}$,
(v) for all $1 \leq j \leq j' \leq m$, make F^j adjacent to all true knees and true shoulders of the clause $\mathcal{C}_{j'}$,
(vi) for all $1 \leq i \leq n$ and all $j, j' \in \Delta_i$ such that $j \leq j'$:

   a) if $v_i = 1$, make $S^{j'}_{\overline{v_i}}$ adjacent to $K^j_{v_i}$, $L^j_{v_i}$ (if exists),
   b) if $v_i = 0$, make $S^{j'}_{v_i}$ adjacent to $K^j_{\overline{v_i}}$, $L^j_{\overline{v_i}}$ (if exists).

Finally, let G^*_σ be constructed from G'_σ by adding the following edges.

(vii) for all $j \in \{1 \ldots m\}$ where $\mathcal{C}_j = X \vee Y \vee Z$:

   a) if $X = 1$, then add edges $F^j L^j_Z$, $K^j_X L^j_Z$, $K^j_Y K^j_Z$, $D^j_2 K^j_Z$, $D^j_2 S^j_Y$, $D^j_3 S^j_Y$
      and make $\{D^j_1, D^j_2, D^j_3, S^j_X, S^j_Z, L^j_Z, K^j_Y\}$ into a complete graph,
   b) if $Y = 1$, then add edges $F^j L^j_X$, $K^j_Y L^j_X$, $K^j_Z K^j_X$, $D^j_3 K^j_X$, $D^j_3 S^j_Z$, $D^j_1 S^j_Z$
      and make $\{D^j_1, D^j_2, D^j_3, S^j_Y, S^j_X, L^j_X, K^j_Z\}$ into a complete graph,
   c) if $Z = 1$, then add edges $F^j L^j_Y$, $K^j_Z L^j_Y$, $K^j_X K^j_Y$, $D^j_1 K^j_Y$, $D^j_1 S^j_X$, $D^j_2 S^j_X$
      and make $\{D^j_1, D^j_2, D^j_3, S^j_Z, S^j_Y, L^j_Y, K^j_X\}$ into a complete graph.

Lemma 2. G'_σ *is a subgraph of every chordal sandwich of* $(G_\sigma, \mathrm{forb}(\mathcal{Q}_I))$.

Lemma 3. *If* σ *is a satisfying assignment for* I, *then* G^*_σ *is a subgraph of every chordal sandwich of* $(G_\sigma, \mathrm{forb}(\mathcal{Q}_I))$.

Lemma 4. *For every chordal sandwich* G' *of* $(\mathrm{int}^*(\mathcal{Q}_I), \mathrm{forb}(\mathcal{Q}_I))$, *there is* σ *such that* G_σ *is a subgraph of* G', *and such that* σ *is a satisfying assignment for* I.

Lemma 5. *If* σ *is a satisfying assignment for* I, *then* G^*_σ *is chordal.*

Proof. (Sketch) Assume that σ is a satisfying assignment for I, i.e., in each clause $\mathcal{C}_j$ exactly one literal evaluates to 1 by the assignment.

Consider the partition $V_1 \cup V_2 \cup V_3 \cup V_4 \cup V_5$ of $V(G^*_\sigma)$ where $V_1 = \{$false knees, D-vertices$\}$, $V_2 = \{$false shoulders$\}$, $V_3 = \{A$-vertices$\}$, $V_4 = \{H$-vertices, F-vertices$\}$, and $V_5 = \{$true knees, true shoulders, the vertex $B\}$.

Let π be an enumeration of $V(G_\sigma^*)$ constructed by listing the elements of V_1, V_2, V_3, V_4, V_5 in this order such that:

($\bullet$) the elements of V_1 are listed by considering each clause $\mathcal{C}_j = X \vee Y \vee Z$ and listing vertices (based on the truth assignment) as follows:

    a) if $X = 1$, then list $K_{\overline{X}}^j$, K_Z^j, L_Y^j, L_Z^j, D_1^j, K_Y^j, D_3^j, D_2^j in this order,

    b) if $Y = 1$, then list $K_{\overline{Y}}^j$, K_X^j, L_Z^j, L_X^j, D_2^j, K_Z^j, D_1^j, D_3^j in this order,

    c) if $Z = 1$, then list $K_{\overline{Z}}^j$, K_Y^j, L_X^j, L_Y^j, D_3^j, K_X^j, D_2^j, D_1^j in this order,

($\bullet$) the elements of V_2 (the false shoulders) are listed by listing the false shoulders of the clauses $\mathcal{C}_1$, $\mathcal{C}_2$, ..., $\mathcal{C}_m$ in this order,

($\bullet$) the elements of V_4 are listed as follows: first H_{v_1}, $H_{\overline{v_1}}$, H_{v_2}, $H_{\overline{v_2}}$, ... H_{v_n}, $H_{\overline{v_n}}$ in this order, then F^m, F^{m-1}, ..., F^1 in this order,

($\bullet$) the elements of V_3 and V_5 are listed in any order.

A simple but tedious analysis shows that π is a perfect elimination ordering of the vertices of G_σ^*. This proves that G_σ^* is indeed a chordal graph (see [14]). $\square$

Proof of Theorem 8. Let G' be a minimal chordal sandwich of $(\text{int}^*(\mathcal{Q}_I), \text{forb}(\mathcal{Q}_I))$. By Lemma 4, there exists σ, a satisfying assignment for I, such that G_σ is a subgraph fo G'. Thus, G' is also a chordal sandwich of $(G_\sigma, \text{forb}(\mathcal{Q}_I))$, and hence, G_σ^* is a subgraph of G' by Lemma 3. But by Lemma 5, G_σ^* is chordal, and so G' is isomorphic to G_σ^* by the minimality of G'.

Conversely, if σ is a satisfying assignment for I, then the graph G_σ^* is chordal by Lemma 5. Moreover, $\text{int}^*(\mathcal{Q}_I)$ is a subgraph of G_σ^*, by definition, and G_σ^* contains no edges of $\text{forb}(\mathcal{Q}_I)$, also by definition. Thus, G_σ^* is a chordal sandwich of $(\text{int}^*(\mathcal{Q}_I), \text{forb}(\mathcal{Q}_I))$, and it is minimal by Lemma 3.

This proves that by mapping each satisfying assignment σ to the graph G_σ^*, we obtain the required bijection. That concludes the proof. $\square$

6   Conclusion

In this paper, we have shown that determining whether a given phylogenetic tree represents the unique evolution of given species is an NP-hard problem. This implies that the problem is actually $CoNP$-complete, as it can be defined by the formula "for every pair of trees, if they are solutions, they are isomorphic". Moreover, the problem clearly remains NP-hard even if the tree is not provided and we only want to test whether there is a unique solution. (For this, note that isomorphism of trees and testing if a tree is a solution takes polynomial time.)

In addition, we proved that the unique chordal sandwich problem is NP-hard. Following this direction, it would be interesting to consider the complexity of uniqueness of other sandwich problems, for instance, interval sandwich (DNA physical mapping) or cograph sandwich (genome comparison); the decision problem for the former is NP-hard [16] while it is polynomial for the latter [5, 15].

References

1. Agarwala, R., Fernández-Baca, D.: A polynomial-time algorithm for the perfect phylogeny problem when the number of character states is fixed. SIAM Journal of Computing 23, 1216–1224 (1994)
2. Böcker, S., Bryant, D., Dress, A.W.M., Steel, M.A.: Algorithmic aspects of tree amalgamation. Journal of Algorithms 37, 522–537 (2000)
3. Bodlaender, H.L., Fellows, M.R., Warnow, T.J.: Two strikes against perfect phylogeny. In: Kuich, W. (ed.) ICALP 1992. LNCS, vol. 623, pp. 273–283. Springer, Heidelberg (1992)
4. Bonet, M.L., Linz, S., John, K.S.: The complexity of finding multiple solutions to betweenness and quartet compatibility. CoRR abs/1101.2170 (2011), http://arxiv.org/abs/1101.2170
5. Bui-Xuan, B.M., Habib, M., Paul, C.: Competitive graph searches. Journal Theoretical Computer Science 393, 72–80 (2008)
6. Buneman, P.: A characterization of rigid circuit graphs. Discrete Mathematics 9, 205–212 (1974)
7. Camin, J., Sokal, R.: A method for deducing branching sequences in phylogeny. Evolution 19, 311–326 (1965)
8. Creignou, N., Hermann, M.: Complexity of generalized satisfiability counting problems. Information and Computation 125, 1–12 (1996)
9. Dekker, M.C.H.: Reconstruction methods for derivation trees. Master's thesis, Vrije Universiteit, Amsterdam (1986)
10. Estabrook, G.F.: Cladistic methodology: a discussion of the theoretical basis for the induction of evolutionary history. Annual Review of Ecology and Systematics 3, 427–456 (1972)
11. Estabrook, G.F., Johnson Jr., C.S., McMorris, F.R.: An idealized concept of the true cladistic character. Mathematical Biosciences 23, 263–272 (1975)
12. Estabrook, G.F., Johnson Jr., C.S., McMorris, F.R.: An algebraic analysis of cladistic characters. Discrete Mathematics 16, 141–147 (1976)
13. Estabrook, G.F., Johnson Jr., C.S., McMorris, F.R.: A mathematical foundation for the analysis of cladistic character compatibility. Mathematical Biosciences 29, 181–187 (1976)
14. Golumbic, M.C.: Algorithmic Graph Theory and Perfect Graphs, 2nd edn. North Holland (2004)
15. Golumbic, M.C., Kaplan, H., Shamir, R.: Graph sandwich problems. Journal of Algorithms 19, 449–473 (1995)
16. Golumbic, M.C., Shamir, R.: Complexity and algorithms for reasoning about time: a graph-theoretic approach. Journal of the ACM 40, 1108–1133 (1993)
17. Gordon, A.D.: Consensus supertrees: The synthesis of rooted trees containing overlapping sets of labeled leaves. Journal of Classification 3, 335–348 (1986)
18. Gusfield, D.: Efficient algorithms for inferring evolutionary trees. Networks 21, 19–28 (1991)
19. Habib, M., Stacho, J.: Unique perfect phylogeny is NP-hard. CoRR abs/1011.5737 (2010), http://arxiv.org/abs/1011.5737
20. Juban, L.: Dichotomy theorem for the generalized unique satisfiability problem. In: Ciobanu, G., Păun, G. (eds.) FCT 1999. LNCS, vol. 1684, pp. 327–337. Springer, Heidelberg (1999)
21. Kannan, S.K., Warnow, T.J.: Triangulating 3-colored graphs. SIAM Journal on Discrete Mathematics 5, 249–258 (1992)

22. Lam, F., Gusfield, D., Sridhar, S.: Generalizing the four gamete condition and splits equivalence theorem: Perfect phylogeny on three state characters. In: Salzberg, S.L., Warnow, T. (eds.) WABI 2009. LNCS, vol. 5724, pp. 206–219. Springer, Heidelberg (2009)

23. LeQuesne, W.J.: Further studies on the uniquely derived character concept. Systematic Zoology 21, 281–288 (1972)

24. LeQuesne, W.J.: The uniquely evolved character concept and its cladistic application. Systematic Zoology 23, 513–517 (1974)

25. LeQuesne, W.J.: The uniquely evolved character concept. Systematic Zoology 26, 218–223 (1977)

26. McMorris, F.R., Warnow, T., Wimer, T.: Triangulating vertex colored graphs. SIAM Journal on Discrete Mathematics 7, 296–306 (1994)

27. Semple, C., Steel, M.: A characterization for a set of partial partitions to define an X-tree. Discrete Mathematics 247, 169–186 (2002)

28. Semple, C., Steel, M.: Phylogenetics. Oxford lecture series in mathematics and its applications. Oxford University Press, Oxford (2003)

29. Steel, M.: Personal webpage, `http://www.math.canterbury.ac.nz/~m.steel/`

30. Steel, M.: The complexity of reconstructing trees from qualitative characters and subtrees. Journal of Classification 9, 91–116 (1992)

31. West, D.: Introduction to Graph Theory. Prentice-Hall, Englewood Cliffs (1996)

32. Wilson, E.O.: A consistency test for phylogenies based upon contemporaneous species. Systematic Zoology 14, 214–220 (1965)

Fast Error-Tolerant Quartet Phylogeny Algorithms

Daniel G. Brown and Jakub Truszkowski

David R. Cheriton School of Computer Science
University of Waterloo
Waterloo ON N2L 3G1 Canada
{browndg,jmtruszk}@uwaterloo.ca

Abstract. We present a quartet-based phylogeny algorithm that returns the correct topology for n taxa in $O(n \log n)$ time with high probability, assuming each quartet is inconsistent with the true tree topology with constant probability, independent of other quartets. Our incremental algorithm relies upon a search tree structure for the phylogeny that is balanced, with high probability, no matter the true topology. In experiments, our prototype was as fast as the fastest heuristics, but because real data do not typically satisfy our probabilistic assumptions, its overall performance is not as good as our theoretical results predict.

1 Introduction

Incremental phylogenetic reconstruction algorithms add new taxa to a topology until all n taxa have been added. In our algorithm, each insertion requires $O(\log n)$ runtime with high probability. The probability that the algorithm makes any mistakes is $o(1)$ in a simple error model. Thus, our randomized algorithm has runtime $O(n \log n)$ and returns the true topology, both with high probability (regardless of the true topology). We believe it is the first $O(n\text{poly} \log n)$-runtime algorithm with such guarantees. An $o(n \log n)$-runtime algorithm cannot return all topologies, so our algorithm is asymptotically optimal.

Our error-tolerant algorithms offer the possibility of producing a phylogenetic tree in runtime smaller than that of producing even the input matrix to a distance method like neighbour joining, while still having high probability of reconstructing the true tree.

2 Related Work

Phylogenetic quartet methods reconstruct trees from sets of four taxa and combine these phylogenies into the overall tree. Quartet puzzling [21] is one of the first algorithms in this line of research. Many heuristic algorithms also operate on this principle (*e.g.* [18,19]).

Some quartet algorithms find the correct phylogeny with high probability under a certain model of evolution. Erdös *et al.* [8] give an $O(n^4 \log n)$ algorithm that reconstructs the phylogeny with high probability, assuming that the

R. Giancarlo and G. Manzini (Eds.): CPM 2011, LNCS 6661, pp. 147–161, 2011.

sequences evolve according to the Cavender-Farris model of evolution, for sufficiently long sequences. The runtime of their algorithm is $O(n^2)$ for most trees. Csűros [5] provided a practical $O(n^2)$ algorithm with similar performance guarantees. Recent papers [11,6] give similar algorithms to identify parts of the tree that can be reconstructed. These approaches choose queries so that, in the assumed model of evolution, all queried quartets are correct with high probability. The only sub-quadratic time algorithm with guarantees on reconstruction accuracy is by King *et al.* [16]; its running time is $O(n^2 \frac{\log \log n}{\log n})$ provided that the sequences are long enough.

Wu *et al.* [22] gave a simple error model where each quartet query independently errs with fixed probability p. They gave an $O(n^4 \log n)$ algorithm that errs with constant probability under this model. This model has also been used for evaluating algorithms for maximum quartet consistency [23].

We improve on Wu *et al.* in runtime and accuracy with an $O(n \log n)$ algorithm that errs with probability $o(1)$. To our knowledge, it is the first provably error-tolerant, substantially sub-quadratic time algorithm for phylogenetic reconstruction. (Recently, a heuristic algorithm has been proposed with a claimed runtime of $O(n^{1.5} \log n)$ [17].)

Fast algorithms have been proposed for error-free data. Kannan *et al.* [13] use error-free *rooted triples* in an $O(n \log n)$ algorithm. Rooted triples reduce to quartets if we pick one taxon as an outgroup and always ask quartet queries for sets with that taxon, so that algorithm works for error-free quartets.

Our algorithm uses ideas from work on noisy binary search in which comparisons have fixed error probability, by Feige *et al.* [9] and Karp and Kleinberg [15].

The data structure used here appears similar to one used by Brodal *et al.* [3] for computing quartet distance between two phylogenies. While both structures represent a partition of the tree into hierarchically nested sets, the goal of their data structure is to aid enumerating certain sets of quartets for a given tree. Our structure is dynamic and supports different queries. Some ideas used in the proofs in this paper are similar to the results of Kao *et al.* [14] on randomized tree splitting.

3   Definitions

We begin with definitions about the two trees we will focus on: the phylogeny we are reconstructing and the search tree that allows us to do the insertions efficiently.

A *phylogeny* T is an unrooted binary tree with n leaves in 1-to-1 correspondence with a set S of terminal taxa. Removing internal node v, and its incident edges, from a phylogeny yields three subtrees, $t_i(T, v)$ for $i = 1, 2, 3$. The tree $t_i(T, v)$ joined with its edge to v is the *child subtree* $c_i(T, v)$. Phylogeny T' is *consistent* with T if its taxa are a subset of those of T, and T' is the union of all paths in T between taxa in T', with internal nodes of degree 2 removed. A *border node* of subtree T' in T is any internal node of T that is a leaf in T'.

A *quartet* is a phylogeny of four taxa. A *quartet query* $q(a, b, c, d)$, returns one of three possible quartet topologies: $ab|cd$, $ac|bd$ and $ad|bc$: in $ab|cd$, if we remove the internal edge, we disconnect $\{a, b\}$ from $\{c, d\}$. We assume a quartet query can be done in $O(1)$ time. In Section 5 our error model considers how often quartet queries for four taxa of T are inconsistent with T. A *node query* $N(T, v, x)$ for internal node v of phylogeny T and new taxon x is a quartet query $q(x, a_1, a_2, a_3)$, where a_i is a leaf of T in $t_i(T, v)$. Such a query identifies the $c_i(T, v)$ where taxon x belongs, if it is consistent with the true topology.

3.1 Search Tree

A natural algorithm to add taxon x to phylogeny T begins at an internal node v and uses node query $N(T, v, x)$ to identify the $t_i(T, v)$ where taxon x belongs. We move to the neighbour of v in that subtree, and repeat the process until the subtree into which x is to be placed is only one edge e, which we break into two edges and hang x onto; see Figure 1. We follow the path from v to an endpoint of e and identify the other endpoint with one more query. The number of node queries equals this path length plus one. For a balanced tree with diameter $\Theta(\log n)$, this gives a $\Theta(n \log n)$ incremental phylogeny algorithm. For trees with $\Theta(n)$ diameter, this algorithm requires $\Theta(n^2)$ queries. We give a

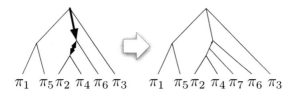

$$\pi_1 \quad \pi_5 \pi_2 \quad \pi_4 \pi_6 \pi_3 \qquad \pi_1 \quad \pi_5 \pi_2 \quad \pi_4 \pi_7 \pi_6 \pi_3$$

Fig. 1. Natural incremental algorithm: start at root and search to find place for new taxon π_7 by asking queries down the path. Break an edge to insert the new taxon.

search tree structure to manage the expected number of queries on the search path, regardless of the underlying tree topology.

Definition 1. *A search tree $Y(T)$ for a phylogeny T is a rooted ternary tree satisfying the following conditions:*

1. *Each node y in $Y(T)$ is associated with a distinct subtree $r(y)$ of T.*
2. *The root of $Y(T)$ is associated with the full tree T.*
3. *For each internal node y in $Y(T)$, there exists an internal node $s(y)$ in T such that the three subtrees associated with the children of y are the child subtrees $c_i(r(y), s(y))$. There are also three nonempty lists $\ell_i(y)$ stored at each internal node y; each element of $\ell_i(y)$ is a terminal taxon in $t_i(T, y)$.*
4. *For each node y in $Y(T)$, $r(y)$ has at most two border nodes in T.*

$Y(T)$ is *complete* if each leaf in $Y(T)$ is associated with a single edge of T, and each edge of T has a corresponding leaf in $Y(T)$. For a given node y in the search tree, its associated node $s(y)$ in T may be picked so the three child subtrees are balanced; this gives expected $O(\log n)$ insertion time. See Figure 2 for an example.

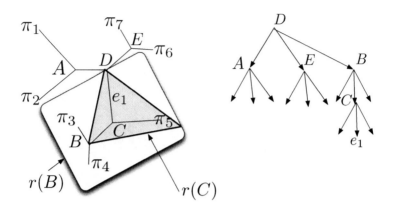

Fig. 2. A phylogeny and a corresponding search tree. Internal nodes of the search tree correspond to subphylogenies: two, for $r(B)$ and $r(C)$, are indicated. Leaves of the search tree correspond to edges of the phylogeny.

4   An Algorithm for Error-Free Data

Using our search tree structure gives a straightforward incremental phylogeny algorithm if quartets are all consistent with T, the true topology.

We pick a random permutation π of the taxa, and start with the unique topology T_3 for $\{\pi_1, \pi_2, \pi_3\}$, and a search tree $Y(T_3)$ with four nodes: a root w with $r(w) = T_3$ and $s(w)$ the internal node of T_3, and with one leaf for each edge of T_3. We also store $\ell_i(w) = \{\pi_i\}$; we also use $\ell_i(w)$ to represent the unique member of this set. This fits our requirements for a complete search tree of T_3.

Now, assuming T_i is consistent with T, and $Y(T_i)$ is a valid search tree for T_i, we add π_{i+1}, to produce T_{i+1} and $Y(T_{i+1})$. We start at the root w of $Y(T_i)$ and ask the node query $N(T_i, s(w), \pi_{i+1})$ using the quartet $q(\pi_{i+1}, \ell_1(w), \ell_2(w), \ell_3(w))$; this tells us which child of w we should move to next. We continue until we reach a leaf y of $Y(T_i)$; this corresponds to the edge e of T_i where the new taxon π_{i+1} belongs. We break edge e into two parts, creating a new node u and a new edge from u to the new leaf π_{i+1}. The new tree is T_{i+1}.

To update $Y(T_i)$, we create three edges from y to a new node for each of the three newly created edges and let $\ell_1(y)$ be $\{\pi_{i+1}\}$, and set $\ell_2(y)$ and $\ell_3(y)$ to contain the taxon closest to π_{i+1} in the final quartet query and one of the two taxa that was not closest to π_{i+1} in that query. Since node y was a leaf in $Y(T_i)$, these nodes are in proper configuration with respect to y in T_{i+1}. See Figure 3. Assuming the quartet queries all are consistent with the true topology T, we discover in this way the proper place in the tree to insert each new taxon and maintain the invariants required for a complete search tree. In particular, the only subtrees whose border nodes need to be considered are those created by the new node addition, and as they are all either single edges or derived from a single edge in $Y(T_i)$, they continue to have at most two border nodes.

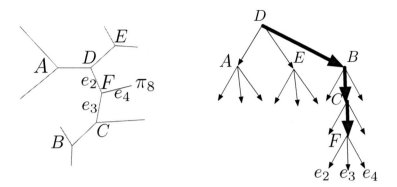

Fig. 3. Inserting into a phylogeny. To insert π_8 into the phylogeny from Figure 2, we follow the path through the search tree indicated with bold lines. We find the correct edge to break to add π_8 to the tree, and modify the search tree locally to accommodate the change.

Theorem 1. *If all quartet queries made by this algorithm are consistent with T, then this algorithm returns T. Its runtime is $O(n \log n)$ with probability $1 - o(1)$.*

Proof. We have seen that the algorithm returns T. In the next subsection, we show that inserting taxon π_i requires $O(\log n)$ queries with high probability. Each query requires $O(1)$ time, and the work to create a new edge requires constant time. The overall runtime is $O(n \log n)$ with high probability.

4.1   The Height of the Search Tree

To prove Theorem 1, we need to know the height of the search tree $Y(T)$. We now show that this tree is almost surely balanced.

Lemma 1. *For any phylogeny T, with n taxa, there exist two disjoint child subtrees A and B of the form $c_i(T, v)$ and $c_j(T, u)$ with at least $n/6$ and at most $n/3$ taxa.*

Proof. We first show there exists a node u where all $t_i(T, u)$ have at most $n/2$ taxa. Pick an internal node u in T; if all $t_i(T, u)$ have at most $n/2$ taxa, we are done. Otherwise, move to the its neighbour in the $t_i(T, u)$ with the most taxa. This process terminates at a node u satisfying the property. Let $n_1 \geq n_2 \geq n_3$ be the numbers of taxa in the trees $t_i(T, u)$ at some step. Call the largest of these trees T_1. If $n_1 > \frac{n}{2}$, we move to the neighbour u^* of u in T_1; trees $t_i(T, u^*)$ have n_{11}, n_{12} and $n_2 + n_3$ taxa respectively where n_{11}, n_{12} are the numbers of taxa in the subtrees T_{11}, T_{12} of T_1 created by removing u^*. Since $n_2 + n_3 < \frac{n}{2}$, the only components that can have size over $\frac{n}{2}$ must be either T_{11} or T_{12}, which have fewer taxa than T_1 since they are its subtrees. Thus the number of taxa in the largest component decreases at each step of the process, which proves the claim.

Now, consider the node u we have found by this process, and let t_1 and t_2 be the two largest $t_i(T, u)$ subtrees, both of which have between $n/4$ and $n/2$ taxa. If t_1 has more than $n/3$ taxa, consider the three child subtrees in t_1 of the neighbour of u in t_1; the subtree that contains u has zero taxa, so the largest of the subtrees three must have at least $n/6$ taxa. If this tree has at most $n/3$ taxa, we have found our subtree A; if not, we move one step more away from u in the direction of the subtree having the most taxa until we find a subtree small enough. We analogously find B as a subtree of t_2.

Lemma 2. *The number of node queries asked by the phylogeny algorithm to assign taxon π_{i+1} to its place in the tree is at most $37(\log_{6/5} i) \approx 203 \ln i$, with probability $1 - o(1/i^4)$, and at most $37(\log_{6/5} n)$ with probability $1 - o(1/n^4)$.*

Proof. Consider the process of adding π_{i+1} to the tree. We consider a sequence $y_1 \ldots y_k$ of nodes in the search tree Y, each corresponding to a subtree $r(y_j)$ of the existing phylogeny. We divide the y_j into phases: phase t corresponds to the period in which $r(y_j)$ contains between $\frac{5}{6}^t i$ and $\frac{5}{6}^{t-1} i$ leaves; after $\log_{6/5} i$ phases, the algorithm has found where to put π_i. We show that the distribution of the length of each phase is bounded above by the sum of three geometrically-distributed random variables.

Each phase corresponds to taking a subtree and shrinking it by a factor of $5/6$. This happens either if the largest of the three subtrees of the phylogeny descendant from the current search tree node y_j has at most $5/6$ of the number of taxa we had at the beginning of the current phase, or earlier if π_i belongs in a tree with fewer than that many taxa. We concern ourselves only with the first of these ways of ending a phase, so we upper bound the length of a phase.

Let A and B be the subtrees of $r(y_j)$ satisfying Lemma 1. The queries asked include taxa found in $r(y_j)$, in the order that they occur in permutation π. In particular, we will ask a node query including a node of A with probability at least $1/6$ at step, independently, until we finally do ask a query of a node from A. (Since our queries always include at least $5/6$ of the taxa, and we have not queried any members of A, we always have all members of A available.) After querying a member of A, for the phase to continue, we must choose the subtree containing all of B. Now, we ask queries corresponding to the current subtree, until we see a taxon from B, which will happen with probability $1/6$ or greater at each step. Now, we arrive in a state where the current subtree of the phylogeny includes border nodes inside A and B, since we must have cut off parts of A and of B, but cannot have cut off all of either without ending the phase. Now, we ask queries until we see a node from neither A nor B; this happens with probability at least $1/3$ at each step. Then, the current search tree node y_j must correspond to a node on the path from A to B in the phylogeny, since otherwise one of its subtrees would have three border nodes.

Thus, the length of a phase is at most the sum of three geometric random variables, with expectations 6, 6 and 3; we then move to a new tree with at most $5/6n$ taxa. However, it may have two border nodes as well; we label these with a taxon from their neighbouring subtrees (thereby adding two taxa to the current subtree) and perform a single quartet query (removing at least two taxa). This gives a new subtree in which we can perform the next phase.

Thus, if $G(i)$ are independent geometric random variables with mean i, then the length of one phase is bounded above by $G(6) + G(6) + G(3) + 1$, and the expected total number of queries is at most $19(\log_{6/5} i) + 1$, where for simplicity, we let the $G(i)$ all have mean 6.

Moreover, this variable is rarely above $37 \log_{6/5} i$. In particular, let $Q(n, r)$ be the negative binomial random variable that is the sum of n geometrically distributed variables with mean r. Then $\Pr[Q(n, r) > knr] = \Pr[B(knr, 1/r) < n]$, where $B(n, p)$ is a binomial random variable that results from the sum of n independent Bernoulli trials, each with mean p. By standard Chernoff methods ([7], p. 6), this probability is bounded above by $\exp(\frac{-kn(1-1/k)^2}{2})$. So, $\Pr[Q(3 \log_{6/5} i, 6) > 36 \log_{6/5} i] \leq i^{-4}$, meaning that the probability we use more than $37 \log_{6/5} i$ queries for taxon π_{i+1} is $o(1/i^4)$; similarly, the probability that we use more than $37 \log_{6/5} n$ queries for taxon π_{i+1} is $o(1/n^4)$.

We emphasize: $Y(T)$ is almost surely balanced regardless of the topology of T. Even if the diameter of T is $\Theta(n)$, its search tree almost surely has height $O(\log n)$. We expect that the actual values of the constants are much smaller than in the above lemma.

5   Accounting for Errors

Our search tree algorithm adapts to the case of error-prone quartets where each quartet query independently errs with probability $p > 0$. We assume that $(1 - p)^3 > 0.5 + \epsilon$ for some $\epsilon > 0$; we relax this assumption at the end of the section.

5.1   Random Walk in the Search Tree

Let $Y(T')$ be a complete search tree for T' and let x be a taxon not in T'. We will perform a random walk on $Y(T')$ to place x into its proper place in T', where each step of the random walk is determined by at most 3 quartet queries.

Let y_i be the location of the random walk after i steps, with y_0 the root of $Y(T')$. If $y(i)$ is not a leaf node, query the border nodes of $r(y_i)$. If any border node queries gives answer $x \notin r(y_i)$, go to the parent node of y_i. If all border nodes give answers consistent with $x \in r(y_i)$, query the node y_i and descend to the child of y_i indicated.

If y_i is a leaf, corresponding to an edge of T', let it have counter variable c initially set to 0. Query its border nodes as before; if each is consistent with $x \in r(y)$, increment c. Otherwise, decrement it if it is greater than 0; if $c = 0$, move to the parent node of y. After a number of queries we will soon compute, we are at a node in $Y(T')$: if it is a leaf, add x to that node of the search tree as for the insertion algorithm with error-free data. If not, signal failure.

The algorithm finds the proper place in the tree with high probability. Let y_x be the leaf in the search tree where we should insert taxon x. After i steps in the random walk, let the random variable d_i be the distance in the search tree between y_i and y_x. Let the random variable g_i have value $-c$ if $y_x = y_i$, $d_i + c$ if $y_x \neq y_i$ and y_i is a leaf of $Y(T')$, and d_i if y_i is not a leaf. If $g_i \leq 0$, then the current node of the random walk is the correct place to put x. The following simple observation is essential to proving the correctness of our algorithm.

Lemma 3. *Consider the random variables g_i defined above.*

1. $E[g_i] \leq d_0 + (1 - 2(1-p)^3)i$.
2. *If $i > \frac{-d_0}{1-2(1-p)^3}$, then* $\Pr[y_i \neq y_x] < \exp(\frac{-(d_0 + i(1-2(1-p)^3)^2}{2i})$

Proof. At each step of the random walk, there are at most two border nodes, so at most three queries. If each gives a correct answer, g_i decreases by 1; if any incorrect queries occur g_i increases by at most one, though it might still decrease by 1. In the worst case, the probability that g_i decreases is at least $(1-p)^3$, so $E[g(i+1) - g(i)] \leq -(1-p)^3 + (1 - (1-p)^3) = 1 - 2(1-p)^3$. The result follows from linearity of expectation, since $g_0 = d_0$. The second claim follows from the Chernoff bound, as the queries are independent.

Now, we have a straightforward taxon insertion algorithm. For each taxon π_{i+1}, we run the random walk long enough to handle the case that $g_0 = 203 \ln i$. To make the error probability at most $(1/i^2)$, we require that the random walk have j steps, where $\exp(\frac{-(203 \ln i + j(1-2(1-p)^3))^2}{2j}) \leq \frac{1}{i^2}$. The minimum value of j to make this guarantee is $j \geq k \ln n$, for $k = \frac{-203(1-2(1-p)^3) + 2 + 2\sqrt{1 - 203((1-2(1-p)^3)}}{(1-2(1-p)^3)^2}$.

We can now state the taxon insertion procedure in detail. Assuming that the

---

Algorithm 1. InsertTaxon$(x, T, Y(T))$

Initialize the random walk at the root of $Y(T)$.
for $i = 1$ to $k \log n$ **do**
    Simulate the next step of the random walk.
end for
Let $y_{k \log n}$ be the current node of the random walk.
if $y_{k \log n}$ is a leaf **then**
    Attach x to $r(y_{k \log n})$ in T and update $Y(T)$.
else
    return Failure.
end if

---

tree T_{i-1} is correct, then, this algorithm adds a new taxon in $O(\log i)$ queries, with error or failure probability $O(1/i^2)$.

5.2   Finding Quartets to Ask

We must ensure that we can always find a quartet that has not been queried before in $O(1)$ time. This requires two separate conditions to hold: first, that enough such quartets exist, and second, that we can find them in $O(1)$ time per query.

The first of these is easy if we start with a constant-sized guide tree T_S on a set S of at least m taxa, where m is the smallest number such that $k \log m < m - 2$, with k equal to the multiple of $\log i$ found using the formula in the previous section. In each insertion, we use at most $k \log i$ quartets at any node of the search tree; the extreme case is where the three child subtrees of the current tree T have 1, 1, and $i - 2$ taxa in them.

The latter is more complicated. Assume that for each node y in Y, $\ell_j(y)$ is the list of all taxa in the child subtree $t_j(r(y), s(y))$ (for $j = 1, 2, 3$). To find the next quartet in $O(1)$ time, we must fetch the next taxon in $t_j(T, s(y))$ in $O(1)$ time. We first enumerate taxa in $\ell_j(y)$. Once all taxa in $\ell_j(y)$ have been used, we pick the border node $b_j(y)$ of y in $t_j(T, s(y))$ (if it exists). The node $b_j(y)$ is associated with some ancestor y_1 of y and we have $r(y) \subseteq t_i(r(y_1), b_j(y))$ for some i. Taxa in $\ell_{(i+1)mod3}(y_1) \cup \ell_{(i+2)mod3}(y_1)$ are also in $t_j(T, s(y))$ so we enumerate them. Once they have been used, we find border nodes of $r(y_1)$ such that two of their taxa lists contain taxa in $t_j(T, s(y))$ that have not been used so far. Once all taxa from a node y_i have been used, we look at border nodes of $r(y_i)$. This process can be thought of as breadth first search on a directed graph where an arc denotes the relationship of being a border node.

Now, we give the complete algorithm. First, pick a constant-sized set $S_0 \subset S$ of m taxa and find the phylogeny for S_0 consistent with the most quartets. Then iteratively add taxa to the tree using the procedure InsertTaxon described above. The running time of this algorithm is $O(n \log n)$ with high probability. The error

---

Algorithm 2. Reconstruct($\mathcal{S}$,m)

Pick a subset $S_0 \subset S$ with m taxa
Find phylogeny T on S_0 consistent with the most quartets by exhaustive search.
Build a search tree $Y(T)$ for T.
for all $s \in \mathcal{S} \backslash S_0$ **do**
   insertTaxon(s,T,Y(T))
end for

---

probability can be bounded by $\mu(m) + \sum_{i=m}^{n} \frac{1}{i^2}$, where $\mu(m)$ is the probability that the maximum quartet compatibility tree on a random set of m taxa is not consistent with T. This quantity is constant for constant m; in the next section we show how to make the total error probability $o(1)$ as n grows.

The remaining case where $(1 - p)^3 \leq \frac{1}{2}$ can be solved by redefining node queries. Each node query is now implemented by asking c_p queries and returning the majority direction, with constants c_p and m chosen appropriately. We defer details to the longer version of this paper.

6   Shrinking the Error Probability to $o(1)$

The algorithm presented in the previous section errs with constant probability, since it starts with a constant-sized tree that may have errors, and since the additions to this tree also have constant probability of error.

If we start with a non-constant-sized guide tree, we can reduce the error probability.

Theorem 2. *The algorithm Reconstruct($\mathcal{S}$, max($\lceil \log \log n \rceil$, m)) both returns the correct tree and runs in $O(n \log n)$ time with probability $1 - o(1)$.*

Proof. The exhaustive search step requires enumerating all $O((\log \log n)^4)$ quartets, on all $O((\log \log n)! \log n)$ topologies on $\log \log n$ taxa; the product of these is $O((\log \log n)^{4 + \log \log n} \log n)$, which is sublinear in n. We have already shown that the rest of the algorithm requires $O(n \log n)$ time with high probability.

We will show below that $\mu(\log \log n)$, the failure probability of the guide tree algorithm, is $o(1)$. The failure probability of the insertion procedure is at most $\sum_{i=\log \log n}^{n} \frac{1}{i^2}$, which is $O(\frac{1}{\log \log n})$, and so $o(1)$. As such, the overall failure probability is $o(1)$.

We note that the guide tree could have more or fewer than $\log \log n$ taxa; we merely require that the brute-force guide tree construction requires $O(n \log n)$ time and has $o(1)$ error probability.

6.1   Maximum Quartet Consistency Is Consistent

Here, we show that the *maximum quartet consistency* approach is consistent for our error model. This result (which may be of independent interest, as our error model has been studied before [23]), shows that $\mu(n) \to 0$ as n grows.

Theorem 3. *Let T_{mqc} be the phylogeny compatible with the most quartet queries for a set of n taxa and let T^* be the true phylogeny. If each quartet query errs independently with probability p, then $\mu(n) = \Pr[T_{mqc} \neq T^*] = o(1)$ as $n \to \infty$.*

To prove this theorem, we first show a few properties of quartets.

Definition 2. *The quartet distance $d_Q(T, T')$ of phylogenies T and T' on the same set of taxa is the number of quartets on which T and T' differ.*

This distance was studied in [3,4] among others.

Lemma 4. *The quartet distance between distinct phylogenies is at least $n - 3$.*

Proof. Let T and T' be distinct phylogenies. Let (S_1, S_2) be a split in T not present in T'. Let (S_1', S_2') be a split in T' not present in T where none of the sets $A = S_1 \cap S_1', B = S_1 \cap S_2', C = S_2 \cap S_1', D = S_2 \cap S_2'$ is empty; such a split exists since T and T' are distinct. Choose taxa a, b, c, d from sets A, B, C, D, respectively. The quartet induced by T is $ab|cd$, whereas in T' it is $ac|bd$. This gives $\phi = |A||B||C||D|$ conflicting quartets; ϕ is at least $n-3$ since $|A|+|B|+|C|+|D| = n$, and the product is minimized when $|A| = n-3$ and $|B| = |C| = |D| = 1$.

The number of trees with small quartet distance from a fixed tree T is small.

Definition 3. *A taxon reinsertion (TR) operation consists of deleting a taxon from a phylogeny and attaching it to a remaining edge, creating three new edges.*

Lemma 5. *Let T and T' be phylogenies such that $d_Q(T, T') < n \log^2 n$. The number of TR operations required to transform T into T' is at most $c \log^4 n$ for some constant c.*

Proof. Let (S_1, S_2) be a split of T not present in T'. Let (S_1', S_2') be some split in T' that is not present in T that minimizes $\phi = |A||B||C||D|$ as defined earlier. Without loss of generality, assume that A is the largest of the sets. Observe that each of the sets B, C, D must have at most $\log^2 n$ taxa: otherwise $\phi > n \log^2 n$, so $d_Q(T, T') > n \log^2 n$. We delete all taxa in B and C from both T and T' to create trees $T^{(1)}$ and $T'^{(1)}$. By Lemma 4, this erases at least $n - 3$ conflicting quartets. We pick splits (S_1, S_2) and (S_1', S_2') in $T^{(1)}$ and $T'^{(1)}$ as we previously did for the original trees and repeat the process to obtain trees $T^{(2)}$ and $T'^{(2)}$, this time removing at least $n - 2 \log^2 n - 3$ discordant quartets.

We iterate the process until $T^{(i)} = T'^{(i)}$ for some i, which is $O(\log^2 n)$ since the total number of conflicting quartets is at most $n \log^2 n$, and each iteration erases $\Omega(n)$. The sets B and C have at most $\log^2 n$ taxa at each step of the algorithm. Therefore, at most $O(\log^4 n)$ taxa are deleted from both trees.

Let R be the taxa removed. The restrictions of both T and T' to $S - R$ are the same. To transform T to T', we move all nodes in R to a new side of the tree T, and then move each to the proper place in T' in $O(\log^4 n)$ TR operations.

Corollary 1. *For any phylogeny T, the number of phylogenies T' such that $d_Q(T, T') < n \log^2 n$ is at most $n^{b \log^4 n}$ for a large enough constant b.*

Proof. Each T' with distance from T at most $n \log^2 n$ can be obtained from T by $c \log^4 n$ TR operations. For any tree, the number of ways to perform a TR operation is less than $2n^2$ since we can choose any of the n taxa and reinsert it at any of the $2n - 5$ edges other than the one at which it was before the operation. This gives fewer than $(2n^2)^{c \log^4 n}$ phylogenies that can be created by repeating the operation $c \log^4 n$ times. Taking $b = 4c$ finishes the proof.

Now we can prove the maximum quartet compatibility consistency theorem.

Proof. Suppose some tree T' is consistent with more quartets than T^*, and $d_Q(T', T^*) = q$. At least half of the q quartets where T^* and T' differ must be erroneous; since they are independent errors, this has probability at most $\exp(-q(\frac{(1-2p)^2}{2}))$ by the Chernoff bound.

Let $\mathcal{T}_0$ be the set of all incorrect phylogenies with quartet distance from T^* less than $n \log^2 n$. Then $|\mathcal{T}^0| \leq n^{b \log^4 n}$, and for trees in $\mathcal{T}_0$, Lemma 4 gives that $q \geq n - 3$. The probability that any tree in $\mathcal{T}_0$ is consistent with more queries than T^* is bounded by $n^{b \log^4 n} \exp(-(n-3)(\frac{(1-2p)^2}{2}))$, which is $o(1)$ as n grows.

Now, consider the incorrect phylogenies $\mathcal{T}_1$ that are not in $\mathcal{T}_0$. There are fewer than $2^n n! < 2^{n(1+\log n)}$ such topologies, and for each, $d_q(T, T^*) \geq n \log^2 n$. The probability that any tree in $\mathcal{T}^1$ is consistent with more quartets than T^* is bounded above by $2^{n(1+\log n)} \exp(-n \log^2 n(\frac{(1-2p)^2}{2}))$, which is $o(1)$ as n grows.

So the probability that any incorrect tree is consistent with more quartets than T^* converges to 0 as n grows.

The following corollary, which may be of independent interest, is a consequence of Theorem 2 and Theorem 3.

Corollary 2. *For the error model assumed above, the algorithm Reconstruct($\mathcal{S}$, max($\lceil \log \log n \rceil, m$)) reconstructs the maximum quartet consistency tree with probability $1 - o(1)$.*

Finding the maximum quartet consistency tree is NP-hard in general [2], though polynomial-time approximation schemes exist [12].

7 Experiments

We have developed a prototype implementation of our algorithm to investigate its running time and accuracy. We have tested the algorithm in four scenarios. First, we tested the performance of the algorithm for the case with no errors. Second, we tested the performance of the random walk algorithm when the data was generated according to the model with independent errors. Third, we ran the random walk algorithm on real biological datasets to observe the running times. Finally, we repeatedly ran the algorithm on a simulated alignment to investigate its reconstruction accuracy where ground truth is known.

The tree topologies used in the synthetic data sets were chosen at random from the uniform distribution. In the iid error case, we perturbed every quartet query to give one of the two possible wrong answers with probability p. We supplied the algorithm with the correct initial guide tree of 150 taxa - only the insertion phase was performed. In our experiments, we set $p = 0.1$.

The algorithm for error-free data is very fast even for reasonably large phylogenies. For data sets having 10^4 taxa or less, constructing the tree takes less than a second. For $2 \cdot 10^4$ taxa, it takes roughly 2 seconds.

The random walk algorithm is roughly 5 times slower than the algorithm for error-free data. Constructing a tree having 10^4 taxa takes about 5 seconds, whereas a tree with $2 \cdot 10^4$ taxa requires 9 seconds. In all experiments on the iid error data set, our algorithm managed to reconstruct the correct tree.

Table 1. The running times of the algorithm for the error-free and iid data sets

Algorithm	1000	5000	10000	20000
Error-free	$< 1s$	$< 1s$	$< 1s$	$2s$
Random walk	$< 1s$	$2s$	$5s$	$9s$

We ran the algorithm on several protein families from the Pfam database [1]. Quartet queries were answered with the Four-Point method [10] based on estimated evolutionary distances between sequences. Distances were estimated based on pairwise BLOSUM62 scores using a method by Sonnhammer and Hollich [20]. We used neighbor-joining trees on a subset of 150 sequences (chosen at random from the whole set of sequences) as our initial guide trees. Our prototype implementation was able to process a dataset of around 12000 sequences in about 16 minutes (see Table 2).

Table 2. The running times of the algorithm for several Pfam families

Protein family	Sequences	Average length	running time
Maf(PF02545)	1980	189.60	38s
2Oxoacid_dh(PF00198)	3701	225.10	1m49s
PALP(PF00291)	11815	294.40	15m42s

7.1   Accuracy on Simulated Alignments

To test the accuracy of the topologies inferred by our algorithm, we used one of the simulated alignments developed by the authors of FastTree [17]. The COG840 alignment has 1250 protein sequences of average length 347 amino acids, with a known true topology. For this alignment, we randomly sampled 200 sequences and computed the neighbour joining tree for this set. We then ran the random walk algorithm starting with this guide tree. For the resulting tree, we calculated the proportion of splits that are present in the true tree. We repeated the procedure 100 times.

The results are shown in Table 3. The random walk often did not end at a leaf of the search tree, which prevents the algorithm from inserting the taxon into the phylogeny. This occured in roughly half of the insertions, so the trees produced by our algorithm only contain around 700 taxa. This suggests that for quartets inferred from sequences, the distribution of errors might be very different from the one assumed in our model. The accuracy of the trees produced by the algorithm is lower than that of neighbour joining on the full data set.

The accuracy of neighbour joining trees for subsets of taxa is lower than for that of the NJ tree for the full data set (49.8% vs. 62.6%). This explains some of the difficulty for the random walk algorithm since it has to insert taxa into an already highly erroneous guide tree. To investigate this issue, we ran the algorithm again using the true topology of randomly chosen 200 taxa as the guide tree. This produced trees whose accuracy was about 3.1% higher. This suggests that guide tree errors have modest impact on the quality of the overall tree. Neighbour joining trees inferred for the taxa that were present in the trees from the above experiment had an average accuracy of 66.2%.

Table 3. Performance of the random walk algorithm on the synthetic alignment

method	taxa inserted	guide accuracy	overall accuracy
RW+NJ guide tree (200 taxa)	704 ± 31	49.8 ± 4.4	46.3 ± 4.6
RW+true guide tree (200 taxa)	716 ± 26	100	49.4 ± 5.0
NJ	1250	n/a	62.6

In all our experiments, the height of search trees constructed by the algorithm was less than 40. This supports our view that the constants in Lemma 2 can be improved.

8   Conclusion

We have presented a fast algorithm that is guaranteed to reconstruct the correct phylogeny with high probability under an error model where each quartet query errs with a fixed probability, independently of others. The algorithm runs in $O(n \log n)$ time, which is the lower bound for any phylogeny reconstruction algorithm. Our prototype implementation seems reasonably fast on both real and simulated datasets, but its accuracy is lower than neighbour joining, and it produces a phylogeny with only a fraction of the input taxa. The experimental results on a simulated alignment suggest that the characteristics of errors in quartet trees inferred from aligned sequences may be quite different from the iid model assumed here.

It remains to be seen whether the algorithm can be improved to yield better practical performance. Our implementation does not use any form of quartet weighting, which greatly improves the performance of many quartet methods used in practice (*e.g.* [21,19]). This would enable the algorithm to distinguish between more and less credible queries, which may lead to an overall performance improvement. Another way to improve our algorithm would be to allow several rounds of taxa reinsertions. We plan to investigate these directions in the near future.

From a theoretical perspective, it is interesting whether there exist fast algorithms that offer similar performance guarantees under commonly studied models of sequence evolution, such as Jukes-Cantor or Cavender-Farris.

References

1. Bateman, A., Birney, E., Cerruti, L., Durbin, R., Etwiller, L., Eddy, S.R., Griffiths-Jones, S., Howe, K.L., Marshall, M., Sonnhammer, E.L.L.: The Pfam protein families database. Nucleic Acids Research 30(1), 276–280 (2002)
2. Berry, V., Jiang, T., Kearney, P.E., Li, M., Wareham, H.T.: Quartet cleaning: Improved algorithms and simulations. In: Nešetřil, J. (ed.) ESA 1999. LNCS, vol. 1643, pp. 313–324. Springer, Heidelberg (1999)
3. Brodal, G., Fagerberg, R., Pedersen, C.: Computing the quartet distance between evolutionary trees in time O(n log n). Algorithmica 38, 377–395 (2003)
4. Bryant, D., Tsang, J., Kearney, P.E., Li, M.: Computing the quartet distance between evolutionary trees. In: Proceedings of SODA 2000, pp. 285–286 (2000)
5. Csűrös, M.: Fast recovery of evolutionary trees with thousands of nodes. J. Comp. Biol. 9(2), 277–297 (2002)
6. Daskalakis, C., Mossel, E., Roch, S.: Phylogenies without branch bounds: Contracting the short, pruning the deep. In: Batzoglou, S. (ed.) RECOMB 2009. LNCS, vol. 5541, pp. 451–465. Springer, Heidelberg (2009)
7. Dubhashi, D.P., Panconesi, A.: Concentration of measure for the analysis of randomized algorithms. Cambridge Univ. Press, Cambridge (2009)
8. Erdös, P.L., Steel, M.A., Székely, L.A., Warnow, T.: A few logs suffice to build (almost) all trees: Part II. Theor. Comput. Sci. 221(1-2), 77–118 (1999)
9. Feige, U., Peleg, D., Raghavan, P., Upfal, E.: Computing with unreliable information. In: Proceedings of STOC 1990, pp. 128–137 (1990)

10. Felsenstein, J.: Inferring Phylogenies. Sinauer (2001)
11. Gronau, I., Moran, S., Snir, S.: Fast and reliable reconstruction of phylogenetic trees with very short edges. In: Proceedings of SODA 2008, pp. 379–388 (2008)
12. Jiang, T., Kearney, P., Li, M.: Orchestrating quartets: Approximation and data correction. In: Proceedings of FOCS 1998, pp. 416–425 (1998)
13. Kannan, S.K., Lawler, E.L., Warnow, T.J.: Determining the evolutionary tree using experiments. J. Algorithms 21(1), 26–50 (1996)
14. Kao, M.Y., Lingas, A., Östlin, A.: Balanced randomized tree splitting with applications to evolutionary tree constructions. In: Meinel, C., Tison, S. (eds.) STACS 1999. LNCS, vol. 1563, pp. 184–196. Springer, Heidelberg (1999)
15. Karp, R.M., Kleinberg, R.: Noisy binary search and its applications. In: Proceedings of SODA 2007, pp. 881–890 (2007)
16. King, V., Zhang, L., Zhou, Y.: On the complexity of distance-based evolutionary tree reconstruction. In: Proceedings of SODA 2003, pp. 444–453 (2003)
17. Price, M.N., Dehal, P.S., Arkin, A.P.: FastTree: Computing large minimum evolution trees with profiles instead of a distance matrix. Molecular Biology and Evolution 26(7), 1641–1650 (2009)
18. Ranwez, V., Gascuel, O.: Quartet-based phylogenetic inference: Improvements and limits. Molecular Biology and Evolution 18(6), 1103–1116 (2001)
19. Snir, S., Warnow, T., Rao, S.: Short quartet puzzling: A new quartet-based phylogeny reconstruction algorithm. Journal of Computational Biology 15(1), 91–103 (2008)
20. Sonnhammer, E.L.L., Hollich, V.: Scoredist: A simple and robust protein sequence distance estimator. BMC Bioinformatics 6, 108 (2005)
21. Strimmer, K., von Haeseler, A.: Quartet puzzling: a quartet maximum-likelihood method for reconstructing tree topologies. Mol. Biol. E 13(7), 964–969 (1996)
22. Wu, G., Kao, M.Y., Lin, G., You, J.H.: Reconstructing phylogenies from noisy quartets in polynomial time with a high success probability. Alg. Mol. Biol. 3 (2008)
23. Wu, G., You, J.H., Lin, G.: Quartet-based phylogeny reconstruction with answer set programming. IEEE/ACM Trans. Comput. Biol. Bioinf. 4(1), 139–152 (2007)

Real-Time Streaming String-Matching*

Dany Breslauer[1] and Zvi Galil[2]

[1] Caesarea Rothchild Institute, University of Haifa, Haifa, Israel
[2] College of Computing, Georgia Institute of Technology, Atlanta, Georgia

Abstract. This paper presents a real-time randomized streaming string matching algorithm that uses $O(\log m)$ space. The algorithm only makes one-sided small probability false-positive errors, possibly reporting phantom occurrences of the pattern, but never misses an actual occurrence.

1 Introduction

The string-matching problem is concerned with finding all exact occurrences of a pattern string $\mathcal{P}[1..m]$ in a text string $\mathcal{T}[1..n]$. Numerous algorithms exist, including algorithms that solve the problem in linear time, in real time, and even using only constant auxiliary space in addition to the input strings [3,5,6,7]. However, all these algorithms, including the on-line algorithms, require repeated access to the pattern or the text. In fact, if the pattern is considered part of the streamed input, without sufficient state space to remember the pattern or associated information, it is impossible to precisely identify occurrences of the pattern in the text.

The string matching problem is often viewed as a candidate elimination problem where initially all text positions are candidate occurrences of the pattern and an algorithm's task is to eliminate candidates and to verify which of the remaining text positions are actual occurrences. The classical Knuth-Morris-Pratt [7] algorithm proceeds by scanning the text and matching subsequent text symbols against the pattern. If a mismatch occurs, then the algorithm shifts the pattern ahead to the next viable text occurrence candidate. The shift is the smallest number of text positions that would align the pattern prefix that was matched thus far with the text, with another matching pattern prefix, skipping candidate occurrences that can be ruled out by the transitivity of the pattern's prefix self-overlap (also called *period*). The lengths of all such shifts are pre-computed in the pattern preprocessing phase and take up $O(m)$ space.

The Karp-Rabin [6] randomized string-matching algorithm deploys an entirely different approach. The algorithm computes a so-called *fingerprint* of a sliding text window of size m, the same length as the pattern, and compares this fingerprint to the fingerprint of the pattern, eliminating candidate occurrences with different fingerprints than the pattern's fingerprint. Their algorithm, however, requires access to the last m text symbols to slide the fingerprint window along

---

* Work partially supported by the European Research Council (ERC) project SFEROT and by the Israeli Science Foundation grants 35/05 and 347/09.

R. Giancarlo and G. Manzini (Eds.): CPM 2011, LNCS 6661, pp. 162–172, 2011.

the text. While the fingerprint functions always identify equal strings, different strings are usually mapped to different fingerprints, but with small probability, to identical fingerprints, possibly introducing erroneous *false-positive phantom* occurrences. Such phantom occurrences can be later verified against the text if both the pattern and text are readily accessible in memory.

Porat and Porat [8] recently gave a streaming-model string-matching algorithm that uses a combination of both the periodicity and the fingerprint approaches. Their one-pass streaming algorithm takes $O(\log m)$ time per symbol, or $O(n \log m)$ time overall, and uses only $O(\log m)$ space. Throughout this paper space refers to the number of $O(\log n)$ bit registers and neither the pattern nor any text segment is accessible after appearing in the input stream.

In addition to possibly reporting false-positive phantom occurrences inherent in fingerprinting, Porat and Porat's [8] algorithm may also commit with small probability *false-negative* errors, omitting actual occurrences of the pattern in the text (two-sided errors). Their algorithm also requires the period lengths and period fingerprints of the pattern and various pattern prefixes to be computed in the pattern preprocessing phase. However, no details were provided about how this information is computed. Note that while period lengths are often computed via straighforward application of string matching algorithms to match the pattern against itself, the streaming model's limitations present some obstacles.

In fact, independently of our work, Ergun, Jowhari and Salgan [1] recently studied the problem of computing the period length of a string in the streaming model. They describe an $O(m \log m)$ time one-pass streaming algorithm to compute the period length of a string using $O(\log m)$ space. Their algorithm, that finds the period only if the input string is *periodic* (the period is no longer than half of the string's length), builds on a simplified streaming string-matching algorithm with simpler pattern preprocessing requirements than Porat and Porat's [8] algorithm (still two-sided errors). Moreover, Ergun, Jowhari and Salgan [1] prove that $\Omega(m)$ space is required by any one-pass streaming algorithm that computes the period length of *non-periodic strings*, but two-passes suffice to reduce the space to $O(\log m)$. They also prove that $\Omega(\log m)$ space is required by any streaming string-matching algorithm, for certain choices of the pattern and text lengths.

We present two streaming string-matching algorithms. The first, like Porat and Porat's [8] algorithm, takes $O(\log m)$ time per symbol and uses $O(\log m)$ space, but is conceptually much simpler and has two important advantages: (1) the algorithm only commits small probability false-positive errors and no false-negative errors; in particular it never misses an occurrence (one-sided errors), and, (2) the pattern preprocessing phase is a trivial real-time streaming algorithm that does not compute period lengths. The second algorithm is a *real-time* algorithm, namely worst-case constant-time per symbol, using the same $O(\log m)$ space, while maintaining the one-sided error and the simple real-time streaming pattern preprocessing. Our techniques can be used to speed up Ergun, Jowhari and Salgan's [1] periodicity streaming algorithm to $O(m)$ time.

The paper starts by reviewing some basic properties of fingerprint arithmetic and periodic strings needed to describe the algorithms. The following sections describe each of the two algorithms and the pattern preprocessing. The next section shows how a slightly more complicated preprocessing that still preserves the properties above can avoid the second kind of errors. The paper concludes with a discussion and open problems.

2   Fingerprints and Periods

Porat and Porat [8] used Karp and Rabin's [6] fingerprints, defining for a prime p, a random integer $r \in \mathcal{F}_p$, and a string $s = s_1 s_2 \cdots s_l$ over the alphabet $\mathcal{F}_p$ ($\mathcal{F}_p$ is the field of integers modulo the prime p) the fingerprint function $\phi_{r,p}(s) = \sum_{i=1}^{l} s_i r^i \bmod p$. The *error probability*, the probability that two different strings share the same fingerprint, can be bounded as follows.

Theorem 1. *Let u and v be two different strings of length l, where $l \le n$ and $p \in \Theta(n^{2+\alpha})$, for some $\alpha \ge 0$. Then, the probability that fingerprints $\phi_{r,p}(u) = \phi_{r,p}(v)$, for a random $r \in F_p$, is smaller than $\frac{1}{n^{1+\alpha}}$.*

The fingerprint function can be arithmetically manipulated to compute the fingerprint of two concatenated strings, requiring only the fingerprints and string lengths and not the concatenated strings themselves. (The powers r^k and r^{-k} can be maintained together with the corresponding fingerprints and updated with the fingerprint operations, and do not need to be computed every time.)

Lemma 1. *One can compute the fingerprint of the concatenated strings u and v as:*

$$\phi_{r,p}(uv) = \phi_{r,p}(u) + r^k \phi_{r,p}(v) \bmod p \qquad uv = u_1 u_2 \cdots u_k v_1 v_2 \cdots v_l. \qquad (1)$$

The last Lemma can be used to *cancel out* the fingerprint of the prefix u from the fingerprint of the concatenated string uv to obtain the fingerprint of $v = u^{-1}(uv)$, an operation referred to as *sliding* by Porat and Porat [8]. Similarly, one can also cancel out the fingerprint of the suffix v to get the fingerprint of $u = (uv)v^{-1}$.

Corollary 1. *To extract the fingerprints of u or v from the fingerprint of uv:*

$$\phi_{r,p}(v) = r^{-k}(\phi_{r,p}(uv) - \phi_{r,p}(u)) \bmod p. \qquad (2)$$

$$\phi_{r,p}(u) = \phi_{r,p}(uv) - r^k \phi_{r,p}(v) \bmod p. \qquad (3)$$

The formulae above will be used to maintain the running fingerprints of multiple text blocks starting at various locations of interest and ending at the current text symbol, without having to update all such fingerprints with every text symbol. Instead, the algorithm will maintain one running fingerprint for the text prefix from the very beginning of the text and up to the current text symbol, and only update this one fingerprint with each input text symbol. Now, by keeping for each location of interest the static fingerprint for the text prefix up to that location, the algorithm can obtain the running fingerprint of the text block starting at that location up to and including the current text symbol:

Lemma 2. *The fingerprint of the text block starting at some location of interest and ending at the current text symbol can be computed in constant time whenever needed.*

Properties of periodic strings are often used in efficient string algorithms. A string u is *a period* of a string w if w is a prefix of u^k for some k, or equivalently if w is a prefix of uw. The shortest period of w is called *the period* of w and w is called *periodic* if it is at least twice as long as its period. Consider prefixes of the pattern of increasing length. If u is a prefix and v is a longer prefix, the period of u is said to continue in v if u and v have the same period and otherwise the period of u terminates in v. The following Theorem is due to Fine and Wilf [2].

Theorem 2. *If a string u has periods of length p and q, and its length $|u| \geq p + q - \gcd(p, q)$, then u also has a period of length $\gcd(p, q)$.*

3   The $O(n \log m)$ Time Algorithm

In this section we describe the $O(n \log m)$ time on-line streaming string-matching algorithm, introducing the basic concepts which are refined in the next section to obtain a real-time algorithm. The algorithm runs $\lceil \log_2 m \rceil$ simultaneous stages that filter the remaining viable occurrences. Each stage requires constant space, and takes constant time per input symbol, adding up to $O(\log m)$ time per input symbol, $O(n \log m)$ time overall, and $O(\log m)$ total space. The pattern preprocessing is trivial. It computes a sequence $\mathcal{P}_i$ of $\lceil \log_2 m \rceil$ increasing prefixes of the pattern $\mathcal{P}[1..m]$, and records their fingerprints, where $|\mathcal{P}_i| = 2^i$, and if m is not a power of 2, adding the last $\mathcal{P}_k = \mathcal{P}[1..m]$, $k = \lceil \log_2 m \rceil$. The $\lceil \log_2 m \rceil$ fingerprints are stored in $O(\log m)$ space.

The algorithm maintains and updates viable occurrences of the pattern while the text is being streamed on-line. A *viable occurrence (VO)* is a position in the text where an occurrence has not been ruled out. The *block of the VO* is the block that starts at the VO and ends at the currently last symbol of the text. Each VO belongs to some stage number i, such that the algorithm has verified earlier that the fingerprint of the text block of length $|\mathcal{P}_i|$ starting at the VO is equal to the fingerprint of pattern prefix $\mathcal{P}_i$, but there are insufficient text symbols yet to verify if this VO belongs to stage number $i+1$. As soon as there are sufficient text symbols, $|\mathcal{P}_{i+1}|$ to be precise, to promote a VO to the next stage (always the first, longest VO in the stage), the fingerprint of the block of the VO is compared to the pre-computed fingerprint of the pattern prefix $\mathcal{P}_{i+1}$ and the VO either gets promoted to the next stage or is eliminated. Clearly, an occurrence of each of the pattern prefixes $\mathcal{P}_i$ must start at each occurrence of the pattern and VOs eliminated this way cannot be occurrences of the whole pattern.

When the algorithm maintains a VO it maintains the position and the fingerprint of the block of the VO. The $O(n \log m)$ algorithm maintains this fingerprint directly. The real-time algorithm will maintain the fingerprint of the prefix of the text up to the VO. This fingerprint is the running fingerprint of the text when this position was reached. When needed the real-time algorithm gets the fingerprint of the block of the VO from the running fingerprint of the text and the static fingerprint of the VO using Corollary 1.

Each text position is initially considered a VO. As soon as the next position is reached, the one text symbol fingerprint is verified against the fingerprint of the pattern prefix $\mathcal{P}_0$, before the new VO may enter stage number 0. Note that VOs that start earlier in the text always correspond to longer text blocks, and therefore belong to the same or higher numbered stages. One can envision the VOs climbing the stage ladder from one stage to the next or falling off the ladder in case of fingerprint mismatch, up to the ultimate stage that verifies the fingerprint of the full pattern $\mathcal{P}_k = \mathcal{P}[1..m]$. Since all text positions are considered VOs and are only eliminated as a consequence of fingerprint mismatch, the algorithm commits no false-negative errors.

Similarly to Galil's [4] parallel string-matching algorithm, multiple VOs that get too crowded in some stage imply that there must be a periodic pattern prefix. Specifically, if there are at least three VOs at the same stage number i, then the pattern prefix $\mathcal{P}_i$ must be periodic, all the VOs in this stage must form an arithmetic progression whose difference is the period length of $\mathcal{P}_i$, and these VOs can be represented compactly and processed efficiently. However, there is one important caveat requiring more caution. The streaming algorithm compares fingerprints and not actual strings, and therefore different strings may be identified by the same fingerprint, conflicting with the periodicity *implied* by string equality. The algorithm may conclude that some fingerprint false-match error must have occurred since the periodicity properties have been violated, without precisely identifying the culprit fingerprint error.

Note that the algorithm only uses periodicity to facilitate the space efficient representation and never to eliminate any VO. In case that the algorithm discovers such low probability false-positive error, it must make some hard choices to remain within its strict space bounds while making sure that only false-positive errors are reported and no actual occurrences are omitted. The algorithm will discard some VOs that can not be compactly represented via the periodic arithmetic progression, essentially throwing them off the stage ladder. However, the algorithm will report all these discarded VOs as occurrences of the pattern so that no occurrences are missed. Note that by allowing extra space to store more individual VOs or arithmetic progressions, the algorithm could continue to examine these VOs. The expected space would still be $O(\log m)$.

Recall that the offending VO that revealed the fingerprint-periodicity inconsistency is in the process of being promoted from some stage to the next. To simplify the presentation and avoid cascading the effects of discarded VOs on higher numbered stages, the algorithm will discard and report all earlier

1. Extend the fingerprints of the blocks of the VOs by the current text symbol. (There is sufficient time to do this directly without Lemma 2.)
2. If the block of the first, longest, VO in the stage has precisely $|\mathcal{P}_{i+1}|$ text symbols, then remove this VO from stage number i, and compare the fingerprint of its block to the fingerprint of the pattern prefix $\mathcal{P}_{i+1}$ as candidate for stage number $i + 1$. The next VO in stage number i, if any, becomes the first VO in the stage.
   - If the fingerprints match, promote the VO to stage number $i + 1$.
   - VOs that get promoted to the ultimate stage matched the fingerprint of the whole pattern and are reported as occurrences of the whole pattern. They do not need to be stored.
3. To initialize, each text symbol's fingerprint that is equal to the fingerprint of $\mathcal{P}_0$ adds to stage number 0 a new VO starting at that text position.

Fig. 1. Stage number i of the $O(n \log m)$ algorithm

VOs (in equal or higher numbered stages), excluding the offending VO and the last VO in the arithmetic progression, and keep all VOs beyond these two since the limiting factor here is the algorithm's ability to compactly store and process all the VOs. The up to $O(\log m)$ discarded arithmetic progressions will be compactly written to the output rather than spelled out individually to remain within the $O(\log m)$ bounds.

Thus, the algorithm might now report two classes of erroneous pattern occurrences, those *phantom* occurrences that passed through the entire stage ladder and eventually had their fingerprint verified against the fingerprint of the whole pattern, and those VOs that were thrown off the stage ladder due to some non-specific fingerprint false-match errors conflicting with the implied periodicity and keeping the algorithm from compactly representing crowded VOs. The error probability in both cases is small, since it is either due to fingerprint false-match of the whole pattern (and some stage prefixes) or a detected fingerprint-periodicity conflict that must be due to fingerprint false-match of some pattern prefix. The algorithm is summarized in Figure 1 where all stages are executed in increasing order for each input symbol.

Lemma 3. *Let u and v be strings such that v contains at least three occurrences of u. Let $t_1 < t_2 < \ldots < t_h$ be the locations of all occurrences u in v, and assume that $t_{i+2} - t_i \leq |u|$, for $i = 1, \cdots, h - 2$ and $h \geq 3$. Then this sequence must form an arithmetic progression with difference $d = t_{i+1} - t_i$, for $i = 1, \ldots, h - 1$, that is equal to the period length of u.*

To get the required time and space bounds one has to compactly represent and efficiently process the multiple VOs that might accumulate in the same stage. Multiple VOs imply periodicity, and periodicity properties are used to represent the VOs and their associated information. The following Lemma follows from Lemma 3.

Lemma 4. *Suppose that there are at least three VOs in stage number i. If these are actual occurrences of the pattern prefix $\mathcal{P}_i$, then these VOs must form an arithmetic progression with difference equal to the period length of $\mathcal{P}_i$.*

If there are one or two VOs in a stage, these VOs are stored directly (the position and the fingerprint). If there are three or more, these VOs should form an arithmetic progression. The algorithm stores the first and last VO, the difference between the positions of the first and second VOs (the difference of the arithmetic progression) and the fingerprint of the block between these two positions (which should be equal to the period of $\mathcal{P}_i$), altogether constant space. The locations and the fingerprints of the middle of the arithmetic progression can be verified and reconstructed as follows:

Lemma 5. *Let $t_1, t_2, \ldots, t_h$ be all the VOs in stage number i, and assume that these are all actual occurrences of $\mathcal{P}_i$. Then it is possible to represent the locations and fingerprints of all the text blocks starting at each of these VOs up to the current position by the location and fingerprints of the first and last VOs and the length and the fingerprint of the block between the first and second VOs.*

A progression is generated when a third VO joins the stage. Note that the data maintained for the progression can be easily computed. The only computation needed is to compute the fingerprint of the block between the first and the second VOs using Corollary 1. We can easily maintain the data related to the arithmetic progression in constant time: If a new VO joins the progression when promoted to stage i, it simply replaces the last one. When the first VO of the progression is promoted to stage $i + 1$, the second one becomes the first and if there remain only two VOs in the progression, the progression stops to exist.

Recall that the algorithm did not compare actual symbols, but only fingerprints of the pattern prefixes $\mathcal{P}_i$, and there might be spurious VOs resulting from false-positive fingerprint errors. The algorithm only uses periodicity properties to verify the validity of the compact representation in stage number i when new VOs are added to the representation to ensure that the full representation can be faithfully reconstructed. There are two factors that need to be verified: (1) The VOs must form an arithmetic progression, and (2) the fingerprints can be reconstructed by Corollary 1. These properties are verified as soon as the third VO is added and every time another VO is added. If there are any problems during the verification, the algorithm concludes that some of the VOs are not actual occurrences due to small probability false-positive errors of some fingerprints. To stay within the time and space bounds, the algorithm will discard some of the VOs as outlined above, and to err only on the false-positive side, the algorithm will report the discarded VOs as actual occurrences.

1. The algorithm verifies that the VOs form an arithmetic progression. The difference of the arithmetic progression, the *implied period length* of $\mathcal{P}_i$, is set to the difference between the the first two VOs and verified against the rest. Normally one should expect the implied period length to be equal to the real period length of $\mathcal{P}_i$, but the algorithm does not know the real period length of $\mathcal{P}_i$ and only that the VOs fall into some arithmetic progression.

2. The algorithm verifies that the text block that starts at the last VO in the stage and ends at the newly added VO has the same fingerprint of the text block between the first and the second VOs in the arithmetic progression, the *implied period fingerprint* which we maintain. This is required so that the algorithm does not introduce false-negative errors when extracting fingerprints from the compact representation via sliding by the implied period fingerprint. Note that these text block fingerprints are extracted using either Equation 2 or 3 in Corollary 1 depending on whether VO fingerprints are maintained for text prefixes up to the VOs or for the blocks of the VOs.

Theorem 3. *The algorithm described above reports all occurrences of the pattern in the text in $O(\log m)$ time per text symbol using total $O(\log m)$ space. The algorithm may report false occurrences, and on occasions even detects that it had fingerprint errors, with probability at most $1/n^{\alpha'}$ for $\alpha' < \alpha$.*

Proof. Each stage number i, whether it has at most two VOs or more, takes constant-time to update one or two fingerprints with the current text symbol, and to discard or promote to stage number $i + 1$ at most one VO. The space requirement for each stage is constant. Multiplying by $O(\log m)$ stages we get the desired bounds. The error probability is bounded by multiplying the probability of fingerprint comparison error by the up to $O(n \log m)$ comparisons made.

4   The Real-Time Algorithm

Observe that in the $O(n \log m)$ algorithm above, fingerprints were only used in stage number i when the length of the first (longest) block of a VO in the stage was equal to the length of the next stage's pattern prefix $|\mathcal{P}_{i+1}|$, to verify whether the VO may be promoted to the next stage or eliminated. The key to the *real time* implementation is in (1) eliminating repetitive verification due to small highly repetitive pattern prefixes (e.g. $aa \cdots aaa$), and, (2) evenly spreading the VO stage promotion verification to avoid contentious text locations that might require up to $\lceil \log_2 m \rceil$ verifications. Both problems are solved by using additional $O(\log m)$ space.

Galil's [3] real-time implementation of the Knuth-Morris-Pratt [7] algorithm will be used. Let $f = \lceil \log_2 \log_2 m \rceil + 1$ and consider the pattern prefix $\mathcal{P}_f$, such that $2 \log_2 m < |\mathcal{P}_f| \le 4 \log_2 m$. The pattern preprocessing of the Knuth-Morris-Pratt [7] real-time variant will start as the pattern appears in the input stream, and will be stopped after the pattern prefix $\mathcal{P}_f$ was processed, having used only $|\mathcal{P}_f| = O(\log m)$ extra space.

Lemma 6. *Let u and v be prefixes of a string w, such that $|u| < |v|$, u is periodic and v is the shortest prefix of w such that the periodicity of u terminates in v. Then the period length of $v > |v| -$ the period length of u, and v is not periodic.*

The failure function of the Knuth-Morris-Pratt [7] algorithm and of Galil's [3] real-time implementation consists of the period of each prefix of the pattern. In the preprocessing we compute the period of each prefix of $\mathcal{P}_f$. If $\mathcal{P}_f$ is periodic, the preprocessing will examine further symbols of $\mathcal{P}$ until either the periodicity ends or the pattern ends. In case the periodicity ends, let π be the prefix up to and including the symbol where the periodicity ends. For each proper prefix of π that is longer than $\mathcal{P}_f$, its period is equal to that of $\mathcal{P}_f$. Since π can be compressed, its period can be easily computed by Galil's algorithm without using any additional space. If the pattern ends, i.e. the period of $\mathcal{P}_f$ is the period of $\mathcal{P}$, then Galil's [3] real-time string-matching algorithm can solve the streaming string-matching problem deterministically without any errors using $O(\log m)$ space.

The real-time string matching algorithm will then be used to match in real-time using $O(\log m)$ space occurrences of either a non-periodic $\mathcal{P}_f$ or π, which is non-periodic by Lemma 6. Let $\hat{f}$ be the largest integer such that $\mathcal{P}_{\hat{f}}$ is contained in this non-periodic pattern prefix (if $\mathcal{P}_f$ is not periodic $f = \hat{f}$). The occurrences found must be spaced by more than $|\mathcal{P}_{\hat{f}}|/2 \geq \log_2 m$ text positions apart, must start with $\mathcal{P}_{\hat{f}}$, and will be reported before sufficient symbols are available to verify $\mathcal{P}_{\hat{f}+1}$. These occurrences will be introduced at stage number $\hat{f}$ of the randomized $O(n \log m)$ algorithm in the previous section, skipping all the prior stages. Observe that no arithmetic progressions are forming at stage number $\hat{f}$ because of the spacing between the VOs.

Thus, the randomized real-time algorithm has two parts that are run alongside each other. Galil's [3] real-time string-matching algorithm that feeds stage number $\hat{f}$ of the following real-time adaptation of the $O(n \log m)$ algorithm from the previous section.

The real-time adaptation simulates the $O(n \log m)$ algorithm by maintaining a *cyclic buffer* $\mathcal{F}P[t]$ of size $s = \lceil \log_2 m \rceil$ that gives the running fingerprints of the *last* s text prefixes of locations up to and including t. Specifically, the fingerprints for positions $t, t-1, \ldots, t-s+1$ are stored at $\mathcal{F}P[t \bmod s]$. The round robin algorithm rotates through the stages numbered $i = \hat{f}, \cdots, \lceil \log_2 m \rceil - 2$, in increasing order, processing one stage at each text location using the buffer for the correct fingerprints. Note that the stage processing is delayed by less than s steps and the fingerprint needed to test whether to promote the VO to the next stage is available in the buffer $\mathcal{F}P$. The following concerns need attention:

1. The simulated action may happen out of order with respect to the $O(n \log m)$ algorithm, and even in different order depending on the text location. Note that since VOs in the same stage are at least $\log_2 m$ apart and the delay in the test for promotion is less than $\log_2 m$, the order of tests for promotion is maintained inside each stage. The only case that the different order will lead to a non temporary different computation is the following: Assume in the real-time algorithm x, y are the first and second VOs in stage i and z was just promoted from stage $i - 1$ to stage i as the third VO in the stage and it reveals an inconsistency with the periodicity. It is possible that in the $O(n \log m)$ algorithm x is promoted from stage i to stage $i + 1$ before z is promoted to stage i and therefore in that algorithm y and z are the only

VOs in stage i and there is no inconsistency. We can easily fix the order in such case to be the same by deleting x from stage i first, since stage i will be considered immediately after stage $i-1$ in the round robin algorithm and x can be promoted then. But in fact, this is not necessary since the algorithm is still correct with the different order.

2. The real-time on-line algorithm has to output the pattern occurrences immediately at their end, and delayed promotions to the last stage number are not acceptable. Such delays will be avoided by examining the last stage (number $\lceil \log_2 m \rceil - 1$) at every text location. In case $\mathcal{P}$ is longer than the $\mathcal{P}_i$ of the next to last stage by less than $\log_2 m$, then this stage too receives the same treatment.

3. Discarding and reporting VOs when some fingerprint-periodicity conflict is detected can take time. The simplest solution is to continue the rotation to larger number stages and discard the VOs in each stage until the last stage number $\lceil \log_2 m \rceil - 1$. Discarded arithmetic progressions will be compactly written to the output rather than spelled out individually to remain within the real-time bounds.

Theorem 4. *The algorithm described above reports all occurrences of the pattern in the text in constant time per text symbol using total $O(\log m)$ space. The algorithm may report false occurrences, and on occasion it even detects that it had fingerprint errors with probability at most $1/n^\alpha$.*

Proof. The algorithm updates the running fingerprint buffer with the current text symbol in constant time. Each delayed stage action can be properly done since the $\lceil \log_2 m \rceil$ fingerprint history is available in the buffer $\mathcal{F}P$. The space requirement for each stage is constant or $O(\log m)$ over all stages, and the overall space required for the Galil's [3] real-time string matching algorithm and for the buffer $\mathcal{F}P$ is $O(\log m)$. The error probability is bounded by multiplying the probability of fingerprint comparison error by the up to $O(n)$ comparisons made.

5   The Pattern Preprocessing

The $O(n \log m)$ time algorithm only requires the trivial preprocessing storing the fingerprints of the pattern prefixes $\mathcal{P}_i$. The real-time algorithm needs in addition to store either the short pattern prefix $\mathcal{P}_f$ and its failure function or (if $\mathcal{P}_f$ is periodic) the compressed versions of the longer prefix π and its failure function. The real-time algorithm will need to know the length of the pattern m or its order of magnitude.

Additional pattern preprocessing can be advantageous, though, to try to obtain a "better" fingerprint function that does not cause any fingerprint-periodicity conflict while matching the given pattern with a text string that is exactly equal to the pattern. (Conflicts would repeat in every occurrence of the pattern.) Such fingerprint function can be obtained by trying out several random seeds, either simultaneously while the pattern is streamed in or sequentially if the pattern is available for additional re-processing (i.e. in a nonstreaming fashion).

Theorem 5. *Given the fingerprint function and the pattern, if the pattern is fingerprint-periodicity conflict free, then when the streaming algorithm discards VOs in the text due to fingerprint-periodicity conflict, the discarded VOs do not need to be reported as potential occurrences.*

Acknowledgments

We thank Noga Alon, Roberto Grossi, Gadi Landau, Benny Pinkas and Ely Porat for useful discussions and comments.

References

1. Ergun, F., Jowhari, H., Salgan, M.: Periodicity in Streams (2010) (manuscript)
2. Fine, N.J., Wilf, H.S.: Uniqueness theorems for periodic functions. Proc. Amer. Math. Soc. 16, 109–114 (1965)
3. Galil, Z.: String Matching in Real Time. J. Assoc. Comput. Mach. 28(1), 134–149 (1981)
4. Galil, Z.: Optimal parallel algorithms for string matching. Inform. and Control 67, 144–157 (1985)
5. Galil, Z., Seiferas, J.: Time-space-optimal string matching. J. Comput. System Sci. 26, 280–294 (1983)
6. Karp, R.M., Rabin, M.O.: Efficient randomized pattern matching algorithms. IBM J. Res. Develop. 31(2), 249–260 (1987)
7. Knuth, D.E., Morris, J.H., Pratt, V.R.: Fast pattern matching in strings. SIAM J. Comput. 6, 322–350 (1977)
8. Porat, B., Porat, E.: Exact And Approxiamate Pattern Matching In The Streaming Model. In: Proc. 50th IEEE Symp. on Foundations of Computer Science, pp. 315–323 (2009)

Simple Real-Time Constant-Space String Matching*

Dany Breslauer[1], Roberto Grossi[2], and Filippo Mignosi[3]

[1] Caesarea Rothchild Institute, University of Haifa, Haifa, Israel
[2] Dipartimento di Informatica, Università di Pisa, Pisa, Italy
[3] Dipartimento di Informatica, Università dell'Aquila, L'Aquila, Italy

Abstract. We use a simple observation about the locations of *critical factorizations* to derive a real-time variation of the Crochemore-Perrin constant-space string matching algorithm. The real-time variation has a simple and efficient control structure.

1 Introduction

Numerous string matching algorithms have been published. The classical algorithm by Knuth, Morris and Pratt [24] takes $O(n + m)$ time and uses $O(m)$ auxiliary space, where n and m are the lengths of the text and the pattern, respectively. The algorithm is also *on-line* in the sense that it reports if an occurrence of the pattern ends at any given text location before moving on to examine the next text symbol. In comparison, the "naive" string matching algorithm works on-line and uses only constant auxiliary space, but takes up to $O(nm)$ time.

In some application, it is not sufficient that the output is produced on-line with an *average* amortized constant number of steps spent at each text location, but it is required that the output is produced in *real-time*, that is, worst case constant time spent at each text location (thus giving $O(m + n)$ time). The Knuth-Morris-Pratt algorithm has certain run-time "hiccups" that prevent real-time execution, spending up to $O(\log m)$ time at some text locations. Karp and Rabin's [23] linear-time randomized string matching algorithm requires only constant auxiliary space, but it comes in various randomized flavors and its real-time version may report false occurrences of the pattern with low probability.

In the early 1980s, two intriguing open question about the feasibility of string matching algorithms under certain conditions were settled when (*a*) Galil [15] derived a *real-time* variation of the Knuth-Morris-Pratt [24] algorithm and described a *predictability* condition allowing to transform a compliant on-line algorithms to real-time, and (*b*) Galil and Seiferas [18] discovered a linear-time

* Work partially supported by the European Research Council (ERC) project SFEROT, by the Israeli Science Foundation grant 347/09 and by Italian project PRIN AlgoDEEP (2008TFBWL4) of MIUR. Part of the work of the first author originated while visiting at the University of Palermo and of the second author originated while visiting the Caesarea Rothschild Institute.

R. Giancarlo and G. Manzini (Eds.): CPM 2011, LNCS 6661, pp. 173–183, 2011.
© Springer-Verlag Berlin Heidelberg 2011

string matching algorithms that requires only *constant auxiliary space* in addition to the read-only input storage; in earlier attempts Galil and Seiferas [16] first reduced the space requirements in variations of the Knuth-Morris-Pratt algorithm to $O(\log m)$ space and then further to constant space, but temporarily "borrowing" and later "restoring" parts of the writable input storage [17]. Galil and Seiferas [18] point out that using their constant-space string matching algorithm, their earlier work [15,29] on real-time Turing machine algorithms for string matching, for recognition of squares and palindromes, and for a number of generalization of these problems can be adapted to use constant-space and even to two-way multi-head finite automata in real-time. Jiang and Li [22] proved that one-way multi-head finite automata cannot solve the string matching problem.

Several other linear-time constant-space string matching algorithms were published [2,6,7,8,9,10,11,20,21,28] later using various combinatorial properties of words. The simplest and most elegant of these, perhaps, is the algorithm by Crochemore and Perrin [7] that relies on the *Critical Factorization Theorem* [5,25,26]. More recently, Gasieniec and Kolpakov [19] studied the auxiliary space utilization in real-time string matching algorithms and derived a real-time variation of the constant-space algorithm by Gasieniec, Plandowski and Rytter [20], that is based on the partial representation of the "next function," using only $O(m^\epsilon)$ auxiliary space, for any fixed constant $\epsilon > 0$.

In this paper we revisit the Crochemore-Perrin [7] *two-way* constant-space string matching algorithm and its use of critical factorizations. We observe that if instead of verifying the pattern suffix in the algorithm's *forward scan* and only then verifying the remaining pattern prefix in the algorithm's *back fill*, we embark on the back fill simultaneously and at the same rate as the forward scan, then for most patterns the algorithm completes the back fill by the time the forward scan is done. We then prove that by deploying a *second instance* of the algorithm to match some pattern substring and by carefully choosing which critical factorizations are used in both instances, the two instances together can verify complementary parts of the pattern, and therefore, we can match the whole pattern in real-time. Thus, we derive a *real-time* variation of the Crochemore-Perrin algorithm circumventing the authors' conclusion that the very nature of their two-way algorithm would not allow it to operate in real-time. The new real-time algorithm has a very simple and efficient control structure, maintaining only three synchronized text pointers that move one way in tandem and induce homologous positions on the pattern, making the variation a good candidate for efficient hardware implementation in deep packet inspection network intrusion detection systems (e.g. [1]) and in other applications. Our observations about the choices of critical factorizations might be of independent interest. Breslauer, Jiang and Jiang [4] also use critical factorizations in the first half of the rotation of a periodic string to obtain better approximation algorithms for the *shortest superstring* problem. It is perhaps worthwhile to emphasis that this paper is concerned with the *auxiliary space* utilization where the input pattern and text are accessible in read-only memory, which is very different from the *streaming model* where space is too scarce to even store the whole inputs. In the streaming model

Breslauer and Galil [3] recently described a randomized real-time string matching algorithm using overall $O(\log m)$ space, improving on previous streaming algorithms [14,27] that take $O(n \log m)$ time.

We start by reviewing critical factorizations and give the basic real-time variation of the Crochemore-Perrin algorithm in Section 2. We then show in Section 3 how to use two instances of the basic algorithm to match any pattern and describe the pattern preprocessing in Section 4. We conclude with some remarks and open questions in Section 5.

2 Basic Real-Time Algorithm

We need the following definitions. A string u is *a period* of a string x if x is a prefix of u^k for some integer k, or equivalently if x is a prefix of ux. The shortest period of x is called *the period* of x and its length is denoted by $\pi(x)$. A *substring* or a *factor* of a string x is a contiguous block of symbols u, such that $x = x'ux''$ for two strings x' and x''. A *factorization* of x is a way to break x into a number of factors. We consider factorizations (u, v) of a string $x = uv$ into two factors: a *prefix u* and a *suffix v*. Such a factorization can be represented by a single integer and is *non-trivial* if neither of the two factors is equal to the empty string.

Given a factorization (u, v), a *local period* of the factorization is defined as a non-empty string z that is consistent with both sides u and v. Namely, (i) z is a suffix of u or u is a suffix of z, and (ii) z is a prefix of v or v is a prefix of z. The shortest local period of a factorization is called *the local period* and its length is denoted by $\mu(u, v)$. A non-trivial factorization (u, v) of a string $x = uv$ is called a *critical factorization* if the local period of the factorization is of the same length as the period of x, i.e. $\mu(u, v) = \pi(uv)$. See Figure 1.

After the preprocessing to find a critical factorization (u, v) of the pattern x, the Crochemore-Perrin algorithm [7] first verifies the suffix v in the what we call the *forward scan* and then verifies the prefix u in what we call the *back fill*. The key to the algorithm is that the algorithm always advances in the forward scan while verifying the suffix v and never needs to back up. The celebrated *Critical Factorization Theorem* is the basis for the Crochemore-Perrin two-way constant-space string matching algorithm.

Theorem 1. *(Critical Factorization Theorem, Cesari and Vincent [5,25]) Given any $|\pi(x)| - 1$ consecutive non-trivial factorizations of a string x, at least one is a critical factorization.*

```
a | b a a a b a        a b | a a a b a        a b a | a a b a
b a   b a               a a a b  a a a b       a       a
    (a)                        (b)                    (c)
```

Fig. 1. The local periods at the first three non-trivial factorizations of string abaaaba. In some cases the local period overflows on either side; this happens when the local period is longer than either of the two factors. The factorization (b) is a critical factorization with (local) period aaab.

We introduce the basic algorithm and assume throughout this section that all the critical factorizations and period lengths that are used were produced in the pattern preprocessing step. The basic idea is to find candidate occurrences of the pattern in real-time by repeatedly interleaving the comparisons in the (Crochemore-Perrin) back fill simultaneously with the comparisons in the forward scan. Note that the basic algorithm might interrupt some of these back fills before they come to completion (while the forward scans are always completed). As a result, the basic algorithm never misses an occurrence of the pattern, but might not fully verify that the produced candidates are real occurrences of the whole pattern, as stated below.

Lemma 1. *Given a pattern x with critical factorization (u, v), there exists a real-time constant-space algorithm that finds candidate occurrences of $x = uv$ such that:*

1. *The actual occurrences of the pattern x are never missed.*
2. *Candidate occurrences end with an occurrence of the pattern suffix v.*
3. *Candidate occurrences contain a specified substring of the pattern prefix u of length up to $|v|$.*

We now prove Lemma 1. Let $x = uv$ be the pattern with period of length $\pi(x)$ and critical factorization (u, v), such that $|u| < \pi(x)$. Such critical factorization always exists by Theorem 1. The basic algorithm aligns the pattern starting at the left end of the text and tries to verify the pattern suffix v (forward scan) and *simultaneously* also some other specified part of the skipped prefix u (back fill), advancing one location in each step.

Suppose that the algorithm has successfully verified some prefix z of v and some part of u, but failed to verify the next symbols, either in the suffix v or in the substring of u. The key observation in the Crochemore-Perrin algorithm is that the pattern may be shifted ahead by $|z| + 1$ text location, thus always moving forward in the text past the prefix of v that has been compared so far. To see this we must convince ourselves that any shorter shift can be ruled out as an occurrence of the pattern. We need the following lemma in our proof.

Lemma 2. *(Crochemore and Perrin [7]) Let (u, v) be a critical factorization of the pattern $x = uv$ and let z be any local period at this factorization, such that $|z| \leq \max(|u|, |v|)$. Then $|z|$ is a multiple of $\pi(x)$, the period length of the pattern.*

Since (u, v) is a critical factorization, by Lemma 2, any shift by $|z|$ or fewer symbols cannot align the pattern with occurrence in the text unless the shift is by a multiple of the period length $\pi(x)$. If the comparison that failed was verifying the suffix v, then shifts by the period length $\pi(x)$ can be ruled are since any pattern symbol $\pi(x)$ locations apart are identical. If the comparison that failed was verifying the prefix u, recalling that $|u| < \pi(x)$, there can be no such multiples of $\pi(x)$. In either case, the pattern may be shifted ahead with respect to the text by $|z| + 1$ locations without missing any occurrences. See Figure 2.

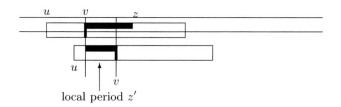

local period z'

Fig. 2. If z is a prefix of v verified in the forward scan, and z' is a prefix of z and a suffix of u, then the critical factorization (u, v) should have a local period z' shorter than the period $\pi(x)$, which gives a contradiction

If the algorithm has finished verifying the whole suffix v, then it has also verified a substring of u of up to the same length, and therefore, if $|u| \leq |v|$ and the algorithm has started at the beginning of u, it has matched the whole pattern $x = uv$. However, if $|u| > |v|$, then the algorithm has only verified the some substring of u of length at most $|v|$, but has not verified the remainder of the pattern (unfortunately, some patterns, e.g. $aa \cdots aaab$, only have critical factorizations close to their right end).

By Lemma 2, subsequent overlapping occurrences of the pattern must be at locations that are multiple of $\pi(x)$ locations apart, allowing the algorithm to shift the pattern ahead by $\pi(x)$ locations using the pattern periodicity to avoid going back. Observe that since $|u| < \pi(x)$, there is no need to go back to verify the substring of the prefix u after shifting the pattern ahead by $\pi(x)$ locations. To simplify the exposition, we give up on this optimization that was discussed by Crochemore and Perrin [7] and further optimized in [2].

We have therefore demonstrated that the algorithm has not missed occurrences of the pattern, and has verified that the candidate occurrences it produced end with v and contained an occurrence of a specified substring of u of length up to $|v|$. Also, after the last symbol of a pattern occurrence is read, it takes just $O(1)$ time to report an occurrence candidate, so the basic algorithm is real-time; the fact it uses $O(1)$ additional space, derives from [7]. The reason we specified a substring of u rather then a prefix of u will become apparent later, where two instances of the algorithm will be used to cover complementary parts of u and one of the instances does not even require the substring of u. This completes the proof of Lemma 1.

3   Real-Time Variation

Consider the basic algorithm from Lemma 1 that finds candidate occurrences of the pattern $x = uv$ with critical factorization (u, v). As mentioned in the proof of Lemma 1, when $|u| \leq |v|$, the basic algorithm is able to verify all the actual occurrences of x. What if $|u| > |v|$? We have to refine the basic algorithm and consider the pattern as $x = uvw$: the main idea to obtain the real-time constant-space variation of the Crochemore-Perrin algorithm is to use *two aligned instances* (see case (2) of Lemma 3 below). When taken together, the complementary parts in x that are verified by the two instances *cover* the whole pattern occurrences, thus identifying the actual occurrences of x. The variation relies on the following observation about the locations of critical factorizations.

Lemma 3. *Given any string* $x = uvw$, *there either exists:*

1. *a critical factorization* (uv, w) *such that* $|uv| \leq |w|$, *or*
2. *a critical factorization* (uv, w) *and another critical factorization* (u, vv') *of a prefix* uvv' *of* x, *such that* $|u| \leq |vv'|$.

Proof. Let (uv, w) be the leftmost critical factorization of $x = uvw$. If $|uv| \leq |w|$, then we have proved case (1). Otherwise, consider all factorizations (u, vw) to the left of (uv, w) that further satisfy the condition that $|u| \leq \mu(u, vw)$. If also $|vw| \leq \mu(u, vw)$, then the local period covers both extremes and (u, vw) is a critical factorization, contradicting that (uv, w) was the leftmost critical factorization. Therefore, $\mu(u, vw) < |vw|$. By Theorem 1, let (u, v) be a critical factorization of uv, such that $|u| \leq \pi(uv) = \mu(u, v)$, and let vv' be the prefix of vw whose length is $\max(|v|, \mu(u, vw)) < |vw|$ (in fact, vv' can be chosen such that uvv' is the longest prefix with period length $\mu(u, vw)$). But then (u, vv') is a critical factorization of uvv' and $|u| \leq \mu(u, v) \leq \mu(u, vv') = \mu(u, vw) \leq |vv'|$. □

Note that case (1) of Lemma 3 holds for most patterns strings: as previously mentioned, one instance of the basic real-time algorithm in Lemma 1 solves the string matching problem. The second instance of the basic algorithm is only required in case (2). There are many different ways to divide up the work between the two instances. We discuss a simple one.

Theorem 2. *There exists a real-time constant-space string matching algorithm that identifies all the occurrences of a pattern* x, *and uses only two instances of the basic algorithm mentioned in Lemma 1.*

To prove Theorem 2, let the first instance use the critical factorization (uv, w) of the whole pattern $x = uvw$ (see Lemma 3) and verify only the suffix w in the forward scan (no back fill). Let the second instance match occurrences of the pattern prefix uvv' (case (2) of Lemma 3) using the critical factorization (u, vv'). The correctness of the algorithm follows from the observation that aligned occurrences of w (partially) overlapping with occurrences of uvv' ending $|w| - |v'|$ text locations earlier, identify occurrences of the whole pattern $x = uvw$. The complexity is twice that of the basic algorithm, and the resulting algorithm is real-time since we repeatedly interleave $O(1)$ steps of the two instances. The variation makes fewer than $3n - m$ symbol comparisons.

We give the pseudocode of the real-time constant-space algorithm in Figure 3, where we focus on the case of the two factorizations mentioned in case (2) of Lemma 3 (since one factorization is straigthforward). We denote the pattern by $x \equiv x[0..m-1]$ and its two factorizations by $(u, vv') \equiv (x[0..a-1], x[a, c-1])$ and

```
for (t = 0, p = t-m+c-a, q = t-b; t < n; t++) {
    s = t-m+c;
    is_pref = false;       /* prefix instance */
    if (T[s]==x[s-p] && T[s-a]==x[s-p-a]) {
        if (s==p+c-1) {
            is_pref = true;
            p += pi_pref;
        }
    } else
        p = s-a+1;
    is_pat = false;        /* pattern instance */
    if (T[t]==x[t-q]) {
        if (t==q+m-1) {
            is_pat = true;
            q += pi_pat;
        }
    } else
        q = t-b+1;
    if (is_pref && is_pat)
        report an occurrence ending at text position t;
}
```

Fig. 3. The real-time constant-space variation of the Crochemore-Perrin algorithm for a pattern x, where $u \equiv x[0..a-1]$, $v \equiv x[a..b-1]$, $v' \equiv x[b..c-1]$, and $w \equiv x[b..m-1]$

$(uv, w) \equiv (x[0..b-1], x[b, m-1])$, whose periods $\pi(uvv') \equiv \pi(x[0..c-1])$ and $\pi(x) \equiv \pi(x[0..m-1])$ are precomputed during the pattern preprocessing (see Section 4). We denote these periods by pi_pref and pi_pat respectively. (Note that the two factorizations are displaced each other by $|v| = b - a$ positions.)

We denote the text by $T[0..n-1]$ and the location of the current text symbol by t, where $s = t - m + c$ is the end of the text as perceived by the independent prefix instance matching uvv'. The algorithm has a very simple control structure using only three aligned text position $s - a < s < t$ that advance in tandem.

Consider the prefix instance $(u, vv') \equiv (x[0..a-1], x[a, c-1])$: the text position for the current candidate occurrence of vv' is denoted by p (and so the candidate position for u is $p - a$). We apply the forward scan to $vv' \equiv x[a, c-1]$ and the back fill to $u \equiv x[0..a-1]$ in the first half of the for loop, so we try to match one symbol from vv' and one from u.

Consider the pattern instance $(uv, w) \equiv (x[0..b-1], x[b, m-1])$: the text position for the current candidate occurrence of $w \equiv x[b, m-1]$ is denoted by q. We just apply the forward scan to $w = x[b, m-1]$, since $uv \equiv x[0..b-1]$ is covered by the prefix instance. We just need to check for the occurrences of w, and we try to match one symbol of it in the second half of the for loop.

As it can be verified in Figure 3, we report an occurrence of pattern x ending at position t after $O(1)$ time and we use just $O(1)$ additional memory words.

4   Pattern Preprocessing

We now describe the pattern preprocessing, where the task is that of computing at most two critical factorizations and the corresponding periods. Namely, given a pattern x, the pattern preprocessing finds at most two factorizations (uv, w) and (u, vv') (if needed) as stated in Lemma 3, where $x = uvw$, and the corresponding period lengths $\pi(x)$ and $\pi(uvv')$.

We build upon the Crochemore-Perrin preprocessing algorithms [7] for the above task. Consider the computation of the periods. Crochemore and Perrin gave a pattern preprocessing algorithm that computes the length of the period $\pi(x)$, when the pattern x is periodic ($\pi(x) \leq |x|/2$), and gave a variation of their algorithm that works when the period is longer than half the pattern length. Other authors have shown how to compute the period length exactly when the period is longer than half of the pattern length [2,8,10]. Our real-time variation can use either approach and either compute the period lengths precisely, or use half the pattern length as an approximation for the period length (i.e. if the period is longer than $|x|/2$, then shift the pattern x by $|x|/2$ positions instead of shifting pattern by the unknown long period length).

Consider now the computation of critical factorizations. Crochemore and Perrin gave a novel constructive proof of the Critical Factorization Theorem (Theorem 1) using properties of lexicographically maximal suffixes and Lyndon words, and a variation of an algorithm by Duval [13] to find the lexicographically maximal suffix of a string. Their proof shows that given any arbitrary order $\leq$ on the input alphabet and its reverse order $\leq^R$, the shorter between the lexicographically maximal suffix of the pattern x by $\leq$ and the lexicographically maximal suffix of x by $\leq^R$, provides a critical factorization in the first $\pi(x) - 1$ positions. The readers are referred to the original paper [7] for the elegant proof.

We will only use the fact that critical factorizations of growing prefixes of the pattern x, within their first period, can be efficiently computed on-line. Before going on, we need a few more definitions and properties of critical factorizations. A factorization (u, v) of a string $x = uv$ is *left external* if $|u| \leq \mu(u, v)$ and similarly, *right external* if $|v| \leq \mu(u, v)$. A factorization is called *external* if it is both left and right external, and *internal* if it is neither left or right external. Any external factorization has a local period length that is equal to the global period length, i.e. $\mu(u, v) = \pi(uv)$, and is therefore critical. Any internal factorization has a *square* centered at the factorization.

We define the set of all the left external non-trivial factorizations $\mathcal{L}(x) = \{(u, v) \| x = uv$ and $1 \leq |u| \leq \mu(u, v)$ and $1 \leq |v|\}$. This set is not empty since it always contains the first factorization (a, v), where a is an alphabet symbol, and it also contains at least one critical factorization by Theorem 1. We now give two properties to characterize $\mathcal{L}(x)$, the first of which can also be partially found in [25, Ch.8].

Lemma 4. *Let (u_1, v_1) and (u_2, v_2) be two factorizations in $\mathcal{L}(x)$, such that $|u_1| \leq |u_2|$. Then the local period lengths satisfy $\mu(u_1, v_1) \leq \mu(u_2, v_2)$.*

Proof. Let x' be a prefix of $x = u_1v_1 = u_2v_2$ of length $|x'| = \min(|x|, |u_1| + \mu(u_1, v_1), |u_2| + \mu(u_2, v_2))$. Then, x' has periods of lengths $\mu(u_1, v_1)$ and $\mu(u_2, v_2)$. Assuming by contradiction that $\mu(u_1, v_1) > \mu(u_2, v_2)$, then the shorter prefix of length $|u_1| + \mu(u_2, v_2)$ has a period of length $\mu(u_2, v_2)$ and therefore (u_1, v_1) has a local period of length at most $\mu(u_2, v_2) < \mu(u_1, v_1)$, establishing that $\mu(u_1, v_1) \leq \mu(u_2, v_2)$. □

Lemma 5. *Let (u, v) be a factorization in $\mathcal{L}(x)$. Then, there exist a prefix x' of x and a left external critical factorization (u_0, v_0) of x', such that $x' = u_0v_0$ and $\mu(u_0, v_0) = \mu(u, v)$.*

Proof. If (u, v) is a critical factorization of $x = uv$ then simply take $(u_0, v_0) = (u, v)$. Otherwise, (u, v) cannot be right external. Let x' be the prefix of x of length $|u| + \mu(u, v) < |x|$. Then x' has period length $\pi(x') = \mu(u, v)$ and left external critical factorization (u_0, v_0), such that $\mu(u_0, v_0) = \pi(u_0v_0) = \mu(u, v)$. □

Given any on-line algorithm that computes left external critical factorizations of growing prefixes of the pattern x (e.g. the Crochemore-Perrin pattern preprocessing algorithm), consider the factorizations induced in x by the critical factorizations of the pattern prefixes. By Lemma 4, the local period lengths of such factorizations are non-decreasing for increasing lengths prefixes, and by Lemma 5, all local period lengths in $\mathcal{L}(x)$ are represented. The constuctive algorithm for finding the factorization in Lemma 3 boils down to the prefix critical factorizations that are representative of the largest two local period lengths in $\mathcal{L}(x)$. We can now describe how to obtain the pattern preprocessing.

Theorem 3. *There exist an on-line constant-space pattern preprocessing algorithm that produces the critical factorizations required in Lemma 3 and the corresponding periods.*

Proof. We already discussed how to find the periods at the beginning of this section. As for the desired critical factorizations, take any on-line constant-space algorithm that computes the left external critical factorizations of growing prefixes of the pattern x and record the last two different critical factorizations (u_0, v_0) and (u_1, v_1) of the pattern prefixes $x_0 = u_0v_0$ and $x_1 = u_1v_1$, such that $|x_0| < |x_1|$ and $|u_0| < |u_1|$ and stop when x_1 is the first critical factorization in the second half of the pattern x, such that $|u_1| \geq |x|/2$, or when the end of the pattern is reached. Since any left external critical factorization in the second half of x is also right external in x and therefore a critical factorization, the last prefix critical factorization x_1 that was found is critical for the whole pattern x. If either of these last two prefix critical factorizations is a critical factorization of the whole pattern x in the first half of x, then we found one critical factorization satisfying case (1) in Lemma 3. Otherwise, (u_0, v_0) induces a non-critical factorization (u_0, v_0v') of u_0v_0v', where u_0v_0v' is the longest prefix of x having period $\mu(u_0, v_0)$ and v_0 is a prefix of v_0v'. Notice that since above prefix critical factorization (u_0, v_0) is left external then $|u_0| \leq |v_0| \leq |v_0v'|$. The second factorization (u_1, v_1) induces a critical factorization in the whole pattern $x = u_1v_1$ since $|u_1| \geq |x|/2$. Letting $u = u_0$, $u_1 = u_0v$ and $x = uvw$, we get the desired three way factorization satisfying case (2) of Lemma 3. □

While the Crochemore-Perrin preprocessing algorithm is works in constant space and in linear time, it does not poroduce the period lengths and critical factorizations in real time. However, the $O(m)$ time preprocessing results of Theorem 3 are produced in time so that the ensuing text scanning works in real-time.

5   Concluding Remarks

We have shown that the Crochemore-Perrin constant-space algorithm [7] can be directly transformed into a real-time string matching algorithm by interleaving its back fill with its forward scan. Other on-line constant-space string matching algorithms can be transformed into real-time algorithms as well, provided that they report the locations of pattern occurrences in increasing order of their locations and no later than cm locations after the end the occurrence, for some fixed constant c and any length m pattern. The general transformation searches for occurrences of a pattern prefix and uses such occurrences as an anchor while catching up with the remaining pattern suffix, utilizing periodicity properties to count highly repetitive pattern prefixes.

The pattern preprocessing of the Crochemore-Perrin algorithm [2,7] computes periods and critical factorizations using symbol comparisons that test for the relative order of symbols under an arbitrary alphabet order. It is an interesting open question to find a critical factorization in linear-time using only equality comparisons without an alphabet order, even if more space is allowed. Duval et al. [12] find all local periods of a string and therefore, also all critical factorizations, in linear time over integer alphabets, but require super-linear time if only symbol comparisons are used.

Acknowledgments

We are grateful to Danny Hermelin, Gadi Landau and Oren Weimann for several discussions.

References

1. Baker, Z.K., Prasanna, V.K.: Time and area efficient pattern matching on FPGAs. In: Tessier, R., Schmit, H. (eds.) FPGA, pp. 223–232. ACM, New York (2004)
2. Breslauer, D.: Saving Comparisons in the Crochemore-Perrin String Matching Algorithm. Theoret. Comput. Sci. 158(1), 177–192 (1996)
3. Breslauer, D., Galil, Z.: Real-Time Streaming String-Matching (2011), this conference proceedings
4. Breslauer, D., Jiang, T., Jiang, Z.: Rotations of periodic strings and short superstrings. J. Algorithms 24(2), 340–353 (1997)
5. Césari, Y., Vincent, M.: Une caractérisation des mots périodiques. C.R. Acad. Sci. Paris 286(A), 1175–1177 (1978)
6. Crochemore, M.: String-matching on ordered alphabets. Theoret. Comput. Sci. 92, 33–47 (1992)

7. Crochemore, M., Perrin, D.: Two-way string-matching. J. Assoc. Comput. Mach. 38(3), 651–675 (1991)
8. Crochemore, M., Rytter, W.: Periodic Prefixes in Texts. In: Capocelli, R., Santis, A.D., Vaccaro, U. (eds.) Proc. of Sequences II: Methods in Communication, Security and Computer Science, pp. 153–165. Springer, Heidelberg (1993)
9. Crochemore, M., Rytter, W.: Text Algorithms. Oxford University Press, Oxford (1994)
10. Crochemore, M., Rytter, W.: Sqares, Cubes, and Time-Space Efficient String Searching. Algorithmica 13(5), 405–425 (1995)
11. Crochemore, M., Rytter, W.: Jewels of stringology. World Scientific, Singapore (2002)
12. Duval, J.P., Kolpakov, R., Kucherov, G., Lecroq, T., Lefebvre, A.: Linear-time computation of local periods. Theor. Comput. Sci. 326(1-3), 229–240 (2004)
13. Duval, J.: Factorizing Words over an Ordered Alphabet. J. Algorithms 4, 363–381 (1983)
14. Ergun, F., Jowhari, H., Salgan, M.: Periodicity in Streams (2010) (manuscript)
15. Galil, Z.: String Matching in Real Time. J. Assoc. Comput. Mach. 28(1), 134–149 (1981)
16. Galil, Z., Seiferas, J.: Saving space in fast string-matching. SIAM J. Comput. 2, 417–438 (1980)
17. Galil, Z., Seiferas, J.: Linear-time string-matching using only a fixed number of local storage locations. Theoret. Comput. Sci. 13, 331–336 (1981)
18. Galil, Z., Seiferas, J.: Time-space-optimal string matching. J. Comput. System Sci. 26, 280–294 (1983)
19. Gasieniec, L., Kolpakov, R.M.: Real-time string matching in sublinear space. In: Sahinalp, S.C., Muthukrishnan, S.M., Dogrusoz, U. (eds.) CPM 2004. LNCS, vol. 3109, pp. 117–129. Springer, Heidelberg (2004)
20. Gasieniec, L., Plandowski, W., Rytter, W.: Constant-space string matching with smaller number of comparisons: Sequential sampling. In: Galil, Z., Ukkonen, E. (eds.) CPM 1995. LNCS, vol. 937, pp. 78–89. Springer, Heidelberg (1995)
21. Gasieniec, L., Plandowski, W., Rytter, W.: The zooming method: A recursive approach to time-space efficient string-matching. Theor. Comput. Sci. 147(1&2), 19–30 (1995)
22. Jiang, T., Li, M.: K one-way heads cannot do string-matching. J. Comput. Syst. Sci. 53(3), 513–524 (1996)
23. Karp, R., Rabin, M.: Efficient randomized pattern matching algorithms. IBM J. Res. Develop. 31(2), 249–260 (1987)
24. Knuth, D., Morris, J., Pratt, V.: Fast pattern matching in strings. SIAM J. Comput. 6, 322–350 (1977)
25. Lothaire, M.: Combinatorics on Words. Addison-Wesley, Reading (1983)
26. Lothaire, M.: Algebraic Combinatorics on Words. Cambridge University Press, Cambridge (2002)
27. Porat, B., Porat, E.: Exact And Approxiamate Pattern Matching In The Streaming Model. In: Proc. 50th IEEE Symp. on Foundations of Computer Science, pp. 315–323 (2009)
28. Rytter, W.: On maximal suffices and constant-space linear-time versions of kmp algorithm. In: Rajsbaum, S. (ed.) LATIN 2002. LNCS, vol. 2286, pp. 196–208. Springer, Heidelberg (2002)
29. Seiferas, J.I., Galil, Z.: Real-time recognition of substring repetition and reversal. Mathematical Systems Theory 11, 111–146 (1977)

Space Lower Bounds for Online Pattern Matching

Raphaël Clifford[1], Markus Jalsenius[1], Ely Porat[2], and Benjamin Sach[1]

[1] University of Bristol, Dept. of Computer Science, Bristol, UK
[2] Bar-Ilan University, Dept. of Computer Science, Ramat-Gan, Israel

Abstract. We present space lower bounds for online pattern matching under a number of different distance measures. Given a pattern of length m and a text that arrives one character at a time, the online pattern matching problem is to report the distance between the pattern and a sliding window of the text as soon as the new character arrives. We require that the correct answer is given at each position with constant probability. We give $\Omega(m)$ bit space lower bounds for L_1, L_2, L_∞, Hamming, edit and swap distances as well as for any algorithm that computes the cross-correlation/convolution. We then show a dichotomy between distance functions that have wildcard-like properties and those that do not. In the former case which includes, as an example, pattern matching with character classes, we give $\Omega(m)$ bit space lower bounds. For other distance functions, we show that there exist space bounds of $\Omega(\log m)$ and $O(\log^2 m)$ bits. Finally we discuss space lower bounds for non-binary inputs and show how in some cases they can be improved.

1 Introduction

We combine existing results with new observations to present an overview of space lower bounds for online pattern matching. Given a pattern that is provided in advance and a text that arrives one character at a time, the online pattern matching problem is to report the distance between the pattern and a sliding window of the text as soon as the new character arrives. In this formulation, the pattern is processed before the first text character arrives and once processed, the pattern is no longer available to the algorithm unless a copy is explicitly made.

This problem has recently gained a great deal of interest with breakthrough results given for exact matching and pattern matching under bounded Hamming distance (k-mismatch) [13]. For both problems it was shown that space sublinear in the size of the pattern is sufficient to give the correct answer at every alignment with high probability. These remarkable results immediately raise a number of significant unresolved questions. The first is for which other distance measures between strings might sublinear space randomised online algorithms be achievable and it is this question which we address here.

Our presentation is divided between what we term local and non-local online pattern matching problems. In the former case the distance function between

R. Giancarlo and G. Manzini (Eds.): CPM 2011, LNCS 6661, pp. 184–196, 2011.

a pattern P of length m and an m-length substring of the text T, starting at position i, is defined by

$$\text{LOCALPM}_{(\oplus,\Delta)}(P,T) = \bigoplus_{j=0}^{m-1} \Delta(P[j], T[i+j]),$$

where $\oplus$ and Δ are both binary operators. In Section 4 we show $\Omega(m)$ bit space lower bounds for online pattern matching for the local problems of L_1, L_2, and Hamming distance as well as for any algorithm that computes the cross-correlation/convolution.

We then go on to show in Section 5 a space dichotomy for local online pattern matching problems of the form $d(i) = \bigwedge_{j=0}^{m-1} \Delta(P[j], T[i+j])$ where the range of Δ is $\{\text{TRUE}, \text{FALSE}\}$. Where the distance function Δ has wildcard-like properties (qv. Section 5), we give an $\Omega(m)$ space lower bound. Where it does not, we have $\Omega(\log m)$ and $O(\log^2 m)$ space bounds. This implies, for example, that online pattern matching with character classes [8] requires linear space.

In Section 6 we go on to consider all eight possible binary Boolean associative operators and give a complete classification in terms of their known upper and lower space bounds. One consequence is that determining if there is an exact "non-match", where the Hamming distance is the same as the pattern length, requires linear space in our online model. This bound also holds if, for example, only the parity of the Hamming distance is required. In Section 7 we then show how our techniques can be used to give linear space lower bounds for L_∞ online pattern matching. In Section 8 we discuss a possible approach to space lower bounds for inputs with large alphabets, focussing on the Hamming distance problem. Finally, in Section 9 we explore non-local problems and show $\Omega(m)$ bit space lower bounds for both online edit and swap distance.

2   Preliminaries and Related Work

Let Σ_P and Σ_T denote the pattern and text alphabet, respectively. We say that $\text{LOCALPM}_{(\oplus,\Delta)}$ is *text independent* with respect to the pattern P if the value of $\text{LOCALPM}_{(\oplus,\Delta)}$ is a constant independent of T. We say that $\text{LOCALPM}_{(\oplus,\Delta)}$ is *pattern independent* with respect to a pattern P if there is a function Δ' such that $\Delta(x, y) = \Delta'(y)$ for all $(x, y) \in P \times \Sigma_T$.

Example 1. Let $\Sigma_P = \{x, y, z\}$, $\Sigma_T = \{a, b, c\}$, $\oplus$ be the Boolean AND-operator and Δ be defined according to the table in Fig. 1, where 1 is TRUE and 0 is FALSE. We can see that $\text{LOCALPM}_{(\wedge,\Delta)}$ is text independent with respect to the pattern $P = xxyyxzxx$ as it always outputs 0. It is also pattern independent with respect to $P = yyzyyzzy$ as $\Delta(y, \alpha) = \Delta(z, \alpha)$ for all $\alpha \in \Sigma_T$. In fact, for this particular definition of Δ, $\text{LOCALPM}_{(\wedge,\Delta)}$ is either text or pattern independent with respect to any pattern P.

Suppose that $\text{LOCALPM}_{(\oplus,\Delta)}$ is text independent with respect to a pattern P. Then any algorithm for $\text{LOCALPM}_{(\oplus,\Delta)}$ on P requires at most $O(1)$ space after

$$\begin{array}{c|ccc} \Delta & a & b & c \\ \hline x & 0 & 0 & 0 \\ y & 0 & 1 & 1 \\ z & 0 & 1 & 1 \end{array}$$

Fig. 1. An example of Δ such that $\text{LOCALPM}_{(\wedge,\Delta)}$ is invalid (either text or pattern independent with respect to any pattern P)

preprocessing P. If $\text{LOCALPM}_{(\oplus,\Delta)}$ is pattern independent with respect to P then $\text{LOCALPM}_{(\oplus,\Delta)}$ does not depend on the pattern and is outside the scope of this paper.

We say that $\text{LOCALPM}_{(\oplus,\Delta)}$ is *invalid* if, for every pattern P, it is either text or pattern independent with respect to P. $\text{LOCALPM}_{(\oplus,\Delta)}$ is *valid* if it is not invalid. The problem $\text{LOCALPM}_{(\wedge,\Delta)}$ in the previous example is therefore invalid. We will only consider from this point pattern matching problems $\text{LOCALPM}_{(\oplus,\Delta)}$ which are valid, and ignore patterns for which $\text{LOCALPM}_{(\oplus,\Delta)}$ is pattern or text independent.

Our focus is on online pattern matching algorithms which output correct answers with constant probability. We are not aware of previous work that considers randomised lower bounds for this specific type of problem. There is however now a considerable literature on communication complexity and on streaming algorithms for single input streams, including those that process a sliding window of the input (see e.g. [4]). This previous streaming work has typically focussed on deterministic or randomised bounds for finding approximate rather than exact solutions. Quantum lower and classical upper bounds for the communication complexity of Hamming distance in more general models than we consider were given previously [5]. A linear lower bound for the randomised communication complexity of the inner product of two binary vectors is given in [3]. The dichotomy presented in Section 5 and in particular the concept of a matching relation that includes wildcard matching, although in a different setting and with different terminology, is similar to a time complexity dichotomy given previously by Muthukrishnan and Ramesh [9]. On the topic of swap matching in Section 9, we note that in [1], the existence of a reduction for time rather than space, from Boolean convolutions to string matching with swaps is claimed without proof.

3   Communication Complexity Problems

Our results are based on reductions from various one-way randomised communication complexity problems with known lower bounds. We list the relevant problems below. In a one-way randomised communication model, only Alice can send messages to Bob and Bob must output the correct answer with probability at least 2/3. Note that the value 2/3 is inconsequential: any probability strictly greater than 1/2 can be amplified to a constant arbitrary close to 1. We assume private randomness.

Definition 1. *The* EQUALITY *problem in one-way communication complexity is defined as follows. Alice has a string $X \in \{0,1\}^m$ and Bob has a string $Y \in \{0,1\}^m$. Bob must determine whether $X = Y$. The communication complexity is $\Theta(\log m)$ bits [14].*

Definition 2. *The* INDEXING *problem in one-way communication complexity is defined as follows. Alice has a string $X \in \{0,1\}^m$ and Bob has an index $n \in \{0, \ldots m-1\}$. Bob must find $X[n]$. The problem is known to have an $\Omega(m)$ bit lower bound (see [6] for an elementary proof).*

4 Addition

In this section we consider the problem LOCALPM$_{(+,\Delta)}$, where $+$ is standard addition and the range of Δ is a subset of the integers. That is, the distance function is

$$d(i) = \sum_{j=0}^{m-1} \Delta(P[j], T[i+j]) \,.$$

Theorem 1. LOCALPM$_{(+,\Delta)}$ *requires $\Omega(m)$ bits of space.*

Proof. Since LOCALPM$_{(+,\Delta)}$ is not text independent, there must exist characters $x \in \Sigma_{\mathrm{P}}$ and $a, b \in \Sigma_{\mathrm{T}}$ such that $\Delta(x, a) \neq \Delta(x, b)$. We reduce from INDEXING: Alice has a string $T = \{a, b\}^m$ and Bob has an index n. Alice initialises a pattern matching algorithm A on the pattern $P = \{x\}^m$ and feeds in her string T. Then she sends the internal state of A to Bob, who feeds in n copies of the symbol a. Let d be the output after those as. Bob then feeds in another a. Let d' be the output. If $d = d'$ then $A[n] = a$. If $d \neq d'$ then $A[n] = b$. If the probability of error per output is bounded by a constant $c < 1/4$, then the union bound for error on two outputs is $2c$, giving the INDEXING problem an error probability of at most $2c < 1/2$. $\square$

Corollary 1. *Computing the L_1, L_2 and Hamming distances, as well as the cross-correlation/convolution, require $\Omega(m)$ bits of space.*

5 Conjunction

In this section we consider LOCALPM$_{(\wedge,\Delta)}$, where $\wedge$ is the Boolean AND-operator and the range of Δ is $\{0,1\}$ (where 0 denotes FALSE and 1 denotes TRUE). There are several natural pattern matching problems that fall under this category, for example, exact matching, matching with wildcards and exact matching with character classes.

The function Δ can be represented with a 0/1-matrix M_Δ, where the rows and columns correspond to the symbols in Σ_{P} and Σ_{T}, respectively. Thus, the entry $(i,j) = \Delta(i,j)$. The 2×2 matrix in Fig. 2 will play an important role, and we call it the *wildcard matrix*.

$$\begin{bmatrix} 1 & 1 \\ 1 & 0 \end{bmatrix} \qquad \begin{bmatrix} 0 & 0 \\ 0 & 1 \end{bmatrix}$$

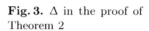

Δ	a	b
$\star$	1	1
x	1	0

Fig. 2. The wildcard matrix (left) and negated wildcard matrix (right)

Fig. 3. Δ in the proof of Theorem 2

M_Δ	a	b	c	d	e	f
v	0	1	0	1	1	0
w	0	0	0	0	0	0
x	1	0	1	0	0	0
y	0	1	0	1	1	0
z	1	0	1	0	0	0

M_Δ'	a	b	c	d	e	f
v	0	1	–	–	–	–
w	–	–	–	–	–	–
x	1	0	–	–	–	–
y	–	–	–	–	–	–
z	–	–	–	–	–	–

Id.	a	b
x	1	0
v	0	1

Fig. 4. An illustration of Lemma 1

We say that M_Δ contains the wildcard matrix if it is a submatrix of M_Δ under some permutation of the rows and columns.

We demonstrate the following dichotomy for $\text{LocalPM}_{(\wedge,\Delta)}$. If M_Δ contains the wildcard matrix, then $\text{LocalPM}_{(\wedge,\Delta)}$ is solvable in $\tilde{\Theta}(m)$ bits of space, otherwise it is solvable in $\tilde{\Theta}(1)$ bits of space. The first class is equivalent to pattern matching with wildcards, and the second class is equivalent to exact matching. Note that both dichotomies are decidable due to the simple characteristic of the function Δ.

Theorem 2. *If M_Δ contains the wildcard matrix, then $\text{LocalPM}_{(\wedge,\Delta)}$ requires $\Omega(m)$ bits of space.*

Proof. Suppose that $\star, x \in \Sigma_\text{P}$ ($\star$ represents a wildcard symbol) and $a, b \in \Sigma_\text{T}$ such that Δ is specified according to Fig. 3. We reduce from the INDEXING problem, in which Alice has an m-length bit string $X \in \{\star, x\}^m$ and Bob has an index $n \in \{0, \ldots m - 1\}$. Let the pattern P be the string X. Let A be any algorithm that solves $\text{LocalPM}_{(\wedge,\Delta)}$ on the pattern P. Alice sends the internal state of A to Bob, who feeds the algorithm with the m-length string that has the symbol a at every position except for at position n where the symbol is b. The output is TRUE iff $X[n] = \star$. $\square$

The following lemma will be useful for the next two theorems (see Fig. 4).

Lemma 1. *Let M_Δ' be the matrix obtained from M_Δ by first removing copies of identical rows and columns, keeping only rows and columns that are distinct in M_Δ, and then removing any row or column that contains only zeros. If M_Δ does not contain the wildcard matrix, then M_Δ' is the identity matrix, under some permutation of rows and columns.*

Proof. Suppose that M_Δ does not contain the wildcard matrix. Let M_Δ' be obtained from M_Δ according to the statement of the lemma. We will show that every column and every row of M_Δ' contains exactly one 1.

$$\begin{array}{c|cc}
\Delta & a & b \\
\hline
x & 1 & 0 \\
y & 0 & 1
\end{array}$$

Fig. 5. Δ in the proof of Theorem 3

First we show that every row of M'_Δ must contain at least one 1. Suppose that some row r of M'_Δ contains only 0s. Since zero-rows of M_Δ were removed and one copy of each column remains after the removal process, it is not possible that all columns in which row r is 1 were removed. We now show that M'_Δ cannot contain a row r with two or more 1s. Without loss of generality, assume that there is a 1 in columns i and j of row r. Since M_Δ does not contain a wildcard matrix, the elements of columns i and j must both be either 0 or 1 in every row. Thus, columns i and j are identical, and one of them must have been removed, contradicting the fact that there are two 1s in row r of M'_Δ. In order to show that every column of M'_Δ contains exactly one 1, we use the exact same argument as for the rows. Thus, M'_Δ is the identity matrix, under some permutation of rows and columns. (See Fig. 4 for an illustration of the lemma). □

Theorem 3. *If M_Δ does not contain the wildcard matrix, then* LOCALPM$_{(\wedge,\Delta)}$ *requires $\Omega(\log m)$ bits of space.*

Proof. We reduce from the EQUALITY problem, where Alice has a string $X \in \{0,1\}^m$ and Bob has a bit string $Y \in \{0,1\}^m$. Since M_Δ doesn't contain the wildcard matrix and as we only consider problems LOCALPM$_{(\wedge,\Delta)}$ that are valid, it follows from Lemma 1 that there must exist $x, y \in \Sigma_P$ and $a, b \in \Sigma_T$ such that Δ is according to Fig. 5. Let P be the m-length pattern obtained from X by replacing every 0 with x and every 1 with y. The m-length text T is obtained similarly from Y by replacing every 0 with a and every 1 with b. For any algorithm A that solves LOCALPM$_{(\wedge,\Delta)}$ on the pattern P, Alice sends the internal state of A on pattern P to Bob, who feeds A with T. The output is TRUE iff $X = Y$. □

Theorem 4. *If M_Δ does not contain the wildcard matrix, then* LOCALPM$_{(\wedge,\Delta)}$ *can be solved in $O(\log^2 m)$ bits of space.*

Proof. We will describe an algorithm for solving LOCALPM$_{(\wedge,\Delta)}$ which uses the exact matching algorithm by Porat and Porat [13], which runs in space $O(\log m)$ words, which is $O(\log^2 m)$ bits of space (under the word-RAM model). In order to use the exact matching algorithm (as a "black box") we must ensure that we do not feed it with distinct symbols that are identical under Δ. In other words, we can think of Δ specifying character classes, and for each class we want to use one representative symbol. We formalise this below.

We make the very reasonable assumption that the alphabets Σ_P and Σ_T are both enumerable and that we can iterate through every symbol of Σ_P and Σ_T, respectively, in no more than $O(\log m)$ bits of space. Let the order by which we iterate through the alphabets describe an ordering of the symbols in Σ_P and Σ_T. We say that the symbol $x \in \Sigma_P$ is *smaller* than $y \in \Sigma_P$ if x appears before y when iterating through Σ_P. We use the same notation for the symbols of Σ_T. We say that two symbols $x, y \in \Sigma_P$ are *equivalent* if $\Delta(x, a) = \Delta(y, a)$ for all $a \in \Sigma_T$. Similarly, $a, b \in \Sigma_T$ are equivalent if $\Delta(x, a) = \Delta(x, b)$ for all $x \in \Sigma_P$. We define the *smallest equivalent* symbol of $x \in \Sigma_P$ to be the symbol $y \in \Sigma_P$ such that y is equivalent to x and no other symbol equivalent to x is smaller than y. The notion of smallest equivalent symbol is defined similarly on Σ_T.

Let $\Sigma_P' \subseteq \Sigma_P$ be the set of all symbols $x \in \Sigma_P$ such that the smallest equivalent symbol of x is x itself. We do not include any symbol x in Σ_P' such that $\Delta(x, a) = 0$ for all $a \in \Sigma_T$. Similarly, let $\Sigma_T' \subseteq \Sigma_T$ be the set of all symbols $a \in \Sigma_T$ such that the smallest equivalent symbol of a is a itself. We do not include any symbol a in Σ_T' such that $\Delta(x, a) = 0$ for all $x \in \Sigma_P$. By Lemma 1 we have that Δ on Σ_P' and Σ_T' is represented by an identity matrix under some permutation of the rows and columns. In the example of Fig. 4, $\Sigma_P' = \{x, v\}$ and $\Sigma_T' = \{a, b\}$. We will ensure that we use the exact matching algorithm of [13] only on Σ_P' and Σ_T' (i.e., normal exact pattern matching).

Given a symbol $x \in \Sigma_P$, we can find its smallest equivalent symbol by iterating through every symbol $y \in \Sigma_P$ and for each y, we iterate through all $a \in \Sigma_T$ to check whether $\Delta(x, a) = \Delta(y, a)$. Similarly we can find the smallest equivalent symbol of any symbol in Σ_T.

Let P be the pattern. We may assume that P does not contain a symbol x for which $\Delta(x, a) = 0$ for all $a \in \Sigma_T$. If it does, the output is always 0. Before we preprocess the pattern, we replace every symbol with its smallest equivalent symbol. Then we preprocess the pattern using the fingerprint technique described in [13]. Now we run the exact matching algorithm with the following additional step. When a new symbol a arrives, we replace it with its smallest equivalent symbol. The only caveat we must take care of is the situation when $\Delta(x, a) = 0$ for all $x \in \Sigma_P$. We can detect this case by iterating through the symbols of Σ_P. As long as a is present in the last m characters of the stream, the output is zero. We use a flag to keep track of this.   □

We now show how these results can be applied to a specific pattern matching problem that has not been considered in the online setting before. The pattern matching with character classes problem allows a set of characters to be defined for each position in the pattern [8]. A character in the text matches a set at a pattern position if it is contained within it. This is a generalisation of exact matching where each set would contain only one character. Using Theorems 2, 3 and 4 we can determine precisely when this problem can and cannot be solved online in sublinear space.

Corollary 2. *Online pattern matching with character classes requires $\Omega(m)$ bits of space in the worst case. However, where the character classes define a matching relation Δ which does not contain the wildcard matrix (see the example in Fig. 4), $O(\log^2 m)$ bits suffice.*

6 Other Boolean Operators

In the previous section we demonstrated a dichotomy for LOCALPM$_{(\oplus,\Delta)}$, where $\oplus$ is the AND-operator. Here we will complete the classification of Boolean operators. There are eight associative Boolean operators $a \oplus b$:

1. TRUE   **2.** FALSE   **3.** a **4.** b **5.** $a \wedge b$ **6.** $a \vee b$ **7.** $a = b$ **8.** $a \neq b$

The operators TRUE and FALSE are trivial; the output is either always TRUE or FALSE. The operator $a \oplus b = b$ is also easy; the output is always $\Delta(P[m-1], t)$, where t is the last received symbol of the text stream.

The operator $a \oplus b = a$ is on the other hand more demanding. Here the output is $\Delta(P[0], t)$, where t is the mth last symbol received from the text stream. The pattern matching algorithm must therefore remember m received characters of the stream. More precisely, we see that $\Omega(m)$ bits of space is necessary by reducing from the INDEXING problem: Alice first feeds her array (text) into the pattern matching algorithm, for which $P[0]$ is a character that can distinguish between the characters of Alice's array. She then sends the internal state to Bob, who feeds in n symbols in order to determine the value at index n of Alice's array.

The OR-operator $\vee$ is equivalent to $\wedge$ under De Morgan's laws: negate the outputs from Δ and negate the output from the pattern matching algorithm. Thus, the dichotomy for $\wedge$ applies to $\vee$ as well, only that we characterise the classes with the wildcard matrix in which each element has been negated. This is called the negated wildcard matrix (see Fig. 2).

We now show that the equality operator "=" requires $\Omega(m)$ bits of space. First note that the output from the pattern matching algorithm is 0 if and only if $\Delta([P[j], T[i+j]]) = 0$ for an odd number of positions j. For example, if M_Δ is the identity matrix, LOCALPM$_{(=,\Delta)}$ gives us the parity of the Hamming distance.

Since LOCALPM$_{(=,\Delta)}$ is valid, there are $x \in \Sigma_{\mathrm{P}}$, $a, b \in \Sigma_{\mathrm{T}}$ such that $\Delta(x,a) = 0$ and $\Delta(x,b) = 1$. We reduce from the INDEXING problem, where Alice has a string in $\{a,b\}^m$ and Bob has an index n. Alice initialises a pattern matching algorithm on the pattern $P = \{x\}^m$ and feeds it with her string. She sends the internal state to Bob, who feeds the algorithm with n copies of the symbol a. The first position of P is now aligned with the nth character of Alice's string. Suppose the output from the algorithm is d. Bob now feeds in another a. Let d' be the new output. If $d = d'$ then the character at position n of Alice's string must have been a. If $d \neq d'$ then the character must have been b.

The operator "$\neq$" is similar to "=" and also requires $\Omega(m)$ bits of space. To see this, note that the output from the pattern matching algorithm is 0 if and

$$
\begin{array}{c|cc}
\Delta & 2 & 3 \\
\hline
1 & 1 & 2 \\
0 & 2 & 3
\end{array}
\qquad
\begin{array}{c|cc}
\Delta' & 2 & 3 \\
\hline
1 & 1 & 1 \\
0 & 1 & 0
\end{array}
$$

Fig. 6. Δ and Δ' in the proof of Theorem 5

only if $\Delta([P[j], T[i+j]) = 1$ for an even number of positions j. We may therefore prove the lower bound using a reduction from the INDEXING problem similar to above.

7   The L_∞ Distance

In this section we consider the L_∞ distance problem which can be defined as LOCALPM$_{(\max,\Delta)}$, where $\Delta(x, y) = |x - y|$ and $\max(a, b)$ is the maximum of a and b. In this section we assume that the pattern and text are integer valued. Here the distance function is the maximum $\Delta(P[j], T[i + j])$ over all j, that is

$$d(i) = \max_{j \in \{0,\dots,m-1\}} \Delta(P[j], T[i + j]) .$$

Theorem 5. *The L_∞ distance problem requires $\Omega(m)$ bits of space.*

Proof. Let $\Sigma_P = \{0, 1\}$ and $\Sigma_T = \{2, 3\}$. Therefore Δ is specified according to Fig. 6. Let $\Delta'(x, y) = 1$ if $\Delta(x, y) < 3$, otherwise $\Delta'(x, y) = 0$. Therefore $M_{\Delta'}$ contains the wildcard matrix and hence by Theorem 2, LOCALPM$_{(\wedge,\Delta')}$ requires $\Omega(m)$ space.

Let $d'(i)$ be the distance under LOCALPM$_{(\wedge,\Delta')}$. If $d'(i) = 1$ then for all j, $\Delta'(P[j], T[i + j]) = 1$, implying that $\Delta(P[j], T[i + j]) < 3$ for all j. Hence $d(i) < 3$. If $d'(i) = 0$ then there exists a j such that $\Delta'(P[j], T[i + j]) = 0$, implying that $\Delta(P[j], T[i+j]) = 3$ and hence $d(i) = 3$. Therefore, if we can solve LOCALPM$_{(\max,\Delta)}$, we can solve LOCALPM$_{(\wedge,\Delta')}$. □

8   Non-binary Alphabets

The space lower bounds we have given so far have been either $\Omega(\log m)$ or $\Omega(m)$ bits. When the pattern or text alphabet is drawn from a large universe, the question arises as to whether even more space is required to perform online pattern matching. We show by way of another different reduction a method that may be applicable to a wider range of pattern matching problems than we consider here. Our approach is to show a reduction from the communication complexity problem DISJOINTNESS [7] to the Hamming distance problem. In DISJOINTNESS Alice and Bob both have sets of m elements each chosen from a universe of size U and Bob wants to determine if their intersection is empty. The lower bound for the space complexity of the Hamming distance problem will then be determined by lower bounds for the one-way randomised communication complexity of the DISJOINTNESS problem with private coins. A result regarded as folklore shows that this complexity is $\Omega(m \log m + \log \log U)$ when U is $\Omega(m^{1+\varepsilon})$ [11,12]. This in turn implies a superlinear lower bound for the space complexity of the online Hamming distance problem with large alphabets.

For an integer n, we write $[n]$ to denote the set $\{0, \ldots, n-1\}$. Alice has a set $A \subseteq [U]$ and Bob has a set $B \subseteq [U]$, and $|A| = |B| = m$. The reduction performs the following steps. We assume for the moment that Alice and Bob both have a shared source of randomness and show later how this assumption can be removed.

1. Alice creates a pairwise independent hash function $h : [U] \to [cm]$, for some constant integer $c > 1$ and creates a pattern P of length cm where each element is initialised to be some unique symbol $\$ \notin [U]$. She then sets $P[h(x)] = x$ for all $x \in A$ by going through A in some arbitrary order. If a position of P is written to multiple times, only the last write is stored.
2. Alice starts the Hamming distance algorithm up until the point at which it has processed the pattern P but none of the text (which is created later) and sends the internal state of the algorithm to Bob.
3. Bob performs the same hashing operation using the same hash function but this time on set B, creating a text T of length cm. Bob uses a different unique symbol $\$' \notin [U]$ for the initialisation of the text.
4. Bob feeds the Hamming distance algorithm with the whole text T. Bob concludes that A and B are disjoint iff the output is cm.

Theorem 6. *Any randomised algorithm for Hamming distance where the symbols are chosen from a universe of size $\Omega(m^{1+\varepsilon})$ uses $\Omega(m \log m + \log \log U)$ bits of space.*

Proof. Considering the reduction above, if A and B are disjoint, then a deterministic Hamming distance algorithm will always output cm. If A and B are not disjoint then a necessary condition for a deterministic Hamming distance algorithm to output cm is if at least two elements are hashed to the same location by either Alice or Bob. We can see that the probability of incorrectly outputting cm is maximised when A and B share exactly one element. Therefore, suppose that $A \cap B = \{x\}$. The element x is hashed to position $h(x)$. By the union bound and the pairwise independence of the hash function, the probability that some other element in either A or B is mapped to $h(x)$ is at most $1/(cm) \cdot m \cdot 2 = 2/c$. If we assume our randomised Hamming distance algorithm is correct with probability at least $2/3$, then the overall process falsely reports disjointness with probability at most $2/c + 1/3$ (union bound). The space complexity of Hamming distance is therefore lower bounded by the communication complexity of the disjointness problem if Alice and Bob have a shared source of random bits to select their common hash function. By Newman's Theorem [10] the cost of transforming the protocol to work with only private coins is at most an additive $O(\log \log U)$ factor in the asymptotic complexity. Assuming that U grows polynomially in m and so $\log \log U$ is $O(\log m)$, the overall lower bound for the space complexity of the Hamming distance problem is therefore $\Omega(m \log m - \log m) = \Omega(m \log m)$. To finish the proof for larger U, we observe first that a lower bound for smaller universes must still hold for larger ones. The final additive $\Omega(\log \log U)$ term is derived by simply setting $m = 1$ and follows directly from the randomised lower bound for EQUALITY. Therefore the overall lower bound is $\Omega(m \log m + \log \log U)$ as required.   □

9   Non-local Pattern Matching

So far we have focused only on local pattern matching where each position in the alignment contributes to the distance independently of the other positions. Here we take a brief look at space lower bounds for two non-local distance measures: edit distance and swap matching.

In online pattern matching, we define the *edit distance* as the minimum number of single character edit operations (insert, delete and replace) required to transform P into the last m characters of the streamed text. This implies that the number of insertions and deletions are equal.

We show that for binary $\Sigma_P = \Sigma_T = \{0, 1\}$, the online edit distance problem requires $\Omega(m)$ bits of space. For non-binary inputs there is a reduction from the Hamming distance problem [2]. The reduction we give covers the binary alphabet case as well and follows directly from INDEXING, where Alice has a string $P \in \{0, 1\}^m$ and Bob has an index n. Alice initialises a pattern matching algorithm on the pattern P and sends the internal state to Bob, who first feeds in m zeros. Let d be the output and note that d is the number of ones in P. Bob then feeds in the m-length string that consists of zeros at every position except for at position n where it is one. Let d' be the output. Bob can now decide the value of $P[n]$ by comparing d with d': $P[n] = 1$ if $d' < d$, and $P[n] = 0$ if $d' \geqslant d$. The probability of error is therefore upper bounded by the union bound on d and d' being wrong.

Given a string S, a *swap* at position i means that the characters $S[i]$ and $S[i + 1]$ swap positions. We say there is a *swap match* if and only if the pattern P can be transformed into the last m characters of the streamed text through a set of swaps. Each $S[i]$ is swapped at most once.

We show that the online swap distance problem requires $\Omega(m)$ space. Our proof is based on the techniques we have presented in this paper. Specifically, we demonstrate a reduction from LOCALPM$_{(\wedge, \Delta)}$ where M_Δ contains the wildcard matrix, hence the space lower bound is $\Omega(m)$. Suppose we have Δ as in Fig. 7. Let $P \in \{\star, x\}^m$ and $\Sigma_T = \{a, b\}$. From P we obtain $P' \in \{0, 1\}^{5m}$ such that every $\star$ in P is replaced with 00100 and every x is replaced with 00010. When we receive characters from the text, we replace a with 00010 and b with 01000. It follows, under the transformation of the symbols, that there is a swap match if and only if LOCALPM$_{(\wedge, \Delta)}$ outputs TRUE for the original (non-transformed) strings. To see this, note that both a and b, under the transformation, swap match $\star$, but b does not swap match x (see Fig. 7). The transformation of the symbols does not allow swaps between adjacent characters; every possible swap will take place "within" the binary encoding of a symbol. Thus, a swap match directly corresponds to a match under LOCALPM$_{(\wedge, \Delta)}$.

$$\begin{array}{c|cc} \Delta & a & b \\ \hline \star & 1 & 1 \\ x & 1 & 0 \end{array}$$

$a:$ 00**0**10 $b:$ 0**1**000 $b:$ 0**1**000

$\star:$ 00**1**00 $\star:$ 00**1**00 $x:$ 000**1**0

Fig. 7. Δ and alignments under swaps

10   Open Problems

We have considered space lower bounds and discussed how they can be derived from known communication complexity lower bounds. Upper bounds can also be directly derived from existing online pattern matching algorithms. For all the problems we have discussed there is at most a log factor gap between these upper and lower bounds. However, where the known lower bound is sublinear, as is the case for exact matching for example, this gap may still be considered significant. Further, for bounded Hamming distance where the distance is only to be given if it is at most some constant k, the best known randomised online space upper bound is $O(k^3 \text{polylog } m)$ [13]). The best known lower bound, on the other hand, is very different at $\Omega(k)$ [5]. Further, it is known that the lower bounds can not be increased to match the known upper bounds using the one-way communication complexity of the functions between two strings of the same length. Either more space efficient algorithms exist for these problems or novel techniques will be needed to improve the lower bounds.

References

1. Amir, A., Aumann, Y., Landau, G., Lewenstein, M., Lewenstein, N.: Pattern Matching with Swaps. Journal of Algorithms 37, 247–266 (2000)
2. Bar-Yossef, Z., Jayram, T.S., Krauthgamer, R., Kumar, R.: Approximating edit distance efficiently. In: FOCS 2004: Proc. 45th Annual Symp. Foundations of Computer Science, pp. 550–559 (2004)
3. Chor, B., Goldreich, O.: Unbiased bits from sources of weak randomness and probabilistic communication complexity. SIAM Journal on Computing 17(2), 230–261 (1988)
4. Datar, M., Gionis, A., Inkyk, P., Motwani, R.: Maintaining stream statistics over sliding windows. SIAM Journal on computing 31(6), 1794–1813 (2002)
5. Huang, W., Shi, Y., Zhang, S., Zhu, Y.: The communication complexity of the Hamming distance problem. Information Processing Letters 99(4), 149–153 (2006)
6. Jayram, T.S., Kumar, R., Sivakumar, D.: The one-way communication complexity of hamming distance. Theory of Computing 4(1), 129–135 (2008)
7. Kushilevitz, E., Nisan, N.: Communication complexity. Cambridge University Press, Cambridge (1997)
8. Linhart, C., Shamir, R.: Faster pattern matching with character classes using prime number encoding. Journal of Computer and System Sciences 75(3), 155–162 (2009)
9. Muthukrishnan, S., Ramesh, H.: String matching under a general matching relation. Inf. Comput. 122(1), 140–148 (1995)
10. Newman, I.: Private vs. common random bits in communication complexity. Information Processing Letters 39(2), 67–71 (1991)

11. Nisan, N.: Personal communication (2011)
12. Pătraşcu, M.: CC4: One-Way Communication and a Puzzle, 2009 (accessed January 20, 2011), `http://infoweekly.blogspot.com/2009/04/cc4-one-way-communication-and-puzzle.html`
13. Porat, B., Porat, E.: Exact and approximate pattern matching in the streaming model. In: FOCS 2009: Proc. 50th Annual Symp. Foundations of Computer Science, pp. 315–323 (2009)
14. Yao, A.C.-C.: Some complexity questions related to distributive computing. In: STOC 1979: Proc. 11th Annual ACM Symp. Theory of Computing, pp. 209–213 (1979)

Counting Colours in Compressed Strings

Travis Gagie[1] and Juha Kärkkäinen[2]

[1] Aalto University, Finland
travis.gagie@aalto.fi
[2] University of Helsinki, Finland
juha.karkkainen@cs.helsinki.fi

Abstract. Motivated by the problem of counting unique visitors to a website, we consider how to preprocess a string $s[1..n]$ such that later, given a substring's endpoints, we can quickly count how many distinct characters that substring contains. The smallest reasonably fast previous data structure for this problem takes $n \log \sigma + \mathcal{O}(n \log \log n)$ bits and answers queries in $\mathcal{O}(\log n)$ time. We give a data structure for this problem that takes $n H_0(s) + \mathcal{O}(n) + o(n H_0(s))$ bits, where $H_0(s)$ is the 0th-order empirical entropy of s, and answers queries in $\mathcal{O}(\log \ell)$ time, where ℓ is the length of the query substring. As far as we know, this is the first data structure, where the query time depends only on ℓ and not on n. We also show how our data structure can be made partially dynamic.

1 Introduction

Imagine you are in charge of web analytics at a company and your boss has asked you to write an interface to the log files that, given an arbitrary time period, returns the number of unique visitors in that period. This task can be broken into two subtasks: find the first and last entries in the logfiles from that time period, and then count the unique visitors between those entries. The first subtask is an instance of the well-studied predecessor problem (see, e.g., [13]), so you concentrate on the second subtask. Counting the unique visitors in a consecutive subset of entries is a special case of coloured range counting, an important problem with applications in, e.g., computational geometry, database research and bioinformatics.

For the general coloured range counting, we are asked to store a set of n coloured points in $\mathbb{R}^d$ such that later, given an axis-aligned box, we can quickly count the distinct colours it contains. Most papers on this problem have focused on $d \geq 2$ dimensions (see, e.g., [6]), whereas counting unique visitors is a one-dimensional problem. The best solution known for general static one-dimensional coloured range counting is an $\mathcal{O}(n)$-word data structure by Bozanis, Kitsios, Makris and Tsakalidis [1] from 1995 that answers queries in $\mathcal{O}(\log n)$ time. The best dynamic solutions known [7] take, for queries and updates, either $\mathcal{O}(n \log n)$ words and $\mathcal{O}(\log n)$ time or $\mathcal{O}(n)$ words and $\mathcal{O}(\log^2 n)$ time. Recently, however, Gagie, Navarro and Puglisi [3] considered the special case in which the coloured points are the integers $1, \ldots, n$. Storing these points is equivalent to storing a

R. Giancarlo and G. Manzini (Eds.): CPM 2011, LNCS 6661, pp. 197–207, 2011.
© Springer-Verlag Berlin Heidelberg 2011

string $s[1..n]$ over an alphabet whose size σ is the number of distinct colours, such that later, given a substring's endpoints, we can quickly count how many distinct characters that substring contains.

Gagie et al. gave a static data structure that takes $n \log \sigma + \mathcal{O}(n \log \log n)$ bits and answers queries in $\mathcal{O}(\log n)$ time. (In this paper log means $\log_2$.) As far as we know, Gagie et al.'s is the smallest and fastest previously known data structure that can be used for counting unique visitors, in which case s is the sequence of visitors recorded in the log files in chronological order, n is the number of visits recorded, and σ is the number of unique visitors recorded. Nevertheless, since n and σ are likely to be very large for a busy website, it is not clear their solution is small enough. In this paper we describe a new data structure for counting colours in strings, one that takes only $n H_0(s) + \mathcal{O}(n) + o(n H_0(s))$ bits, where $H_0(s)$ is the 0th-order empirical entropy of s. Furthermore, we simultaneouly reduce the query time to $\mathcal{O}(\log \ell)$ time, where ℓ is the size of the query range. As far as we know, no other data structure for coloured range counting has a non-trivial upper bound depending only on ℓ.

Gagie et al.'s solution is built on work by Muthukrishnan [12] about coloured range queries in strings. Muthukrishnan defined $C[1..n]$ to be the array in which each cell $C[q]$ stores the largest value $p < q$ such that $s[p] = s[q]$ (or 0 if no such p exists). He observed that $s[q]$ is the first occurrence of that distinct character in $s[i..j]$ if and only if $i \leq q \leq j$ and $C[q] < i$. Therefore, the number of distinct characters in $s[i..j]$ is the number of values in $C[i..j]$ strictly less than i. Notice that C could quite reasonably be much more compressible than s. For example, if most visitors are unique, then s becomes less compressible but C becomes more compressible, as it consists mostly of 0s. Gagie et al. pointed out that, if we store C in a wavelet tree [5], which takes $n \log n + o(n \log n)$ bits, then we can count all such values in $\mathcal{O}(\log n)$ time; for details, see [10]. This is already a slight improvement over the bounds we achieve with Bozanis et al.'s data structure [1], but Gagie et al. showed it can be reduced to $n \log \sigma + \mathcal{O}(n \log \log n)$ by modifying the wavelet tree. Our own work also uses the C array and wavelet trees, but we achieve compression by modifying the representation of C rather than the wavelet trees. Apart from counting the unique visitors, both Gagie et al.'s data structure and our own can support other interesting queries. For example, we can easily count the first-time visitors in an interval $s[i..j]$ by counting the number of 0s in $C[i..j]$. More generally, we can easily count the visitors in an interval $s[i..j]$ whose last visit was in another interval $s[i'..j']$, by counting the number of values in $C[i..j]$ that are between i' and j'. Finally, we can use three instances of our data structure so that we can count the visitors that visit exactly once in $s[i..j]$ (i.e., the number of unique visitors whose visits are unique, as well). To do this, we build the first instance normally, we build the second instance replacing even occurrences of each character by a special filler character #, and we build the third instance replacing odd occurrences of each character by this filler; e.g., if $s = $ abracadabra, then the three instances are for abracadabra, abr#cad###a and ###a###abr#, respectively. (Since the second and third instances are for complementary strings, we could merge them fairly easily; we consider them separately for the sake of simplicity.) Given $s[i..j]$, we use the first instance to find the total number d_{all} of distinct characters in $s[i..j]$, we use the second instance to find the number d_{odd} of distinct characters that have an odd occurrence in $s[i..j]$, and we use the third instance to find the number d_{even} of distinct characters that have an even occurrence in $s[i..j]$; the number of distinct characters that have both an odd and an even occurrence is $d_{odd} + d_{even} - d_{all}$, so the number of characters that have only an odd or only an even occurrence — i.e., exactly one occurrence — is $2d_{all} - d_{odd} - d_{even}$.

Muthukrishnan and Gagie et al. were motivated by problems in document retrieval. Given a collection of documents, they build the suffix array for the documents' concatenation and then an array indicating in which document each suffix begins. To compute the document frequency of a given pattern — i.e., how many documents contain it — they find the interval in the suffix array corresponding to that pattern, then count the distinct documents in the same interval in the document array. We note that Sadakane [14] gave a faster and more space-efficient data structure for computing the document frequencies of single patterns, but his solution cannot be used for coloured range counting in arbitrary strings. As Gagie et al. pointed out, their approach can be used in some cases where Sadakane's cannot; e.g., when we want to compute the total document frequency of patterns in a lexicographic range. Our own data structures are functionally equivalent to Gagie et al.'s and, thus, can also be used in such cases.

In Section 2 we describe a simple data structure that achieves essentially the same bound as Gagie et al.'s. In Section 3 we extend the ideas from Section 2 to build a data structure that takes $nH_0(s) + \mathcal{O}(n) + o(nH_0(s))$ bits and answers queries in $\mathcal{O}(\alpha(n) \log n \log \log n)$ time, where α is the inverse Ackermann function. We adjust our data structure and analysis slightly in Section 4, so that our time bounds are in terms of ℓ, the length of the substring whose distinct colours we are counting, rather than in terms of n. In Section 5, we reorganize our data structure and improve query time to $\mathcal{O}(\log \ell)$. For this result, we need a couple of simple but non-standard tricks in implementing wavelet trees; previous sections use standard wavelet trees as black box. In Section 6 we show how our data structure can be made partially dynamic. Specifically, we first show how to achieve the same time bound for querying and a space bound of $\mathcal{O}(n(H_0(s) + 1))$ bits while supporting an $\mathcal{O}(\log n)$-time append operation, which is the most natural update when, e.g., maintaining log files. We then show how to support colour substitutions and deletions, at the cost of using $\mathcal{O}(\log^2 n)$ time for both queries and updates. For the current version of this paper, in both cases we ignore the resources needed to perform a constant number of rank/select queries on s.

2   Simple Blocking

In this section we give a simple proof that, using two normal wavelet trees and a straightforward encoding of C, we need store only $(1+o(1))(n \log \sigma + n \log \log n)$

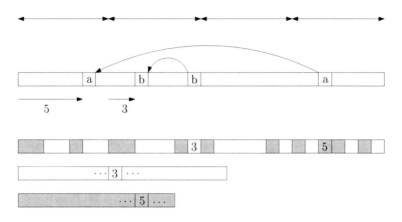

Fig. 1. The array s, with lines above indicating blocks and arcs indicating characters' previous occurrences; our representation of the C array overlaid on the bitvector, with white indicating intra-block pointers and grey indicating inter-block pointers; and the contents of the two wavelet trees — intra-block pointers in one and inter-block pointers in the other. Notice that, since both copies of b are contained within one block, the distance 3 is measured from the beginning of that block.

bits to answer queries in $\mathcal{O}(\log n)$ time. Without loss of generality, assume $\sigma = o(n/\log n)$; otherwise, we achieve our desired bound by simply storing C in a single, normal wavelet tree. Our idea is to break s into blocks of length $b = \sigma \log n$ and encode the entry $C[q]$ differently depending on whether the previous occurrence $s[p]$ of the character $s[q]$ is contained in the same block. If $s[p]$ is contained in the same block as $s[q]$, then we write $C[q]$ as the $\lceil \log b \rceil$-bit offset of p within the block; otherwise, we write it as the $\lceil \log n \rceil$-bit binary representation of p. Notice that, for each block, there are at most σ entries of C encoded as $\lceil \log n \rceil$-bit numbers.

We build a bitvector indicating how each entry of C is encoded, which takes $n + o(n)$ bits. We build one wavelet tree storing all the $\lceil \log b \rceil$-bit encodings, which takes at most $n \log b + o(n \log b) = (1 + o(1))(n \log \sigma + n \log \log n)$ bits, and another storing all the $\lceil \log n \rceil$-bit encodings, which takes at most $\sigma \lceil n/b \rceil \log n + o(\sigma \lceil n/b \rceil \log n) = n + o(n)$ bits. This is illustrated in Figure 1. Notice that, if $s[q]$ is the first occurrence of that distinct character in $s[i..j]$ and $C[q]$ is encoded in $\lceil \log b \rceil$ bits, then $s[q]$ must be between $s[i]$ and the end of the block containing $s[i]$. This is because, if $s[q]$ were in a later block, then $C[q] < i$ would be encoded using $\lceil \log n \rceil$ bits. We can count all such characters in $\mathcal{O}(\log b) = \mathcal{O}(\log \sigma + \log \log n)$ time using the bitvector and the first wavelet tree. We can count all the other first occurrences in $\mathcal{O}(\log n)$ time using the bitvector and the second wavelet tree.

Theorem 1. *Given a string $s[1..n]$, we can build a data structure that takes $(1 + o(1))(n \log \sigma + n \log \log n)$ bits such that later, given a substring's endpoints, in $\mathcal{O}(\log n)$ time we can count how many distinct characters it contains.*

Notice that, if $\sigma \geq \log n$, then Gagie et al.'s data structure is within a constant factor of being succinct and the data structure we just presented is within a factor of 2 of being succinct. If $\sigma < \log n$, then we can store s in a multiary wavelet tree [2], which takes $nH_0(s)+o(n)$ bits, and answer any query by enumerating the characters in the alphabet and, for each one, using two $\mathcal{O}(1)$-time rank queries to see whether it occurs in the given substring.

Corollary 1. *Given a string $s[1..n]$, we can build a data structure that takes $2n \log \sigma + o(n \log \sigma)$ bits such that later, given a substring's endpoints, in $\mathcal{O}(\log n)$ time we can count how many distinct characters it contains.*

3   Multi-size Blocking

In this section we extend our idea from the previous section so that, instead of encoding entries of C differently for only two block sizes — i.e., $\sigma \log n$ and n — we use many block sizes. In particular, we use $\mathcal{O}(\log \log n / \log(1 + \delta))$ different block sizes,

$$2^{1+\delta}, 2^{\max\left((1+\delta)^2, 2\right)}, 2^{\max\left((1+\delta)^3, 3\right)}, 2^{\max\left((1+\delta)^4, 4\right)}, \ldots, n ,$$

where $\delta \in (0,1]$ is a value we will specify later. Also, for each block size b, we consider s to consist of about $2n/b$ evenly overlapping blocks,

$$s[1..b], s[b/2..3b/2], s[b+1..2b], s[3b/2+1..5b/2], \ldots, s[n-b+1, n] .$$

If $C[q] = p$ and the smallest block containing both $s[p]$ and $s[q]$ has size b, then we write $C[q]$ as the $\lceil \log b \rceil$-bit offset of p within the lefthand block of size b containing $s[q]$ (there are at most two such blocks and, if there are two, then they overlap). Since

$$2^{\max\left((1+\delta)^{i-1}, i-1\right)-1} < q - p + 1 \leq b = 2^{\max\left((1+\delta)^i, i\right)} ,$$

we have $\lceil \log b \rceil < (1 + \delta) \log(q - p + 1) + 3$. In other words, if $s[p]$ and $s[q]$ are occurrences of a character a that does not occur in $s[p + 1..q - 1]$, then we use fewer than $(1+\delta) \log(q-p+1)+3$ bits to store $C[q]$. By Jensen's Inequality, since the logarithm is concave, the total number of bits we use to store the offsets for occurrences of a is maximized when those occurrences are evenly spaced and, thus, at most

$$(1 + \delta) \sum_a \mathrm{occ}(a, s) \log \left(\frac{n}{\mathrm{occ}(a, s)} + 1 \right) + 3n = (1 + \delta)nH_0(s) + \mathcal{O}(n) ,$$

where $\mathrm{occ}(a, s)$ is the number of occurrences of a in s.

Let t be a string indicating whether each entry of $C[q]$ is 0 and, if not, the block size used for it. We build a multiary wavelet tree [2] storing t. Notice we can always encode a block size $b = 2^{\max\left((1+\delta)^i, i\right)}$ in $\mathcal{O}(\log i) = \mathcal{O}(\log \log b)$ bits.

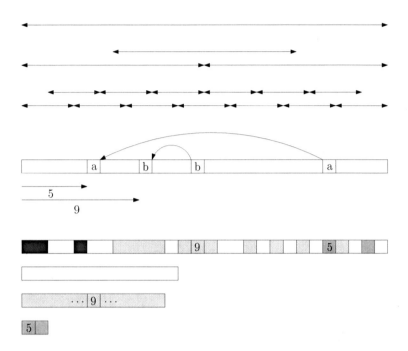

Fig. 2. The array s, with lines above indicating the overlapping block structure (with blocks of three different sizes, in this case) and arcs indicating characters' previous occurrences; our representation of the C array overlaid on the string t, with shades of grey indicating which encoding length is used for each pointer (black for 0s); and the contents of the three wavelet trees — pointers contained in short blocks, pointers contained in medium-length blocks, and pointers contained in long blocks. Notice that, although 9 is larger than 5, the pointer with value 9 has a shorter encoding because both copies of b are contained within the same medium-length block, while the two copies of a are not contained in any single block except the one long block, which contains the whole string.

By the calculations in the paragraph above and another application of Jensen's Inequality, $H_0(t) = \mathcal{O}(\log(H_0(s) + 1))$. It follows that, if $H_0(s)$ grows without bound as n goes to infinity, then the size of the tree is $o(nH_0(s))$ bits; otherwise, it is $\mathcal{O}(n)$ bits. Using the tree, in $\mathcal{O}(1)$ time we can count all the characters whose first appearance in s is in $s[i..j]$.

For each block size b, we build a wavelet tree storing all the $\lceil \log b \rceil$-bit encodings. By the same calculation as before, these wavelet trees take a total of $(1+\delta)nH_0(s)+\mathcal{O}(n)+o(nH_0(s))$ bits. This is illustrated in Figure 2. Notice that, for any block size b, if $s[q]$ is the first occurrence of that distinct character in $s[i..j]$ and $C[q]$ is encoded in $\lceil \log b \rceil$ bits, then $s[q]$ must be between $s[i]$ and the end of the righthand block of size b containing $s[i]$. Using the multiary wavelet tree and the wavelet tree for block size b, in $\mathcal{O}(\log b)$ time we can count all such characters in the right halves of both the lefthand and the righthand blocks of size b containing $s[i]$. Since these are the only blocks of size b containing $s[i]$ and the right half of the lefthand block is the left half of the righthand block, the sum is the total number of such characters. That is, in $\mathcal{O}(\log b)$ time, we can count all the first occurrences $s[q]$ of distinct characters in $s[i..j]$ such that $C[q]$ is encoded in $\lceil \log b \rceil$ bits. Repeating this for each of the $\mathcal{O}(\log \log n / \log(1 + \delta))$ block sizes, in $\mathcal{O}(\log n \log \log n / \log(1 + \delta)) = \mathcal{O}(\log n \log \log n / \delta)$ time we can count the distinct characters in $s[i..j]$. Choosing $\delta = 1/\alpha(n)$, for example, where α is the inverse Ackermann function, yields the a space bound of $(1 + 1/\alpha(n))nH_0(s) + \mathcal{O}(n) + o(nH_0(s)) = nH_0(s) + \mathcal{O}(n) + o(nH_0(s))$ and a time bound of $\mathcal{O}(\alpha(n) \log n \log \log n)$.

Theorem 2. *Given a string $s[1..n]$, we can build a data structure that takes $nH_0(s) + \mathcal{O}(n) + o(nH_0(s))$ bits such that later, given a substring's endpoints, in $\mathcal{O}(\alpha(n) \log n \log \log n)$ time we can count how many distinct characters it contains.*

4   Time Independent of n

Suppose we are to count the distinct colours in $s[i..j]$ using the data structure from Section 3, and let $\ell = j - i + 1$. Notice that, if $C[q] = p$ is encoded using a block size larger than the size $b_{\max}$ of the smallest block that completely contains $s[i..j]$, then $p < i$ or $q > j$. Therefore, any such entries in $C[i..j]$ indicates the first occurrence of some distinct character in $s[i..j]$. We can compute how many such entries there are by using the multiary wavelet tree to count all the entries in $C[i..j]$ that are encoded with block sizes at most $b_{\max}$, then subtracting from ℓ. Since there are $\mathcal{O}(\log \log b_{\max} / \log(1 + \delta)) = \mathcal{O}(\alpha(n) \log \log(\ell + 1))$ block sizes up to $b_{\max}$, this takes $\mathcal{O}(\alpha(n) \log \log(\ell + 1))$ time. Also notice that, using the technique described in the last paragraph of Section 3, we can count the entries $C[q] < i$ there are in $C[i..j]$ that are encoded using a block size at most $b_{\max}$. Therefore, we can replace the $\mathcal{O}(\alpha(n) \log n \log \log n)$ time bound in Theorem 2 by $\mathcal{O}(\alpha(n) \log \ell \log \log(\ell + 1))$. Since $\alpha(n)$ grows very, very slowly as n increases, our time bound is now almost independent of n.

To make our time bound completely independent of n, we adjust our block sizes: the first block size b_1 is 2; for $i \geq 2$, the kth block size is

$$b_k = 2^{\max\left(\prod_{h=1}^{k-1}(1+1/\alpha(b_h)), k\right)} .$$

If the smallest block containing both $s[p]$ and $s[q]$ has size b_k then, since

$$2^{\max\left(\prod_{h=1}^{k-2}(1+1/\alpha(b_h)), k-1\right)-1} < q - p + 1 \leq 2^{\max\left(\prod_{h=1}^{k-1}(1+1/\alpha(b_h)), k\right)} ,$$

we have $\log(q - p + 1) < \lceil \log b_k \rceil < (1 + 1/\alpha(b_{k-1})) \log(q - p + 1) + 3$. Also notice that, since b_{k-1} can be bounded from below in terms of b_k and b_k can be bounded from below in terms of $q - p$, $\alpha(b_{k-1})$ increases without bound (albeit very, very slowly) as $q - p$ goes to infinity. Therefore, we use fewer than $\log(q - p + 1) + o(\log(q - p + 1))$ bits to store $C[q]$. By calculations similar to those in Section 3, we still use $nH_0(s) + \mathcal{O}(n) + o(nH_0(s))$ bits in total. Now, however, since $\alpha(b_1) \leq \cdots \leq \alpha(b_k)$, more calculation shows that the number of block sizes up to b_k is $\mathcal{O}(\log \log b_k / \log(1 + 1/\alpha(b_k)))$, from which it follows that our new time bound is $\mathcal{O}(\alpha(\ell) \log \ell \log \log(\ell + 1))$.

Theorem 3. *Given a string $s[1..n]$, we can build a data structure that takes $nH_0(s)+\mathcal{O}(n)+o(nH_0(s))$ bits such that later, given a substring's endpoints i and j, in $\mathcal{O}(\alpha(\ell) \log \ell \log \log(\ell + 1))$ time we can count how many distinct characters it contains, where $\ell = j - i + 1$.*

5   Improving Time

We now modify the data structure so that instead of having one wavelet tree for each block size, we have a separate wavelet tree for each block. If $C[q] = p$ is encoded using a block size b then one or two blocks of size b contain both p and q, and we store the encoding in the wavelet tree of the leftmost block. Notice that q is always in the second half of the block. The total number of bits in the encodings does not change.

A standard wavelet tree implementation technique is to represent each level of a wavelet tree with a single bitvector, which is the concatenation of the bitvectors for individual nodes over that level [2,9]. Here we can similarly use a single bitvector to represent a level over *all* wavelet trees for a given block size. As in the standard case, given the location of the bitvector for a node, we can easily locate the bitvectors for the children. For each block size b_k, we provide two additional bitvectors to directly locate nodes, one for the root level and one for the level at height k (with leafs at height 0). The size of such a bitvector is $n_k + v + o(n_k + v)$, where n_k is the length of the level bitvector, which equals the number of entries encoded with block size b_k, and v is the number of nodes on the given level. Since $v = \mathcal{O}(n/2^k)$ for height k and less or equal for the root level, the size of the locating bitvectors for block size b_k is $\mathcal{O}(n_k + n/2^k)$, which is $\mathcal{O}(n)$ over all block sizes.

When counting the number of distinct colors in $s[i..j]$, we handle block sizes larger than b_{max} as before using the multiary wavelet tree in $\mathcal{O}(\alpha(\ell) \log \log(\ell + 1))$ time. For each block size $b \leq b_{max}$, we need to query the wavelet trees for the two blocks that contain i. For block size b_{max} we do this as before in $\mathcal{O}(\log b_{max}) = \mathcal{O}(\log \ell)$ time. Block sizes smaller than b_{max} are handled differently.

If B is a block of size $b_k < b_{max}$ that contains i, it does not contain j. If $C[q] = p$ is stored in the wavelet tree for B, then $q < j$. We want to count an entry $C[q] = p$ in B if (i) $q \geq i$ and (ii) $p < i$. Since $p < q$, both conditions cannot be violated simultaneously. Thus we count entries that violate (i) and entries that violate (ii) and subtract the sum from the total number of entries for block B. Notice that this does not work for larger block sizes because we need an additional condition $q \leq j$. Using the multiary wavelet tree we can count in constant time all entries $C[q] = p$ that are encoded using block size b_k and have q in a given range. This is sufficient to count all entries in B as well as those that violate (i). Counting (ii) can be done by locating the leaf that represents the position i in the wavelet tree for block B, which we do by locating its ancestor at height k in constant time and traversing down in $\mathcal{O}(k)$ time. Thus block size b_k can be processed in $\mathcal{O}(k)$ time and all block sizes smaller than $b_{\max}$ in $\mathcal{O}\big((\alpha(\ell)\log\log(\ell+1))^2\big)$ time.

Theorem 4. *Given a string $s[1..n]$, we can build a data structure that takes $nH_0(s) + \mathcal{O}(n) + o(nH_0(s))$ bits such that later, given a substring's endpoints i and j, in $\mathcal{O}(\log \ell)$ time we can count how many distinct characters it contains, where $\ell = j - i + 1$.*

6   Partial Dynamism

Suppose we want to append a character $s[n+1]$ to s. To maintain C, we must append $C[n+1] = p$ to it, where p is the position of the last occurrence of $s[n+1]$ in $s[1..n]$, or 0 if there is no such occurrence. In the current version of this paper, we do not worry about how we find p (which can be done with, say, one rank query and one select query on s; a visitor to a website might also provide p directly via, e.g., a cookie) and focus only on how to append $C[n+1]$ to our representation of C stored in the data structure we gave in Section 3 (as appending it to the data structure from Section 2 is similar and simpler).

Our first concern is to append to the string t a character indicating whether p is 0 and, if not, the block size used for it. Instead of storing t with a multiary wavelet tree, we now store it with a Huffman-shaped binary wavelet tree [8], with the bitvectors at the internal nodes stored separated from each other (i.e., not concatenated, as is usual). As long as these bitvectors are each stored with at most linear redundancy, they take a total of at most $\mathcal{O}(n(H_0(t)+1)) \subseteq \mathcal{O}(n\log(H_0(s)+1) + n)$ bits. Also, since t is over an alphabet of size $\mathcal{O}(\log\log n / \log(1+\delta))$, which is $\mathcal{O}\big((\log\log n)^2\big)$ with our choice of $\delta = 1/\alpha(n)$, we can store the shape of the tree using $\mathcal{O}(\log n)$-bit pointers at each internal node without increasing our overall space bound.

To append a character to t, we append a bit to each bitvector on the path from the root of the wavelet tree to the leaf labelled with the character we append (we create this leaf if it does not already exist). Each bit indicates whether the next node on the path is the current node's left child or its right child. Many implementations of bitvectors are based on breaking them into blocks (see, e.g., [10] for more discussion) and, thus, make appending relatively easy. Since we allow ourselves linear redundancy, whenever a bitvector outgrows the space allocated to it, we double that space; we use background processing to copy the bitvector into its new location, so that our time bounds are still worst-case. Appending to t takes a total of $\mathcal{O}(\log\log\log n)$ time.

Our other concern is to append $C[n+1]$ to a sequence of values encoded with the same block size b, all of which are stored in a wavelet tree. We use essentially the same approach as when appending a character to t. One complication is that the sequence of values is no longer guaranteed to be over a small alphabet, so it is not immediately clear how we can use $\mathcal{O}(\log n)$-bit pointers at the internal nodes. If b is small, at most $n/\log n$, then, as with t, there is no problem: calculation shows that using pointers in all the wavelet trees for small block sizes increases our space bound by at most $\mathcal{O}(n)$. For the case when $b > n/\log n$, we replace the standard trie shape of wavelet trees with a Patricia trie shape. From the standard wavelet tree, we remove all nodes associated with an empty sequence. If any remaining node has exactly one child, the associated bitvector is all 0s or all 1s and can be encoded with a single bit stored in the closest existing decendant of the node. The resulting wavelet tree shape is a Patricia trie [11], where the number of internal nodes is less than the number of leaves, which is equal to the number of distinct values in the sequence and, thus, at most the length of the sequence. Recall that, if we use $\log b$ bits for each value stored in the wavelet tree for block size b, for every b, then we use a total of $(1 + \delta)nH_0(s) + \mathcal{O}(n) + o(nH_0(s))$ bits. Therefore, if we use $\mathcal{O}(\log n) \subseteq \mathcal{O}(\log b)$ bits for pointers at each internal node, then we use $\mathcal{O}(n(H_0(s) + 1))$ bits altogether. Appending to the sequence stored in a wavelet tree for a block size takes $\mathcal{O}(\log n)$ time.

Theorem 5. *We can modify the data structure from Theorem 2 such that we achieve the same time bound for querying and a space bound of $\mathcal{O}(n(H_0(s) + 1))$ bits while supporting an $\mathcal{O}(\log n)$-time append operation.*

If we modify the data structure from Section 3 by replacing all the wavelet trees (including the multiary wavelet tree) with dynamic wavelet trees [4], which support queries, insertions and deletions in $\mathcal{O}(\log^2 n)$ time, then we still use $nH_0(s) + \mathcal{O}(n) + o(nH_0(s))$ bits, but $\mathcal{O}(\log^2 n \log \log n)$ time for queries and appends. This data structure can also support colour substitutions and deletions in $\mathcal{O}(\log^2 n \log \log n)$ time. In order to replace a character $s[q] = a$ by a', we find the last occurrences $s[p]$ and $s[p']$ of a and a' strictly before $s[q]$, and the first occurrences $s[r]$ and $s[r']$ of a and a' strictly after $s[q]$. Again, these can be found using a constant number of rank and select queries on s. We can also find p and r using the wavelet trees, and p' given r' (or vice versa), in $\mathcal{O}(\log^2 n \log \log n)$ time. We update C such that $C[q] = p'$, $C[r] = p$ and $C[r'] = q$, again using $\mathcal{O}(\log^2 n \log \log n)$ time. To delete a character from s, we replace it with a special null character not in the alphabet (which we search for and exclude when performing queries).

Theorem 6. *We can modify the data structure from Theorem 2 such that it again takes $nH_0(s) + \mathcal{O}(n) + o(nH_0(s))$ bits, but supports queries, appends, colour substitutions and deletions in $\mathcal{O}(\log^2 n \log \log n)$ time.*

Acknowledgments

Many thanks to Veli Mäkinen, Giovanni Manzini, Gonzalo Navarro, Simon Puglisi and Jorma Tarhio, for helpful discussions.

References

1. Bozanis, P., Kitsios, N., Makris, C., Tsakalidis, A.K.: New upper bounds for generalized intersection searching problems. In: Proceedings of the 22nd International Colloquium on Automata, Language and Programming (ICALP), pp. 464–474 (1995)
2. Ferragina, P., Manzini, G., Mäkinen, V., Navarro, G.: Compressed representations of sequences and full-text indexes. ACM Transactions on Algorithms 3(2) (2007)
3. Gagie, T., Navarro, G., Puglisi, S.J.: Colored range queries and document retrieval. In: Chavez, E., Lonardi, S. (eds.) SPIRE 2010. LNCS, vol. 6393, pp. 67–81. Springer, Heidelberg (2010)
4. González, R., Navarro, G.: Rank/select on dynamic compressed sequences and applications. Theoretical Computer Science 410(43), 4414–4422 (2009)
5. Grossi, R., Gupta, A., Vitter, J.S.: High-order entropy-compressed text indexes. In: Proceedings of the 14th Symposium on Discrete Algorithms (SODA), pp. 636–645 (2003)
6. Kaplan, H., Rubin, N., Sharir, M., Verbin, E.: Efficient colored orthogonal range counting. SIAM Journal on Computing 38(3), 982–1011 (2008)
7. Lai, Y.K., Poon, C.K., Shi, B.: Approximate colored range and pointer enclosure queries. Journal of Discrete Algorithms 6(3), 420–432 (2008)
8. Mäkinen, V., Navarro, G.: Succinct suffix arrays based on run-length encoding. In: Apostolico, A., Crochemore, M., Park, K. (eds.) CPM 2005. LNCS, vol. 3537, pp. 45–56. Springer, Heidelberg (2005)
9. Mäkinen, V., Navarro, G.: Implicit compression boosting with applications to self-indexing. In: Ziviani, N., Baeza-Yates, R. (eds.) SPIRE 2007. LNCS, vol. 4726, pp. 229–241. Springer, Heidelberg (2007)
10. Mäkinen, V., Navarro, G.: Rank and select revisited and extended. Theoretical Computer Science 387(3), 332–347 (2007)
11. Morrison, D.R.: PATRICIA — practical algorithm to retrieve information coded in alphanumeric. Journal of the ACM 15(4) (1968)
12. Muthukrishnan, S.: Efficient algorithms for document retrieval problems. In: Proceedings of the 13th Symposium on Discrete Algorithms (SODA), pp. 657–666 (2002)
13. Patrascu, M.: Predecessor search. In: Kao, M.-Y. (ed.) Encyclopedia of Algorithms. Springer, Heidelberg (2008)
14. Sadakane, K.: Succinct data structures for flexible text retrieval systems. Journal of Discrete Algorithms 5(1), 12–22 (2007)

On Wavelet Tree Construction*

German Tischler

Institut für Informatik, Universität Würzburg, Am Hubland,
97074 Würzburg, Germany
tischler@informatik.uni-wuerzburg.de

Abstract. The wavelet tree is a compact data structure allowing fast rank, select, access and other queries on non binary sequences. It has many applications in indexed pattern matching and data compression. In contrast to applications of wavelet trees their construction has so far been paid little attention. In this paper we discuss time and space efficient algorithms for constructing wavelet trees.

1 Introduction

The wavelet tree was introduced by Grossi, Gupta and Vitter in 2003 (cf. [5]). Augmented with binary rank and select dictionaries it allows rank, select and access queries on strings over an alphabet of size σ in time $O(\log \sigma)$. Such queries are used in many compressed index structures such as the compressed suffix array (cf. [5]) and FM type indexes (cf. [2]). The space required by the uncompressed wavelet tree is the same as the space required for the input string. Note that this statement refers only to the tree itself. Additional space is necessary for the rank and select dictionaries. We neglect this space in this paper, as these dictionaries are not required for the construction of the tree. Apart from rank, select and access queries, wavelet trees also efficiently support orthogonal range queries (counting and enumerating points in a given rectangle on the plane, cf. [9]) and range quantile queries (finding the k'th smallest element in a subinterval of a sequence, cf. [4]). Furthermore, wavelet trees facilitate data compression (cf. e.g. [1]). While many papers discuss functionality and applications of the wavelet tree, the construction of wavelet trees has so far gotten little attention. It is simple to set up an algorithm for constructing the wavelet tree in time $O(n \log \sigma)$ which is linear in the size of the number of bits in the output, but such a straight-forward algorithm will require at least an amount of memory which is three times as large as the size of the input. For a large input this may be prohibitive. In this paper we discuss algorithms for producing the wavelet tree of a string with various time and space bounds. We mainly focus on balanced wavelet trees. Most of our approaches can be translated to Huffman shaped wavelet trees. We omit most details about Huffman shaped wavelet trees though due to lack of space. Throughout this paper we assume a word RAM model with a word size of w bits and log denotes the base 2 logarithm.

---

* Part of this work was done while the author was a Newton Fellow at King's College London.

R. Giancarlo and G. Manzini (Eds.): CPM 2011, LNCS 6661, pp. 208–218, 2011.

The paper is structured as follows. In Section 2 we present efficient algorithms for sorting integers with respect to one of their bits. In Section 3 we present several algorithms for the generation of wavelet trees from strings. We conclude the paper in Section 4, where we give some open problems.

2   Stable Sorting of Bit Key Sequences

It is well known that a finite sequence of length m can be sorted stably and in place in time $O(m \log m)$ in the comparison model (cf. [10]). In the following we provide simple and practical algorithms for the case of binary keys in the word RAM model. By binary keys we mean that we are sorting integers of $k \in O(w)$ bits (not necessarily binaries), but for the comparison of two numbers only a certain single bit of the binary representation of the integers is relevant. In particular the keys are part of the integers. The length m of the sequence is assumed to be such that $\log m \in O(w)$

The following observation is derived from the permutation operation by Kronrod (cf. [8]). In its formulation we consider sequences of binaries, the generalisation to bit key sequences is straight forward.

Observation 1. *Let X and Y be sorted bit sequences of length $|X|$ and $|Y|$ concatenated in a sequence Z of length $|Z| = |X| + |Y|$. More precisely let $X = 0^{x_0} 1^{x_1}$, $Y = 0^{y_0} 1^{y_1}$ and $Z = XY$. Then X and Y can be merged stably and in place in time $O(|X| + |Y|) = O(|Z|)$.*

Proof. The stably merged sequence is obtained by computing

$$X[0 \mathinner{.\,.} x_0 - 1]((X[x_0 \mathinner{.\,.} x_0 + x_1 - 1])^R (Y[0 \mathinner{.\,.} y_0 - 1])^R)^R Y[y_0 \mathinner{.\,.} y_0 + y_1 - 1]$$

where R denotes the reversal operator. As reversal can be implemented in place in linear time, we obtain the given runtime bound.

An $O(m \log m)$ time stable in-place sorting algorithm for binary keys can be obtained by implementing a common merge sort approach while using the method given in the proof of Observation 1 for merging. Note that this approach requires only a constant number of words of additional memory. A pseudo-code version of the algorithm is shown in Algorithm 1. The parameter l denotes the length of pre sorted sub arrays, i.e. we would set it to 1 for an array we have no further information about. For $l > 1$ the algorithm works on the assumption that the portions $A[il \mathinner{.\,.} \min((i+1)l, m) - 1]$ are already sorted for $i = 0, 1, \ldots$ (if the left bound is larger than the right bound this notation denotes an empty sub array).

An interesting and important feature of this sorting algorithm consists in the fact that it still works if we do not move the bits we sort by in the reversal operations. This means we can sort the information attached to the sorted bits by the bits while keeping the keys in their original place. We name this algorithm working without moving the keys BITMERGESORTKEYS. To see why this works note that the values b_l and o_r computed in line 4 and 5 of the algorithm remain

---

Algorithm 1. Bit comparison based merge sort algorithm

BITMERGESORT(A, m, b, l)

```
1   while l < m do
2       for i ← 0 to ⌈m/2l⌉ − 1 do
3           (l_l, r_l, l_r, r_r) ← (2il, (2i + 1)l), (2i + 1)l, min(2(i + 1)l, m))
4           b_l ← |{j | l_l ≤ j < r_l and A[j]&b = b}|
5           o_r ← |{j | l_r ≤ j < r_r and A[j]&b ≠ b}|
6           REVERSE(A[r_l − b_l .. r_l − 1])
7           REVERSE(A[l_r .. l_r + o_r − 1])
8           REVERSE(A[r_l − b_l .. l_r + o_r − 1])
9       l ← 2l
```

---

the same even if we do not move the keys. This is due to the fact that for the counting of the zeroes and ones it is unimportant where they are in the considered interval. The considered intervals though remain the same, as they do not depend on the sorted data.

If we are given another stable bit key sorting algorithm $\mathcal{A}$ sorting portions of length $\frac{m}{c}$ in time $t(\frac{m}{c})$ (with or without moving the keys), then we can obtain a bit key sorting algorithm (with or without moving the keys respectively) running in time $O(ct(\frac{m}{c}) + m \log c)$ and requiring the same space as $\mathcal{A}$ by first sorting portions of length $\frac{m}{c}$ and finally merging the portions using $\log c$ merging stages based on Observation 1.

Another interesting property of the algorithm BITMERGESORTKEYS consists in the fact that it can easily be modified to perform the reverse operation, i.e. restoring the unsorted from the sorted sequence while keeping the keys in place. For this purpose we only need to reverse the order of lines 6–8 and let l run from its maximal value down to 1 in the while loop. In fact this reversibility holds for the whole wavelet tree construction process such that we can obtain algorithms transforming a wavelet tree to the string it represents satisfying the same time and space constraints as in the string to wavelet tree direction. We omit the details due to lack of space.

The translation of the algorithm BITMERGESORTKEYS to the case of a Huffman coded sequence is not quite as simple as the algorithm for the case of block code. We need to know each bit of the Huffman code of a symbol to be able to determine the length of the code. We thus switch from an iterative bottom up formulation of the recursive structure of the algorithm to a traversal of the recursion tree in depth first order. We use a stack for the control of the recursion tree traversal. The algorithm thus requires $O(\log n)$ additional words of memory instead of a constant number of words. During the depth first traversal of the recursion tree we visit each node twice, first while descending into the tree and second while returning from the bottom of the tree. During the first visit we need to determine the code length assigned to the left and right child, as we need to pass on the relevant information for the handling of the respective child nodes. During the second visit we merge the partial results produced for the child nodes. Whenever we have finished handling a node we need to return the number of relevant zero and one bits used for sorting.

The algorithms BITMERGESORT and BITMERGESORTKEYS can be used in combination with other sorting approaches. More precisely they can be used to combine partial results produced by other stable bit key sorting algorithms. The algorithm BITMERGESORT can be combined with a bucket sorting approach in a recursive scheme which produces free space within the array by compressing sorted portions as in [3]. This yields an $O(m \log k)$ time stable in place bit key sorting algorithm moving the keys. We omit the details here because this algorithm is in this form not suitable for our application of wavelet tree construction as we want the keys to stay in place while we sort. The algorithm may be interesting in it's own right though. Here we consider two approaches, where only the first one seems easily translatable to the case of sorting Huffman encoded sequences.

The first approach we consider is a combination of the algorithm BITMERGE-SORTKEYS and bucket sorting. It requires $\frac{(k-1)m}{c}$ bits of additional space where $c \geq 1$. We can choose c as a function of m. For values of about $c > \frac{m}{2}$ the algorithm turns into a pure run of BITMERGESORTKEYS, as we do not have sufficient space to sort more than one element using bucket sort. Assume we have $d(k-1)$ bits of additional space, where $d = \lfloor \frac{m}{c} \rfloor > 1$. Then we sort each sub array $A[id \mathinner{\ldotp\ldotp} \min((i+1)d, m) - 1]$ using bucket sort. Note that there is no need to copy the key bits to the buckets, we just leave them where they are while we first copy the attached information to the correct of the two buckets and then copy it back in sorted order (i.e. first the bucket for zero, then the bucket for one). After all the sub arrays have been handled, we merge the partial results via the algorithm BITMERGESORTKEYS using $l = d$. We call this algorithm BITBUCKET-SORTKEYS. It takes time $O(m \log c)$, i.e. by choosing the amount of additional memory we provide we can get any runtime between $O(m)$ and $O(m \log m)$. The linear time version requires additional space within a constant of $m(k-1)$ bits, the $O(m \log m)$ time version requires only a constant number of additional memory words, i.e. $O(\log m)$ additional bits.

The second approach uses the sorting method proposed in appendix B of [7]. It uses additional space in the order of $O(\sqrt{m}(\log m + k))$ additional bits to stably sort a sequence of k bit numbers of length m according to binary keys in time $O(m)$. As in the simpler approach BITBUCKETSORTKEYS, there is no need to move the keys during the procedure. The algorithm works in two phases. In the first phase, the integers are sorted into buckets, where each bucket is represented as a list of blocks. Each block has space for $O(\sqrt{m})$ elements, i.e. it requires $O(\sqrt{m}k)$ bits. In the case of binary keys we need $O(1)$ blocks in addition to the array to be sorted. These blocks are called the external buffer. The array is scanned from left to right and elements are distributed to the buckets. This effectively means they are copied to the external buffer. As soon as a block in the external buffer is full, it is copied to some free block in the original array.

Due to the allocation strategy of the algorithm it is guaranteed that such a free block exists. The block is not necessarily copied into the right position of the array, thus we need to note where it should be. This is stored in an additional array of size $O(\sqrt{m} \log \sqrt{m}) = O(\sqrt{m} \log m)$ bits. In the second phase the blocks are permuted to obtain the sorted sequence. This is done by following cycles in the permutation stored in the additional array. The interested reader can find more details in [7]. We call the variant of this algorithm leaving the keys in place KSBBITBUCKETSORTKEYS.

The algorithm BITBUCKETSORTKEYS can be translated to the case of a Huffman coded array. Like for the translation of the algorithm BITMERGESORTKEYS we use a stack facilitating a depth first traversal of the recursion tree. When we have reached a sufficient depth at which the code length of the considered sub array is reduced enough we switch to bucket sorting and truncate the recursion tree. From the bucket sorting we need to return the number of 0 and 1 key bits found as if we would return from a recursive call to BITMERGESORTKEYS. It is unclear, whether the approach used in KSBBITBUCKETSORTKEYS can easily be applied to Huffman coded sequences.

A simple modification of BITMERGESORT allows us to transform a sequence $A = a_0, b_0, a_1, b_1, \ldots, b_{m-1}$ such that the a_i are bits and the b_i numbers of $k-1$ bits in place in time $O(m \log m)$ to the sequence $B = a_0, a_1, a_2, \ldots, a_{m-1}, b_0, b_1, \ldots, b_{m-1}$. For this purpose we may imagine the a_i are assigned key 0, the b_i key 1 and we perform the sorting procedure according to these keys. The reversal operations are easily modified for taking into account that the elements of the sequence do not all have the same length in this case. We name this algorithm TRANSPOSE$_k$.

If we are given another method which is able to transpose sequences of length $\frac{m}{c}$ for some $c \geq 1$ in time $O(t(\frac{m}{c}))$, then we can use an adapted version of TRANSPOSE$_k$ and this other method to transpose an array in time $O(ct(\frac{m}{c}) + m \log c)$ by first transposing portions of length $\frac{m}{c}$ and finally merging the portions using the adapted version of TRANSPOSE$_k$.

As for sorting we can consider several methods for transposition using additional space. A first simple method for transposing a sequence of length $\frac{m}{c}$ in time $O(\frac{m}{c})$ can be implemented easily if we have $\frac{m}{c}$ additional bits of space. We first copy the bits moved to the front to the additional space, move the other numbers into place by copying them to the back and copy the bits from the additional space back to the front. A second method can be based on the sorting scheme used in KSBBITBUCKETSORTKEYS. The only adaption necessary here is to make sure we fill blocks of a fixed size, i.e. $O(\sqrt{m}(\delta - 1))$ single bits make one block and $O(\sqrt{m})$ integers (the rest of the integers without the keys) make one block, so the second stage can handle blocks of equal size in bits. As above the justification for describing the first simpler method is that it can easily be applied to Huffman coded sequences, while this is not clear for the second method.

In the presence of available additional memory we will mean these hybrid approaches instead of the pure merge sort based variant below if we use the name

TRANSPOSE$_k$ (i.e. using the second method similiar to KSBBITBUCKETSORTKEYS for block code and the first method for Huffman code).

As the transposition method given is an adaption of the sorting method, the translation to the Huffman code case can formulated analogously.

3   Generating Wavelet Trees

3.1   Definitions

Let Σ be a finite alphabet equipped with an injective function $r : \Sigma \mapsto \mathbb{N}$ assigning a rank to each symbol. We assume that $|\Sigma| \geq 2$, otherwise all problems handled in the following become trivial. Let $\sigma = \max\{r(a)|a \in \Sigma\}$ be the maximal rank of a symbol in the alphabet. Each rank and thus each alphabet symbol can be represented in $\delta = \lceil \log_2(\sigma + 1) \rceil$ bits. We assume that $\delta \in O(w)$, i.e. our word RAM can access the representation of each symbol in constant time.

Let $b : \Sigma \times \mathbb{N} \mapsto \{0,1\}$ be defined by $b(a, i) = 1$ iff the i'th least significant bit in $r(a)$ is 1, i.e. $r(a) = \sum_{i=0}^{\delta-1} b(a, i)2^i$.

Throughout this section we consider an input sequence $S \in \Sigma^n$ of length $n \in \mathbb{N}$, which we assume is given as a bit sequence

$$B = b(S[0], \delta - 1)b(S[0], \delta - 2) \ldots b(S[0], 0)b(S[1], \delta - 1) \ldots b(S[n-1], 0)$$

of length $|B| = n\delta$, where we assume $\log n \in O(w)$. We identify the symbol $S[i]$ with the block of bits $B[\delta i] \ldots B[\delta(i + 1) - 1]$ in B.

Let B_k for $0 \leq k < \delta$ denote the subsequence given by

$$B_k = b(S[0], \delta - k - 1)b(S[0], \delta - k - 2) \ldots b(S[0], 0)$$
$$b(S[1], \delta - k - 1) \ldots b(S[n-1], 0) \;,$$

i.e. the k most significant bits are masked from each symbol.

In analogy to S and B we define for $0 \leq k < \delta$ the sequence S_k of length n by setting $S_k[i] = B_k[i(\delta - k)] \ldots B_k[(i + 1)(\delta - k) - 1]$. Let $S^{(i,k)}$ denote the subsequence of S containing exactly those symbols a such that $\sum_{j=0}^{k-1} b(a, \delta - j - 1)2^{k-j-1} = i$, i.e. the binary interpretation of the top k bits is i. We use S_k^i as a more convenient notation for $(S^{(i,k)})_k$ below and use S^i as a short form of $S^{(i,1)}$.

The *balanced wavelet tree* of the string S is a binary tree such that the root of the tree contains the bit sequence $B[0], B[\delta], B[2\delta], \ldots, B[(n - 1)\delta]$, i.e. the most significant bit of each character's rank. If $\delta = 1$, then the root of the balanced wavelet tree is a leaf. Otherwise, if $\delta > 1$, then the left child of the root is the balanced wavelet tree of the subsequence S_1^0 (i.e. the subsequence S^0 with the most significant bit removed. Accordingly we substitute δ by $\delta - 1$ in the construction.) and the right child is the balanced wavelet tree of S_1^1. The tree is represented by the concatenation of the bit-vectors found in the tree nodes traversed in breadth first order. The motivation for storing the nodes in breadth

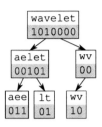

Fig. 1. Wavelet tree for the string *wavelet* represented by 101000000101000110110

first order is that this representation augmented with a rank dictionary allows us to navigate in the tree without using pointers, as the start of the left child of a node always equals the start of a node plus n bits and the start of the right child can be determined by the start of the left child plus the number of 0 bits assigned to the node. The width of a left (right) child is determined by the number of zero (one) bits in its parent node.

For the computation of the rank, select, access, etc. functions using the balanced wavelet tree the reader is referred to the respective literature (cf. [4,5,9]).

As an example consider the string T = wavelet over the alphabet $\Gamma = \{a, e, l, t, v, w\}$ equipped with the rank function given by the pairs $(a, 0 = 000)$, $(e, 1 = 001)$, $(l, 2 = 010)$, $(t, 3 = 011)$, $(v, 4 = 100)$ and $(w, 5 = 101)$. The balanced wavelet tree of T is shown in Figure 1. The tree only consists of the bit sequences shown with a grey background, the letters are provided for demonstration purposes only.

For a precise definition of Huffman shaped wavelet trees the reader is referred to the literature (cf. [2]). In short the balanced structure of the block code based wavelet tree is replaced by a tree structure based on the tree of a Huffman code (cf. [6]) derived from the input sequence. This reduces the space used from δn bits to Mn bits, where M denotes the average code length used by the code per symbol for the complete input sequence. In contrast to the balanced wavelet tree navigation in the Huffman shaped wavelet tree is not easily done without pointers between the nodes. We consider the pre order depth first concatenation of the nodes' binary sequences as the representation of the tree. For the construction of Huffman shaped wavelet trees we assume that the input sequence is Huffman coded and we have supporting data structures allowing fast encoding and decoding of the underlying code.

3.2   Wavelet Tree Construction Algorithms

The algorithms we will present generate the tree top down. They are based on the sorting schemes KSBBITBUCKETSORTKEYS in the balanced case (where we assume we have $0 \leq O(\frac{\sqrt{n}(\delta + \log n)}{c}) \leq O(\sqrt{n}(\delta + \log n))$ free bits) and BITBUCKETSORTKEYS in the Huffman case (where we assume we have $0 \leq \frac{n(M-1)}{c} \leq n(M-1)$ free bits).

We first discuss the case of balanced wavelet trees. We consider one breadth and one depth first approach. The first algorithm which we call WTBFS is based on a breadth first traversal, i.e. it generates the tree level per level, while it uses only a constant number of words of additional space for the control of the tree traversal. It consists of $\delta - 1$ repetitions of two phases. Consider iteration k, where $0 \le k < \delta - 1$. The first phase sorts 2^k arrays of total length n using the algorithm KSBBITBUCKETSORTKEYS. These arrays are $S_k^0, \ldots, S_k^{2^k - 1}$, i.e. the data relevant for the levels k to $\delta - 1$ of the tree. The second phase uses the algorithm TRANSPOSE$_k$ to move the bits corresponding to the k'th level into place. Both phases take $O(n \log c)$ time while we have $\delta - 1$ such phases, thus the overall time for the sort and transposition operations is $O(\delta n \log c)$. As an example consider the string `wavelet` represented by 5041213 which is $(101)(000)(100)(001)(010)(001)(011)$ in binary where $\delta = 3$ (the parentheses are inserted to improve readability). We first sort by the most significant bit while keeping the keys in place. This yields $(100)(001)(110)(001)(011)(001)(000)$. The first bit in each triple is still the same. The second and third bit of each triple is obtained by taking the second and third bit of the original sequence sorted by the first bit (which would be $(000)(001)(010)(001)(011)(101)(100)$, this sequence is however not computed explicitly). Then we apply TRANSPOSE$_3$ to $(100)(001)(110)(001)(011)(001)(000)$, which yields the intermediate result $(1010000)((00)(01)(10)(01)(11)(01)(00))$. The first $n = 7$ bits are the first n bits of the output. The rest is the concatenation of the sequences S_1^0 and S_1^1. Subsequently for level $k > 0$ below the root we are confronted with the problem of determining where the sequences S_k^i for $0 \le i < 2^k$ start and end in the bit sequence starting from index $k\delta$. To find the number of symbols in the input which equal i in the k most significant bits, we use a top down scan of the existing output. The starting point of the sequence S_k^i is obtained by accumulating the number of symbols assigned to sequences S_k^j for $j < i$. These operations can be performed in time $O(\sigma \delta n)$ using a constant number of additional words of memory for one level. The overall runtime of the algorithm is thus $O(\delta n \log c + \delta \sigma \delta n) = O(\delta n(\log c + \sigma \delta))$. Consider for instance our example above. We have produced the output for the first level of the tree. The next level has two inner nodes. We consider them from low to high value. For $i = 0$ there is no smaller value, thus the left bound is 0. The number of values which equal $i = 0$ is 5 (the number of zero bits on the first level). Thus the range of S_1^0 extends from bit $1 \cdot n + 0 \cdot (3 - 1)$ to $1 \cdot n + (0 + 5) \cdot (3 - 1)$ (where $1 \cdot n$ denotes the start of the input to be processed (we have processed one level, thus the 1) and $3 = \delta$). The range for S_1^1 extends from the right bound of S_1^{1-1} to $1 \cdot n + (0 + 5 + 2) \cdot (3 - 1)$ (where the 2 denotes the number of 1 bits on the first level).

The second algorithm which we call WTDFS performs the sorting along a pre order depth first traversal of the tree using a stack of depth $O(\delta)$, where each stack element consists of an index interval and a depth and thus requires $O(\log n + \log \delta)$ bits. In the beginning the stack contains the pair $([0, n), 0)$ representing the root of the tree. All the transposition operations are postponed until the end of the algorithm. A slight adaption of the sorting algorithm is required due to the fact that the sorted data is not contiguous (i.e. when we perform the sorting for the lower levels, the bits corresponding to the levels above will still be between the bits relevant for sorting), this however does not change the runtime of the algorithm. The algorithm uses $O(\delta)$ additional words of memory and has a runtime of $O(\delta n \log c)$. During the stack based traversal we need to take care not to push empty intervals onto the stack, otherwise the term $n \log c$ in the runtime of the algorithm is replaced by $\max\{n \log c, \sigma\}$. As an example again consider the string `wavelet` represented by $(101)(000)(100)(001)(010)(001)(011)$. On the stack we find $([0,7),0)$. We first sort the interval $[0,7)$ at level 0 by the most significant bit while keeping the keys in place. As above this yields $(1(00))(0(01))(1(10))(0(01))(0(11))(0(01))(0(00))$. Now we count the number of zeroes found in the most significant bits. This number is 5. Thus we push the tuples $([5,7),1)$ and $([0,5),1)$ onto the stack as to be handled after we have popped the tuple $([0,7),0)$. The interval $[0,5)$ denotes the left child node, $[5,7)$ the right child node and the final 1 denotes the depth of the respective child nodes. The next node handled is the left child of the root denoted by $([0,5),1)$. We sort the subsequence $(00)(01)(10)(01)(11)$ (the interval $[0,5)$ without the output for level zero of the tree) by the most significant bit while keeping the keys in place. This yields $(00)(01)(11)(00)(11)$ and the global bit string becomes $(1(00))(0(01))(1(11))(0(00))(0(11))(0(01))(0(00))$. The node has no children, so we push no elements onto the stack. The next node handled is $([5,7),1)$. We sort $(01)(00)$ by the most significant bit while leaving the keys in place and obtain $(01)(00)$. Globally we obtain $(1(00))(0(01))(1(11))(0(00))(0(11))(0(01))(0(00))$. The node has no children, so we push no elements onto the stack. The stack is now empty. It remains to perform the postponed transpositions. We first perform TRANSPOSE_3 and then TRANSPOSE_2, where TRANSPOSE_3 handles all the bits and TRANSPOSE_2 does not touch the front n bits. This gives us our final result $(1010000)(0010100)(0110110)$.

The following table shows a comparison of the two algorithms.

algorithm	runtime	additional space (bits)
WTBFS	$O(\sigma\delta^2 n + \delta n \log c)$	$O(\frac{\sqrt{n}(\delta+\log n)}{c}) + O(\log n)$
WTDFS	$O(\delta n \log c)$	$O(\frac{\sqrt{n}(\delta+\log n)}{c}) + O(\delta(\log n + \log \delta))$

For the generation of Huffman shaped wavelet trees we use a pre order depth first traversal of the tree and perform the transposition operations directly after a node has been handled. For the sorting we use an additional space of $\frac{(M-1)n}{c} + O(\log^2 n)$ bits for some $c \geq 1$ which may be a function of n. As above M denotes the average code length used by the Huffman code per symbol for the complete input sequence. The term $\frac{(M-1)n}{c}$ may be reduced to zero by using a pure merge sort approach. The $O(\log^2 n)$ term stems from the stack of depth $\log n$ used for sorting. Let m denote the maximum code length of the Huffman code used. Then we need a stack of size $O(m)$ to facilitate the depth first traversal through the wavelet tree during construction. The elements on the stack consist of the left and right bound of the relevant sub array (each stored using $O(\log(Mn))$ bits) and the code prefix assigned to the represented node (which has a length of up to m bits. Each stack element stores only a single bit of the prefix though). Thus we need $O(m \log(Mn))$ bits for the stack. Overall the space required for the algorithm thus is $\frac{(M-1)n}{c} + O(\log^2 n) + m \log(Mn)$ bits. Assuming the encoding and decoding routines for the Huffman code are fast enough, we can expect the algorithm to run in time $O(Mn \log c)$.

4   Conclusion

In this paper we have discussed time and space efficient algorithms for constructing wavelet trees from strings. We have given algorithms which can be parametrised by the amount of additional space they are allowed to use for both balanced and Huffman shaped wavelet trees. For the case of balanced wavelet trees this additional space can be between constant additional space (a fixed number of words or $O(\log n)$ additional bits) and $O(\sqrt{n}(\delta + \log n))$ bits. On the assumption of a constant sized alphabet we can get any runtime between $O(n)$ and $O(n \log n)$. Interesting problems include whether the $O(n \log n)$ time bound can be broken, if we do not allow an algorithm to use additional memory and if we can obtain a linear runtime in $O(\delta n)$ using an additional space of $o(\sqrt{n}(\delta + \log n))$ bits.

The author would like to thank the anonymous reviewers for helpful comments and Juha Kärkkäinen and Roberto Grossi for interesting discussions on the topic of this paper, in particular concerning the hint to the linear time sorting algorithm given by Kärkkäinen et al. in [7].

References

1. Ferragina, P., Giancarlo, R., Manzini, G.: The myriad virtues of wavelet trees. Inf. Comput. 207(8), 849–866 (2009)
2. Ferragina, P., Manzini, G., Mäkinen, V., Navarro, G.: An alphabet-friendly FM-index. In: Apostolico, A., Melucci, M. (eds.) SPIRE 2004. LNCS, vol. 3246, pp. 150–160. Springer, Heidelberg (2004)
3. Franceschini, G., Muthukrishnan, S.M., Pătraşcu, M.: Radix sorting with no extra space. In: Arge, L., Hoffmann, M., Welzl, E. (eds.) ESA 2007. LNCS, vol. 4698, pp. 194–205. Springer, Heidelberg (2007)
4. Gagie, T., Puglisi, S., Turpin, A.: Range quantile queries: Another virtue of wavelet trees. In: Karlgren, J., Tarhio, J., Hyyrö, H. (eds.) SPIRE 2009. LNCS, vol. 5721, pp. 1–6. Springer, Heidelberg (2009)
5. Grossi, R., Gupta, A., Vitter, J.S.: High-order entropy-compressed text indexes. In: SODA, pp. 841–850 (2003)
6. Huffman, D.A.: A method for the construction of minimum-redundancy codes. Proceedings of the Institute of Radio Engineers 40(9), 1098–1101 (1952)

7. Kärkkäinen, J., Sanders, P., Burkhardt, S.: Linear work suffix array construction. J. ACM 53(6), 918–936 (2006)
8. Kronrod, M.A.: Optimal ordering algorithm without operational field. Soviet Math. Dokl. 10, 744–746 (1969)
9. Mäkinen, V., Navarro, G.: Rank and select revisited and extended. Theoretical Computer Science 387(3), 332–347 (2007); The Burrows-Wheeler Transform
10. Salowe, J., Steiger, W.: Simplified stable merging tasks. J. Algorithms 8(4), 557–571 (1987)

Lightweight BWT Construction for Very Large String Collections

Markus J. Bauer[1], Anthony J. Cox[1], and Giovanna Rosone[2]

[1] Illumina Cambridge Ltd., United Kingdom
{mbauer,acox}@illumina.com
[2] University of Palermo, Dipartimento di Matematica e Informatica,
Via Archirafi 34, 90123 Palermo, Italy
giovyros@virgilio.it

Abstract. A modern DNA sequencing machine can generate a billion or more sequence fragments in a matter of days. The many uses of the BWT in compression and indexing are well known, but the computational demands of creating the BWT of datasets this large have prevented its applications from being widely explored in this context.

We address this obstacle by presenting two algorithms capable of computing the BWT of very large string collections. The algorithms are lightweight in that the first needs $O(m \log m)$ bits of memory to process m strings and the memory requirements of the second are constant with respect to m.

We evaluate our algorithms on collections of up to 1 billion strings and compare their performance to other approaches on smaller datasets. Although our tests were on collections of DNA sequences of uniform length, the algorithms themselves apply to any string collection over any alphabet.

Keywords: BWT, text indexes, next-generation sequencing.

1 Introduction

We consider the problem of computing the Burrows-Wheeler transform (BWT) of very large datasets. Our interest in this topic arises from DNA sequencing, a field that has been transformed by the advent of 'next-generation' sequencing technologies [15]. Whole-genome sequencing of human DNA may sample the 3 billion base-pairs of the human genome to 30× redundancy or higher. A modern DNA sequencing machine can produce this quantity of data in just a few days and datasets of 100Gbases or larger have therefore become commonplace.

The BWT of such a dataset would be useful to have for the purposes of data compression or for creating self-indexing data structures such as the FM-index [4]. However, generating the BWT for datasets on this scale is the challenge that we address in this paper.

Computing the BWT of a string from its suffix array (SA) is simple and fast, and much effort has been devoted to devising algorithms for SA construction that

R. Giancarlo and G. Manzini (Eds.): CPM 2011, LNCS 6661, pp. 219–231, 2011.

are efficient in both space and CPU time [9–11]. However, all share the need for the entire SA to be held in RAM, which becomes problematic for datasets of the size in which we are interested. For instance, Simpson and Durbin [19] extrapolate from their experiences with a variant of the SA construction algorithm by Nong *et al.* [16] to estimate 700Gbytes of RAM would be necessary to build the FM-index of a 20× oversampling of the human genome (60Gbases) by this route.

A divide-and-conquer approach is one way to address this bottleneck: aiming to compute the compressed suffix array (CSA) of large collections of texts, Sirén [20] divides the collection into batches, computes the BWT of each in a distributed fashion and then merges the results to obtain the BWT of the entire collection. Nevertheless, the RAM requirements of the final merge remain considerable - Sirén quotes 32Gbytes to index 42.03Gbytes of English text if 8 parallel threads are used, and double that if 16 threads are used.

Some studies [8, 18] showed that one can compute the BWT without a SA by implicitly adding suffixes from the shortest ones to longest ones. However, these algorithms are slow due to the large constant factor.

External memory algorithms are an alternative way of tackling large datasets. Ferragina *et al.* [3] show how to build the BWT of a large string T by logically partitioning it into r blocks $T_r \cdots T_1$. Working from right to left in T, r passes through the data are made. At pass h, the suffix array of T_{h+1} is used to update $\mathsf{bwt}(T_h \cdots T_1)$ to give $\mathsf{bwt}(T_{h+1} \cdots T_1)$, following the idea of [7]. At any point, the RAM usage is dominated by the suffix array of the block T_{h+1} being merged, so r can be adjusted to 'tune' the RAM usage of the algorithm to fit the memory available.

In this paper, we present two lightweight algorithms for constructing the BWT in external memory. With our application to DNA data in mind, our algorithms assume the BWT is to be generated from a dataset comprising a large collection of short strings.[1] We therefore need to consider how the notion of the BWT should be extended from a single string to a collection of strings. Mantaci *et al.* [12] gave the first such generalization (cf. also [13, 14]), but this can be defined in more than one way, the differences centering around how string comparisons across text boundaries are handled. Here we follow the approach taken by Sirén and others, assuming each string is appended with a unique terminating character ordered lexicographically so as to match the ordering of the strings in the collection.

For a collection of m strings of length k, our algorithms are lightweight in the sense that one uses only $O(m \log(mk))$ bits of space and $O(k \, \mathrm{sort}(m))$ time, where $\mathrm{sort}(m)$ is the time taken to sort m integers, and the other works almost entirely in external memory, taking $O(km)$ time with RAM usage that is constant for collections of any size, dependent only on the alphabet size and therefore negligible for DNA data. The overall I/O volume is $O(k^2 m)$. Our algorithms are

---

[1] In addition, our implementations expect each string in the collection to be of equal size, since many sequencing technologies produce data that is of this form. However this limitation is not intrinsic to the algorithms themselves.

scan-based in the sense that all files are accessed in a manner that is entirely sequential and either wholly read-only or wholly write-only.

We carry out computational experiments to compare our code to the Sirén and Ferragina *et al.* algorithms and to assess how the performance of our algorithms scales to string collections as large as one billion 100-mers. Our experiments raise some interesting points about the behaviour of external memory algorithms in practice and suggest areas for future improvement.

2 Preliminaries

Let $\Sigma = \{c_1, c_2, \ldots, c_\sigma\}$ be a finite ordered alphabet with $c_1 < c_2 < \ldots < c_\sigma$, where $<$ denotes the standard lexicographic order. Given a finite string $w = w_0 w_1 \cdots w_{k-1}$ with each $w_i \in \Sigma$, a *substring* of a string w is written as $w[i, j] = w_i \cdots w_j$. A substring of type $w[0, j]$ is called a *prefix*, while a substring of type $w[i, k-1]$ is called a *suffix*. We denote by j-*suffix* the suffix of w that has length j. The *concatenation* of two words w and v, written wv, is simply the string consisting of the symbols of w followed by the symbols of v.

Let $S = \{S_1, S_2, \ldots, S_m\}$ be a collection of m strings, each comprising k symbols drawn from Σ. Each S_i can be imagined to have appended to it an end marker symbol \$ that satisfies $\$ < c_1$. We define lexicographic order among the strings in the usual way, except that each end marker \$ is considered a different symbol, so that every suffix of every string is unique in the collection. The (implicit) end marker is in position k, i.e. $S_i[k] = S_j[k] = \$$, and we define $S_i[k] < S_j[k]$, if $i < j$.

We define the 0-suffix as the suffix that contains only the end marker \$.

The suffix array SA of a string w is the permutation of integers giving the starting positions of the suffixes of w in lexicographical order. The BWT of w is a permutation of its symbols such that the i-th element of the BWT is the symbol preceding the first symbol of the suffix starting at position $SA[i]$ in w, for $i = 0, 1, \ldots, k$ (we assume that the symbol preceding the suffix starting at the position 0 is \$). We refer interested readers to [17] and [1] for further reading on suffix arrays and the Burrows-Wheeler Transform respectively.

3 Lightweight BWT Construction for a Collection of Strings

In this section we describe two algorithms, inspired by [3, 12], that compute the BWT of a collection of strings without concatenating the strings and without needing to compute their suffix array. We assume that $j = 1, 2, \ldots, k$ and $i = 1, 2, \ldots, m$.

Our algorithms compute the BWT of the collection S incrementally via k iterations. At each of the iterations $j = 1, 2, \ldots, k-1$, the algorithms compute a partial BWT string $\text{bwt}_j(S)$ by inserting the symbols preceding the j-suffixes of S at their correct positions into $\text{bwt}_{j-1}(S)$. Each iteration j simulates the insertion of the j-suffixes in the SA. The string $\text{bwt}_j(S)$ is a 'partial BWT' in the sense

that the addition of m end markers in their correct positions would make it the
BWT of the collection $\{S_1[k - j - 1, k], S_2[k - j - 1, k], \ldots, S_m[k - j - 1, k]\}$.

A trivial 'iteration 0' sets the initial value of the partial BWT by simulating
the insertion of the end markers \$ in the SA. Since their ordering is entirely
determined by the position in S of the string they belong to, $\mathsf{bwt}_0(\mathsf{S})$ is just the
concatenation of the last non-\$ symbol of each string, that is $S_1[k - 1]S_2[k - 1] \cdots S_m[k - 1]$.

Finally, iteration k inserts m end markers into $\mathsf{bwt}_{k-1}(\mathsf{S})$ at their correct
positions. This simulates the insertion of the suffixes corresponding to the entire
strings into the SA.

Like in [3], the fundamental observation is that going from $\mathsf{bwt}_{j-1}(\mathsf{S})$ to
$\mathsf{bwt}_j(\mathsf{S})$ at iteration j requires only that we insert m new symbols and does
not affect the relative order of the symbols already in $\mathsf{bwt}_{j-1}(\mathsf{S})$. We can think
of $\mathsf{bwt}_j(\mathsf{S})$ as being partitioned into $\sigma + 1$ strings $B_j(0), B_j(1), \ldots, B_j(\sigma)$, with
the symbols in $B_j(h)$ being those that are associated with the suffixes of S that
are of length j or less and begin with $c_0 = \$$ and $c_h \in \Sigma$, for $h = 1, \ldots, \sigma$. We
note that $B_j(0)$ is constant for all j and, at each iteration j, we store $B_j(h)$ in
$\sigma + 1$ external files that are sequentially read one-by-one.

During the iteration $j = 1, \ldots, k$, we must insert the symbol associated with
the new suffix $S_i[k - j, k]$ of each string $S_i \in \mathsf{S}$ (this symbol is $S_i[k - j - 1]$
for $j < k$, or \$ at the final iteration) into the BWT segment $B_j(z)$, where
$c_z = S_i[k - j]$ (we recall that $B_j(z)$ contains all symbols associated with suffixes
starting with the symbol c_z). Our main idea is that the position in $B_j(z)$ where
this symbol needs to be inserted can be computed from the position r where,
in the previous step, the symbol c_z has been inserted into the BWT segment
$B_{j-1}(v)$, where $c_v = S_i[k - (j - 1)]$ (we recall that $B_{j-1}(v)$ contains all symbols
associated with suffixes that have already been inserted and that start with the
symbol c_v).

To do this, we need to retain the BWT segments $B_{j-1}(h)$, for $0 \le h \le \sigma$, and
keep track of the positions within them of the symbols that correspond to the
$(j-1)$-suffixes of S, which we do by associating to each $B_{j-1}(h)$ an array $P_{j-1}(h)$
of integers that stores the absolute positions of the $(j - 1)$-suffixes starting with
c_h. Each $P_{j-1}(h)$ is in turn associated with an array $N_{j-1}(h)$ that has the same
number of entries and is such that $N_{j-1}(h)[q]$ stores i, the original position in S
of the string S_i whose $(j-1)$-suffix is pointed to by $P_{j-1}(h)[q]$. Here q is a generic
subscript of the array $N_{j-1}(h)$ or (equivalently, since their number of entries is
the same) $P_{j-1}(h)$. The maximum value of q is determined by the number of
$(j - 1)$-suffixes starting with c_h and will therefore vary with both h and j.

Stated formally, at the start of iteration j, we assume the following structures
are available for each $h = 0, \ldots, \sigma$, where $c_0 = \$$ and $c_n \in \Sigma$, for $n = 1, \ldots, \sigma$,
and the maximum value of q depends on the number of the $(j - 1)$-suffixes
starting with c_h:

$B_{j-1}(h)$ is a segment of the partial BWT.
$N_{j-1}(h)$ is an array of integers such that $N_{j-1}(h)[q]$ is associated with the
$(j - 1)$-suffix of the string $S_i \in \mathsf{S}$, where $i = N_{j-1}(h)[q]$.

$P_{j-1}(h)$ is an array of integers such that $P_{j-1}(h)[q]$ is the absolute position of the symbol $S_i[k-j]$, associated with the $(j-1)$-suffix of S_i, in $B_{j-1}(h)$, where $i = N_{j-1}(h)[q]$.

Hence, at the end of the iteration $j-1$, for each element in N_{j-1} and P_{j-1}, we have that the symbol $c_z = S_i[k-j]$, with $i = N_{j-1}(v)[q]$, has been inserted in the position $P_{j-1}(v)[q]$ in $B_{j-1}(v)$, where $c_v = S_i[k-(j-1)]$.

During the iteration j, we have to update these structures for each string $S_i \in \mathsf{S}$. The crucial point is to insert the new symbol associated with the j-suffix of S_i into $B_{j-1}(z)$, where $c_z = S_i[k-j]$, for some $z = 1, \ldots, \sigma$, at its correct position in order to obtain $B_j(z)$. Hence, our task is to compute $P_j(h)$ by considering how many suffixes of S that are of length j or less are smaller than each suffix of length j.

The following lemma (similar to [3, Lemma 1]) is the key to this point and it is based on a function called *LF-mapping* [5] that is also used extensively in compressed self-indexes. This method is based on the count of symbols, from first position to the position of the last inserted symbol of S_i in $\mathsf{bwt}_{j-1}(\mathsf{S})$, that are smaller than $c_z = S_i[k-j]$. It is equivalent to count the number of symbols that are associated with suffixes smaller than $S_i[k-j,k]$. We observe that we do not need to do exactly this, because the suffixes starting with a symbol smaller than c_z are associated with symbols in $B_{j-1}(r)$ for $r = 0, \ldots, z-1$. So, we only need to count how many suffixes of length j or less starting with the symbol c_z are smaller than the suffix $S_i[k-j,k]$.

Lemma 1. *For any iteration $j = 1, 2, \ldots, k$, given a symbol c_h, with $0 \le h \le \sigma$, let q be an index that depends on the number of the $(j-1)$-suffixes starting with c_h. For each string $S_i \in \mathsf{S}$, with $i = N_{j-1}(h)[q]$, we assume that the suffix $S_i[k-(j-1),k]$ is lexicographically larger than precisely $r = P_{j-1}(v)[q]$ suffixes of length $0, 1, \ldots, j-1$ that begin with the symbol $c_v = S_i[k-(j-1)]$. Now, we fix $c_z = S_i[k-j]$. Then the new suffix $S_i[k-j,k] = c_z S_i[k-(j-1),k]$ is lexicographically larger than precisely r' suffixes of length $0, 1, \ldots, j$, where $r' = \mathsf{rank}(c_z, r, c_v)$ and $\mathsf{rank}(c_z, r, c_v)$ denotes the number of occurrences of c_z in $B_{j-1}(0) \cdots B_{j-1}(v-1)B_{j-1}(v)[0, r-1]$.*

Proof. The proof is similar to that of Ferragina *et al.* in [3]. By hypothesis $c_z = S_i[k-j]$ and $c_v = S_i[k-(j-1)]$. Clearly, $S_i[k-j,k]$ is larger than the suffixes starting with a symbol smaller than c_z (they are associated with the symbols in $B_{j-1}(0), \ldots, B_{j-1}(z-1)$), and is smaller than all suffixes starting with a symbol greater than c_z (they are associated with the symbols in $B_{j-1}(z+1), \ldots, B_{j-1}(\sigma)$). Since the suffixes starting with c_z are associated with the symbols in $B_{j-1}(z)$, the correct position of the symbol associated with the suffix $S_i[k-j,k]$ is in $B_{j-1}(z)$. Now, the crucial point is to compute how many suffixes of length j or less starting with c_z are smaller than $S_i[k-j,k]$. The sorting of the rows in BWT implies that counting how many suffixes starting with c_z in $\{S_1[k-j,k], S_2[k-j,k], \ldots, S_m[k-j,k]\}$ that are smaller than $S_i[k-j,k]$ is equivalent to counting the number of occurrences of c_z in $B_{j-1}(0), \ldots, B_{j-1}(v-1)$ and in $B_{j-1}(v)[0, r-1]$. This is precisely $\mathsf{rank}(c_z, r, c_v)$. □

The positions of each j-suffix $S_i[k-j,k]$ are computed using Lemma 1 and stored in P_j according to the symbol $S_i[k-j]$. In other words, if $c_z = S_i[k-j]$, the computed position r' is stored into $P_j(z)$ and i is stored into $N_j(z)$. Moreover, the value r' corresponds to the *absolute* position in $B_j(z)$ where we have to insert the new symbol associated with $S_i[k-j,k]$ starting with c_z. This means that, for each symbol c_h, with $0 \le h \le \sigma$, we consider, in the computation of the new positions, all new j-suffixes in S that begin with c_h. Hence, if $S_r[k-j] = S_s[k-j]$ and $S_r[k-(j-1),k] < S_s[k-(j-1),k]$, for some $1 \le r,s \le m$, it follows that $S_r[k-j,k] < S_s[k-j,k]$. For this reason and since each $B_j(h)$ is stored in an external file, we have to sort each $P_j(h)$ (respectively $N_j(h)$) and insert the new symbols according to the value of their position, from the smallest to the largest. Given this information, we can build $B_j(1), \ldots, B_j(\sigma)$ by using the current files $B_{j-1}(1), \ldots, B_{j-1}(\sigma)$. The idea is very simple; we read —in a sequential way— each external file associated with each $B_j(h)$ once and insert all symbols associated with the j-suffixes starting with the symbol c_h. Once all the symbols are read and copied, $B_{j-1}(0) \cdots B_{j-1}(\sigma)$ form $B_j(0) \cdots B_j(\sigma)$ respectively, i.e. the partial BWT string required by the next iteration. Since we no longer need $P_{j-1}(h)$ and $B_{j-1}(h)$, we can write $P_j(h)$ and $B_j(h)$ over the already processed $P_{j-1}(h)$ and $B_{j-1}(h)$.

The counts for $B_{j-1}(d)$, $c_d < S[k-j]$, are dealt with by keeping a count of the number of occurrences of each symbol for all $B_{j-1}(h)$ in memory, which takes $O(\sigma^2 \log(mk))$ bits of space. For $B_{j-1}(z)$, $c_z = S[k-j]$, the pointer value corresponding to S—which we read from $P_{j-1}(h)$—tells us how far along the count needs to proceed in $B_{j-1}(z)$. So for each B_{j-1} we need $O(\sigma \log(mk))$ bits of space: a trivial amount for DNA data, although potentially an issue for very large alphabets.

We can summarize the steps at the iteration j in the following way:

1. For each symbol c_v, with $0 \le v \le \sigma$ and for each element q (we observe that the maximum value of q depends on the number of the $(j-1)$-suffixes starting with c_v), we know:
   - The number of the sequence $i = N_{j-1}(v)[q]$ (clearly $S_i[k-(j-1)] = c_v$).
   - $r = P_{j-1}(v)[q]$ (it means that $c_z = S_i[k-j]$ has been inserted in the position r of $B_{j-1}(v)$ at the end of the previous step).
   - By using c_z, r and c_v, we compute $r' = \mathsf{rank}(c_z, r, c_v)$ (see Lemma 1), i.e. the position where we have to insert the new symbol in $B_j(z)$. We store r into $P_j(z)$.
   - We store i into $N_j(z)$.
   - We observe that we do not need to store c_z, because we can read the symbol c_z from $B_{j-1}(v)$ when we compute the new position.
2. For each symbol c_z, with $0 \le z \le \sigma$, we sort the pair $(P_j(z), N_j(z))$ in ascending order, where $P_j(z)$ is the primary key.
3. For each symbol c_z, with $0 \le z \le \sigma$, and for each element q (where the maximum value of q depends on the number of the j-suffixes starting with c_z), we insert the new symbol associated with j-suffix of the string S_i, where $i = N_j(z)[q]$, into $B_j(z)$ in the position $P_j(z)[q]$.
4. Return B_j, P_j, N_j.

Now, we are ready to describe our algorithms. Both are based on the above description, but they differ mainly in the data structures used. In particular, the first algorithm uses more internal memory and less time, whereas for small alphabets the second algorithm uses almost no memory at all.

Algorithm 1 - BCR. In the above description, we used $P_j(h)$ and $N_j(h)$ for each symbol c_h, with $0 \leq h \leq \sigma$, whereas in the implementation of the first algorithm, for each iteration j, we allocate a unique array P_j for all $P_j(h)$ and a unique array N_j for all $N_j(h)$ in internal memory. We observe that P_j and N_j contain exactly one integer for each sequence in the collection, P_j uses $O(m \log(mk))$ bits of workspace and N_j uses $O(m \log m)$ bits of workspace. Since $P_j[q]$, for some q, denotes the position into $B_j(z)$ of the new symbol associated with the j-suffix $S_i[k - j, k]$ starting with $c_z = S_i[k - j]$ and $i = N_j[q]$, we need another array Q_j, setting $Q_j[q] = z$. It uses $O(m \log \sigma)$ bits of workspace. We do not want to read the σ external files containing the BWT segments B_j more than once and since the values in P_j are absolute positions (see the above description), we need to sort the values in P_j before inserting the new symbols. The first, second and third keys of the sort are the values in Q_j, P_j and N_j respectively. We do not need to store the associated suffixes in memory, so this algorithm uses $O(m \log(mk))$ bits of workspace and $O(k \, \mathrm{sort}(m))$ of time, where $\mathrm{sort}(m)$ is the time needed to sort Q_j, P_j and N_j. We can observe that if we store Q_j, P_j, N_j in external files and use an external sorting algorithm, we could significantly reduce the workspace.

Algorithm 2 - BCRext. Our second algorithm is based on least-significant-digit radix sort. For this variant, sorting of arrays is not required because the sequences themselves are sorted externally. At the start of iteration j, the elements of S are assumed to be ordered by $(j - 1)$-suffix, this ordering being partitioned into external files $T_{j-1}(1), \ldots, T_{j-1}(\sigma)$ according to the first characters of the $(j - 1)$-suffixes. Files $P_{j-1}(1), \ldots, P_{j-1}(\sigma)$ are such that $P_{j-1}(h)$ contains the positions of the $(j - 1)$-suffixes in $B_{j-1}(h)$, ordered the same way.

All files are assumed to be accessed sequentially via read-only R() or write-only W() file streams. In the order $h = 1, \ldots, \sigma$, we open read-only file streams to each of $T_{j-1}(h)$ and $P_{j-1}(h)$, while *two* read-only file streams $R_1(B_{j-1}(h))$ and $R_2(B_{j-1}(h))$ reading from each segment of the partial BWT remain open throughout the iteration.

Reading a string $S \in$ S from $R(T_{j-1}(h))$ and its associated pointer P (which points to the position of its $(j - 1)$-suffix in $B_{j-1}(h)$) from $R(P_{j-1}(h))$), each S is then placed into one of σ distinct output files $T_j(1), \ldots, T_j(\sigma)$ according to the value of $S[k - j]$. Once all the sequences are processed, reading these files in the order $T_j(1) \ldots T_j(\sigma)$ forms the j-suffix ordering of the collection S that is needed for the next iteration.

The key observation here is that since the strings of S are presented in $(j - 1)$-suffix order, so also must be the subset whose $(j - 1)$-suffixes share a common first symbol c_h. Thus we use $R_1(B_{j-1}(h))$ to count the number of occurrences of each symbol seen so far in $B_{j-1}(h)$, keeping track of how far into $B_{j-1}(h)$ we have read so far. We then read forward to the position pointed to by P, updating the counts as we go. Since the strings are processed in $(j-1)$-suffix order, we never need to backtrack.

Having determined where to put the new BWT symbol $S[k-j-1]$ in $B_{j-1}(z)$, where $c_z = S[k-j]$, we use $\mathsf{R}_2(B_{j-1}(z))$ to read up to that position, then write those symbols plus the appended $S[k-j-1]$ to $\mathsf{W}(B_j(z))$. All strings S' whose symbols need to be inserted into $B_{j-1}(z)$ arrive in $(j-1)$-suffix order and also satisfy $S'[k-j] = c_z$. They are therefore j-suffix ordered so, again, we never need to backtrack.

Finally, we must write to $\mathsf{W}(P_j(z))$ the entry that corresponds to S. To do this, we need to keep count of the number of additional symbols that have so far been inserted between the symbols from $B_{j-1}(z)$ and sent to $\mathsf{W}(B_j(z))$. This provides an offset that must be added to the number of symbols read from $\mathsf{R}_2(B_{j-1}(z))$ so far to create the value we need.

Once the last element of $T_{j-1}(\sigma)$ has been read, we update the cumulative count values to reflect any symbols not yet read from each $\mathsf{R}_1(B_{j-1}(h))$ and send any symbols not yet read from $\mathsf{R}_2(B_{j-1}(h))$ to $\mathsf{W}(B_j(h))$.

Figure 1 uses a simple example to illustrate how the data structures associated with both variants of the algorithm are updated during an iteration.

4   Computational Experiments

We tested our approach on subsets of a collection of one billion human DNA sequences, each one 100 bases long, sequenced from a well-studied African male individual [2] (available from the Sequence Read Archive [6] using the accession number ERA015743[2]). To prove the low resource needs of our algorithms, all tests were carried out on one of two identical machines, each having 16Gbytes memory and two Intel Xeon X5450 (Quad-core) 3GHz processors (we only used one processor for our tests). Each machine was directly connected to its own array of 146Gbytes SAS disks in RAID6 configuration, each array having a Hewlett-Packard P6000 RAID controller with 512Mbytes cache. We had exclusive access to both test machines and their associated disk arrays for the duration of our experiments.

We developed prototypical implementations BCR and BCRext of the two algorithms described in Section 3. The code is available upon request from the authors.

For smaller input instances, we compared these programs to bwte from the bwtdisk toolkit (version 0.9.0[3]), which implements the blockwise BWT construction algorithm described in [3]. Since bwte constructs the BWT of a single string, we concatenated our string collections into this form using 0 as a delimiter, choosing 0 because it is lexicographically smaller than any A, C, G, T, or N. An entirely like-for-like comparison would use a different end marker for each string, however the bwte implementation does not support the many millions of distinct end marker symbols this would require.

---

[2] Available at ftp://ftp.sra.ebi.ac.uk/vol1/ERA015/ERA015743/srf/
[3] Available at http://people.unipmn.it/manzini/bwtdisk/

$B_5(0)$	Suffixes
0	C $\$_1$
1	C $\$_2$
2	T $\$_3$

$B_6(0)$	Suffixes
0	C $\$_1$
1	C $\$_2$
2	T $\$_3$

BCR :
Read in sequences
in Q and P order

$B_5(1)$	Suffixes
0	C $AAC\$_1$
1	A $AC\$_1$
2	**G** **$AGCTC\$_2$**

$P_5 = [2, 3, 4]$ $\qquad P_6 = [0, 1, 2]$
$N_5 = [2, 1, 3] \Rightarrow N_6 = [2, 1, 3]$
$Q_5 = [1, 2, 2]$ $\qquad Q_6 = [3, 3, 4]$

$B_6(1)$	Suffixes
0	C $AAC\$_1$
1	A $AC\$_1$
2	G $AGCTC\$_2$

$B_5(2)$	Suffixes
0	A $C\$_1$
1	T $C\$_2$
2	C $CAAC\$_1$
3	**G** **$CCAAC\$_1$**
4	**T** **$CGCTT\$_3$**
5	G $CTC\$_2$
6	G $CTT\$_3$

$\Rightarrow$

$B_6(2)$	Suffixes
0	A $C\$_1$
1	T $C\$_2$
2	C $CAAC\$_1$
3	G $CCAAC\$_1$
4	T $CGCTT\$_3$
5	G $CTC\$_2$
6	G $CTT\$_3$

$B_5(3)$	Suffixes
0	A $GCTC\$_2$
1	C $GCTT\$_3$

BCRext :
Read in sequences
in 5-suffix order

$B_6(3)$	Suffixes
0	**T** **$GAGCTC\$_2$**
1	**A** **$GCCAAC\$_1$**
2	A $GCTC\$_2$
3	C $GCTT\$_3$

$P_5(0) = []$ $\qquad P_6(0) = []$
$P_5(1) = [2]$ $\qquad P_6(1) = []$
$P_5(2) = [3, 4] \Rightarrow P_6(2) = []$
$P_5(3) = []$ $\qquad P_6(3) = [0, 1]$
$P_5(4) = []$ $\qquad P_6(4) = [2]$

$B_5(4)$	Suffixes
0	T $T\$_3$
1	C $TC\$_2$
2	C $TT\$_3$

$B_6(4)$	Suffixes
0	T $T\$_3$
1	C $TC\$_2$
2	**G** **$TCGCTT\$_3$**
3	C $TT\$_3$

Fig. 1. Iteration 6 of the computation of the BWT of the collection $S = \{TGCCAAC,$ $AGAGCTC, GTCGCTT\}$ on the alphabet $\{A, C, G, T\}$. The two columns represent the partial BWT before and after the iteration and, in between, we see how the auxiliary data stored by the two variants of the algorithm changes during the iteration. The positions of the new symbols corresponding to the 6-suffixes (shown in bold on the right) are computed from the positions of the 5-suffixes (in bold on the left), which were retained in the arrays P after the previous iteration. For clarity, we give distinct subscripts to the end markers of each of the sequences in the collection.

Intuitively, the BWT is more work to compute for a string of size km than for a collection of m strings of length k, since the number of symbol comparisons needed to decide the order of two suffixes is not bounded by k. In our particular case, however, the periodic nature of the concatenated string means that 99 out of 100 suffix/suffix comparisons will still terminate within 100 symbols, because one suffix will hit 0 but the other will not, the only exception being the case where both suffixes start at the same position in different strings. The problem bwte is being asked to solve is therefore of comparable complexity to ours. We ran bwte using 4Gbytes of memory, the maximum amount of memory the program allowed us to specify.

We also constructed the compressed suffix array (CSA) on smaller instances using Sirén's program rlcsa [20][4]. On those input instances rlcsa poses an interesting alternative, especially since this algorithm is geared towards indexing

---

[4] Available at http://www.cs.helsinki.fi/group/suds/rlcsa/rlcsa.tgz, version tested was downloaded on 8th December 2010.

text collections as well. We split the input data into 10 batches, constructing a separate index for each batch then merging the indexes afterwards. With increasing data volumes, however, the computational requirements for constructing the CSA become prohibitive on our testing environment. In [20, Section 6], 8 threads and up to $36 - 37$Gbytes of memory are used to construct the CSA of a text collection 41.48Gbytes in size, although we note the author describes other variants of the algorithm that would use less RAM than this.

Table 1 gives the results for all the input instances that we generated. The first two (0043M and 0085M) were sized to match the largest datasets considered in [3]. We created the latter three to show the effectiveness of our approach on very large string collections. rlcsa and bwte show efficiency (defined as user CPU time plus system CPU time as a fraction of wallclock time) approaching 100% in all our experiments (not only on our test machines, but on other servers as well), whereas both BCR and BCRext exhibit a drop in efficiency for large datasets. Even so, BCR and BCRext are both able to process the 1000M dataset at a rate that exceeds the performance of bwte on the 0085M dataset, which is less than one tenth of the size. This efficiency loss, which we believe is due to the internal cache of the disk controller becoming saturated, starts to manifest itself on smaller datasets for BCRext than for BCR, which is likely to be a consequence of the greater I/O demands of BCRext . Since the I/O of BCRext is dominated by the repeated copying of the input sequences during the radix sort, we modified BCRext to minimise the data read to and from disk during this activity.

In initial experiments with the zlib[5] library, the CPU overhead of on-the-fly compression and decompression of the input sequences to and from gzip format more than outweighed any possible efficiency gain that could arise from the reduced file sizes. We had more success by using a 4-bits-per-base encoding and by observing that, during a given iteration, we do not need to copy the entire input sequences but only the prefixes that still remain to be sorted in future iterations. The resulting new version BCRext++ was otherwise identical to BCRext but reduced the processing time for the 1 billion read dataset by 47%, with even greater gains on the smaller datasets.

To see how performance scales with respect to sequence length, we concatenated pairs of sequences from our collection of 100 million 100-mers to create a set of 50 million 200-mers. While this collection contains the same number of bases, BCRext and BCRext++ both needed a similar proportion of additional time to create the BWT (69% and 67% respectively), whereas the time needed by BCR was only 29% more than was required for the original collection. The likely explanation for the difference is that the radix sort performed by BCRext and BCRext++ requires twice as much I/O for the 200-mer dataset than for the original collection.

To look further at the relationship between disk hardware and efficiency, we also performed some tests on a machine whose CPU was identical to those used for our previous tests but that was also equipped with a solid state hard drive

---

[5] Available at http://www.zlib.net

Table 1. The input string collections were generated on an Illumina GAIIx sequencer, all reads are 100 bases long. We chose the first two instances to have data sets comparable in size to the largest ones tested in [3]. Size is the input size in gigabytes, wall clock time—the amount of time that elapsed from the start to the completion of the instance—is given as microseconds per input base, and memory denotes the maximal amount of memory (in gigabytes) used during execution. `BCRext` and `BCRext++` need to store only a constant and (for the DNA alphabet) negligibly small number of integers in RAM regardless of the size of the input data, we therefore state a -. The efficiency column gives the CPU efficiency values, *i.e.* the proportion of time for which the CPU was occupied and not waiting for I/O operations to finish, as taken from the output of the `/usr/bin/time` command. Some of the tests were repeated on a solid-state hard drive (SSD), the results from these are shown last. For all tests, the best wall clock time achieved is marked in bold.

instance	size	program	wall clock	efficiency	memory
0043M	4.00	bwte	5.00	0.99	4.00
	4.00	rlcsa	2.21	0.99	7.10
	4.00	BCR	0.99	0.84	0.57
	4.00	BCRext	2.15	0.58	−
	4.00	BCRext++	**0.93**	0.66	−
0085M	8.00	bwte	7.99	0.99	4.00
	8.00	rlcsa	2.44	0.99	13.40
	8.00	BCR	1.01	0.83	1.10
	8.00	BCRext	4.75	0.27	−
	8.00	BCRext++	**0.95**	0.69	−
0100M	9.31	BCR	**1.05**	0.81	1.35
	9.31	BCRext	4.6	0.28	−
	9.31	BCRext++	1.16	0.61	−
0800M	74.51	BCR	**2.25**	0.46	10.40
	74.51	BCRext	5.61	0.22	−
	74.51	BCRext++	2.85	0.29	−
1000M	93.13	BCR	5.74	0.19	13.00
	93.13	BCRext	5.89	0.21	−
	93.13	BCRext++	**3.17**	0.26	−
0085M	8.00	bwte	8.11	0.99	4.00
(SSD)	8.00	rlcsa	2.48	0.99	13.40
	8.00	BCR	0.78	0.99	1.10
	8.00	BCRext	0.89	0.99	−
	8.00	BCRext++	**0.58**	0.99	−
1000M	93.13	BCR	**0.98**	0.91	13.00
(SSD)	93.13	BCRext++	1.24	0.64	−

(SSD)[6]. Since both `rlcsa` and `bwte` already operate at close to maximum efficiency, we would not expect the run time of these programs to benefit from the faster disk access speed of the SSD and their performance when the 0085M dataset was stored on the SSD was in line with this expectation. However the

---

[6] We used an OCZ Technology R2 p88 Z-Drive with 1Tbyte capacity and a claimed maximum data transfer rate of 1.4Gbytes per second.

SSD greatly improved the efficiency of our algorithms, reducing the run times of BCRext and BCRext++ on the 1000M dataset by more than 5-fold and 2-fold respectively, meaning that the BWT of 1 billion 100-mers was created in just over 27 hours using BCR, or 34.5 hours with BCRext++.

5 Discussion

The algorithms we describe here represent a step towards making the BWT a practical tool for processing of the vast collections of strings that are generated by modern DNA sequencers. Their effectiveness on very large datasets was discussed in Section 4, but a further aspect of practical relevance is that our transformations are reversible: the inversion procedure closely follows that of the original BWT and that in [13]. This is especially important if one wants to use the (compressed) BWT as an archival format that allows us to extract the original strings. The idea is to define a permutation π on bwt(S) and F, that is the symbols of bwt(S) in lexicographic order. We are able to decompose the permutation π into disjoint cycles: $\pi = \pi_1 \pi_2 \cdots \pi_m$. Each cycle π_i corresponds to a conjugacy class of a string in S.

Following the same reasoning, we can delete any string S in S and obtain the BWT of $S \setminus \{S\}$. Adding new strings to an existing BWT is feasible as well. We simply use Lemma 1 and obtain the BWT of $S \cup \{S\}$. In both cases, there is no need to construct the BWT from scratch.

Finally, note that BCR allows a certain degree of parallelization. The computation of the new positions is independent of each other and is thus easily parallelizable. Inserting the new symbols into the partial BWTs can be done in parallel as well. This allows us to use multiple processors on multi-core servers that are commonplace nowadays while keeping the computational requirements low.

Acknowledgments. The authors would like thank Kimmo Kallio for setting up the test environment and many helpful discussions on various test-related issues, Jochen Singer for contributing at an early stage of this project and Dirk Evers and the anonymous referees for their constructive comments on earlier versions of this manuscript. The third author wishes to thank Illumina Cambridge Ltd for financial support.

References

1. Adjeroh, D., Bell, T., Mukherjee, A.: The Burrows-Wheeler Transform: Data Compression, Suffix Arrays, and Pattern Matching, 1st edn. Springer, Heidelberg (2008)
2. Bentley, D.R., et al.: Accurate whole human genome sequencing using reversible terminator chemistry. Nature 456(7218), 53–59 (2008)
3. Ferragina, P., Gagie, T., Manzini, G.: Lightweight data indexing and compression in external memory. In: López-Ortiz, A. (ed.) LATIN 2010. LNCS, vol. 6034, pp. 697–710. Springer, Heidelberg (2010)
4. Ferragina, P., Manzini, G.: Opportunistic data structures with applications. In: Proceedings of the 41st Annual Symposium on Foundations of Computer Science, Washington, DC, USA, pages 390. IEEE Computer Society, Los Alamitos (2000)

5. Ferragina, P., Manzini, G.: Indexing compressed text. J. ACM 52, 552–581 (2005)
6. National Center for Biotechnology Information. Sequence Read Archive, http://trace.ncbi.nlm.nih.gov/Traces/sra/sra.cgi?
7. Gonnet, G.H., Baeza-Yates, R.A., Snider, T.: New indices for text: PAT Trees and PAT arrays, pp. 66–82. Prentice-Hall, Inc., Upper Saddle River (1992)
8. Hon, W.K., Lam, T.W., Sadakane, K., Sung, W.K., Yiu, S.M.: A space and time efficient algorithm for constructing compressed suffix arrays. Algorithmica 48, 23–36 (2007)
9. Kärkkäinen, J., Sanders, P., Burkhardt, S.: Linear work suffix array construction. J. ACM 53, 918–936 (2006)
10. Kim, D., Sim, J., Park, H., Park, K.: Linear-time construction of suffix arrays. In: Baeza-Yates, R., Chávez, E., Crochemore, M. (eds.) CPM 2003. LNCS, vol. 2676, pp. 186–199. Springer, Heidelberg (2003)
11. Ko, P., Aluru, S.: Space efficient linear time construction of suffix arrays. Journal of Discrete Algorithms 3(2-4), 143–156 (2005)
12. Mantaci, S., Restivo, A., Rosone, G., Sciortino, M.: An extension of the burrows wheeler transform and applications to sequence comparison and data compression. In: Apostolico, A., Crochemore, M., Park, K. (eds.) CPM 2005. LNCS, vol. 3537, pp. 178–189. Springer, Heidelberg (2005)
13. Mantaci, S., Restivo, A., Rosone, G., Sciortino, M.: An extension of the Burrows-Wheeler Transform. Theor. Comput. Sci. 387(3), 298–312 (2007)
14. Mantaci, S., Restivo, A., Rosone, G., Sciortino, M.: A new combinatorial approach to sequence comparison. Theory Comput. Syst. 42(3), 411–429 (2008)
15. Metzker, M.L.: Sequencing technologies – the next generation. Nature Reviews Genetics 11(1), 31–46 (2009)
16. Nong, G., Zhang, S., Chan, W.H.: Linear time suffix array construction using d-critical substrings. In: Kucherov, G., Ukkonen, E. (eds.) CPM 2009 Lille. LNCS, vol. 5577, pp. 54–67. Springer, Heidelberg (2009)
17. Puglisi, S.J., Smyth, W.F., Turpin, A.H.: A taxonomy of suffix array construction algorithms. ACM Comput. Surv. 39 (July 2007)
18. Walenz, B.P., Lippert, R.A., Mobarry, C.M.: A Space-Efficient Construction of the Burrows-Wheeler Transform for Genomic Data. Journal of Computational Biology 12(7), 943–951 (2005)
19. Simpson, J.T., Durbin, R.: Efficient construction of an assembly string graph using the FM-index. Bioinformatics 26(12), i367–i373 (2010)
20. Sirén, J.: Compressed suffix arrays for massive data. In: Karlgren, J., Tarhio, J., Hyyrö, H. (eds.) SPIRE 2009. LNCS, vol. 5721, pp. 63–74. Springer, Heidelberg (2009)

Palindrome Pattern Matching

Tomohiro I[1], Shunsuke Inenaga[2], and Masayuki Takeda[1]

[1] Department of Informatics, Kyushu University
[2] Graduate School of Information Science and Electrical Engineering,
Kyushu University
744 Motooka, Nishiku, Fukuoka 819–0395, Japan
{tomohiro.i,takeda}@inf.kyushu-u.ac.jp,
inenaga@c.csce.kyushu-u.ac.jp

Abstract. A palindrome is a string that reads the same forward and backward. For a string x, let $Pals(x)$ be the set of all maximal palindromes of x, where each maximal palindrome in $Pals(x)$ is encoded by a pair (c, r) of its center c and its radius r. Given a text t of length n and a pattern p of length m, the palindrome pattern matching problem is to compute all positions i of t such that $Pals(p) = Pals(t[i : i + m - 1])$. We present linear-time algorithms to solve this problem.

1 Introduction

A palindrome is a symmetric string that reads the same forward and backward. Namely, a string w is a palindrome if $w = xax^R$ where x is a string, x^R is a reversal of x, and a is either a single character or the empty string.

Recently, palindromic structures in strings have been extensively studied: A string of length n is called *palindromic rich* (or simply *rich*) if it contains $n + 1$ distinct palindromes (including the empty string). It is known that any string of length n can contain at most $n + 1$ distinct palindromes [6]. A unified study of palindromic richness of finite and infinite strings was initiated in [7]. A close relationship between palindromic richness and the Burrows-Wheeler transform [5] was recently discovered in [16]. Another concept regarding palindromic structures is *palindrome complexity* [1,4,2] of a string, which is the number of palindromic substrings of a given length in the string.

There exist several efficient algorithms that solve interesting problems on palindromes: A linear-time algorithm to check if a given string is palindromic rich or not, is presented in [8]. One can compute the set of all maximal palindromes of a given string in linear time [13]. The reverse engineering problem of computing a string from a given set of maximal palindromes is solvable in linear time [11], and its closely related problem is also considered in [14].

In this paper, we introduce a new paradigm of pattern matching based on palindromes in strings. Two strings of same length m are said to be *pal-equivalent* iff the length of the maximal palindrome at every center in the strings is equal [11]. Given a text string t and a pattern string p, we are interested in finding all text positions i ($1 \leq i \leq n$) such that p and $t[i : i + m - 1]$ are

R. Giancarlo and G. Manzini (Eds.): CPM 2011, LNCS 6661, pp. 232–245, 2011.
© Springer-Verlag Berlin Heidelberg 2011

pal-equivalent, where n and m are text and pattern lengths, respectively. This problem is called the *palindrome pattern matching*.

It is not difficult to see that the palindrome pattern matching problem can be solved in $O(nm)$ time: We pre-compute all maximal palindromes for t and p using linear time algorithms [13,9]. For every text position i, we compare the length of the maximal palindromes of t at position $i + j - 1$ and that of p at position j for every $1 \leq j \leq m$. If a maximal palindrome of the text "goes over" the interval $[i : i+j-1]$, then the left and right arms of the maximal palindrome are trimmed accordingly for comparison.

There exists a linear-time algorithm for small alphabets. In [11] it was shown that if the alphabet size is at most 3, then two strings are pal-equivalent iff those strings parameterized match [3]. Hence the palindrome pattern matching can be solved in $O(n + m)$ time for ternary and smaller alphabets.

In this paper, we present efficient solutions for larger alphabets. Firstly, we present an algorithm which solves the problem in $O(n+m)$ time for *arbitrary* alphabets. This algorithm is a palindrome-pattern-matching version of the Morris-Pratt [15] pattern matching algorithm. Secondly, we propose another algorithm that uses a new text indexing structure called the *palindrome suffix trees*. We show that palindrome suffix trees can be constructed in $O(n \log \sigma)$ time, where σ is the alphabet size. Using the palindrome suffix tree, we can solve the problem in $O(m \log \sigma + r)$ time, where r is the number of text positions to report.

The algorithms of this paper are applicable to several practical problems, e.g., in bioinformatics. For instance, similar palindromic sequences often need to be identified in DNA and RNA sequence analysis [9]. Sequences having similar palindromic structures may code for similar 3-D structures of the respective molecules, leading to possible functional interpretation of the identified sequences. Due to the size of genomes, efficiency of search methods is of great importance.

2   Preliminaries

Let Σ be a finite *alphabet*. An element of Σ^* is called a *string*. The length of a string w is denoted by $|w|$. The empty string ε is a string of length 0, that is, $|\varepsilon| = 0$. Let $\Sigma^+ = \Sigma^* - \{\varepsilon\}$. For a string $w = xyz$, x, y and z are called a *prefix*, *substring*, and *suffix* of w, respectively. The i-th character of a string w is denoted by $w[i]$ for $1 \leq i \leq |w|$, and the substring of a string w that begins at position i and ends at position j is denoted by $w[i : j]$ for $1 \leq i \leq j \leq |w|$. For convenience, let $w[i : j] = \varepsilon$ if $j < i$.

For any string w, let w^R denote the reversed string of w, that is, $w^R = w[|w|] \cdots w[2]w[1]$. A string w is called a *palindrome* if $w = w^R$. If $|w|$ is even, then w is called an *even palindrome*, that is, $w = xx^R$ for some $x \in \Sigma^*$. If $|w|$ is odd, then w is called an *odd palindrome*, that is, $w = xax^R$ for some $x \in \Sigma^*$ and $a \in \Sigma$. The *radius* of a palindrome w is $\frac{|w|}{2}$.

The *center* of a palindromic substring $w[i : j]$ of a string w is $\frac{i+j}{2}$. A palindromic substring $w[i : j]$ is called the *maximal palindrome* at the center $\frac{i+j}{2}$ if no other palindromes at the center $\frac{i+j}{2}$ have a larger radius than $w[i:j]$, i.e., if $w[i-1] \neq w[j+1]$, $i = 1$, or $j = |w|$. In particular, a maximal palindrome $w[i:|w|]$ is called a *suffix palindrome* of w.

Let $Pals(w)$ be the set of all center-distinct maximal palindromes where each element is encoded by a pair of its center and radius, namely,

$$Pals(w) = \left\{ (c,r) \; \middle| \; \begin{array}{l} w[c-r+0.5:c+r-0.5] \text{ is a maximal palindrome} \\ \text{at center } c = 1, 1.5, 2, \ldots, n \end{array} \right\},$$

Also, let

$$SPals(w) = \{(c,r) \mid (c,r) \in Pals(w), c + r - 0.5 = |w|\},$$

namely, $SPals(w)$ represents the set of all suffix palindromes of w.

For example, let $w = \mathtt{abbacabbba}$. Then

$$\begin{aligned} Pals(w) = \{ & (1,0.5),(1.5,0),(2,0.5),(2.5,2),(3,0.5),(3.5,0),(4,0.5),(4.5,0), \\ & (5,3.5),(5.5,0),(6,0.5),(6.5,0),(7,0.5),(7.5,1),(8,2.5),(8.5,1), \\ & (9,0.5),(9.5,0),(10,0.5) \} \text{ and} \end{aligned}$$
$$SPals(w) = \{(8,2.5),(10,0.5)\}.$$

Theorem 1 ([13]). *For any string w of length m, $Pals(w)$ can be computed in $O(m)$ time.*

Throughout this paper, we assume that the elements of $Pals(w)$ are sorted in increasing order of centers c. Actually, the algorithm of [13] computes the elements of $Pals(w)$ in this order.

In this paper, we tackle the following problem.

Problem 1 (Palindrome pattern matching, pal-matching in short). Given a text string t of length n and a pattern string p of length m, compute all positions i of t such that $Pals(p) = Pals(t[i:i+m-1])$.

3   Linear-Time Palindrome Pattern Matching Algorithm

To achieve a linear time solution to Problem 1, we design a pal-matching version of the Morris-Pratt algorithm [15].

Definition 1. *A palindrome border (pal-border in short) of a string p of length m is any integer j s.t. $0 \leq j < m$ and $Pals(p[1:j]) = Pals(p[m-j+1:m])$.*

For example, the set of pal-borders of string $p = \mathtt{aabcdaacdbcc}$, is $\{7,2,1,0\}$, since $Pals(\mathtt{aabcdaa}) = Pals(\mathtt{aacdbcc})$, $Pals(\mathtt{aa}) = Pals(\mathtt{cc})$, $Pals(\mathtt{a}) = Pals(\mathtt{c})$, and $Pals(\varepsilon) = Pals(\varepsilon)$.

Let $\mathcal{N}$ be the set of non-negative integers. For any string p of length m, let $Pal\_Border_p : \mathcal{N} \to \mathcal{N}$ be the function such that $Pal\_Border_p(m)$ equals the largest pal-border of string p. When clear from the context, we abbreviate Pal_Border_p as Pal_Border. Since $Pal\_Border(m)$ is strictly smaller than m, we finally obtain 0 by iteratively applying the function Pal_Border to m. For any function $f : \mathcal{N} \to \mathcal{N}$ and any $m, k \in \mathcal{N}$, we define $f^k(m)$ as follows: $f^k(m) = f(m)$ if $k = 1$, and $f^k(m) = f(f^{k-1}(m))$ if $k \geq 2$. Similar to a standard border of a string [15], the following lemma holds.

Lemma 1. *For any string p of length m, let k be the smallest integer such that $Pal\_Border^k(m) = 0$. Then*

$$Pal\_Border(m), Pal\_Border^2(m), \dots, Pal\_Border^k(m)$$

are all the pal-borders of p with $m > Pal\_Border(m) > Pal\_Border^2(m) > \cdots > Pal\_Border^k(m) = 0$.

Definition 2. *The* palindrome border array *(pal-border array) β_p of a string p of length m is an integer array of length m such that $\beta_p[i] = Pal\_Border_{p[1:i]}(i)$ for each $1 \leq i \leq m$.*

For example, for string $p = $ aabbaa, we have $\beta_p = [0, 1, 1, 2, 3, 4]$. When it is clear from the context, we abbreviate β_p as β.

In what follows, we present how to compute the pal-border array β_p of a given string p in linear time.

For any string w of length $m \geq 1$, let $Lpal_w$ be an integer array of length m such that

$$Lpal_w[i] = \max\{i - k + 1 \mid w[k : i] = w[k : i]^R, 1 \leq k \leq i\}.$$

That is, the value of $Lpal_w[i]$ is equal to the length of the longest palindrome that ends at position i in w, for every $1 \leq i \leq m$[1]. Note that the above palindrome $w[k : i]$ is not necessarily a maximal palindrome at center $\frac{k+i}{2}$ in w.

For example, for string $w = $ abbacabbba, $Lpal_w = 1\ 1\ 2\ 4\ 1\ 3\ 5\ 7\ 3\ 5$.

The following lemma is a key to solve Problem 1 of pal-matching.

Lemma 2. *For any strings $w, z \in \Sigma^+$, $Pals(w) = Pals(z)$ iff $Lpal_w = Lpal_z$.*

Proof. ($\Longrightarrow$) We prove the claim by contradiction. Assume for contrary that $Lpal_w \neq Lpal_z$. Then there exists position i such that $Lpal_w[i] \neq Lpal_z[i]$. Assume w.l.o.g. that $Lpal_w[i] < Lpal_z[i]$. Let $k = (Lpal_z[i])/2$. The radius of the maximal palindrome centered at position $i - k + 0.5$ of w is less than k, however, that of the maximal palindrome centered at position $i - k + 0.5$ of z is at least k. This contradicts the assumption that $Pals(w) = Pals(z)$. Hence if $Pals(w) = Pals(z)$, then $Lpal_w = Lpal_z$.

($\Longleftarrow$) We prove the claim by contradiction and infinite descent. Assume for contrary that $Pals(w) \neq Pals(z)$. Then there exists center c such that $(c, r) \in Pals(w)$, $(c, u) \in Pals(z)$, and $r \neq u$. Assume w.l.o.g. that $r < u$.

In what follows, we consider position $j = \lceil c + u \rceil - 1$.

1. When $Lpal_w[j] < 2u$. Since $(c, u) \in Pals(z)$, $Lpal_z[j] \geq 2u$. This contradicts the assumption that $Lpal_w = Lpal_z$.

---

[1] The notion of $Lpal_w[i]$ was previously introduced in [8], denoted LPS[i] therein.

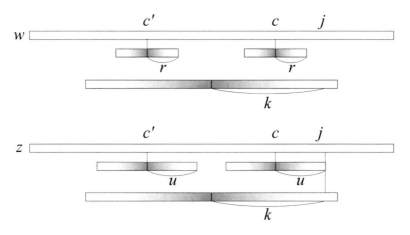

Fig. 1. Illustration for infinite descent in the proof of Lemma 2

2. When $Lpal_w[j] \geq 2u$. Let $k = (Lpal_w[j])/2$. Then clearly w has a palindrome that is centered at $j - k + 0.5$ and is of radius k. Also z has a palindrome that is centered at $j - k + 0.5$ and is of radius k, since otherwise it contradicts the assumption that $Lpal_w = Lpal_z$. Then there exists center $c' < c$ such that $(c', r) \in Pals(w)$, $(c', u) \in Pals(z)$, and $r < u$. (See also Fig. 1.)

The same must hold for those smaller centers, ad infinitum. However, this is impossible since w and z are finite strings.

Hence if $Lpal_w = Lpal_z$, then $Pals(w) = Pals(z)$.   □

It is shown in [8] that $Lpal_w$ can be computed in linear time from $Pals(w)$. The following lemma is essentially the same as what is claimed in [8], but is more specifically tailored for our needs.

Lemma 3. *Let w be any string of length m. Given $Pals(w)$, $Lpal_w$ can be computed in $O(m)$ time, in an on-line fasion, from $Lpal_w[1]$ to $Lpal_w[m]$.*

Proof. For any position i of w with $1 \leq i \leq m$, the value of $Lpal_w[i]$ is equal to $2(i-c)+1$ where c is the smallest center of a maximal palindrome $(c, r) \in Pals(w)$ such that $c + r \geq i$. Hence we process the given string w from left to right.

Assume that we have computed $Lpal_w[1 : i]$ and let $(c, r) \in Pals(w)$ with $Lpal_w[i] = 2(i - c) + 1$. Now we compute $Lpal_w[i + 1]$. If $c + r \geq i + 1$, then $Lpal_{w[1:i+1]} = 2((i + 1) - c) + 1$. Otherwise, we increment the value of c by 0.5 until satisfying $c + r \geq i + 1$, where r is the radius of the maximal palindrome with the updated center c.

A pseudo-code of the algorithm is shown in Algorithm 1. The correctness should be clear from the above arguments. Note that the value of c does not decrease and does not exceed the value of i. Also, (c, r) can be picked up from $Pals(w)$ in constant time at each step, since $Pals(w)$ is sorted in increasing order of c. Consequently the time complexity is linear in m.   □

Algorithm 1. On-line algorithm to compute $Lpal_w$ of w

Input: String w of length m.
Output: $Lpal_w[1:m]$.
1 compute $Pals(w)$;
2 $c \leftarrow 1$; let $(c,r) \in Pals(w)$;
3 **for** $i \leftarrow 1$ **to** m **do**
4 $\quad$ **while** $c + r < i$ **do**
5 $\quad\quad$ $c \leftarrow c + 0.5$; let $(c,r) \in Pals(w)$;
6 $\quad$ $Lpal_w[i] \leftarrow 2(i - c) + 1$;

7 **return** $Lpal_w[1:m]$;

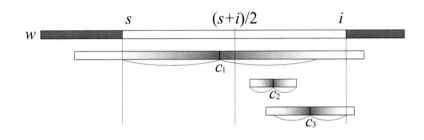

Fig. 2. If (c_3, r_3) is the maximal palindrome in $Pals(w)$ such that c_3 is the smallest center satisfying $c_3 \geq (s + i)/2$ and $c_3 + r_3 \geq i$, c_3 is the active center for s and i, and $Lpal_{w[s:i]}[i - s + 1] = 2(i - c_3) + 1$. Note that c_1 is not the active center for s and i since $c_1 < (s + i)/2$.

Let w be any string of length m, and let s and i be any integers with $1 \leq s \leq i \leq m$. Here we consider computing $Lpal_{w[s:i]}[i - s + 1]$ from $Pals(w)$. By the definition of $Lpal$, the value of $Lpal_{w[s:i]}[i - s + 1]$ is equal to $2(i - c) + 1$, where (c, r) is the maximal palindrome in $Pals(w)$ such that c is the smallest center satisfying $c \geq (s + i)/2$ and $c + r \geq i$ (See also Fig. 2). We call this center c *the active center* for s and i w.r.t. w, and denote it by $AC_w(s, i)$. It holds that $Lpal_{w[s:i]}[i - s + 1] = 2(i - AC_w(s, i)) + 1$.

Lemma 4. *Let w be any string of length m. For any integers s, i, s', i' with $1 \leq s \leq i \leq m$ and $1 \leq s' \leq i' \leq m$, if $s \leq s'$ and $i \leq i'$, then $AC_w(s, i) \leq AC_w(s', i')$.*

Proof. Assume for contrary that $AC_w(s, i) > AC_w(s', i')$. Since $AC_w(s, i) \leq i$, $AC_w(s', i') < i$. Let $(AC_w(s', i'), r) \in Pals(w)$. It follows from $AC_w(s', i') \geq (s' + i')/2 \geq (s + i)/2$ and $AC_w(s', i') + r \geq i' \geq i$ that $AC_w(s', i') \geq (s + i)/2$ and $AC_w(s', i') + r \geq i$. However this contradicts that $AC_w(s, i)$ is the active center for s and i w.r.t. w. $\quad\square$

In the algorithms which follow, we will need to know the value of $Lpal_{w[s:i]}[i - s + 1]$ for some s and i. It seems difficult to compute $Lpal_{w[s:i]}[i - s + 1]$ in constant time for "randomly" chosen s and i, with $O(m)$-time preprocessing.

Algorithm 2. Algorithm to compute β_p of a given string p

Input: String p of length m.

Output: Pal-border array $\beta_p[1 : m]$.

1 compute $Pals(p)$ and $Lpal_p[1 : m]$;

2 $\beta_p[1] \leftarrow 0$;

3 $j \leftarrow 0; c \leftarrow 0$;

4 **for** $i \leftarrow 2$ **to** m **do**

5   **while true do**

6          $c \leftarrow \max\{c, i - j/2\}$; let $(c, r) \in Pals(p)$;

7   **while** $c + r < i$ **do** /* Shift c to $AC_p(i - j, i)$.   */

8              $c \leftarrow c + 0.5$; let $(c, r) \in Pals(p)$;

        /* $2(i - c) + 1 = Lpal_{p[i-j:i]}[j + 1]$.   */

9   **if** $Lpal_p[j + 1] = 2(i - c) + 1$ **then** break;

10         $j \leftarrow \beta_p[j]$;

11     $j \leftarrow j + 1$;

12     $\beta_p[i] \leftarrow j$;

13 **return** $\beta_p[1 : m]$;

Nevertheless, Lemma 4 suggests that, if s and i monotonically increase from 1 to m, then the total cost for computing $Lpal_{w[s:i]}[i - s + 1]$ for all s and i never exceeds the number of the centers in w, which is $2m - 1$. The point is that all the following algorithms only require to compute $Lpal_{w[s:i]}[i - s + 1]$ for monotonically increasing positions s and i, with $1 \leq s \leq i \leq m$.

Lemma 5. *For any string p of length m, β_p can be computed in $O(m)$ time.*

Proof. Algorithm 2 describes our algorithm. This algorithm is mostly the same as the linear-time algorithm for computing a standard border array of a string [15], except that we match the values of $Lpal$ instead of characters.

We firstly compute $Pals(p)$ and $Lpal_p[1 : m]$. This takes $O(m)$ time by Theorem 1 and Lemma 3. Then we compute $\beta_p[1 : m]$ in ascending order. Consider the i-th iteration of the **for** loop of Line 2. Here we have computed $\beta_p[1 : i - 1]$, and variable j is set to be $\beta_p[i-1]$. Next we compute $Lpal_{p[i-j:i]}[j+1]$ by shifting the current center c right to $AC_p(i - j, i)$. If $Lpal_p[j + 1] = Lpal_{p[i-j:i]}[j + 1]$, $\beta_p[i] = j + 1$. Otherwise, we set j to be $\beta_p[j]$ and check again if $Lpal_p[j + 1] = Lpal_{p[i-j:i]}[j + 1]$ or not. The above procedure is repeated until j, such that $Lpal_p[j + 1] = Lpal_{p[i-j:i]}[j + 1]$, is found. Note that we break this loop at the latest when $j = 0$, since $Lpal_p[1] = Lpal_{p[i:i]}[1] = 1$.

In each iteration of the **for** loop of Line 2, the value of j increases by at most 1. Since each execution of the **while** loop of Line 2 decreases the value of j at least 1 and $j \geq 0$, the **while** loop of Line 2 is executed at most m times in total. Moreover, since the value of c does not decrease and does not exceed the value of i, the total cost of the **while** loop of Line 2 is $O(m)$. Therefore Algorithm 2 runs in time linear in m. □

Algorithm 3. Algorithm to solve pal-matching problem in linear time

---

Input: Text string t of length n and pattern string p of length m.
Output: All positions i of t such that $t[i : i + m - 1]$ pal-matches p.
1   compute $Pals(t)$, $Lpal_p[1 : m]$, and $\beta_p[1 : m]$;
2   $j \leftarrow 0$; $c \leftarrow 0$;
3   **for** $i \leftarrow 1$ **to** n **do**
4   **while true do**
5         $c \leftarrow \max\{c, i - j/2\}$; let $(c, r) \in Pals(t)$;
6   **while** $c + r < i$ **do** /* Shift c to $AC_t(i - j, i)$.   */
7           $c \leftarrow c + 0.5$; let $(c, r) \in Pals(t)$;
        /* $2(i - c) + 1 = Lpal_{t[i-j:i]}[j + 1]$.   */
8   **if** $Lpal_p[j + 1] = 2(i - c) + 1$ **then** break;
9         $j \leftarrow \beta_p[j]$;
10      $j \leftarrow j + 1$;
11   **if** $j = m$ **then**
12         $j \leftarrow \beta_p[j]$; **report** $i - m + 1$;

Theorem 2. *The pal-matching problem (Problem 1) can be solved in $O(n + m)$ time.*

Proof. Algorithm 3 describes our algorithm. This algorithm is a pal-matching version of the Morris-Pratt algorithm [15].

We firstly compute $Pals(p)$ by Algorithm 1 and $Lpal_p[1 : m]$ by Algorithm 2 in $O(m)$ time, and $Pals(t)$ in $O(n)$ time. Consider the i-th iteration of the **for** loop of Line 3. Here variable j represents an integer such that $p[1 : j]$ and $t[i - j : i - 1]$ pal-match. Next we compute $Lpal_{t[i-j:i]}[j + 1]$ by shifting the current center c right to $AC_t(i - j, i)$. If $Lpal_p[j + 1] = Lpal_{t[i-j:i]}[j + 1]$, we break the **while** loop of Line 3. Otherwise, we set j to be $\beta_p[j]$ and check again if $Lpal_p[j + 1] = Lpal_{t[i-j:i]}[j + 1]$ or not. The above procedure is repeated until j, such that $Lpal_p[j + 1] = Lpal_{t[i-j:i]}[j + 1]$, is found. Note that we break this loop at the latest when $j = 0$, since $Lpal_p[1] = Lpal_{t[i:i]}[1] = 1$. After breaking the **while** loop of Line 3, we increment j by 1, and if j becomes m, the algorithm reports that $t[i - m + 1 : i]$ and $p[1 : m]$ pal-match.

In each iteration of the **for** loop of Line 3, the value of j increases by at most 1. Since each execution of the **while** loop of Line 3 decreases the value of j at least 1 and $j \geq 0$, the **while** loop of Line 3 is executed at most n times in total. Moreover, since the value of c does not decrease and does not exceed the value of i, the total cost of the **while** loop of Line 3 is $O(n)$. Therefore Algorithm 3 runs in $O(n + m)$ time.   □

4   Palindrome Suffix Trees

In this section, we consider an indexing structure for pal-matching. We propose a new data structure called *palindrome suffix trees* (*pal-suffix trees* in short).

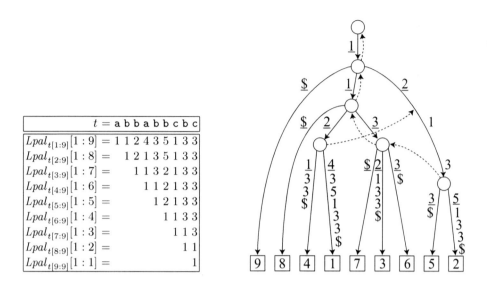

$t = \mathtt{a\,b\,b\,a\,b\,b\,c\,b\,c}$
$Lpal_{t[1:9]}[1:9] = 1\ 1\ 2\ 4\ 3\ 5\ 1\ 3\ 3$
$Lpal_{t[2:9]}[1:8] = \quad 1\ 2\ 1\ 3\ 5\ 1\ 3\ 3$
$Lpal_{t[3:9]}[1:7] = \qquad 1\ 1\ 3\ 2\ 1\ 3\ 3$
$Lpal_{t[4:9]}[1:6] = \qquad\quad 1\ 1\ 2\ 1\ 3\ 3$
$Lpal_{t[5:9]}[1:5] = \qquad\qquad 1\ 2\ 1\ 3\ 3$
$Lpal_{t[6:9]}[1:4] = \qquad\qquad\quad 1\ 1\ 3\ 3$
$Lpal_{t[7:9]}[1:3] = \qquad\qquad\qquad 1\ 1\ 3$
$Lpal_{t[8:9]}[1:2] = \qquad\qquad\qquad\quad 1\ 1$
$Lpal_{t[9:9]}[1:1] = \qquad\qquad\qquad\qquad 1$

Fig. 3. Illustration of $Pal\_ST(t)$ for string $t = \mathtt{abbabbcbc}$. The solid arrows represent the edges, and the broken arrows do the suffix links. The path from the root to each leaf s spells out $Lpal_{t[s:9]}[1:s]\$$.

The pal-suffix tree of a string t, denoted $Pal\_ST(t)$, is a compacted trie which represents $Lpal_{t[s:n]}[1 : n - s + 1]$ for all the suffixes $t[s : n]$ of t, where n is the length of t and $1 \le s \le n$. Each internal node of $Pal\_ST(t)$ has at least two children, and the labels of two distinct out-going edges of each internal node must start with distinct non-negative integers. Moreover, for $Pal\_ST(t)$ to have exactly n leaves, we use the following convention: Each leaf of $Pal\_ST(t)$ is uniquely labeled with integer s ($1 \le s \le n$) in such a way that the path from the root to leaf s spells out $Lpal_{t[s:n]}[1 : n - s + 1]\$$, where $\$$ is a special end-marker. The *length* of a node v, denoted $len(v)$, is the length of $Lpal$ represented by v. Fig. 3 illustrates $Pal\_ST(\mathtt{abbabbcbc})$.

Notice that there are $O(n)$ distinct values for the elements of $Lpal_t[1 : n]$. For instance, consider $t = (\mathtt{ab})^{\frac{n}{2}}$. Then $Lpal_t[1 : n] = 1\ 1\ 3\ 3 \cdots n-1\ n-1$. This suggests that an internal node of $Pal\_ST(t)$ might have $O(n)$ children. However, the following lemma holds.

Lemma 6. *For any string t, each node of $Pal\_ST(t)$ has at most σ children, where σ is the alphabet size.*

Proof. For any string w, let $S(w) = SPals(w) \cup \{(|w| + 0.5, 0)\} - \{(|w|/2 + 0.5, |w|/2)\}$. To show the lemma, we consider the following claim.

Claim. Let w and z be any strings of length i s.t. $Pals(w) = Pals(z)$. For any integers j, k with $1 \le j \le i$, $1 \le k \le i$ and $(\frac{i+j+1}{2}, \frac{i-j}{2}), (\frac{i+k+1}{2}, \frac{i-k}{2}) \in S(w)$, if $w[j] = w[k]$ then $z[j] = z[k]$.

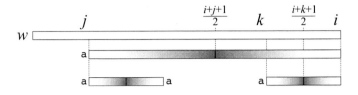

Fig. 4. Illustration for the proof of Claim in Lemma 6

Proof of Claim. When $j = k$, it is clear the claim holds. Then we consider the case $j \neq k$. Assume w.l.o.g. that $j < k$. Since $w[j+1 : i+j-k] = w^R[k+1 : i] = w[k+1 : i] = w^R[j+1 : i+j-k]$, $w[j+1 : i+j-k]$ is a palindrome. It follows from $w[j] = w[k]$ and $w[j+1 : i] = w^R[j+1 : i]$ that $w[j] = w[k] = w[i+j+1-k]$. Putting $w[j+1 : i+j-k] = w^R[j+1 : i+j-k]$ and $w[j] = w[i+j+1-k]$ together, we get $w[j : i+j+1-k]$ is a palindrome (See also Fig. 4). Since $Pals(w) = Pals(z)$, $z[j : i+j+1-k]$ and $z[j+1 : i]$ are palindromes, and thus $z[j] = z[i+j+1-k] = z[k]$. Hence the claim holds.

Consider any substring w of length i of t. We introduce an equivalence relation on $S(w)$ such that

$$\left(\frac{i+j+1}{2}, \frac{i-j}{2}\right) \equiv \left(\frac{i+k+1}{2}, \frac{i-k}{2}\right) \iff w[j] = w[k],$$

where $1 \leq j \leq i, 1 \leq k \leq i$, and $\left(\frac{i+j+1}{2}, \frac{i-j}{2}\right), \left(\frac{i+k+1}{2}, \frac{i-k}{2}\right) \in S(w)$. By definition, there are at most σ equivalence classes w.r.t. $\equiv$. Consider any substring z of t with $Pals(z) = Pals(w)$. Due to the above claim, the equivalence classes on $S(z)$ are identical to those on $S(w)$.

Let v be any node of $Pal\_ST(t)$, and assume that the path from the root to v spells out $Lpal_w$. Note that every substring z of t that pal-matches w is represented by the same node v in $Pal\_ST(t)$, since it has the same $Lpal$ values as w, i.e., $Lpal_w = Lpal_z$. Therefore, the number of children of v is at most $d+1$, where d is the number of equivalence classes on $S(w)$, which is bounded by σ. Hence the lemma holds. □

In order to implement $Pal\_ST(t)$ with $O(n)$ space, we encode the label of each edge as follows. Assume that there is an edge of $Pal\_ST(t)$ labeled with x, where x is a sequence of positive integers. We encode x by a triple $(x[1], q, |x|)$, where $x[1]$ is the first element of x, q is a position of text t such that $x = Lpal_{t[s:n]}[q-s+1 : q-s+|x|]$ for some $1 \leq s \leq n$, and $|x|$ is the length of the edge label. See Fig. 3 and focus on the edge which is labeled with 2 1 3. Choosing $s = 2$, the label is encoded by $(2, 3, 3)$ as $q = 3$, $|x| = 3$, and $Lpal_{t[2:9]}[2 : 4] = 2$ 1 3. In Fig. 3, the first element of each edge label is shown underlined.

Theorem 3. *Provided that $Pal\_ST(t)$ and $Pals(t)$ are already computed, the pal-matching problem (Problem 1) can be solved in $O(m \log \sigma + r)$ time, where r is the output size.*

Proof. We compute $Lpal_p$ using Algorithm 1 in $O(m)$ time. Then we search $Pal\_ST(t)$ for $Lpal_p[1:m]$. Assume that $Lpal_p[1:j]$ matches the label of an outgoing edge of the root node of $Pal\_ST(t)$, with some $1 \le j < m$. Assume the edge label is encoded as $(Lpal_{t[q:n]}[1], q, j)$, where $Lpal_{t[q:n]}[1:j] = Lpal_p[1:j]$. Let v be the node that represents $Lpal_{t[q:n]}[1:j]$. Assume that there is an out-going edge of v, which is labeled with $(Lpal_{t[q'-j:n]}[j+1], q', j')$, where $Lpal_{t[q'-j:n]}[j+1] = Lpal_p[j+1]$ and $j' \ge 2$. This edge can be found in $O(\log \sigma)$ time by Lemma 6. Now we have to check whether $Lpal_{t[q'-j:n]}[j+2] = Lpal_p[j+2]$. Although q' is *not* necessarily equal to $q+j$, we can compute $Lpal_{t[q'-j:n]}[j+2]$ as follows: By the definition of $Pal\_ST(t)$ it holds that $Lpal_{t[q'-j:n]}[1:j+1] = Lpal_{t[q:n]}[1:j+1]$, which implies that $AC_t(q'-j, q') = AC_t(q, q+j) + q' - (q+j)$. As described in Section 3, we can compute $Lpal_{t[q'-j:n]}[j+2]$ by shifting the current center from $AC_t(q'-j, q')$ to $AC_t(q'-j, q'+1)$. Moreover, $Lpal_{t[q'-j:n]}[j+2] = Lpal_p[j+2]$ iff $AC_t(q'-j, q'+1) - AC_t(q'-j, q') = AC_p(1, j+2) - AC_p(1, j+1)$. In light of this, the total cost for computing such values of $Lpal$ is bounded by the cost for computing $Lpal_p$, which is $O(m)$. We continue the above procedure until either we find $Lpal_p$ in $Pal\_ST(t)$ or we find a mismatch. This takes $O(m \log \sigma)$ time. If $Lpal_p$ is found, then we traverse the sub-tree rooted at the (possibly implicit) node that represents $Lpal_p$, and report the id of the leaves in the sub-tree, in $O(r)$ time. □

4.1   Constructing Palindrome Suffix Trees

We employ Ukkonen's on-line construction techniques for suffix trees [17]. Here let us briefly review the behavior of the Ukkonen's algorithm. The algorithm processes the characters of a given string t of length n in ascending order. After processing the $(i-1)$-th character of t, the algorithm has constructed the suffix tree of $t[1:i-1]$. Now the algorithm waits for the next i-th character on the location which represents the longest suffix $t[s:i-1]$ of $t[1:i-1]$ that matches a substring of $t[1:i-2]$, with some $2 \le s \le i$. Let us call this location on the path *the active point* for $i-1$. Next the algorithm obtains the i-th character $t[i]$. If we can transit from the active point for $i-1$ with $t[i]$, then the active point for i is the location that represents $t[s:i]$. Otherwise, the algorithm creates a new edge from the active point for $i-1$ leading to a new leaf node, with edge label $t[i:n]$. After that, the algorithm finds the location which represents $t[s+1:i-1]$ by using a suffix link, in amortized constant time. The above procedure is repeated until the active point for i is found. Readers are referred to [17] for more details of the Ukkonen algorithm.

In the sequel, we show main technical issues of our algorithm to construct $Pal\_ST(t)$.

Suffix Links. Let v be any node of $Pal\_ST(t)$, and assume that the path from the root to v spells out $Lpal_w$ for some substring w of t. The *suffix link* of node v is an auxiliary edge from node v to node u, such that the path from the root to u spells out $Lpal_{w[2:|w|]}$. For example, see Fig. 3, and focus on the node which

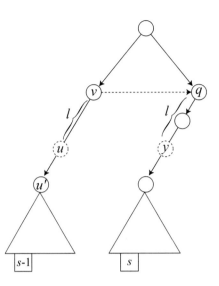

Fig. 5. Illustration of maintenance of the active point. u is the active point for $i - 1$, and y is a candidate for the active point for i.

represents 1 2 1 3. The suffix link of this node points to the node which represents 1 1 3. This is because there exists a substring bbab with $Lpal_{\text{bbab}} = 1\ 2\ 1\ 3$, and $Lpal_{\text{bab}} = 1\ 1\ 3$.

Unlike the case of suffix trees, the node u, which is to be pointed by the suffix link of some node v, is *not* always explicit in $Pal\_ST(t)$. For example, see Fig 3. The suffix link of the node which represents 1 1 2 is illustrated to point to the implicit node which represents 1 2. In such a case, we set the suffix link of node v to the child node u' of implicit node u, and record the length of the partial edge label from u to u'. This way we can access from node v to the location for u in constant time. In the above example, the suffix link of node 1 1 2 is implemented to point to node 1 2 1 3, with auxiliary value 2 which is the length of the partial label from implicit node 1 2 to node 1 2 1 3. The same technique was used in [3] to implement the suffix links of *parameterized suffix trees*.

Maintaining Active Point. Assume that we have constructed $Pal\_ST(t[1 : i - 1])$ for given string t, for some $1 \le i \le n$. Assume that the active point for $i - 1$ is on an implicit node u. Let v be the explicit parent node of u, and let u' be the explicit child node of v, i.e., u is on the edge from v to u'. Let x be the label of the edge from v to u', and let ℓ be the length of the partial edge label from v to u. Then, the active point for $i - 1$, the implicit node u, is represented by $(v, x[1], s - 1 + len(v), \ell)$, where $x[1]$ is the first element of x and $s - 1$ is a position of t such that $Lpal_{t[s-1:n]}[len(v) + 1 : len(v) + \ell] = x[1 : \ell]$.

Similarly to construction of suffix trees, we look for the active point for i from the active point for $i - 1$, i.e., the implicit node u. See Fig. 5. In so doing, we use the suffix link of node v. Consider any leaf $s - 1$ in the subtree rooted

244 T. I, S. Inenaga, and M. Takeda

at v. Let q be the node we have reached by the suffix link of node v. Now we want to look for a (possibly implicit) child y of q such that the subtree rooted at y has leaf s and $len(y) = len(u) - 1 = len(q) + \ell$. The difficulty we face is that $x[1 : \ell] = Lpal_{t[s-1:n]}[len(v) + 1 : len(v) + \ell]$ may not be equal to $Lpal_{t[s:n]}[len(q) + 1 : len(q) + \ell]$. This happens when there exists an integer k, $len(v) + 1 \le k \le len(v) + \ell$, such that $Lpal_{t[s-1:n]}[k] = k$. For example, see Fig 3. The edge leading to leaf 2 is labeled with 5 1 3 3 \$, while the edge leading to leaf 3 is labeled with 2 1 3 3 \$. This is because $Lpal_{t[2:9]}[5] = 5$.

Nevertheless, we can efficiently locate y starting from q, as follows. Since $x[1] = Lpal_{t[s-1:n]}[len(v) + 1]$, we can calculate $AC_t(s - 1, s - 1 + len(v))$ in constant time. Since $len(q) = len(v) - 1$, we can compute $Lpal_{t[s:n]}[len(q) + 1 : len(q) + \ell]$ in $O(AC_t(s, s + len(q)) - AC_t(s - 1, s + len(q)) + \ell)$ time, as described in Section 3. Then we can find y in $O(\ell \log \sigma)$ time, since there can be at most $\ell - 1$ explicit nodes in the path from q to y. We check whether y is the active point for i or not, and if not, we repeat the above procedure until the active point for i is found. The total cost of the above operations, after constructing $Pal\_ST(t)$, is $O(n \log \sigma)$.

Consequently, we obtain the following result.

Theorem 4. *For any string t of length n, $Pal\_ST(t)$ can be constructed in $O(n \log \sigma)$ time, where σ is the alphabet size.*

5 Conclusions and Future Work

Palindromes in strings have widely been studied both in theoretical and practical contexts, such as in word combinatorics and in bioinformatics. In this paper, we presented linear-time algorithms to solve a new problem called the palindrome pattern matching problem. The first algorithm is a Morris-Pratt type algorithm, and the second one is a suffix-tree type algorithm.

In practical applications such as DNA and RNA sequence analysis, it is desired to cope with *gapped palindromes* which have a spacer between the left and right arms of the palindromes. Several versions of gapped palindromes have been introduced and studied [9,12,10]. Our future work includes development of efficient solutions to a gapped-palindromes version of the palindrome pattern matching problem.

References

1. Allouche, J.P., Baake, M., Cassaigne, J., Damanik, D.: Palindrome complexity. Theoretical Computer Science 292(1), 9–31 (2003)
2. Anisiu, M.C., Anisiu, V., Kása, Z.: Total palindrome complexity of finite words. Discrete Mathematics 310(1), 109–114 (2010)
3. Baker, B.S.: Parameterized pattern matching: Algorithms and applications. Journal of Computer and System Sciences 52(1), 28–42 (1996)
4. Brlek, S., Hamel, S., Nivat, M., Reutenauer, C.: On the palindromic complexity of infinite words. International Journal of Foundations of Computer Science 15(2), 293–306 (2004)

5. Burrows, M., Wheeler, D.J.: A block-sorting lossless data compression algorithm. Tech. rep., DIGITAL System Research Center (1994)
6. Droubay, X., Justin, J., Pirillo, G.: Episturmian words and some constructions of de Luca and Rauzy. Theoretical Computer Science 255(1–2), 539–553 (2001)
7. Glen, A., Justin, J., Widmer, S., Zamboni, L.Q.: Palindromic richness. European Journal of Combinatorics 30(2), 510–531 (2009)
8. Groult, R., Prieur, É., Richomme, G.: Counting distinct palindromes in a word in linear time. Information Processing Letters 110(20), 908–912 (2010)
9. Gusfield, D.: Algorithms on Strings, Trees, and Sequences. Cambridge University Press, New York (1997)
10. Hsu, P.H., Chen, K.Y., Chao, K.M.: Finding all approximate gapped palindromes. In: Dong, Y., Du, D.-Z., Ibarra, O. (eds.) ISAAC 2009. LNCS, vol. 5878, pp. 1084–1093. Springer, Heidelberg (2009)
11. I, T., Inenaga, S., Bannai, H., Takeda, M.: Counting and verifying maximal palindromes. In: Chavez, E., Lonardi, S. (eds.) SPIRE 2010. LNCS, vol. 6393, pp. 135–146. Springer, Heidelberg (2010)
12. Kolpakov, R., Kucherov, G.: Searching for gapped palindromes. Theoretical Computer Science 410(51), 5365–5373 (2009)
13. Manacher, G.: A new linear-time "on-line" algorithm for finding the smallest initial palindrome of a string. Journal of the ACM 22(3), 346–351 (1975)
14. Massé, A.B., Brlek, S., Frosini, A., Labbé, S., Rinaldi, S.: Reconstructing words from a fixed palindromic length sequence. Proc. TCS 2008. IFIP 273, 101–114 (2008)
15. Morris, J.H., Pratt, V.R.: A linear pattern-matching algorithm. Tech. Rep. 40, University of California, Berkeley (1970)
16. Restivo, A., Rosone, G.: Burrows-Wheeler transform and palindromic richness. Theoretical Computer Science 410(30–32), 3018–3026 (2009)
17. Ukkonen, E.: On-line construction of suffix trees. Algorithmica 14(3), 249–260 (1995)

Sparse and Truncated Suffix Trees on Variable-Length Codes

Takashi Uemura and Hiroki Arimura

Graduate School of Information Science and Technology, Hokkaido University
Kita 14, Nishi 9, Kita-ku, Sapporo, 060-0814 Japan
{tue,arim}@ist.hokudai.ac.jp

Abstract. The sparse suffix trees (SST), introduced by (Kärkkäinen and Ukkonen, *COCOON 1996*), is the suffix tree for a subset of all suffixes of an input text T of length n. In this paper, we study a special case that an input string is a sequence of k codewords drawn from a regular prefix code $\Delta \subseteq \Sigma^+$ recognized by a finite automaton, and index points locate on the code boundaries. In this case, we present an online algorithm that constructs the sparse suffix tree for an input string T on any variable-length regular prefix code, called the code suffix tree (CST), in $O(n+m)$ time and $O(k)$ additional space for a fixed base alphabet Σ, where m is the size of an automaton for Δ. Furthermore, we present a modified algorithm for ℓ-truncated version of code suffix trees that runs in the same time and space complexities. Hence, these results generalize the previous results (Inenaga and Takeda, *CPM 2006*) for word suffix trees and (Na, Apostolico, Iliopoulos, and Park, *Theor. Comp. Sci.*, 304, 2003) for truncated suffix trees on arbitrary variable-length regular prefix codes, such as Huffman codes and multi-byte codes (e.g. UTF-8).

1 Introduction

Backgrounds. The *sparse suffix trees* (SST), introduced by Kärkkäinen and Ukkonen [11] in 1996, is the suffix tree (ST) [6, 13, 17] for storing a subset consisting of k suffixes in an input text T of length n on a base alphabet Σ, where $k \leq n$. In its most general form, the set $I = \{i_1, \ldots, i_k\} \subseteq \{1, \ldots, n\}$ of *index points* is given as an arbitrary subset of all n text positions. We denote by $\mathsf{SST}^I(T)$ the sparse suffix tree for T with respect to the set I of index points. [11] showed that a sparse suffix tree on a k-evenly indexed string in $O(n)$ worst-case time and $O(k)$ space. Although a sparse suffix tree for a string with an arbitrary index set is well-defined in any sense, interestingly enough, it is still open since its introduction whether a sparse suffix tree for arbitrary set I of k index positions can be constructed in $O(n)$ time and $O(k)$ additional space.

For the problem, recently, a collection of word-based suffix indexes have been introduced [3, 5, 8–10]. To formalize this notion, we introduce the set $\Delta_I(T) \subseteq \Sigma^+$ of words, called the *induced code*, obtained by partitioning an input text T by the index positions in I. Then, a suffix index is called *word-based* if the set $\Delta_I(T)$ is restricted to a set of words in $\Sigma^+ W$, where W is a finite set of symbols,

R. Giancarlo and G. Manzini (Eds.): CPM 2011, LNCS 6661, pp. 246–260, 2011.

called word delimiters, such that $W \cap \Sigma = \emptyset$. In 1999, Andersson, Larsson, and Swanson [3] introduced a word-suffix tree as a $\mathsf{SST}^I(T)$ on a word alphabet, and presented a construction algorithm in $O(n)$ average time and $O(k)$ space. In 2006, Inenaga and Takeda [8] presented the first construction algorithm that runs in $O(n)$ worst-case time and $O(k)$ space by modifying Ukkonen's linear time online construction algorithm for full suffix trees [17]. Their work is most closely related to this work. Ferragina and Fischer [5] introduced word-suffix arrays and presented a construction algorithm with $O(n)$ worst-case time and $O(k)$ space.

Our Contribution. In this paper, we study the sparse suffix tree construction in more general setting than that of the word-based suffix trees [3, 8]. In particular, we consider the sparse suffix tree for a string on an arbitrary regular prefix code $\Delta \subseteq \Sigma^+$ which is recognized by a finite deterministic automaton. The sparse suffix tree of this type is called the *code suffix tree* (CST), and can be regarded as a natural generalization of word suffix trees [3, 8]. As a main result of this paper, we show that the code suffix tree for an input string T of length n on a prefix code Δ can be constructed in $O(n + m)$ worst-case time and $O(k)$ space for a fixed base alphabet Σ, where k is the number of words in T and m is the size of the automaton for Δ (Theorem 2). Thus, the CST can be linearly constructed for texts on regular prefix codes such as Huffman codes or UTF-8.

Key Techniques. To show this, we propose a modified version of Ukkonen's online suffix tree construction algorithm [17] augmented with a DFA, called a *code automaton*, for recognizing Δ, which is similar to the construction in [8]. However, the proofs for correctness and time complexity are not straightforward due to the complex behavior of the algorithm when it traverses inside of a code automaton. To overcome this difficulty, we introduce an extended domain of strings augmented with the erasing element $\bot$, which is the inverse of any codeword in Δ and acting from left. Using $\bot$, we give a general definition of suffix links, and show that most properties of full suffix trees [17], including the existence lemma for suffix links, still remains valid when Δ is a prefix code. Hence, Theorem 2 above gives a partial answer to a natural question: what is the largest class of codes for which the approach of Ukkonen's linear-time construction algorithm [17] is sufficient for constructing sparse suffix trees on a code?

An Extension. For every $\ell \geq 1$, an ℓ-*truncated suffix tree* (ℓ-TST) for T is a variation of suffix trees that stores all factors of T with length ℓ. Na *et al.* [14] introduced ℓ-TST and presented an online construction algorithm for ℓ-TST in $O(n)$ time and space. Generalizing ℓ-TSTs for regular prefix codes, we introduce the ℓ-*truncated code suffix trees* (ℓ-TCST) that stores all factors of T consisting of at most ℓ codewords. Based on our algorithm for CST, we present a modified version of the algorithm that constructs ℓ-TCST for a text on Δ in $O(n)$ worst-case time and $O(k)$ space, where k is the number of words in T (Theorem 3). Finally, we ran experiments on real datasets to evaluate the usefulness of the proposed methods. For example, CSTs are 3 to 5 times smaller than STs on English and UTF-8 texts as shown in Section 5.

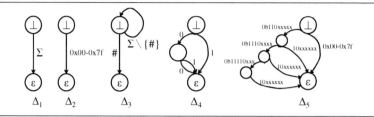

Fig. 1. The code automata for prefix codes $\Delta_1, \Delta_2, \Delta_3, \Delta_4$, and Δ_5 in Example 1, where $\bot$ and ε are the initial and the final states, respectively, and labels with wildcard $x \in \{0, 1\}$, e.g., $0x00 - 0x7f$, represent sets of the corresponding multiple edges

Organization of this Paper. In Sec. 2, we give basic definitions. In Sec. 3, we present our linear-time construction algorithm ConstructCST for code suffix trees. In Sec. 4, we extend the algorithm for ℓ-truncated code suffix trees. In Sec. 5, we show experimental results, and in Sec. 6, we conclude this paper.

2   Preliminaries

Basic Definitions. We introduce basic definitions on suffix trees according to [4, 6, 11, 14, 17]. We assume that the reader has basic knowledge of the linear time construction algorithm by Ukkonen [17]. Let Σ be an alphabet of *base letters*. We denote by ε the *empty string*. Let Σ^* and Σ^+ denote the *sets of all possibly empty finite strings* and *non-empty finite strings* on Σ. For a string T, if $T = xyz$ for some $x, y, z \in \Sigma^*$, then we call x, y, and z a *prefix*, a *factor* (*substring*), and a *suffix* of T, respectively. Let $T = a_1 \cdots a_n \in \Sigma^*$ be a string on Σ of length $|T|_\Sigma = |T| = n$, where $T[i] = a_i \in \Sigma$ is the *i-th letter* for every $i = 1, \ldots, n$. For any $1 \le i \le j \le n$, we denote by $T[i..j] = a_i \cdots a_j$ the *factor from i to j* of T. If $i > j$ then we define $T[i..j] = \varepsilon$. For a set $\mathcal{S} \subseteq \Sigma^*$ of strings, the *sets of the prefixes* and *proper prefixes* of all strings in $\mathcal{S}$ are denoted by $\mathsf{Pre}(\mathcal{S})$ and by $\mathsf{PropPre}(\mathcal{S})$, respectively. $|\mathcal{S}|$ and $||\mathcal{S}||$ denote the *cardinality* and the *total size* of $\mathcal{S}$. For strings x, y, we denote by $\mathsf{lcp}(x, y)$ the *longest common prefix* of x and y.

Prefix Codes. A *code* is a set $\Delta \subseteq \Sigma^+$, where each $w \in \Delta$ is a non-empty string, called a *codeword* (or *word*, for short) of Δ. A *preword* is any prefix $u \in \mathsf{Pre}(\Delta)$ and a *proper preword* is any proper prefix $u \in \mathsf{PropPre}(\Delta)$ of a word in Δ. A code Δ is either infinite or finite. Δ is a *prefix code* if it is *prefix-free*, that is, any codeword is not a prefix of some other codeword of Δ.

Example 1. Let Σ be a letter alphabet and $\mathbb{B} = \{1, 0\}$ be a binary alphabet. A trivial prefix code $\Delta_1 = \Sigma$ and the *ASCII code* $\Delta_2 = [0x00 - 0x7f] \subseteq \mathbb{B}^8$ are prefix codes of fixed-length. For a word delimiter $\sharp \notin \Sigma$, a *word alphabet* $\Delta_3 = \Sigma^+\sharp$ is an example of prefix-codes of variable-length ([3, 5, 8–10]). A *Huffman code* $\Delta_4 = \{00, 01, 1\}$ for the set of symbols $\{A, B, C\}$ with probabilities $p(A) = 1/4, p(B) = 1/4, p(C) = 1/2$ ([4]), and the code $\Delta_5 = (10\mathbb{B}^1) \cup (110\mathbb{B}^2) \cup (1110\mathbb{B}^3)$, called the *three-byte fragment of UTF-8* ([7]), are also examples of prefix-codes of variable-length. As seen later, all of these prefix codes are regular (Fig. 1).

Codeword Strings. A *word string* on Δ (or Δ-*string*) is a string $T \in \Delta^*$. Then, an *input string* or a *prestring* of letter length $n \geq 0$ is any prefix $T = T[1] \cdots T[n] \in \mathsf{Pre}(\Delta^*)$ of a word string on Δ, where $T[i] \in \Sigma$ for $i = 1, \ldots, n$. Since Δ is a prefix code and thus $\mathsf{Pre}(\Delta^*) = \Delta^*\mathsf{PropPre}(\Delta)$, we have the unique Δ-factoring $T = w_1 \cdots w_k w_{k+1} \in \mathsf{Pre}(\Sigma^*)$ of T, where $w_1, \ldots, w_k \in \Delta$ and $w_{k+1} \in \mathsf{PropPre}(\Delta)$. Then, the proper preword w_{k+1} is called the *tail* of T. Clearly, T is a complete word string if and only if the tail w_{k+1} is emtpy. We define the *word length* of T by $|T|_\Delta = k$, and the *letter length* by $|T|_\Sigma = |T| = n$. where $n = \sum_{i=1}^{k+1} |w_i|$.

In this paper, construction of a CST is done in online manner through the stage $i = 0, \ldots, n$ as in [15, 17]. At stage i, we define the current input $T^i = T[1] \cdots T[i] = w_1 \cdots w_k w_{k+1}$, where $k \geq 0$, $w_j \in \Delta$, and $w_{k+1} \in \mathsf{PropPre}(\Delta)$. Let $j = 1, \ldots, k+1$ be any index. The j-th Δ-*suffix* of T is defined as the suffix of T starting at the j-th word boundary, i.e., $\mathsf{suf}_j^\Delta(T) = w_j \cdots w_k w_{k+1}$. For $\ell \geq 1$, the j-th (Δ, ℓ)-*factor* of T is defined as the factor $\mathsf{fac}_j^{\Delta,\ell}(T) = w_j \cdots w_h$ of T, where $h = \min\{j + \ell - 1, k+1\}$. We denote by $\mathsf{Suf}^\Delta(T)$ and $\mathsf{Fac}^\Delta(T, \ell)$ the *sets of all Δ-suffixes* and *all (Δ, ℓ)-factors* of T, respectively.

Code Suffix Trees. A *code suffix tree* (CST, for short) for an input string $T \in \mathsf{Pre}(\Delta^*)$ w.r.t. a prefix code Δ, denoted by $\mathsf{CST}^\Delta(T)$, is a compacted trie [6] that represents all Δ-suffixes of T. Formally, the CST for T is a rooted tree $S = \mathsf{CST}^\Delta(T) = (V, child, root, Label(\cdot), SL(\cdot))$ that satisfies the following properties. V is a finite set of *tree nodes* (or *nodes*). Each directed edge $e = (u, v)$ is labeled with a factor of T, $Label(v) \in \Sigma^+$, stored in v. Every internal node except the root is *branching*, i.e., it has at least two children. For every base letter $a \in \Sigma$, each internal node v has at most one directed edge from v to the a-child, $u = child(v, a)$, whose label starts with a. $SL(\cdot)$ is a *suffix link* function, which will be defined later in Section 3. For each v, we denote by $L(v)$ the string *represented by* v, that is, the string obtained by concatenating all labels on the path from the root to v. All the suffixes are represented by the leaves of $\mathsf{CST}^\Delta(T)$ when T ends with the unique marker $T[n]$ that does not appear elsewhere. All the factors in T that start at word boundaries are represented as the prefixes of all leaves, that is, the elements of the set $\mathsf{Pre}(\mathsf{Suf}^\Delta(T))$. We give the naming function $[\cdot]$ below. For any $\alpha \in \mathsf{Pre}(\mathsf{Suf}^\Delta(T))$, we define the *locus* of α, denoted by $[\alpha]$, to be the unique tree node $v \in Q$ such that $L(v) = \alpha$. This mapping $[\cdot]$ is one-one.

Clearly, $\mathsf{CST}^\Delta(T)$ has at most k leaves and $k - 1$ internal nodes [11]. To save the space, we represent $Label(v)$ by a pair $\langle j, i \rangle \in \mathbb{N}^2$ of the starting and ending positions of the label in T such that $T[j..i] = Label(v)$, while ε is represented by $\langle i + 1, i \rangle$ for some i. Assuming this, $\mathsf{CST}^\Delta(T)$ occupies only $O(k)$ space. For any factor α of T that starts at a word boundary, we represent its location in $\mathsf{CST}^\Delta(T)$ by a triple $p = \langle v, j, i \rangle \in V \times \mathbb{N}^2$, called a *pointer* (or a *reference*) to α, such that $\alpha = L(v) \cdot T[j..i]$ if it exists. A pointer $\langle v, j, i \rangle$ is *canonical* if $T[j..i]$

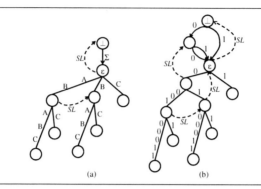

Fig. 2. Examples of (a) an ordinary suffix tree for string $S = ABABC$ on alphabet $\{A, B, C\}$, and (b) a code suffix tree for coded string $T = 000100011$ on prefix code $\Delta = \{A/00, B/01, C/1\}$

is shortest. The *locus* of a factor α is the canonical pointer for α, and denoted by $loc(\alpha)$. We often call p a *virtual node* if $j \leq i$, i.e., $T[j..i] \neq \varepsilon$, and *real node* if $j > i$, i.e., $T[j..i] = \varepsilon$.

3 A Linear-Time Online Algorithm for Code Suffix Trees

In this section, we show our algorithm ConstructCST for constructing a code suffix tree $\mathsf{CST}^\Delta(T)$ on a prefix code Δ, which is based on Ukkonen's online construction algorithm for suffix trees [17]. The only difference is that it is augmented with a *code automaton* and *code suffix links* explained below. Let us fix an input string $T = w_1 \cdots w_k w_{k+1} \in \mathsf{Pre}(\Delta^*)$ on a prefix code $\Delta \subseteq \Sigma^+$ with letter length n and word length k.

3.1 Code Automata and Code Suffix Links

Code Automata. In our problem setting, a prefix code $\Delta \subseteq \Sigma^+$ on Σ is *regular* if Δ is recognized by a finite automaton. A *code automaton* for a prefix code $\Delta \subseteq \Sigma^+$ is a possibly cyclic deterministic finite automaton (DFA) $A = (\Sigma, Q, child, \hat{\perp}, root)$ on a base alphabet Σ, where Q is a finite set of *code nodes* (or *nodes*). $\hat{\perp}$ and *root* are the unique initial and final states, called the *source* and *sink*, respectively. The function $child : Q \times \Sigma \to Q$ is a *transition function* such that for every $u, v \in Q$ and $a \in \Sigma$, $child(u, a) = v$ if and only if there exists an a-edge from u to v labeled with a. We extend $child$ to a mapping $child^* : Q \times \Sigma^* \to Q$ in a standard way [4]. If it is clear from context, we refer to the code automaton A for Δ as $\mathsf{DFA}(\Delta)$. In the treatment of this paper, $\mathsf{DFA}(\Delta)$ need not be minimal in general. As a related work, Takeda, Miyamoto *et al.* [15] used code automata to extend Aho-Corasick pattern matching machines on a prefix code. In Fig. 1, we show the code automata for codes $\Delta_1, \Delta_2, \Delta_3, \Delta_4$, and Δ_5 of Example 1, respectively. It is not hard to see that the automata for Δ_1 and Δ_3 are exactly those automata that are employed by the linear-time construction algorithms of Ukkonen [17] and Inenaga *et al.* [8].

Now, we give the naming function $[\cdot]$ from strings to nodes as follows. We introduce a special element $\perp \notin \Sigma$ as the *inverse* of any word of Δ, i.e., $\perp w = \varepsilon$ for every $w \in \Delta$. We define $\perp w \alpha = \varepsilon \alpha = \alpha$ if $w \in \Delta$ and $\alpha \in \Sigma^*$. For a proper preword $\alpha \in \mathsf{PropPre}(\Delta)$, $\perp \alpha$ is a special element different from α. For any preword $\alpha \in \mathsf{Pre}(\Delta)$, we define the *locus* of α, denoted by $[\perp \alpha]$, to be the unique code node $v \in Q$ such that $child^*(\hat{\perp}, \alpha) = v$. For set S, let $\perp S = \{ \perp \alpha \mid \alpha \in S \}$. Note that the mapping $[\cdot]$ is many-one and naturally induces an equivalence relation $\equiv$ on the set $\mathsf{Pre}(\Delta)$. Since $\mathsf{DFA}(\Delta)$ has the unique final state $root = [\varepsilon]$, we see that $[\perp \alpha] = [\perp \beta] = [\varepsilon]$ holds for any codewords $\alpha, \beta \in \Delta$. We also note that $[\perp \alpha] \neq [\alpha]$ for any string α since the former and the latter are the nodes reachable from $\hat{\perp}$ in $\mathsf{DFA}(\Delta)$ and from $root$ in $\mathsf{CST}^\Delta(T)$, resp. The above notations are just for analysis of our algorithm, and do not affect the behavior and complexity of the algorithm.

By the above encoding, we can represent $\mathsf{DFA}(\Delta)$ as follows: The node set is $Q = \{ [\perp \alpha] \mid \alpha \in \mathsf{PropPre}(\Delta) \}$. The transition function is given by $child([\alpha], a) = [\alpha a]$ for every $\alpha \in \mathsf{PropPre}(\Delta)$ and $a \in \Sigma$. The source and the sink are $[\perp]$ and $[\varepsilon]$, respectively. If $[\alpha]$ is either a tree node or a code node and $a \in \Sigma$ is any letter, then we define $[\alpha] \cdot a = [\alpha \cdot a]$. We define the domains $\mathsf{dom}(\mathsf{code}) = Q - \{ [\varepsilon] \} = \{ [\perp \alpha] \mid \alpha \in \mathsf{PropPre}(\Delta) \}$ of all *code nodes*, $\mathsf{dom}(\mathsf{tree}) = \{ [\alpha] \mid \alpha \in \mathsf{Pre}(\mathsf{Suf}^\Delta(\mathsf{T})) \}$ of all *tree nodes*, and $\mathsf{dom}(\mathsf{pre}) = \{ [\alpha] \mid \alpha \in \mathsf{PropPre}(\Delta) \} \subseteq \mathsf{dom}(\mathsf{tree})$, of all *prenodes*. By definition, $[\varepsilon] \in \mathsf{dom}(\mathsf{tree})$ but $[\varepsilon] \notin \mathsf{dom}(\mathsf{code})$. In what follows, we often use α and $[\alpha]$ interchangeably if no confusion arises.

Code Suffix Links. Next, we introduce the suffix links for $\mathsf{CST}^\Delta(T)$ as follows. Similarly to Ukkonen's algorithm, each internal node $v = [\alpha]$ in the CST has the suffix link of v, denoted by $SL^\Delta(v)$, which is a pointer from v to the internal node u such that $SL^\Delta([\alpha]) = [\perp \cdot \alpha]$, where $\alpha \in \mathsf{Pre}(\mathsf{Suf}^\Delta(T))$. Equivalently, if $v = [w\beta]$ for some $w \in \Delta, \beta \in \perp\mathsf{PropPre}(\Delta) \cup \mathsf{Pre}(\mathsf{Suf}^\Delta(T))$ then $SL^\Delta([w\beta]) = [\beta]$. Any code node $v \in \mathsf{dom}(\mathsf{code})$ does not have a suffix link. The next lemma is crucial to the correctness of our algorithm.

Lemma 1 (existence lemma for code suffix links). *Let $\Delta \subseteq \Sigma^+$ be a prefix-free code and T be any prestring on Δ. Then, (i) any tree node v in $\mathsf{CST}^\Delta(T)$ has the suffix link $SL^\Delta(v)$ pointing to a branching internal node u in either $\mathsf{CST}^\Delta(T)$ or $\mathsf{DFA}(\Delta)$. Furthermore, (ii) v is a preword node if and only if $SL^\Delta(v)$ is a code node in $\mathsf{DFA}(\Delta)$, and (iii) v is not a preword node if and only if $SL^\Delta(v)$ is a tree node in $\mathsf{CST}^\Delta(T)$.*

Proof. There are two cases on the domain of v. (1) If $v = [\alpha] \in \mathsf{dom}(\mathsf{pre})$ is a prenode, then $SL^\Delta(v) = [\perp \alpha]$ belongs to $\mathsf{dom}(\mathsf{code})$ by the definition of $\mathsf{DFA}(\Delta)$. (2) Suppose that $= [w\alpha] \in \mathsf{dom}(\mathsf{tree}) \backslash \mathsf{dom}(\mathsf{pre})$ for some $w \in \Delta, \alpha \in \Sigma^*$. It is shown in [11] that a SST has a branching node v if and only if

---

Algorithm ConstructCST:
input: A preword string $T = w_1 \cdots w_k \in \mathsf{Pre}(\Delta^*)$ on a prefix code $\Delta \subseteq \Sigma$;
output: The sparse suffix tree $\mathsf{CST}^\Delta(T)$ for t w.r.t. Δ. ;
1: { **global variables:** Θ, Θ': word counters //for ℓ-TCST }
2: Create an empty tree CST^Δ with the root node $root = [\varepsilon]$;
3: Build $\mathsf{DFA}(\Delta)$ for Δ with the source $\hat{\bot} = [\bot]$ and the sink $root = [\varepsilon]$;
4: $SL^\Delta(root) = \hat{\bot}$;
5: $\phi \leftarrow \langle root, 1, 0 \rangle$; $\psi \leftarrow \langle root, 1, 0 \rangle$;
6: { $\mathsf{Reset}(\Theta)$; $\mathsf{Reset}(\Theta')$ //for ℓ-TCST }
7: **for** $i = 1, \ldots, n$ **do** //Stage i
8:     $\phi \leftarrow \mathsf{Extend}(\phi, i)$;
        {$\psi \leftarrow \mathsf{Terminate}(\psi, i)$; //for ℓ-TCST}
9: **end for**
10: **return** CST^Δ;

Fig. 3. A construction algorithm for a code suffix tree $\mathsf{CST}^\Delta(T)$ for a text T on a prefix code $\Delta \subseteq \Sigma^+$

$L(v) = \mathsf{lcp}(\mathsf{suf}_i^\Delta(T), \mathsf{suf}_j^\Delta(T))$ for some indexes i and j. Since Δ is prefix-free and $v = [w\alpha]$, we know that both of $\mathsf{suf}_i^\Delta(T)$ and $\mathsf{suf}_j^\Delta(T)$ start with w, and thus, we have $\alpha = \mathsf{lcp}(\mathsf{suf}_{i+1}^\Delta(T), \mathsf{suf}_{j+1}^\Delta(T))$. From the above claim, the lemma follows. $\square$

3.2   Main Algorithm

In Fig. 3, we show the algorithm ConstructCST, and in Fig. 4, the subprocedures Extend and Terminate. The only difference between our algorithm and Ukkonen's algorithm is lines 3 and 4 of ConstructCST that attaches $\mathsf{DFA}(\Delta)$ to the CST. For an input string $T = T[1] \cdots T[n] = w_1 \cdots w_k w_{k+1} \in \mathsf{Pre}(\Delta^*)$ on $\Delta \subseteq \Sigma^+$, the algorithm constructs the CST for $T_i = T[1..i]$ in an online manner for every stage $i = 1, \ldots, n$. At stage 0, the CST consists only of the root node $root = [\varepsilon]$ and $\mathsf{DFA}(\Delta)$. Let $T^i = T[1..i]$ be the current input text and $\mathsf{CST}^\Delta(T^i)$ be the code suffix tree for T^i obtained at the end of stage i. At each step i, the algorithm extends α to the new suffixes αa_i by appending the current base letter $a_i = T[i]$ for all Δ-suffixes in $\mathsf{CST}^\Delta(T^i)$.

This extension process is based on the following idea. Let $S^\Delta(i) = Suf^\Delta(T^i)$ be the set of all Δ-suffixes in T^i. For every stage $i = 0, \ldots, n$, we define the set $Bd^\Delta(i) \subseteq \mathsf{dom}(\mathsf{tree}) \cup \mathsf{dom}(\mathsf{code})$, the *border*, by the following recurrence:

- $Bd^\Delta(0) = \{\bot, \varepsilon\}$,
- $Bd^\Delta(i) = (Bd^\Delta(i-1) \cdot a_i) \cup \{\bot\}$ if $\varepsilon \in (Bd^\Delta(i-1) \cdot a_i)$,
- $Bd^\Delta(i) = Bd^\Delta(i-1) \cdot a_i$ otherwise,

where $S \cdot a = \{\alpha a \mid \alpha \in S\}$ for any set $S \subseteq \Sigma^*$ and $a \in \Sigma$. Then, we have the following lemma. Recall that $\mathsf{dom}(\mathsf{tree})$ is the domain of tree nodes in $\mathsf{CST}^\Delta(T^i)$.

procedure Extend($\phi = \langle s, j, i-1 \rangle, i$):
1: $last \leftarrow NULL$; //oldp
2: **while** $(child(\phi, T[i])$ is not defined) **do**
3:    **if** $j \leq i$ **then begin**
4:      $\phi \leftarrow$ Split(ϕ);
5:      **if** $last \neq NULL$ **then begin**
6:        $SL^\Delta(\phi) \leftarrow last$;
7:        $last \leftarrow \phi$; **end**
8:    **end**
9:    create a leaf q with $label(q) \leftarrow \langle i, \infty \rangle$;
10:   $child(\phi, T[i]) \leftarrow q$;
11:   $\phi \leftarrow$ Canonize$(\langle SL^\Delta(s), j, i-1 \rangle)$;
12:   {Decrement(Θ) //for ℓ-TST}
13: **end while**
14: $\phi \leftarrow$ Canonize$(\langle s, j, i \rangle)$;

15: {ChildTrans$(\Theta, T[i])$; //for ℓ-TST}}
16: {**if** toClose(Θ) **then** SuffixTrans(Θ, ϕ);
   //for ℓ-TST}
17: **return** ϕ; {End of Extend}

procedure Canonize$(\phi = \langle s, j, i \rangle)$:
1: **while** $j \leq i$ **do begin**
2:   $u \leftarrow child(s, T[j])$;
3:   $\langle q, p \rangle \leftarrow label(u)$;
4:   **if** $p - q > i - j$ **then**
5:     **break**;
6:   $j \leftarrow j + (p - q + 1)$;
7:   $s \leftarrow u$;
8: **end**
9: **return** $\langle s, j, i \rangle$; {End of Canonize}

Fig. 4. The subprocedures Extend and Canonize for the code suffix tree construction

Lemma 2. *For every* $i = 0, \ldots, n$, $S^\Delta(i) = Bd^\Delta(i) \cap \mathrm{dom}(tree)$.

Proof. From a similar argument to [15, 17], the following recurrence holds:

(i) $S^\Delta(0) = \{\varepsilon\}$,
(ii) $S^\Delta(i) = (S^\Delta(i-1) \cdot a_i) \cup \{\varepsilon\}$ if $T^i = T[1..i]$ is a complete word string (*),
(iii) $S^\Delta(i) = S^\Delta(i-1) \cdot a_i$ otherwise,

By induction on $i \geq 0$, we then can show that the condition (*) holds iff $w \in S^\Delta(i-1) \cdot a_i$ for some $w \in \Delta$ iff $\varepsilon \in Bd^\Delta(i-1) \cdot a_i$ holds. Furthermore, $\perp \alpha \in Bd^\Delta(i)$ iff $\alpha \in S^\Delta(i)$ for any $\alpha \in \mathrm{Pre}(\Delta)$. Thus, the result follows. □

We call each suffix α in $Bd^\Delta(i)$ the *extension point* at stage i. Next, we consider how we can efficiently find the extension points and extend them. The detection of ε is also crucial to the synchronization of word boundaries. We use a pointer ϕ to keep track to extension points in $Bd^\Delta(i)$ from longer to shorter. Let act_i^Δ be the *active point* at stage i as the pointer ϕ such that $L(\phi)$ is the longest suffix of T^{i-1} that occurs at least twice in T^i. For $i = 1, \ldots, n$, the maintenance of $Bd^\Delta(i)$ proceeds in the following way for all extension points $\phi = [\alpha a_i]$ of three types 1–3. Let $Bd^\Delta(0) = \{\perp, \varepsilon\}$.

- *type 1*: If α occurs only at the end of T^{i-1}, then, α is represented by a leaf, and thus so is αa_i. As in Ukkonen [17], by representing α as an open leaf $\langle j, \infty \rangle$, ∞ is interpreted as the current index i, the extension point α is automatically extended without any management. This correctly extends all extension points α of type 1.
- *type 2*: If α occurs at least twice in T^{i-1}, but αa_i does not occur, then by induction on i, we can show that act_i^Δ is the first node of type 2 satisfying this condition. Then, we create the new node for αa_i extending α by appending

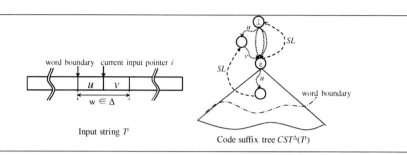

Fig. 5. Proof sketch for the time complexity in the case of cyclic code automata

a to the tail of α, while the parent α is materialized by procedure Split if α is virtual. Repeat this process until we reach the first node ϕ of type 3. This correctly extends all extension points α of type 2.

- *type 3*: If α occurs at least twice in T^{i-1}, and αa_i also occurs, then we can show as in Ukkonen [17] that all extension points on the suffix links from α to some node in dom(code) $\cup \{[\varepsilon]\}$, the end of the border, are already contained in T^{i-1}. Therefore, these extension points are correctly extended in $\mathsf{CST}^{\Delta}(T^i)$ without any explicit extension.

The procedure Extend in Fig. 4 implements the above incremental computation of $Bd^{\Delta}(i)$. From the discussion above, we have the next lemma.

Lemma 3. *For every stage $i = 1, \ldots, n$, The procedure* Extend *in Fig. 4 correctly computes the border $Bd^{\Delta}(i)$ from $Bd^{\Delta}(i-1)$ and $a_i = T[i]$.*

Proof. It is easy to see that Extend correctly implements the extensions of suffixes in $Bd^{\Delta}(i)$ mentioned above. Then, the most part is shown in a similar way to Ukkonen [17]. For extension of all three types, the above procedure correctly extends the original suffix $[\alpha] \in Bd^{\Delta}(i-1)$ to obtain $[\alpha a_i] \in Bd^{\Delta}(i)$. Remaining thing is to show the while-loop from lines 2 to 13 of Extend eventually terminates. If the while-loop is executed repeatedly, the depth of the extension pointer ϕ become smaller, and finally, either it ends with extension of type 3 or it enters the domain dom(code). In the former case, the proof is done. In the latter case, ϕ enters dom(code), and thus immediately ends with extension of type 3, too since $DFA(\Delta)$ accepts any preword. This completes the proof. □

From Lemma 2 and Lemma 3, we show the following theorem.

Theorem 1 (correctness). *For every stage $i = 1, \ldots, n$, the algorithm* ConstructCST *in Fig. 3 correctly constructs* $\mathsf{CST}^{\Delta}(T^i)$.

3.3   Time Complexity

The remaining task is to estimate the time complexity of ConstructCST in Fig. 3. Let N_{tree} and N_{code} be the numbers of tree and code edges traversed during the computation, respectively. Let $N = N_{tree} + N_{code}$. In a special case that Δ is finite and thus $\mathsf{DFA}(\Delta)$ is acyclic, the linear time complexity of $\mathsf{ConstructCST}$ can be easily proved by applying the *telescope argument* on the changes of the depth $D(\phi)$ of ϕ as used in [17] with a little twist that $D(\phi)$ is defined by the number of the code and tree nodes on the path from $[\perp]$ to ϕ.

In the general case that $\mathsf{DFA}(\Delta) = A$ is possibly cyclic, and consequently Δ is infinite, however, it is not straightforward to show a linear bound of $N = N_{tree} + N_{code}$ because the extension pointer ϕ can move inside a cycle in $\mathsf{dom(tree)}$ many times without monotonically increasing the depth parameter $D(\phi)$, and thus, it is not sufficient to linearly bound N_{code} for $\mathsf{dom(code)}$. To overcome this difficulty, we bound the number N_{code} by the total number of letters consumed during the traversal on $\mathsf{dom(code)}$. We have the main theorem of this paper.

Theorem 2 (linear time construction of code suffix trees). *Let $\Delta \subseteq \Sigma^*$ be any regular prefix code on Σ recognized by a code automaton $A = \mathsf{DFA}(\Delta)$. Then, the algorithm $\mathsf{ConstructCST}$ in Fig. 3 constructs $\mathsf{CST}^{\Delta}(T)$ for an input text $T \in \mathsf{Pre}(\Delta^*)$ in $O(n \log |\Sigma| + m)$ time and $O(k)$ space in online manner, where $n = |T|_{\Sigma}$ is the total text size, $k = |T|_{\Delta}$ is the number of codewords, and $m = ||A||$ is the size of the code automaton A.*

Proof. We show that the number $N = N_{tree} + N_{code}$ is bounded by $i = |T^i|$ for every stage i. We can show that the number N_{tree} in $\mathsf{dom(tree)}$ is linearly bounded by $i = |T^i|$. Therefore, we estimate the total number N_{code} of all child and suffix links that the algorithm traverses in $\mathsf{dom(code)}$ through all stages. Let $i = 1, \ldots, n$ be any stage, and let $T^i = T[1] \cdots T[i] = w_1 \cdots w_k u_{k+1} \in \mathsf{Pre}(\Delta^*)$ be the current input string. At stage i, we denote by ∂N_{code}^i the number of suffix and child edge traversals added to N_{code}. Then, there are three cases below when ϕ traverses inside $\mathsf{dom(code)}$: (a) The case that at stage $i-1$, the extension ends at node ϕ in $\mathsf{dom(code)}$ such that $\phi \neq [\varepsilon]$. From the construction of $\mathsf{DFA}(\Delta)$ and Lemma 3, the algorithm executes exactly one extension of type 3 in $\mathsf{dom(code)}$ by going down a child edge. Thus, $\partial N_{code}^i \leq 1$ is immediate. (b) Otherwise, at stage $i-1$, the extension ends at node $\phi \notin \mathsf{dom(code)}$. This implies that $\phi \in \mathsf{dom(tree)}$. In this case, the algorithm repeats extensions of type 2 in the while-loop from lines 2 to 13 of Extend by traversing suffix links in $Bd^{\Delta}(i-1)$, and terminates with extension of type 3. Let $\phi = [\beta]$ be the final extension point of type 2. Then, we have two subclasses below on ϕ: (b.1) The case that $\phi \in \mathsf{dom(tree)}$. Since all the preceding extension for $Bd^{\Delta}(i-1)$ were done in $\mathsf{dom(tree)}$, we have $\partial N_{code}^i = 0$. (b.2) The case that $\phi \notin \mathsf{dom(tree)}$. Then, $\phi = [\beta] \in \mathsf{dom(code)}$ for a preword β. Let $\phi = \langle s, j, i-1 \rangle$ be the canonical pointer of $[\beta]$. From Lemma 3, we can show that $\beta = u_{k+1}$ holds, that is, the string label $\beta = L(\phi)$ coincides the current tail preword u_{k+1} of T^i being scanned. In Extend, we then move from $\phi = [\beta]$ to $SL^{\Delta}(\beta) = [\perp\beta]$ by firstly following one suffix link from the real node s to $SL^{\Delta}(s)$, and by successively going down at most $|\beta|$ child edges by applying $\mathsf{Canonize}(SL^{\Delta}(s), j, i-1)$ at line 11 (See Fig. 5). Therefore, ∂N_{code}^i is at most $|\beta| + 1$. On the other hand, suppose that we are scanning a prefix $\beta = u_{k+1}$ of some complete word $w_{k+1} \in \Delta$. Then, it is not hard to show that the extension in the case (b.2), where a jump from $\mathsf{dom(tree)}$ into $\mathsf{dom(code)}$ is performed, can occur at most once per complete codeword w_{k+1} during the whole scan, because once the case (b.2) occurs, only the case (b.3) can occurs iteratively in successive stages until ϕ reaches dom(tree). Thus, we can amortize the cost for the case (b.2) over the whole computation. Combining the above arguments, we have the number of edge traversals bounded by:

$$N_{code} \leq \sum_{i=0}^{n} \partial N_{code}^i \leq (\sum_{i=0}^{n} 1) + (\sum_{j=0}^{k+1} |w_j|) \leq 2n,$$

where we used the equality $\sum_{j=0}^{k+1} |w_j| = |T| = n$. From similar arguments as in [17], we can show the remaining part that $N_{tree} \leq 2n$. Hence, the total number of edge traversals is given by $N = N_{tree} + N_{code} \leq 4n$. Space complexity is obvious since $\mathsf{CST}^{\Delta}(T)$ has at most $O(n)$ real nodes. For the total time complexity, the algorithm takes $O(m)$ time for the preprocessing $A = \mathsf{DFA}(\Delta)$. It takes $O(1)$ time per suffix link traversal and $O(\log|\Sigma|)$ time per child edge traversal with an appropriate dictionary structure. Hence, we have the result. $\square$

From Theorem 2, the algorithm runs in $O(n + m)$ time and $O(k)$ space for a fixed base alphabet Σ.

4   Application to Truncated Code Suffix Trees

Let $T \in \mathsf{Pre}(\Delta^*)$ $T = T[1] \cdots T[n] = w_1 \cdots w_k w_{k+1} \in \mathsf{Pre}(\Delta^*)$ be an input prestring on Δ and $\ell > 0$ be a fixed integer. Then, the ℓ-*truncated code suffix tree* (ℓ-*TCST*) of T on Δ, denoted by ℓ-$\mathsf{TCST}^{\Delta}(T)$, is a compacted trie that represents the set $\mathsf{Fac}^{\Delta}(T, \ell)$ of all (Δ, ℓ)-factors of T. It is easy to see that the number of nodes in ℓ-$\mathsf{TCST}^{\Delta}(T)$ is linear in the number $k' = |\mathsf{Fac}^{\Delta}(T, \ell)|$ of unique (Δ, ℓ)-factors of T. Since k' is smaller than $k = |\mathsf{Suf}^{\Delta}(T)|$ in real data sets, we expect that ℓ-TCST is more space efficient than CST for small values of ℓ.

The modified algorithm ConstructTCST for ℓ-TCST is obtained from the original ConstructCST of Fig. 3 by inserting Terminate in Fig. 6 after Extend of line 8. The main difference of the new algorithm from the old one is the use of the *termination pointer* ψ for closing suffixes in addition to the extension pointer ϕ for opening suffixes. At every stage $i = 1, \ldots, n$, Extend first extends each $\phi = \alpha$ of type 2 in T^{i-1} to αa_i by attaching the i-th letter $a_i = T[i] \in \Sigma$. At the same time, Terminate keeps track of termination point ψ, which is an open leaf $\psi = \langle j, \infty \rangle$ with word depth at most $\ell - 1$, and terminates it whenever ϕ reaches the depth ℓ by replacing $\langle j, \infty \rangle$ with $\langle j, i \rangle$, where i is the current index. A key observation is that there exists at most one open leaf to be closed at every stage i.

To implement this idea, we have to count the length of the open suffixes in the number of codewords to detect when $|L(\psi)|$ exceeds the limit ℓ. To do this we use a data structure $\Theta = \langle \eta, wc \rangle$, where $\Theta.\eta = \eta \in \mathsf{dom}(code)$ is a *boundary pointer* to a code node and $\Theta.wc = wc \in \mathbb{N}$ is a *word counter*, with the following operations, where $a \in \Sigma$:

Algorithm Terminate($\psi = \langle s, j, i - 1 \rangle, i$):
Input: A terminating point ψ and $i \geq 0$;
1: $\psi \leftarrow$ Canonize($\langle s, j, i \rangle$); {type 1 extension by letter a_i}
2: ChildTrans($\Theta', T[i]$);
3: **if** toClose(Θ') **then**
4:   $v := child(s, T[j])$;
5:   **if** $label(v) = \langle j, \infty \rangle$ **then** $label(v) \leftarrow \langle j, i \rangle$;
6:   SuffixTrans(Θ', ψ);
7: **return** ψ;

Fig. 6. Terminating truncated word suffix trees

- Reset(Θ) $\equiv$ **begin** $\Theta.\eta \leftarrow [\bot]$; $\Theta.wc \leftarrow 0$; **end.**
- Decrement(Θ) $\equiv$ $\Theta.wc \leftarrow \Theta.wc - 1$;
- ChildTrans(Θ, a) $\equiv$
    1: $\Theta.\eta \leftarrow child_{\mathsf{DFA}(\Delta)}(\Theta.\eta, a)$;
    2: **if** $\Theta.\eta = [\varepsilon]$ **then begin** $\Theta.\eta \leftarrow \bot$; $\Theta.wc \leftarrow \Theta.wc + 1$ **end**;
- SuffixTrans($\Theta, \phi = \langle s, j, i \rangle$) $\equiv$
    1: $\Theta.wc \leftarrow \Theta.wc - 1$; $\phi \leftarrow$ Canonize($\langle SL^{\Delta}(s), j, i \rangle$); **return** ϕ;
- toClose(Θ) $\equiv$
    1: **return** $\Theta.wc = k$ and $\Theta.\eta = [\varepsilon]$;

The meaning of the above operations will be easily understood. Using these operations, we modify the algorithm ConstructCST and procedures Extend and Canonize by adding comment lines with "for ℓ-TCST." In Canonize, we replace the sentence "**if** $p - q > i - j$ **then**" at line 4 with "**if** $p - q > i - j$ **or** u is a leaf **then**." For time complexity, ChildTrans takes $O(\log |\Sigma|)$ time, and all the other operations except SuffixTrans take constant time. By analysis similar to one in the previous section, we can show that SuffixTrans requires amortized constant time per operation. From a similar discussion in Sec. 3 and in Na *et al.* [14], we have the following theorem.

Theorem 3 (linear time construction of a TCST on prefix code). *If Δ is a fixed, possibly infinite prefix code, then the modified algorithm* ConstructTCST *constructs a truncated code suffix tree ℓ-TCST$^{\Delta}(T)$ for an input text T in $O(n)$ time and $O(k)$ space in online manner, where $k = |T|_{\Delta}$ and $n = |T|_{\Sigma}$.*

5 Experimental Results

We ran experiments on real datasets. Input data were an English text from the Pizza & Chili Corpus[1], where the delimiters are spaces SPC and LF, and a UTF-8 text from the Mainichi Newspaper Corpus 1991 in Japanese[2]. The length of

---

[1] http://pizzachili.dcc.uchile.cl/
[2] http://www.nichigai.co.jp/sales/corpus.html

Table 1. Node count (10^6 nodes), where ℓ is the length of code factors

Data	ST	CST	IST	HST	IHST	ℓ	TST	TCST	TIST	THST	TIHST
English	87	17	14	105	20	2	28	3.6	2.6	39	4.9
						5	56	11	8.3	74	14
						10	58	11	8.7	76	14
UTF-8	85	29	25	105	35	2	2.5	0.43	0.35	4.0	0.70
						5	43	13	11	58	17
						10	70	23	20	89	30

Table 2. Query time (microseconds per query)

Data	ST	CST	IST	HST	IHST	ℓ	TST	TCST	TIST	THST	TIHST
English	1.03	0.827	0.686	1.295	0.624	2	0.889	0.718	0.577	1.139	0.546
	1.233	0.952	0.827	1.497	0.733	5	1.170	0.905	0.764	1.435	0.718
	1.263	0.952	0.827	1.514	0.749	10	1.185	0.904	0.827	1.467	0.718
UTF-8	0.537	0.47	0.441	0.787	0.434	2	0.421	0.341	0.305	0.608	0.304
	0.88	0.772	0.803	1.138	0.718	5	0.827	0.72	0.684	1.077	0.677
	0.886	0.774	0.767	1.138	0.723	10	0.858	0.755	0.787	1.108	0.705

each text was 50MB. The English text has 336,578 different words of the average length 5.20 (byte). The UTF-8 text has 4054 different codes of the average length 2.96 (byte).

We implemented several types of sparse and truncated suffix trees. ST is the suffix tree, CST is the code suffix tree in Chapter 3, IST is the suffix tree over the code alphabet Δ using four-byte integers as base letters, HST is the code suffix tree over the (letter-based) Huffman code, and IHST is the code suffix tree over the word-based Huffman code for Δ. TST, TCST, TIST, THST and TIHST are their truncated versions with the factor length $\ell = 2, 5$, and 10 (words). These programs were written in C++ and compiled by Microsoft Visual Studio 2010. We ran the programs on an Intel Core i7 920 and 12GB of RAM, running Windows 7 Professional 64bit.

Tables 1 and 2 show the node counts of the suffix trees and the average query time for 10^6 strings of the code lengths $\ell = 2, 5$, and 10 (words), respectively. In the experiments, we observed that the sparse suffix trees were more space-efficient and faster than their non-sparse versions, roughly, by the factor of $O(n/k)$. Truncated suffix trees also improved space efficiency and query time in both full and truncated versions. IHST was the fastest in the algorithms, even though HST was slowest. CST was slightly slower than IST, however, it is comparable because it does not need any additional space and preprocessing.

6   Conclusion

In this paper, we presented an online construction algorithm for CSTs and ℓ-TCSTs on regular, variable-length prefix codes that runs in $O(n + m)$ time and

$O(k)$ space, where n is the total text size, k is the number of codewords, and m is the size of a code automaton.

As future works, extensions of this approach to other suffix indexes, e.g., DAWG [9], CDAWG [10], and suffix arrays [5], and application to enhanced suffix arrays [1, 12] and property suffix trees [2, 16] would be interesting. Also, it would be an interesting future problem to study lowerbounds of the worst case time complexity of construction of sparse suffix trees with an arbitrary index set of size k when $O(k)$ space is allowed.

Acknowledgements

The authors are grateful to anonymous referees for many useful comments and suggestions that significantly improve the quality of this paper. They also would like to thank Masayuki Takeda, Ayumi Shinohara, Shunsuke Inenaga, Thomas Zeugmann, Makoto Haraguchi, Shin-ichi Minato, Takuya Kida, and Osamu Watanabe for their discussions and valuable comments, This research was partly supported by MEXT Grant-in-Aid for Scientific Research (A), 20240014, FY2008–2011, and MEXT/ JSPS Global COE Program, FY2007–2011.

References

1. Abouelhoda, M.I., Kurtz, S., Ohlebusch, E.: Replacing suffix trees with enhanced suffix arrays. J. of Discrete Algorithms 2(1), 53–86 (2004)
2. Amir, A., Chencinski, E., Iliopoulos, C.S., Kopelowitz, T., Zhang, H.: Property matching and weighted matching. In: Lewenstein, M., Valiente, G. (eds.) CPM 2006. LNCS, vol. 4009, pp. 188–199. Springer, Heidelberg (2006)
3. Andersson, A., Larsson, N.J., Swanson, K.: Suffix trees on words. Algorithmica 23(3), 246–260 (1999)
4. Crochemore, M., Rytter, W.: Jewels of Stringology: Text Algorithms (2002)
5. Ferragina, P., Fischer, J.: Suffix arrays on words. In: Ma, B., Zhang, K. (eds.) CPM 2007. LNCS, vol. 4580, pp. 328–339. Springer, Heidelberg (2007)
6. Gusfield, D.: Algorithms on Strings, Trees, and Sequences, – Computer Science and Computational Biology, Cambridge (1997)
7. IETF, UTF-8, a transformation format of ISO 10646, RFC 3629 (2003), http://tools.ietf.org/html/rfc3629
8. Inenaga, S., Takeda, M.: On-line linear-time construction of word suffix trees. In: Lewenstein, M., Valiente, G. (eds.) CPM 2006. LNCS, vol. 4009, pp. 60–71. Springer, Heidelberg (2006)
9. Inenaga, S., Takeda, M.: Sparse directed acyclic word graphs. In: Crestani, F., Ferragina, P., Sanderson, M. (eds.) SPIRE 2006. LNCS, vol. 4209, pp. 61–73. Springer, Heidelberg (2006)
10. Inenaga, S., Takeda, M.: Sparse compact directed acyclic word graphs. In: Holub, J., Zdarek, J. (eds.) Proc. PSC 2006, pp. 197–211 (2006)
11. Kärkkäinen, J., Ukkonen, E.: Sparse suffix trees. In: Cai, J.-Y., Wong, C.K. (eds.) COCOON 1996. LNCS, vol. 1090, pp. 219–230. Springer, Heidelberg (1996)

12. Kasai, T., Lee, G.H., Arimura, H., Arikawa, S., Park, K.: Linear-time longest-common-prefix computation in suffix arrays and its applications. In: Amir, A., Landau, G.M. (eds.) CPM 2001. LNCS, vol. 2089, pp. 181–192. Springer, Heidelberg (2001)
13. McCreight, E.M.: A space-economical suffix tree construction algorithm. J. ACM 23, 262–272 (1976)
14. Na, J.C., Apostolico, A., Iliopoulos, C.S., Park, K.: Truncated suffix trees and their application to data compression. Theoretical Computer Science 304, 87–101 (2003)
15. Takeda, M., Miyamoto, S., Kida, T., Shinohara, A., Fukamachi, S., Shinohara, T., Arikawa, S.: Processing text files as is: Pattern matching over compressed texts, multi-byte character texts, and semi-structured texts. In: Laender, A.H.F., Oliveira, A.L. (eds.) SPIRE 2002. LNCS, vol. 2476, p. 170. Springer, Heidelberg (2002)
16. Uemura, T., Arimura, H.: A linear-time off-line construction of property suffix trees. IEICE Trans. Inf. & Syst. J91-D(3), 595–607 (2008) (in Japanese). An English version appears in Chapter 4, T. Uemura, Efficient Construction of Constrained Suffix Trees, Ph.D thesis, IST, Hokkaido Univ. (February 2011) (submitting)
17. Ukkonen, E.: On-line construction of suffix-trees. Algorithmica 14(3), 249–260 (1995)

On the Weak Prefix-Search Problem[*]

Paolo Ferragina

Dipartimento di Informatica, Università di Pisa, Italy
ferragina@di.unipi.it

Abstract. The weak-prefix search problem asks for the strings in a dictionary S that are prefixed by a pattern $P[1,p]$, if any, otherwise it admits any answer. Strings in S have average length ℓ, are n in number, and are given in advance to be preprocessed, whereas pattern P is provided on-line. In this paper we solve this problem in the cache-oblivious model by using the optimal $O(n \log \ell)$ bits of space and $O(p/B + \log_B n)$ I/Os. The searching algorithm is of Monte-Carlo type, so its answer is correct with high probability. We also extend our algorithmic scheme to the case in which a probability distribution over the queried prefixes is known, and eventually address the deterministic case too.

1 Introduction

Searching for a pattern $P[1,p]$ as a prefix of a set S of n strings of average length ℓ is a classic problem in the string-matching field, with plenty of solutions available. (See e.g. [9] for a survey and a discussion of the literature.) All known solutions require $O(n\ell)$ bits in the worst case, because they need to store the set S. Recently [1] introduced the *weak* variant of the problem that asks for a *one-side* answer, which is requested to be correct only in the case that P prefixes some of the strings in S; otherwise, it leaves to the algorithm the possibility to return an un-meaningful answer. Formally,

Problem 1 (Bellazougui et al. *2010).* Let $S = \{s_1, s_2, \ldots, s_n\}$ be a prefix-free set of binary strings which have variable length and are sorted alphabetically. Each string consists of at most L bits, and it has average length ℓ. The *weak* prefix-search problem requires, given a binary query-string $P[1,p]$, to return the range of strings of S having P as prefix. The range is expressed in terms of string-ids, and it may be *any* range if P does not prefix any string of S.

The *weak*-feature allowed [1] to reduce the space occupancy to $O(n \log \ell)$ bits, and indeed they proved that this is a space lower-bound. Their solution takes $O(p/w + \log_2 p)$ memory transfers for searching in a RAM model with memory-word of w bits. The key issue here is not to store the string set S, but use an index just taking $O(\log \ell)$ bits per string. This improvement is significant for very-large string sets, and we refer the reader to [1] for applications.

---

[*] Supported by PRIN MadWeb 2008 and FIRB Linguistica 2006.

R. Giancarlo and G. Manzini (Eds.): CPM 2011, LNCS 6661, pp. 261–272, 2011.

In the large-set setting it is more appropriate to evaluate algorithms in the External-Memory model in which B is the disk-page size and M is the size of the internal memory [13]. In this context the above result can be rephrased as $O(p/B + \log_2 p)$ I/Os, by setting $w = B$, still within the optimal $O(n \log \ell)$ bits of space occupancy. Noteworthy the I/O-bound holds also in the more powerful Cache-Oblivious model where the two model's parameters M and B are unknown to the running algorithm (see e.g. [10]).

The best known result for the *classic* prefix-search problem [3,8,9] uses $O(n\ell)$ bits of space and $O(p/B + \log_B n)$ I/Os. This is always worse in space than what is obtained for the weak prefix-search, but it is faster whenever $\log_B n < \log_2 p$. This latter inequality is realistic in practice even for moderate pattern's lengths, given that it corresponds to $n < p^{\log_2 B} \approx p^{15}$. In this paper, we provide a (randomized) solution to the weak-prefix search problem which matches the best of the two solutions above by obtaining $O(p/B + \log_B n)$ I/Os within $O(n \log \ell)$ bits of space occupancy (see Theorem 2). The searching algorithm is of Monte-Carlo type, so its answer is correct with high probability. Our solution has the nice feature of being algorithmically simpler than [1], and thus worth to be implemented. Eventually we will extend our algorithmic scheme to the case in which a *probability distribution* over the queried prefixes is known (see Theorem 3), and then address the deterministic case too.

We conclude this section by noticing that our results are stated in terms of binary strings but they generalize easily to strings drawn from an arbitrary alphabet Σ. This is obvious for a constant-sized alphabet, whereas the case of a larger alphabet can be managed by re-mapping its symbols occurring in $\mathcal{S}$'s strings to the range $[0, n\ell - 1]$, the second extreme denoting the total length of the strings in $\mathcal{S}$.

2   Background

A *compacted trie* $\mathcal{T}_\mathcal{S}$ built on the string set $\mathcal{S}$ is a tree in which the root may be unary, but all other internal nodes are binary. $\mathcal{T}_\mathcal{S}$ consists of n leaves (one per string in $\mathcal{S}$) and no more than n internal nodes, hence $O(n)$ nodes in total. Each edge $e = (u, v)$ is labeled with a substring $s(e)$ of the strings in $\mathcal{S}$, each leaf l is labeled with an integer in the range $[1, n]$ denoting the rank of its associated string in $\mathcal{S}$, and each internal node u is labeled with an integer denoting the length of the string spelled out by the root-to-u path in $\mathcal{T}_\mathcal{S}$. We denote that spelled string by $s(u)$, and observe that in the case u is a leaf it is $s(u) \in \mathcal{S}$.

A *Patricia trie* $\mathcal{PT}_\mathcal{S}$ is derived from the compacted trie $\mathcal{T}_\mathcal{S}$ by substituting each substring labeling a tree edge with its first character, commonly called *branching character*. Figure 1 shows an example. In the following we will use $\mathcal{PT}_\mathcal{S}$ in combination with a special prefix-search procedure introduced in [8] and called *blind search*. The key idea is to search for the lexicographic position of P in $\mathcal{S}$ by percolating a root-to-leaf path in $\mathcal{PT}_\mathcal{S}$, matching only the available branching characters. Two cases may occur: (i) either P is exhausted and a node u is reached (possibly a leaf); or (ii) a mismatch occurred when branching

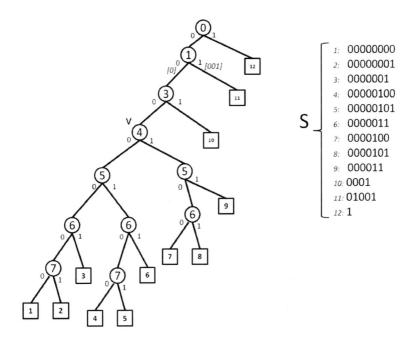

Fig. 1. An example of compacted trie and its Patricia's variant built over a set S of 12 binary strings. Bits between brackets are the part of an edge-label that has been dropped when turning the compacted trie into the corresponding Patricia trie. In this example, this dropping occurs just twice.

out from a node u. In both cases, the authors in [8] suggest to take any leaf l descending from u, and they prove that $s(l)$ is one of the strings in S sharing the *longest common prefix* with P [8, Lemma 1, pag 253]. This lcp-value together with another percolation of PT_S was used in [8] to find P's position in S. The nice fact was that just one string of S was compared.

This single string-comparison is not possible in our weak-prefix search problem because S is not available. Nevertheless that result can be rephrased in our problem as follows: case (ii) immediately leads us to say that P does not prefix any string of S; if we are in case (i), then either the string $s(u)$ has prefix P or P does not prefix any string in S. As an example take $P = 0010$, the blind search over the Patricia Trie in Figure 1 reaches the node v which is *not* the correct locus for the string P because there is a mismatch at the third bit that was not catched by the blind search since that bit was missing in the corresponding edge label. Anyway, this is not a problem because P is not a prefix of any string in S, and so the answer to the weak-prefix search can be arbitrary. On the other hand if $P = 0000$ then the blind search stops again at the node v, but this is now correct since P prefixes all 9 strings descending from v.

The advantage of using the combination between blind search *and* PT_S, with respect to the classical approach over compacted tries, is that the blind search identifies one single node in PT_S whose spelled string is the unique candidate for being prefixed by P, and this node can be found by deploying only the information available in $\mathcal{PT}_S$, thus without any access to the string set $\mathcal{S}$.[1]

On the other hand, there are two shortcomings in adopting this scheme to solve our weak-prefix search problem over $\mathcal{S}$. The first one is that the overall space is $\Omega(n \log n)$ bits, which is possibly $\omega(n \log \ell)$; the second one is that the packing of $\mathcal{PT}_S$ into pages of size B is a difficult problem in its generality, as shown in [6]. We address the first issue in the next Section 3, and the second issue in Section 4 where we will propose a solution tailored to deploy the specialities of the blind-search procedure over Patricia tries.

3   A 2-Level Indexing Scheme for RAM

Our first step is to propose a 2-level indexing scheme whose goal is to reduce the space occupancy of the classic Patricia trie's approach from $O(n \log n)$ to $O(n \log \ell)$ bits. Of course it is $\ell = \Omega(\log_2 n)$ because all n strings are distinct.

A proper sampling of $\mathcal{S}$. Recall that the strings in $\mathcal{S}$ are lexicographically sorted. For the sake of presentation we assume that $s_1 = 0^+$ and $s_n = 1^+$, so that P is lexicographically included in $\mathcal{S}$, and let n be a multiple of $\log n$.

We partition $\mathcal{S}$ into $g = n/\log n$ groups of (contiguous) strings defined as follows: $\mathcal{S}_i = \{s_{1+i \log n}, s_{2+i \log n}, \ldots, s_{(i+1) \log n}\}$ for $i = 0, 1, 2, \ldots, g-1$. We then construct a subset of $\mathcal{S}$, call it $\mathcal{S}'$, that consists of $2g$ strings obtained by taking the smallest (first) and the largest (last) string in each of those groups. In some sense this *sampling* process recalls the one adopted to design the String B-tree [8] but it is restricted to just two levels and assumes that the block size is $\log n$. Subtly we cannot prefix-search as done in String B-trees because we do not have the strings of $\mathcal{S}$ available, and thus we cannot compute the lcp's between P and one of those strings when needed by the blind search. Hence we need to devise some further properties about Patricia Tries and blind searches, in order to implement an efficient weak prefix search over this 2-level index.

The data structures. For the sake of presentation, let us forget the I/O-issues and concentrate on the design of a solution for the RAM model. We denote by $[s_l, s_r]$ the two strings in $\mathcal{S}$ which delimit the range of strings prefixed by P and thus the integer-pair (l, r) is the solution to the weak-prefix search problem. We recall that these two strings are arbitrary in the case that P does not prefix any string of $\mathcal{S}$. Given our notation we construct two types of Patricia Tries:

- $\mathcal{PT}'$ is the Patricia Trie built over the strings in $\mathcal{S}'$ with the speciality that we store in each node u of $\mathcal{PT}'$ a fingerprint of $O(\log n)$ bits computed for the string-prefix $s(u)$ (spelled by the root-to-u path). Since $\mathcal{PT}'$ consists of $O(n/\log n)$ nodes, this additional information needs $O(n)$ bits overall and can be efficiently computed by using the Karp-Rabin-scheme [5] which ensures distinct substrings collide with polynomially small probability.

---

[1] The compacted trie needs to access $\mathcal{S}$'s strings to resolve the edge labels during a string search.

– For each edge $e = (u, v)$ of $\mathcal{PT}'$ we build the Patricia trie $\mathcal{PT}_e$ which indexes a group of $O(\log n)$ strings defined as follows. Assume that each node v of $\mathcal{PT}'$ points to its leftmost/rightmost descending leaves, denoted by $L(v)$ and $R(v)$. We denote by $\mathcal{S}_{L(v)}$ and $\mathcal{S}_{R(v)}$ the two groups of strings, from the grouping above, that contain $s_{L(v)}$ and $s_{R(v)}$. The Patricia trie $\mathcal{PT}_e$ is built on the string set $\mathcal{S}_{L(v)} \cup \mathcal{S}_{R(v)}$. Since the strings in $\mathcal{S}_{L(v)}$ or $\mathcal{S}_{R(v)}$ are contiguous in $\mathcal{S}$ we can encode their string-ids relatively to the first string-id of each group, and thus use $O(\log n + (\log n) \times (\log \log n))$ bits. Each node encodes the string-len info within $O(\log \ell)$ bits on average, thus $O((\log n) \times (\log \ell))$ average bits overall. Since $\ell = \Omega(\log_2 n)$ (see above) and since the number of edges of $\mathcal{PT}'$ is $O(n/\log n)$, all Patricia tries $\mathcal{PT}_e$s take $O(n \log \ell)$ bits.

Lemma 1. *The data structures $\mathcal{PT}'$ and $\mathcal{PT}_e$, for all edges e in $\mathcal{PT}'$, occupy a total of $O(n \log \ell)$ bits.*

Few useful properties. Let us start with a property that can be easily derived from [8] and has been sketched in Section 2.

Fact 1. *Let $\mathcal{PT}_X$ be the Patricia trie built on a string set X. The execution of the blind search for a pattern P over $\mathcal{PT}_X$ stops at a node u such that either the string $s(u)$ is prefixed by P, or P does not prefix any string in X.*

The impact of this Fact cannot be overestimated, in the sense that it can be used in the second-level of our data structure to perform a weak-prefix search over the (contiguous) groups of strings indexed by some $\mathcal{PT}_e$; but it cannot be used over the first-level Patricia trie $\mathcal{PT}'$ to determine the lexicographic position of P in $\mathcal{S}'$, as it was done originally in [8]. This is because the set $\mathcal{S}'$ is unavailable. An example is given in Figure 2 for a pattern $P = 01001$. The classic blind-search procedure would reach the right child of node w, thus identifying the group of strings $\mathcal{S}_3$. This is not correct since $P \in \mathcal{S}_4$.

We circumvent this drawback by devising a new structural property of Patricia Tries.

Fact 2. *If P prefixes some of the strings in $\mathcal{S}$, then it does exist an edge $e = (u, v)$ in $\mathcal{PT}'$ such that s_l and s_r can be identified by looking at $\mathcal{S}_{L(v)} \cup \mathcal{S}_{R(v)}$.*

Proof. For ease of exposition we denote by $s_{L(i)}$ and $s_{R(i)}$ the leftmost and rightmost strings sampled from $\mathcal{S}_i$. We consider two cases: (a) P does not prefix anyone of the strings in $\mathcal{S}'$, and (b) P prefixes at least one string of $\mathcal{S}'$. Case (a) implies that the solution-range $[s_l, s_r]$, if not empty, is totally included in one group $\mathcal{S}_i$, so $s_{L(i)} < s_l \le s_r < s_{R(i)}$. Case (b) implies that the solution-range $[s_l, s_r]$ spans one or more groups, say $\mathcal{S}_x, \mathcal{S}_{x+1}, \ldots, \mathcal{S}_y$.

The Fact is trivially true in Case (a) because it is enough to take as edge e the one incident into the leaf $v = s_{L(i)}$ (or equivalently, $v = s_{R(i)}$) so that $\mathcal{S}_{L(v)} = \mathcal{S}_{R(v)}$. For Case (b), we have that $s_l \in \mathcal{S}_x$ and $s_r \in \mathcal{S}_y$, with $x \le y$. Given that P prefixes all strings in $[s_l, s_r]$, it prefixes the leftmost/rightmost strings of $\mathcal{S}_{x+1}, \ldots, \mathcal{S}_{y-1}$ as well as it prefixes $s_{R(x)}$ (which lies on the right of s_l,

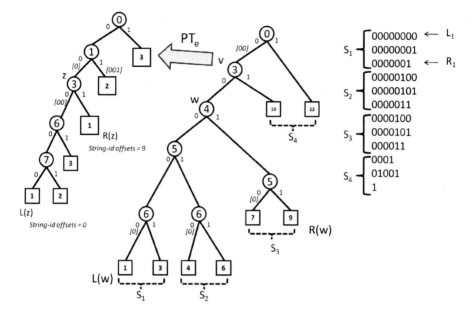

Fig. 2. The Patricia Trie built on the subset $\mathcal{S}'$, in which each group is formed by 3 strings. On the left it is shown $\mathcal{PT}_e$, where e is the edge incident into v, with its leaves encoded *relatively* to the first two string-ids of groups $\mathcal{S}_1$ and $\mathcal{S}_4$, namely 1 and 10. The figure also indicates the leftmost $s_{L(w)}$ and the rightmost $s_{R(w)}$ strings descending from node w.

possibly it is equal) and $s_{L(y)}$ (which lies on the left of s_r, possibly it is equal). In addition, if $s_l = s_{L(x)}$, then P also prefixes that string; similarly for $s_r = s_{R(y)}$. We now concentrate on s_l because the proof for s_r is symmetric; and we prove that $s_l \in \mathcal{S}_{L(v)}$ for some node v by distinguishing two sub-cases: (b.1) $s_{L(x)} < s_l$ and so P does not prefix $s_{L(x)}$, (b.2) $s_{L(x)} = s_l$ and thus P prefixes $s_{L(x)}$.

Case (b.1) implies that $lcp(s_{L(x)}, s_l) < p$, and since we know that P prefixes $s_{R(x)}$, the blind search over $\mathcal{PT}'$ will exhaust P stopping at a node v such that $|s(v)| \geq p$. Moreover we have that $s_{R(x)}$ descends from v but its adjacent string on-the-left in $\mathcal{S}'$, namely $s_{L(x)}$, does not. Therefore the leftmost string in $\mathcal{S}'$ descending from v is $s_{R(x)}$. So we can conclude that $s_l \in \mathcal{S}_{L(v)}$.

Case (b.2) follows a similar argument to prove that $s_{L(x)} = s_l$ descends from v but its adjacent string on-the-left in $\mathcal{S}'$, namely $s_{R(x-1)}$, does not (given that P does not prefix it). So the leftmost string in $\mathcal{S}'$ descending from v is $s_{L(x)} = s_l$, and thus again $s_l \in \mathcal{S}_{L(v)}$.

As we mentioned above, the proof for s_r follows similar lines hinging on the same vertex v, and shows that $s_r \in \mathcal{S}_{R(v)}$. □

Fact 2 relaxes the requirement that $\mathcal{PT}'$ can identify the lexicographic position of P in $\mathcal{S}'$, and aims for an *approximate solution* of it: namely, it shows that the position of P in the original set $\mathcal{S}$ can be determined within a distance $O(\log n)$.

It is the second round of the weak-prefix search procedure, executed on $\mathcal{PT}_e$, to resolve the ambiguity thanks to Fact 1.

As an illustrative example, refer to Figure 2 and consider a pattern $P = 0100$ which prefixes the single string s_{11} internal into $\mathcal{S}_4$ (Case a). The edge identified by Fact 2 is the one incident in $s_{L(4)} = s_{10}$, so the candidate group of strings to be searched for s_l, s_r is $\mathcal{S}_4$. On the hand, assume that $P = 00$ which prefixes $[s_1, s_{10}]$ (Case b). The edge e is the one connecting the root of $\mathcal{PT}'$ to v. It identifies the groups $\mathcal{S}_1 \cup \mathcal{S}_4$ and indeed $s_1 \in \mathcal{S}_1$ and $s_{10} \in \mathcal{S}_4$. Both groups are indexed by the Patricia Trie $\mathcal{PT}_e$ shown in the figure. We observe that, if we blind search P in $\mathcal{PT}_e$, we identify the node z such that $s_{L(z)} = s_1$ and $s_{R(z)} = s_{10}$ as stated in Fact 1.

The algorithmic solution. Now we are left with the problem of showing how the Patricia Trie $\mathcal{PT}'$ can be used to efficiently identify the edge e characterized by Fact 2, and how the search proceeds in $\mathcal{PT}_e$ to detect s_l, s_r among the strings of $\mathcal{S}_{L(v)} \cup \mathcal{S}_{R(v)}$ (Fact 1).

Subtly enough, the identification of edge e specified in Fact 2 is not immediate. As an example let us assume that P does not prefix any string in $\mathcal{S}'$, but it prefixes one string in some sub-group $\mathcal{S}_i$. Given that the Patricia trie $\mathcal{PT}'$ is built over $\mathcal{S}'$, the blind search could lead to a node which is completely far away from the leaf $s_{L(i)}$ we are searching for (because it points to $\mathcal{S}_i$). The problem here arises because $\mathcal{PT}'$ contains a "reduced" set of branching nodes (wrt $\mathcal{PT}_{\mathcal{S}}$), which are indeed the lowest-common-ancestors of the leaves associated to the sampled strings of $\mathcal{S}'$. So the path leading to $s_{L(i)}$ is surely followed by the blind-search procedure up to the first $lcp(P, s_{L(i)}) < p$ characters, but then it may diverge from that path when matching the pattern's characters following $lcp(P, s_{L(i)})$. Figure 2 shows this dangerous case for $P = 0100$. Here P prefixes $s_{11} \in \mathcal{S}_4$ but no string in $\mathcal{S}'$ is prefixed by P. The path followed by the blind search in $\mathcal{PT}'$ leads to the node w and thus the application of Fact 2 would lead to search in $\mathcal{S}_1 \cup \mathcal{S}_3$, which is incorrect. The problem here is that the mismatch-bit between P and the first traversed edge resides at position $lcp(P, s(v)) = 1$ which is internal into the edge-label and thus has not been compared by the blind search. The next compared bit, namely $P[4]$, matches the branching bit leading to w and thus drives the search far from node v, and thus far from $R(v) \in \mathcal{S}_4$.

In order to circumvent this problem we have to empower the blind-search procedure for detecting the edge of the mismatch bit. The pseudo-code in Figure 3 details our approach, which hinges on the deployment of the fingerprints $f(u)$ available at each node u of $\mathcal{PT}'$. These fingerprints maintain a succinct encoding of the substrings $s(u)$, taking $O(\log n)$ bits each. We highlight that, for efficiency reasons, $f(u)$ does not represent the label of the edge leading to u, but it rather represents the entire string-prefix $s(u)$, as spelled out by the root-to-u path. This way, we need to compute and store only $O(p)$ fingerprints for P, which can be done in linear time according to the Karp-Rabin's scheme.

It is evident in the pseudo-code of Figure 3 that fingerprints will be equal for all the ancestors a of v (and thus for all pattern prefixes of length $\leq |s(a)|$), after that they'll be different with high probability. The crucial twist here is that we

1. Compute the Karp-Rabin's fingerprint of all prefixes of the pattern P, according to the function $f()$ used for the nodes in $\mathcal{PT}'$.
2. Execute a blind search for P over the Patricia trie $\mathcal{PT}'$. Each time a node v is reached, we check whether $f(v)$ equals the fingerprint of the corresponding pattern prefix $f(P[1, |s(v)|])$. If the two fingerprints match and $|s(v)| < p$, the blind search proceeds branching out of v with the bit $P[|s(v)| + 1]$, otherwise we stop at v and call e the edge of $\mathcal{PT}'$ entering into v.
3. Repeat the blind-search procedure over $\mathcal{PT}_e$ and call v' the node where it stops. Return $s_l = L(v')$ and $s_r = R(v')$.

Fig. 3. The weak-prefix search algorithm

are not able to identify the position of the mismatching bit, but we are able to identify the edge $e = (u, v)$ containing that mismatching bit. This is enough to conclude that P lies lexicographically either to the left of the subtree descending from v (hence to the left of $L(v)$) or to its right (hence to the right of $R(v)$). This is what has been stated in Fact 2 and implemented in Step 2. Finally, Fact 1 guarantees that Step 3 correctly identifies s_l and s_r among the strings of $\mathcal{S}_{L(v)} \cup \mathcal{S}_{R(v)}$ as the leftmost/rightmost descendants of node v' in $\mathcal{PT}_e$.

Theorem 1. *The combination of $\mathcal{PT}'$ and the set of $\mathcal{PT}_e$ solves the weak-prefix search problem in $O(n \log \ell)$ bits. The search for a binary pattern $P[1, p]$ percolates two downward paths of length at most p, and uses local information stored at each node in $O(\log n + \log \ell) = O(w)$ bits. The search is correct with high probability because of the use of the Karp-Rabin's fingerprintings.*

4   I/O-Packing of the Patricia Tries

Packing trees of t nodes into pages of size B is difficult if there is no additional restriction either on the tree structure or the type of tree traversals. Surprisingly enough, even if we restrict the tree traversals to root-to-leaf paths of length L, the type of I/O-bounds we are aiming for— namely $O(L/B + \log_B t)$— cannot be guaranteed in general [6]. In particular, when the path length is $L = O(\log t)$ then the bound $\frac{L}{\log B} = O(\log_B t)$ I/Os is possible. But in order to extend this I/O-bound to all other p, we need to dig into the structural properties of tries and prefix searches as we do next.[2]

The centroid tree. We resort a known decomposition of trees proposed in [2] for the static cache-oblivious String B-tree. We apply it to our Patricia Trie $\mathcal{PT}'$, and then modify accordingly the blind-search procedure of Figure 3. This will be enough to get our final I/O-bounds because the other Patricia Tries $\mathcal{PT}_e$ have size $O(\log n)$, thus their depth is $L = O(\log n)$, and hence they can be searched in $O(\log_B n)$ I/Os as observed above.

---

[2] For other cache-oblivious or I/O-efficient mappings of tries please have a look at [3,4,11,12].

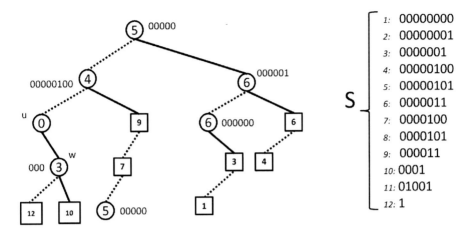

Fig. 4. The centroid tree of the Patricia Trie $\mathcal{PT}'$ of Fig.2

So let us recall how to pack $\mathcal{PT}'$. The key idea is that there is a *centroid node* z in $\mathcal{PT}'$ that has at least $t/3$ and at most $2t/3$ descendants, where $t = |\mathcal{PT}'| = O(n/\log n)$. The centroid tree of $\mathcal{PT}'$ is obtained by making z the root and attaching as z's children the recursively defined centroid trees of z's up and down tries. At every level of the recursion we eliminate a constant fraction of trie nodes from consideration, so the centroid tree has depth $O(\log n)$. Thus, as observed above [6], there exists a packing for which any root-to-leaf path traverses $O(\log_B n)$ disk pages (hence I/Os). By packing we mean not only the tree-structure of the centroid tree, but also the fingerprinting information that is associated to the nodes of $\mathcal{PT}'$.

Figure 4 shows an example of centroid tree computed for the Patricia Trie of Figure 1. Each internal node is labeled with a string s, and has a *solid* edge linking it to the root of its down trie (the leaves of which have s as a prefix), and a *dotted* edge linking it to the root of its up trie (the leaves of which do not have s a prefix). Each internal node maintains a pointer to the leftmost and rightmost descending leaf from the entire $\mathcal{PT}'$ (not shown in the picture). In Figure 4 we use circles to denote internal nodes of $\mathcal{PT}'$ and squares to denote leaves of $\mathcal{PT}'$. Notice that leaves of $\mathcal{PT}'$ can become internal nodes of the centroid tree, and vice versa.

Prefix-searching over the centroid tree. We are left with showing how Step 2 of Figure 3 is implemented (recall that Step 3 acts on $\mathcal{PT}_e$ which takes $O(\log_B n)$ I/Os to be percolated). Let z be the node of the centroid tree currently visited, initially z is the root. We determine whether $s(z)$ is a prefix of P by comparing the corresponding fingerprints. If it does, then we follow the solid edge, otherwise we follow the dotted edge. The ratio is that, in the case of a match, the prefix search must proceed in the subtrie of $\mathcal{PT}'$ that descends from z (previously called down-trie) and containing all strings of $\mathcal{S}'$ that have $s(z)$ as a prefix; otherwise, the prefix search does not pass through z but lies into its up-trie. It should be evident that we are doing a sort of *binary search* over the entire structure of $\mathcal{PT}'$ which eventually identifies the deepest node u whose fingerprint $f(u)$ equals the fingerprint of the corresponding pattern prefix $f(P[1, |s(u)|])$. Now if $|s(u)| < p$, we take v as the child of u in $\mathcal{PT}'$ whose branching bit equals $P[|s(u)| + 1]$; otherwise it is $|s(u)| = p$ and thus we can set $v = u$. This node v is exactly the one identified in Step 2 of the pseudo-code of Figure 3. We remind the reader that false-positive matches between Karp-Rabin's fingerprints occur with small probability.

As an illustrative example, take $P = 0100$ and start matching the centroid-tree's nodes in Figure 4. We percolate the leftmost downward path up to node u, since its ancestors have longer labeling substrings which therefore do not prefix-match P (so dotted edges are followed). Since $s(u)$ is the empty string, it prefix-matches P; but its solid child w does not prefix P. Given that u has not dotted outgoing edges, u is the deepest node in $\mathcal{PT}'$ whose label prefix-matches P. So we jump[3] on u's "copy" in $\mathcal{PT}'$ and take the child whose branching bit is $P[1] = 0$. This is correctly the node v of Figure 2.

Theorem 2. *The weak-prefix search problem can be solved in the cache-oblivious model by using $O(n \log \ell)$ bits of space and $O(p/B + \log_B n)$ I/Os. The correctness holds with high probability.*

We conclude the paper by observing that our algorithmic scheme can be turned into a *query-distribution* aware approach that changes the term $O(\log_B n)$ into a term that depends on the probability of querying the pattern P. This improvement can be significant in the case that some patterns are more frequent than others. One (common) way to proceed is to start from a probability distribution over the strings of $\mathcal{S}$ (and thus over the leaves of $\mathcal{PT}_\mathcal{S}$), and then derive a distribution over their prefixes: for every node $v \in \mathcal{PT}_\mathcal{S}$, we define $\wp(v)$ as the sum of the probabilities of the descending leaves (i.e. strings having prefix $s(v)$). Then, we can adapt the centroid decomposition to work on the probability mass of the nodes in $\mathcal{PT}_\mathcal{S}$ rather than on the number of their descending leaves. Actually, to avoid the limiting cases of very rare strings s such that $\wp(s) < 1/n$, we divide the nodes of $\mathcal{PT}_\mathcal{S}$ into two subsets: one composed by rare nodes whose probability is $< 1/n$, and the other set formed by the rest. It is clear that the rare nodes form subtrees of $\mathcal{PT}_\mathcal{S}$ that we store using the scheme which is distribution's UNaware, and thus get the I/O-bound $O(p/B + \log_B n)$ for searches into them. Conversely, for the former set which forms a single trie, we proceed by adding to it all roots of the rare subtries, possibly splitting some edges, and then storing the overall trie using the distribution-aware approach.

The search starts in this latter trie, and identifies an edge (u, v) taking $O(|s(u)|/B + \log_B 1/\wp(u))$ I/Os. Since $s(u)$ prefixes P it is $|s(u)| \le p$ and $\wp(u) \ge \wp(P)$. Now, two situations may occur. Either v is not rare, and thus the search stops at this edge and the above bound can be written as $O(p/B + \log_B \frac{1}{\wp(P)})$.

---

[3] Clearly the "jump" is logical in that v can bring the two pointers within its occurrence in the centroid tree, thus saving one I/O, and the duplicate storage of $\mathcal{PT}'$.

Or v is rare, and if the prefix of P matches $s(v)$, then the search continues in the rare subtrie descending from v, using the distribution UN-aware scheme. This second search takes $O(p/B + \log_B n)$ I/Os. Overall this second case (which actually refers to a rare P) takes $O(p/B + \log_B 1/\wp(u) + \log_B n) = O(p/B + \log_B n)$ I/Os, where the second bound derives from the fact that u is not rare.

If we apply the above query-distribution aware approach to the entire trie $\mathcal{PT}_S$, we would get the aimed I/O-bound but with an overall space occupancy of $O(n\ell)$ bits. To get the following theorem we just use the 2-level indexing scheme of Section 3 and apply the query-distribution aware approach to both $\mathcal{PT}'$ and to all $\mathcal{PT}_e$.

Theorem 3. *The weighted weak-prefix search problem, for which a query-distribution is known in advance, can be solved in the cache-oblivious model using $O(n \log \ell)$ bits of space and $O(p/B + \log_B \min\{n, \frac{1}{\wp(P)}\})$ I/Os. The correctness holds with high probability.*

As a conclusive note we point out that our algorithms can be changed from Montecarlo's to *deterministic* ones in the reasonable applicative scenario (mentioned in [1]) that the strings reside on disk and the weak-prefix-search data structure is stored in internal memory. In this case, our approach needs only to store on disk the sampled set $\mathcal{S}'$ and uses just one additional disk I/O to determine v in Step 2 (in practice we argue that $p/B \leq 1$). Conversely the String B-tree [8] requires the same number of I/Os but with a storage cost of $\Omega(n \log n)$ bits in internal memory, and the need to keep the whole string set $\mathcal{S}$ on disk.

As a future research we would like to deploy our algorithms and data structures in order to address the energy-issues raised in [7].

Acknowledgements. I thank Djamal Belazzougui and the anonymous referees for their useful comments.

References

1. Belazzougui, D., Boldi, P., Pagh, R., Vigna, S.: Fast prefix search in little space, with applications. In: de Berg, M., Meyer, U. (eds.) ESA 2010. LNCS, vol. 6346, pp. 427–438. Springer, Heidelberg (2010)
2. Bender, M., Farach-Colton, M., Kuszmaul, B.: Cache-oblivious string B-trees. In: Procs. ACM PODS, pp. 233–242 (2006)
3. Brodal, G., Fagerberg, R.: Cache-oblivious string dictionaries. In: ACM-SIAM SODA, pp. 581–590 (2006)
4. Clark, D.R., Munro, J.I.: Efficient suffix trees on secondary storage. In: Procs. ACM-SIAM SODA, pp. 383–391 (1996)
5. Cormen, T.H., Leiserson, C.E., Rivest, R.L.: Introduction to Algorithms. MIT Press, Cambridge (1990)
6. Demaine, E.D., Iacono, J., Langerman, S.: Worst-case optimal tree layout in a memory hierarchy (2004), available on arXiv:cs.DS/0410048 (manuscript)
7. Ferragina, P.: Data structures: Time, I/Os, entropy, joules! In: de Berg, M., Meyer, U. (eds.) ESA 2010. LNCS, vol. 6347, pp. 1–16. Springer, Heidelberg (2010)

8. Ferragina, P., Grossi, R.: The string B-tree: A new data structure for string search in external memory and its applications. J. ACM 46(2), 236–280 (1999)
9. Ferragina, P., Grossi, R., Gupta, A., Shah, R., Vitter, J.S.: On searching compressed string collections cache-obliviously. In: Procs. ACM PODS, pp. 181–190 (2008)
10. Frigo, M., Leiserson, C.E., Prokop, H., Ramachandran, S.: Cache-oblivious algorithms. In: Procs. IEEE FOCS, pp. 285–298 (1999)
11. Ko, P., Aluru, S.: Obtaining Provably Good Performance from Suffix Trees in Secondary Storage. In: Lewenstein, M., Valiente, G. (eds.) CPM 2006. LNCS, vol. 4009, pp. 72–83. Springer, Heidelberg (2006)
12. Maheshwari, A., Zeh, N.: A survey of techniques for designing I/O-efficient algorithms. In: Meyer, U., Sanders, P., Sibeyn, J.F. (eds.) Algorithms for Memory Hierarchies. LNCS, vol. 2625, pp. 36–61. Springer, Heidelberg (2003)
13. Vitter, J.: Algorithms and Data Structures for External Memory. Series on Foundations and Trends in Theoretical Computer Science. Now Publishers (2008)

Quick Greedy Computation for Minimum Common String Partitions

Isaac Goldstein and Moshe Lewenstein[*]

Department of Computer Science, Bar-Ilan University, Ramat Gan 52900, Israel
goldshti@gmail.com, moshe@cs.biu.ac.il

Abstract. In the *minimum common string partition* problem one is given two strings S and T with the same character statistics and one seeks the smallest partition of S into substrings so that T can also be partitioned into the same substring multiset. The problem is fundamental in several variants of edit distance with block operations, e.g. signed reversal distance with duplicates and edit distance with moves.

The minimum common string partition problem is known to be NP-complete and the best approximation known is of order $O(\log n \log^* n)$. Since this problem is of utmost practical importance one seeks a heuristic that will (1) usually have a low approximation factor and (2) will run fast.

A simple greedy algorithm is known and it has been well-studied from an approximation point of view. It has been shown to have a bad worst case approximation factor. However, all the bad approximation factors presented so far stem from complicated recursive construction. In practice the greedy algorithm seems to have small approximation factors. However, the best current implementation of greedy runs in quadratic time.

We propose a novel method to implement greedy in linear time.

1 Introduction

The classical edit distance is the number of edit operations, insertions, deletions, character exchanges and (sometimes) swaps, needed to transform one string into another. All the edit operations work on single characters, besides swap which operates on two. Motivation for block operations has stimulated a large collection of new problems, some of which have a completely different flavor than the original edit distance problem.

Kruskal and Sankoff [12] and Tichy [16] were first to deal with block operations together with simple character operations. Lopresti and Tomkins [13] gave several distance measures for block operations in the same spirit as they were defined for the case of single characters. The subject of block edit distance gained an increasing interest in the last decade especially due to its important impact

---

[*] This research was partially supported by an Israel Science Foundation (ISF) grant # 1484/08.

R. Giancarlo and G. Manzini (Eds.): CPM 2011, LNCS 6661, pp. 273–284, 2011.

on the computational biology field. In the research of the homology of structures there is an interest in finding organisms' characteristics that are derived
from a common ancestor. The task of identifying the orthologs, especially direct
descendants of ancestral genes in current species is a fundamental problem in
computational biology. A promising approach to solving this problem is to take
into account not only local mutations but also global genome rearrangements
events. This approach is strongly connected to the area of edit distance where
block operations are allowed, since those genes can be represented as sequences
of characters with block operations applied to them.

1.1   Sorting by Reversals and MCSP

Chen et al. [1] studied the problem of signed reversal distance with duplicates
(SRDD) which is a slight extension of the sorting by reversals (SBR) problem.
They showed that this problem is NP-hard even for $k = 2$ (it is worth noting
that in the unsigned case the problem is NP-hard even if $k = 1$. However, for
the signed case there is a polynomial algorithm for solving SBR for $k = 1$,
see [7]). Moreover, they pointed out that this problem is closely related to the
minimum common string partition problem (MCSP). In the MCSP we are given
two strings S and T and we need to find a partition of each of these strings to
substrings, so that we can match all of S's parts to those of T.

1.2   Edit Distance with Moves and MCSP

The problem of edit distance with moves was studied by Cormode and Muthukrishnan [4]. In this variant of the classical edit distance, moving a substring is
allowed in addition to deleting and inserting single characters. They showed the
problem to be NP-Complete and suggested and approximation algorithm with
an approximation factor of $O(\log n \log^* n)$ and running time $O(n \log^* n)$. This is
still the best approximation algorithm given for the problem.

    Shapira and Storer [15] observed that the problem, while recursive, can be
transformed into a non- recursive version with a constant-factor cost in the
approximation. Moreover, they showed that the problem can be transformed
into a version in which moves only are allowed. It has been observed that this is
in essence is a reduction to the MCSP problem, although not specifically noted so
in the paper. Using Shapira and Storer's observations one can use Cormode and
Muthukrishnan [4] approximation algorithm for MCSP as well. This is currently
the best known approximation.

1.3   Restricted Versions of MCSP

Goldstein et al. [6] proved that even the simple case of 2-MCSP in which each
character occurs at most twice is NP-hard. Moreover, they showed that it is APX-
hard. They also gave an 1.1037-approximation algorithm for this problem which
improved the 1.5 ratio showed by Chen et al. [1]. For 3-MCSP a 4-approximation
algorithm was given. Kolman [10] found an $O(k^2)$-approximation algorithm for

k-MCSP with $O(nk)$ running time for general k, where each symbol can occur at most k times. Kolman and Walen [11] improved this result and gave an $O(k)$-approximation algorithm with $O(n)$ running time.

Due to the close relation between the MCSP problem and the SBR problem (as shown by Chen et al. [1]) the approximation factor for these restricted versions of MCSP also applies to parallel restricted versions of SBR with just a constant factor penalty. Instead of restricting the problem by the number of each symbol occurrences, the MCSP problem and the related problem of SBR could be also restricted by the alphabet size. Unsigned SBR for unary alphabet could be solved trivially. However, Christie and Irving [2] showed that for binary alphabet unsigned SBR becomes NP-hard. Similar NP-hardness results for MCSP incorporating alphabet sizes greater than 1 were given by Jiang et al. [8].

1.4   MCSP and the GREEDY Algorithm

A simple greedy algorithm, to be denoted by GREEDY, was first examined by Shapira and Storer [15]. They showed that for many inputs the algorithms has a logarithmic approximation factor.

Chrobak et al. [3] demonstrated that in the general case the approximation ratio of GREEDY for the MCSP is not better than $\Omega(n^{0.43})$. In addition, for the special case of 4-MCSP it was found that it is at least $\Omega(\log n)$. Nevertheless, they showed it has an $O(n^{0.69})$ approximation factor for the general case, and a 3-approximation factor for 2-MCSP. The lower bound on GREEDY's performance for the general case was later increased to $\Omega(n^{0.46})$ by Kaplan and Shafrir [9].

1.5   Our Results

Since the edit distance with moves and the signed reversal distance with duplicates are used in practice it is of interest to construct an algorithm for MCSP that also (1) gives good approximation factors and (2) runs fast.

The approximation algorithm of [4] is the best in the worst-case sense, but the approximation factor is inherent in the method. On the other hand, GREEDY, while it can be bad in the approximation factor sense, as mentioned above, the worst-case examples are constructed from recursive definitions. In practice, GREEDY seems to do quite well in practice.

Shapira and Storer presented an implementation of Greedy that runs in $O(n^2)$ time. We will show a novel algorithm for implementing the greedy that runs in $O(n)$ time.

2   Problem Definitions and Preliminaries

2.1   Preliminary Definitions and Notations

Given a string S, $|S|$ is the length of S. Throughout this paper we denote $n = |S|$. An integer i is a *location* or a *position* in S if $i = 1, \ldots, |S|$. The substring $S[i .. j]$ of S, for any two positions $i \leq j$, is the substring of S that begins at index i and ends at index j. The *suffix* S_i of S is the substring $S[i .. n]$.

The *suffix tree* [5,14,17,18] of a string S, denoted $\text{ST}(S)$, is a compact trie of all the suffixes of $S\$$, i.e. S concatenated with a delimiter symbol $\$ \notin \Sigma$. $\text{ST}(S)$ requires $O(n)$ space. Algorithms for the construction of a suffix tree enable $O(n)$ preprocessing time when $|\Sigma|$ is constant (where Σ is the alphabet set), and $O(n \log \min(n, |\Sigma|))$ time when $|\Sigma|$ is not. A more sophisticated construction of the suffix tree can be constructed in linear time even for alphabets drawn from a polynomially-sized range, see [5].

Let the longest common substring, for short LCS, of S and T be denoted by $LCS(S, T)$. Moreover, if $x = LCS(S, T)$ appears at location i of S and location j of T then we say that it appears at (i, j).

We say that two strings S and T over alphabet Σ are *equi-statistic* if for every $\sigma \in \Sigma$ the number of appearances of σ in S and T is the same. Clearly, this implies that $|S| = |T|$.

A *partition* of a string $S = s_1, \cdots, s_n$ is a sequence of strings $R_1, R_2, \cdots, R_k$ such that $S = R_1 \cdot R_2 \cdot \cdots \cdot R_k$, where $\cdot$ denotes concatenation. A *common string partition* of two strings S and T is a partition $R_1, R_2, \cdots, R_k$ of S such that there is a permutation $\pi\{1, \cdots, k\} \rightarrow \{1, \cdots, k\}$ such that $R_{\pi[1]}, \cdots, R_{\pi[k]}$ is a partition of T. Note that if S and T are equi-statistic there is a simple common string partition of S and T; that is $R_i = s_i$ for every $1 \leq i \leq n$. Moreover, for any two strings that have a common string partition it is easy to verify that they are equi-statistic. Nevertheless, for equi-statistic strings S and T there may be various different common string partitions. The *minimum common string partition* problem is defined as follows.

Input: Two equi-statistic strings S and T.
Output: A common string partition $R_1, \cdots, R_k$ such that k is minimized.

The greedy algorithm for the minimum common string partition problem is presented below. We assume that the input strings S and T are equi-statistic.

Algorithm GREEDY(S,T)

- Set P and Q to be the initial (empty) partition
- Initially all letters in S and T are unmarked and P, Q are empty.
- **while** there are unmarked letters in T do
  - LCST $\leftarrow$ longest common substring of S and T that does not contain marked letters.
  - $LCST_S, LCST_T \leftarrow$ occurrences of LCST in S and T, respectively.
  - Designate $LCST_S$ as a block of P in S and $LCST_T$ as a block of Q in T.
  - Mark all letters in $LCST_S$ and $LCST_T$.
- output (P,Q)

3   Algorithm Outline

When solving the GREEDY method it is natural to use suffix trees. Specifically, start by constructing a generalized suffix tree for S and T, or in other words a suffix tree for $S\#T\$$, where $\#, \$ \notin \Sigma$. To find the LCS of S and T seek for the deepest node v which contains leaves in its subtree which represent suffixes from both S and T. This can be done by a traversal of the suffix tree in inorder. The substring which node v represents is the LCS.

Once the LCS is found one can reiterate the process, but beforehand x needs to be removed. One way to handle this is to replace x (in both S and T) with a new symbol not in the alphabet and to start from scratch. The time this process takes is $O(n^2)$ because in each round the suffix tree needs to be built from scratch and there can be O(n) rounds. In essence this is the implementation of the greedy method in [15].

Since we desire to reduce the time to have a more efficient implementation we do not want to construct such a suffix tree from scratch. Rather we want to fix the given suffix tree to reflect the changes that have been made. This is challenging. We now give the outline of how we shall do this.

Let $x = LCS(S,T)$ and assume that it appears at (i,j). More specifically, $S = S[1 \cdots i-1] \cdot x \cdot S[i + |x| \cdots n]$ and $T = T[1 \cdots j-1] \cdot x \cdot T[j + |x| \cdots n]$ or, in suffix terms, x is the prefix of suffix S_i and x is also the prefix of T_j. Once x is found it is necessary to update the suffix tree to reflect the fact that the appearances of x cannot be used again, not x itself and not any part of it. This means that we need to create a generalized suffix tree for $S = S[1 \cdots i - 1], S[i+|x| \cdots n]T[1 \cdots j-1], T[j+|x| \cdots n]$. However, the problem is that when transforming the generalized suffix tree for S and T into the generalized suffix tree for $S = S[1 \cdots i-1], S[i+|x| \cdots n]T[1 \cdots j-1], T[j+|x| \cdots n]$, all the suffixes $S_1, \cdots, S_{i-1}$ and $T_1, \cdots, T_{j-1}$ are affected. This means that the transformation can be too costly, namely O(n) which is the same as reconstructing the suffix tree from scratch.

Nevertheless, it turns out that the number of suffixes that affect the future runs of GREEDY can be bounded by the following lemma.

Lemma 1. *Let $x = LCS(S,T)$ be such that $k = |x|$. Assume that x appears at (i,j). Let y be any other common substring of S and T. Say, y appears at (i',j'). If $i' \notin [i-k+1, i+k-1]$ then the appearances of y and x do not intersect in S. Likewise, if $j' \notin [j-k+1, j+k-1]$ then the appearances of y and x do not intersect in T.*

Proof. The correctness of the proof follows from the fact that $|y| \leq k = |x|$ because $x = LCS(S,T)$. □

It follows from the lemma that if the LCS x appears at (i,j) and is of length $k = |x|$ then the first $i-k$ suffixes of S and the first $j-k$ suffixes of T, while they technically should be shortened because of the removal of x, they need not be shortened because they will not affect the GREEDY method. Therefore, when preparing the generalized suffix tree for the next step it is sufficient to take care of the $2k-1$ suffixes from S in the range $[i-k+1, i+k-1]$ and the $2k-1$ suffixes from T in the range $[j-k+1, j+k-1]$.

The suffixes of S in the range $[i, i + k - 1]$ need to be removed completely, as they overlap the x removed and the suffixes of S in the range $[i - k + 1, i - 1]$ need to be shortened. Each S_r, where $r \in [i - k + 1; i - 1]$ needs to be shortened to length $i - r$. Likewise, the suffixes of T in the range $[j, j + k - 1]$ need to be removed completely and the suffixes of T in the range $[j - k + 1; j - 1]$ need to be shortened appropriately.

Note that while the discussion until here has been about transforming the generalized suffix tree after finding the LCS in S and T all discussed holds for the next rounds as well. In other words, in the next round we will find an $LCSy$ in the generalized suffix tree for $S[1 \cdots i - 1], S[i + |x|, n], T[1 \cdots j - 1]$ and $T[j + |x|, n]$, which will, once again, be the deepest node v with suffixes from S and T in its subtree. The string represented by v, y will be the LCS (in the 2nd round) and will appear at (i', j') if the leaf representing $S_{i'}$ is in the subtree of v and the leaf representing $T_{j'}$ is also in the subtree of v. Removing y will split $S[1 \cdots i - 1], S[i + |x|, n], T[1 \cdots j - 1]$ and $T[j + |x|, n]$ into (at most) six substrings, (at most) three from S and (at most) three from T. We say "at most" because y splits a fragment into two only if it is in the middle of a fragment. Otherwise, if y is at the end, or beginning there will still only be one fragment after removing y. If y is the whole fragment then there is one fragment less. The suffix tree will need to be updated again. Obviously, Lemma 1 holds also here. This process is repeated iteratively and in round rS and T will each be fragmented into (at most) r substrings and we will need to find the LCS that appears both in the fragments of S and the fragments of T. We denote the LCS in round r with $LCS[r]$. We point out that $\Sigma_r |LCS^{[r]}| = n$. Hence, if we can solve each round r in time $O(|LCS^{[r]}|)$ then we will have overall time $O(n)$.

In lieu of the above define the four distinct sets of suffixes described above to be handled in round r of the GREEDY procedure.

1. $Delete(T)$: all T_i such that i is one of the $|LCS^{[r]}|$ locations where $LCS^{[r]}$ appears in T.
2. $Affects(T)$: all T_i such that i is one of the $|LCS^{[r]}|$ locations preceeding the appearance of $LCS^{[r]}$ in T.
3. $Delete(S)$: all S_i such that i is one of the $|LCS^{[r]}|$ locations where $LCS^{[r]}$ appears in S.
4. $Affects(S)$: all S_i such that i is one of the $|LCS^{[r]}|$ locations preceeding the appearance of $LCS^{[r]}$ in S.

Now we are left with two major tasks. First, we need to update the suffix tree to reflect the changes described for the four above mentioned groups of suffixes. Second, in each round we will need to find $LCS^{[r]}$. The challenging part for the first problem is to fix the suffix tree for $Affects(T)$ and $Affects(S)$ since it requires putting those suffixes in new places within the suffix tree. The challenging part of the second problem is to find $LCS^{[r]}$ in the limited time we have, namely in $O(|LCS^{[r]}|)$ time. This will require a data structure, for otherwise we will need to restart from scratch every time. The data structure needs to adapt according to the changes in the suffix tree.

4   Transforming the Suffix Tree in Round Change

Say we are in round r and moving to round $r + 1$. Currently S and T are each fragmented into at most r substrings and removing $LCS^{[r]}$ fragments the text further into, at most, one more fragment for each of S and T. The task at hand is to update the suffix tree to reflect this situation. Recall, that we will be updating only the suffixes from the four previously described suffix sets.

For any suffix in any of these sets we will need to remove it from the suffix tree. We will later need to reposition those from $Affects(S)$ and $Affects(T)$. Removing a suffix is actually quite straightforward, albeit a bit technical. We reach the appropriate leaf v representing the suffix by a pointer from an array representing the string (S or T). v's parent is then checked whether it has two children or more. If it has more it is sufficient to remove v and the edge connecting it to it's parent. If v's parent has only two sons, one of which is v, in order to maintain the suffix tree in it's compressed format one needs to phase out the parent of v by putting an edge between the grandparent and sibling of v. The labeling of the new edge needs to be fixed, which is easily done from the current labels.

So, what really needs to be handled is entering the truncated suffixes of $Affects(S)$ and $Affects(T)$ into the suffix tree. We treat $Affects(T)$ and $Affects(S)$ in the same manner but separately. The update we will do is done concurrently for all the members of $Affects(T)$.

We will refer to the suffix tree construction of Weiner [18]. Weiner constructs the suffix tree from the shortest suffix to the longest. Say the process is done on string Q of length q. Then the suffixes are entered into the compressed trie of suffixes (which will become a suffix tree) in the order of suffixes $\$, Q[q]\$, Q[q-1 \cdots q]\$, \cdots, Q[2 \cdots q]\$, Q\$$. The suffixes are inserted with the help of suffix pointers, which point from node v representing x to node v' which represent ax, if such a node exists. Weiner shows that the whole process is linear using an amortized argument that goes along the following lines. The amortized process assumes that we have just entered suffix x and need to enter suffix ax. x may not be represented by a node, but then x', the longest prefix of x which is represented by a node is the current parent of the suffix x that was entered. If x' has a pointer to ax' then we use it and we enter the suffix ax as its child. If not, then we walk up the path of root-to-leaf-representing-x until we reach a node v representing x' a prefix of x for which there is a pointer to ax' and do as mentioned before. The argument is that the overall walk is linear in length of the number of suffixes = the length of $Q = q$.

This suits our purpose well. Since we have just found $LCS^{[r]}$ we are allowed to spend $k = |LCS^{[r]}|$ time in work for reinserting all the suffixes in $Affects(T)$. If $x = LCS^{[r]}$ then $y = y_1, \cdots, y_{k-1}$ the substring (of length $k-1$) that appears just before x in T is what $Affects(T)$ is constructed from. Hence, the suffixes of $Affects(T)$, after being truncated, are $y_{k-1}, y_{k-2}y_{k-1}, y_{k-3}y_{k-2}y_{k-1}, \cdots, y$. We traverse the suffix tree and enter these suffix in the described order using the Weiner scheme. The time arguments of Weiner shows that this all these suffixes can be inserted in $O(|LCS^{[r]}|)$ time. We do the same for $Affects(S)$. Therefore, in $O(|LCS^{[r]}|)$ time we can update the suffix tree. What still needs to be done is to actually find the $LCS^{[r]}$. We will do so in the next section.

5   Maintaining Data Structures to Quickly Find the Current Round LCS

Therefore the question that remains open is how to do the updates fast enough so we can find the $LCS^{[r]}$ in time which is at most $O(|LCS^{[r]}|)$. Obviously, even if we would manage to keep the nodes of the suffix tree marked as we did in the beginning this would require us to traverse the whole suffix tree to find them. One approach is to save these nodes in some data structure ordered by their length and to deduce the deepest node from these. An example would be to maintain them in a heap ordered by node depth. However, even if this could be done, and it is not all obvious how to do so dynamically, then one would need to extract suffixes (leaves) S_i and T_j from it's subtree (within the above described time bounds). This can be really difficult. Another option is to save all the (i,j)'s and to update them dynamically, but there may be $O(n^2)$ of these pairs.

Our approach is to do it in a skewed manner. We will treat S and T asymmetrically. S will be our guide in initially finding an appropriate length and a substring for which we will know where it appears in S. Then we will need to work to find a counterpart from T. To this end we define $LP(i)$ to be the length of the longest prefix of S_i that appears in T. If we maintain the $LP()$s in some data structure, say a heap, ordered by their value then we can find the desired rather quickly. We will actually use a more efficient data structure than a heap. However, the two main challenges here are:

1. Maintain the pairs under the dynamic changes of the suffix tree.
2. Find a substring in T that is equal to the $LP(S_i)$-length prefix of S_i.

5.1   Maintaining LP() Dynamically

We are interested in maintaining the values $LP(i)$ for each i which is an original location of S and belongs to one of the fragments. $LP(i)$ needs to be maintained in a way that we can quickly access i^* where $LP(i^*)$ is maximal over all i. An array of length $n(=|S|=|T|)$ called $LPB[]$ (for LP Buckets) maintains all the i's that belong to fragments of S. In each cell ℓ of the array $LPB[\ell]$ contains all i's such that $LP(i) = \ell$. These will be saved in a linked list. Since we will return i^* where $LP(i^*)$ is maximal, we also maintain a special "maximal"-pointer. This pointer is initialized to n and when seeking a maximum $LP()$ value scans the array in a decreasing manner, i.e. moves from cell n to $n-1$ etc. Since the $LP(i)$ values can only decrease (by being in $Affects(S)$ or by changes in T) or disappear (by being in $Deleted(S)$ or by being chosen for an $LCS^{[r]}$) the "maximal"-pointer will never increase and hence, the "maximal"-pointer scan will spend, over the whole algorithm, $O(n)$ time.

The problem with maintaining this data structure is that a small change may lead to many changes in the values of $LP(i)$ and this may be costly. For example, if in round r we choose a substring in T, say x, it may be that for many suffixes S_i the longest prefix y which appears in T intersects x and, hence, $LP(i)$ will need to be decreased. Updating the $LP(i)$s means moving them from one cell (bucket) of the array to a smaller one. Other changes in T can also affect the $LP(i)$s, for example if a suffix of T belongs to $Affects(T)$. Nevertheless, the following observation will help to reduce the amount of work necessary to update these changes.

Observation 1. *Let i and j be two locations in S such that they both belong to fragments of S. If $LP(i) = LP(j)$ and the $LP(i)$-length prefix of S_i and S_j is the same then after any changes in T (removals, shortenings, etc.) it holds that $LP(i) = LP(j)$ (even if $LP(i)$ has changed).*

Proof. $LP(i)$, by definition, is the length of the longest prefix of S_i that appears in T. Since this prefix, say x, is also the prefix of S_j then a change in T affects the longest prefix of S_i that appears in T iff it affects the longest prefix of S_j that appears in T, since it is the same prefix. $\qquad \square$

This leads us to a further refinement that we will make on the linked lists of $LPB[\ell]$. We will group the values i in $LPB[\ell]$ as follows. If i and i' are both in $LPB[\ell]$ then they will be in the same group if the ℓ-length prefix of S_i and $S_{i'}$ is the same, say x. We will implement this by maintaining, for $LPB[\ell]$, a linked list of representatives of groups. Each group will also be represented by a linked list. Actually, it is easy to see that the groups are represented by nodes. A group in which all i's have x as the prefix of their suffix are represented by the locus of x.

5.2 Detailed Data Structure Maintenance

This requires careful handling of the data within the suffix tree. We will outline how this is done.

Initially, at the beginning of the algorithm, we sweep the suffix tree and mark each node if it has leaves (in its subtree) which represent S only, which represent T only, or which represent both. We call nodes essential if their subtree contains leaves that represent both. We will create, from the suffix tree, a skeleton tree by retaining all the essential nodes (note that if a node is essential then so are all its ancestors). Each essential node will be the parent of all its children who are essential. Moreover, we will create two special new children nodes for every essential node. One for S, which we will call an S-node, and one for T, which we will call a T-node. Each suffix represented by leaf u in the original tree will belong to exactly one special node. Let v be the lowest ancestor of u which is an essential node. The leaf u will belong to the T-node of v if the suffix is from T and it will belong to the S-node of v if the suffix is from S.

This grouping needs to be maintained under the various changes to the suffix tree implemented in Section 4. Here is how it is done.

Consider the changes that happen because of a suffix removed (completely or shortened) from T. Changes because of suffixes of S will be done similarly.

Say we have removed a node v that represents a suffix from T. As described in the beginning of Section 4 besides removing the suffix we may need to also remove its parent and to create a new label for the new edge. We'll call this "parent compression".

In the skeleton tree node v will belong to a T-node which may represent several suffixes. Therefore, we (carefully) make changes as follows:

If there are other suffixes in the T-node then we need to do nothing more than remove the suffix. Otherwise, the T-node needs to be removed. If the T-node's parent u has another 2 children (either an S-node and an essential node or 2 or more essential nodes) then we need to remove the T-node and are done. If u has only one other child then there are 2 cases. (1) the other child is an essential node, (2) the other child is an S-node. If the other child is an essential node then the T-node is removed and parent compression is done as described in Section 4. If the other child is an S-node then u was essential up till now and has now become non-essential. This requires a special process done as follows.

Assume w is the parent of u. u will be removed. If w did not have a child S-node then the S-node of u becomes the S-node of w and we are done (because in this situation w must have another child, either a T-node or a different essential node). If w had an S-node then if there is another child of w beside the S-node then we merge the S-nodes (by chaining the linked lists of the S-nodes). If w's only children were the S-node and u (which is being removed) then we merge the S-nodes and now need to remove w. We repeat the process.

If we do not need to do "parent compression" then the time is $O(1)$. If parent compression was done then the time is proportional to the number of parent compressions done. Since there are $O(n)$ nodes to start off with and the nodes added, described in Section 4, are at most $O(n)$ it follows that the number of parent compressions done overall in the whole process is $O(n)$, as a node is compressed (and disappears) each time.

What still needs to be done is to update the relevant LPs in the data structure.

When removing a suffix S_i (completely or by shortening it) it only affect the one value of $LP(i)$ that represents it. Therefore, the update takes $O(1)$ time, for taking the $LP(i)$ value out of its current location in the data structure and moving it to the new value if necessary. On the other hand, the removal of a suffix T_j may affect several LP values. Nevertheless, all these LP values will end up in one group at the end and the time it takes to group them is proportional to the number of parent compressions. Therefore, it will be overall $O(n)$ as well.

5.3   Finding a Matching T-Suffix

Once one has an $LP(i)$ that is maximal and, from it (previous section), a node in the suffix tree v which represents x which is the LP(i)-length prefix of S_i. By definition of $LP(i)x$ appears in T and, hence, there is at least one suffix of T, T_j, that has prefix x. This T_j must also be a child of v. Of course, v is essential because i of S and j of T are leaves in its subtree. However, v has no essential children. Otherwise, there would be an i' with higher $LP(i')$. Therefore, v only has the two special children and it is immediate to choose a T_j.

6   Putting It All Together

Putting all the pieces together the whole algorithm could be sketched by the following procedure:

FAST-GREEDY(S,T)

1. Initialization
   (a) Construct a suffix tree $ST(S,T)$ of the unified string $S\#T\$$.
   (b) Traverse the tree and mark each node whether it has suffixes (in its subtree) from S, T or both (an essential node).
   (c) Create the skeleton tree from the essential nodes.
   (d) Allocate an empty array of pointers of size $n, LPB[]$.
   (e) Traverse the tree again and fill $LPB[\ell]$ with all i such that $LP(i) = \ell$. The data is saved in $LPB[\ell]$ with a linked list, where each elements in the list represents a group of i's such that the S_is have a common prefix of length ℓ.
2. Initialize i to 0.
3. Repeat until both string become empty:
   (a) Find $LCS^{[r]}$ using the maximum-pointer in the data structure $LPB[\ell]$.
   (b) Delete the occurrences of $LCS^{[r]}$ from both S and T after outputting the indices of both occurrences.
   (c) Update the values of the $LP(S_i)$s in the data structure by making the changes needed for each of the following groups:
   $Delete(T), Affects(T), Delete(S), Affects(S)$.
   (d) Increase i by 1.

From the discussion throughout this paper it follows that:

Theorem 1. *The GREEDY algorithm for the Minimum Common String Partition problem can be implemented in $O(n)$ time for strings S and T of length n each.*

References

1. Chen, X., Zheng, J., Fu, Z., Nan, P., Zhong, Y., Lonardi, S., Jiang, T.: Assignment of orthologous genes via genome rearrangement. IEEE/ACM Trans. Comput. Biology Bioinform. 2(4), 302–315 (2005)
2. Christie, D.A., Irving, R.W.: Sorting strings by reversals and by transpositions. SIAM J. Discrete Math. 14(2), 193–206 (2001)
3. Chrobak, M., Kolman, P., Sgall, J.: The greedy algorithm for the minimum common string partition problem. ACM Transactions on Algorithms 1(2), 350–366 (2005)

4. Cormode, G., Muthukrishnan, S.: The string edit distance matching problem with moves. In: Proceedings of the Annual Symposium on Discrete Algorithms (SODA), pp. 667–676 (2002)
5. Farach, M.: Optimal suffix tree construction with large alphabets. In: FOCS 1997: Proceedings of the 38th Annual Symposium on Foundations of Computer Science, Washington, DC, USA, p. 137. IEEE Computer Society, Los Alamitos (1997)
6. Goldstein, A., Kolman, P., Zheng, J.: Minimum common string partition problem: Hardness and approximations. Electr. J. Comb. 12(1) (2005)
7. Hannenhalli, S., Pevzner, P.A.: Transforming cabbage into turnip: Polynomial algorithm for sorting signed permutations by reversals. J. ACM 46(1), 1–27 (1999)
8. Jiang, H., Zhu, B., Zhu, D., Zhu, H.: Minimum common string partition revisited. In: Lee, D.-T., Chen, D.Z., Ying, S. (eds.) FAW 2010. LNCS, vol. 6213, pp. 45–52. Springer, Heidelberg (2010)
9. Kaplan, H., Shafrir, N.: The greedy algorithm for edit distance with moves. Inf. Process. Lett. 97(1), 23–27 (2006)
10. Kolman, P.: Approximating reversal distance for strings with bounded number of duplicates. In: Jedrzejowicz, J., Szepietowski, A. (eds.) MFCS 2005. LNCS, vol. 3618, pp. 580–590. Springer, Heidelberg (2005)
11. Kolman, P., Walen, T.: Reversal distance for strings with duplicates: Linear time approximation using hitting set. Electr. J. Comb. 14(1) (2007)
12. Kruskal, J., Sankoff, D.: Time Warps, String Edits and Macromolecules: The Theory and Practice of Sequence Comparison. Addison-Wesley, Reading (1999)
13. Lopresti, D.P., Tomkins, A.: Block edit models for approximate string matching. Theor. Comput. Sci. 181(1), 159–179 (1997)
14. McCreight, E.M.: A space-economical suffix tree construction algorithm. J. ACM 23(2), 262–272 (1976)
15. Shapira, D., Storer, J.A.: Edit distance with move operations. J. Discrete Algorithms 5(2), 380–392 (2007)
16. Tichy, W.F.: The string-to-string correction problem with block moves. ACM Trans. Comput. Syst. 2(4), 309–321 (1984)
17. Ukkonen, E.: On-line construction of suffix trees. Algorithmica 14(3), 249–260 (1995)
18. Weiner, P.: Linear pattern matching algorithms. In: 14th Annual Symposium on Switching and Automata Theory, pp. 1–11. IEEE, Los Alamitos (1973)

LRM-Trees: Compressed Indices, Adaptive Sorting, and Compressed Permutations*

Jérémy Barbay[1], Johannes Fischer[2], and Gonzalo Navarro[1]

[1] Department of Computer Science, University of Chile
{jbarbay,gnavarro}@dcc.uchile.cl
[2] Computer Science Department, Karlsruhe University
johannes.fischer@kit.edu

Abstract. LRM-Trees are an elegant way to partition a sequence of values into sorted consecutive blocks, and to express the relative position of the first element of each block within a previous block. They were used to encode ordinal trees and to index integer arrays in order to support range minimum queries on them. We describe how they yield many other convenient results in a variety of areas: compressed succinct indices for range minimum queries on partially sorted arrays; a new adaptive sorting algorithm; and a compressed succinct data structure for permutations supporting direct and inverse application in time inversely proportional to the permutation's compressibility.

1 Introduction

Introduced by Fischer [9] as an indexing data structure which supports *Range Minimum Queries* (RMQs) in constant time with no access to the main data, and by Sadakane and Navarro [26] to support navigation operators on ordinal trees, *Left-to-Right-Minima Trees* (LRM-Trees) are an elegant way to partition a sequence of values into sorted consecutive blocks, and to express the relative position of the first element of each block within a previous block.

We describe how the use of LRM-Trees and variants yields many other convenient results in the design of data structures and algorithms:

1. We define three compressed succinct indices supporting RMQs, which use less space when the indexed array is partially sorted, improving in those cases on the $2n + o(n)$ bits usual space [9], and on other techniques of compression for RMQs such as taking advantage of repetitions in the input [10].
2. Based on LRM-Trees, we define a new *measure of presortedness* for permutations. It combines some of the advantages of two well-known measures, *runs* and *shuffled up-sequences*: the new measure is computable in linear time (like the former), but considers sorted sub-*sequences* (instead of only contiguous sub-*arrays*) in the input (similar, yet distinct, to the latter).

---

* First and third author partially funded by Fondecyt grant 1-110066, Chile; second author supported by a DFG grant (German Research Foundation).

R. Giancarlo and G. Manzini (Eds.): CPM 2011, LNCS 6661, pp. 285–298, 2011.
© Springer-Verlag Berlin Heidelberg 2011

3. Based on this measure, we propose a new sorting algorithm and its adaptive analysis, asymptotically superior to sorting algorithms based on runs [2], and on many instances faster than sorting algorithms based on subsequences [19].
4. We design a compressed succinct data structure for permutations based on this measure, which supports the access operator and its inverse in time inversely proportional to the permutation's presortedness, improving on previous similar results [2].

All our results are in the word RAM model, where it is assumed that we can do arithmetic and logical operations on w-bit wide words in $\mathcal{O}(1)$ time, and $w = \Omega(\lg n)$. In our algorithms and data structures, we distinguish between the work performed in the input (often called "data complexity" in the literature) and the accesses to the internal data structures ("index complexity"). This is important in cases where the input is large and cannot be stored in main memory, whereas the index is potentially small enough to be kept in fast main memory. For instance, in the context of compressed indexes like our RMQ structures, given a fixed limited amount of local memory, this additional precision permits identifying the instances whose compressed index fits in it while the main data does not. On these instances, between two data structures that support operators with the same total asymptotic complexity but distinct index complexity, the one with the lowest index complexity is more desirable.

2   Previous Work and Concepts

2.1   Left-to-Right-Minima Trees

LRM-Trees partition a sequence of values into sorted consecutive blocks, and express the relative position of the first element of each block within a previous block. They were introduced under this name as an internal tool for basic navigational operations in ordinal trees [26], and, under the name "2d-Min Heaps," to index integer arrays in order to support range minimum queries on them [9].

Let $A[1, n]$ be an integer array. For technical reasons, we define $A[0] = -\infty$ as the "artificial" overall minimum of the array.

Definition 1 (Fischer [9]; Sadakane and Navarro [26]). *For $1 \leq i \leq n$, let $\mathrm{PSV}_A(i) = \max\{j \in [0..i-1] : A[j] < A[i]\}$ denote the* previous smaller value *of position i. The* Left-to-Right-Minima Tree *(LRM-Tree) $\mathcal{T}_A$ of A is an ordered labeled tree with $n+1$ vertices each labeled uniquely from $\{0, \ldots, n\}$. For $1 \leq i \leq n$, $\mathrm{PSV}_A(i)$ is the parent node of i. The children of each node are ordered in increasing order from left to right.*

See Fig. 1 for an example of LRM-Trees. Fischer [9] gave a (complicated) linear-time construction algorithm with advantages that are not relevant for this paper. The following lemma shows a simpler way to construct the LRM-Tree in at most $2(n-1)$ comparisons within the array and overall linear time, which will be used in Thms. 4 and 5.

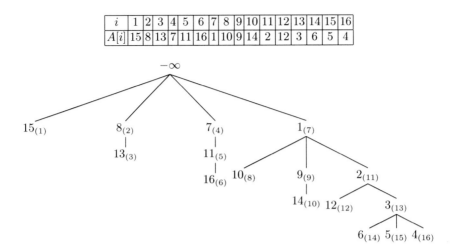

i	1	2	3	4	5	6	7	8	9	10	11	12	13	14	15	16
$A[i]$	15	8	13	7	11	16	1	10	9	14	2	12	3	6	5	4

Fig. 1. An example of an array and its LRM-Tree

Lemma 1. *Given an array $A[1, n]$ of totally ordered objects, there is an algorithm computing its LRM-Tree in at most $2(n - 1)$ comparisons within A and $\mathcal{O}(n)$ total time.*

Proof. The computation of the LRM-Tree corresponds to a simple scan over the input array, starting at $A[0] = -\infty$, building down iteratively the current rightmost branch of the tree with increasing elements of the sequence until an element x smaller than its predecessor is encountered. At this point one climbs the rightmost branch up to the first node v holding a value smaller than x, and starts a new branch with a rightmost child of v of value x. As the root of the tree has value $A[0] = -\infty$ (smaller than all elements), the algorithm always terminates.

The construction algorithm performs at most $2(n - 1)$ comparisons: the first two elements $A[0]$ and $A[1]$ can be inserted without any comparison as a simple path of two nodes (so $A[1]$ will be charged only once). For the remaining elements, we charge the last comparison performed during the insertion of an element x to the node of value x itself, and all previous comparisons to the elements already in the LRM-Tree. Thus, each element (apart from $A[1]$ and $A[n]$) is charged at most twice: once when it is inserted into the tree, and once when scanning it while searching for a smaller value on the rightmost branch. As in the latter case all scanned elements are removed from the rightmost path, this second charging occurs at most once for each element. Finally, the last element $A[n]$ is charged only once, as it will never be scanned: hence the total number of comparisons of $2n - 2 = 2(n - 1)$. Since the number of comparisons within the array dominates the number of other operations, the overall time is also in $\mathcal{O}(n)$. □

2.2   Range Minimum Queries

We consider the following queries on a static array $A[1, n]$ (parameters i and j with $1 \le i \le j \le n$):

Definition 2 (Range Minimum Queries). $\text{RMQ}_A(i, j) = $ *position of a minimum in $A[i, j]$.*

RMQs have a wide range of applications for various data structures and algorithms, including text indexing [11], pattern matching [7], and more elaborate kinds of range queries [6].

For two given nodes i and j in a tree T, let $\text{LCA}_T(i, j)$ denote their *Lowest Common Ancestor* (LCA), that is, the deepest node that is an ancestor of both i and j. Now let $\mathcal{T}_A$ be the LRM-Tree of A. For arbitrary nodes i and j in $\mathcal{T}_A$, $1 \le i < j \le n$, let $\ell = \text{LCA}_{\mathcal{T}_A}(i, j)$. Then if $\ell = i$, $\text{RMQ}_A(i, j)$ is i, otherwise, $\text{RMQ}_A(i, j)$ is given by the child of ℓ that is on the path from ℓ to j [9].

Since there are succinct data structures supporting the LCA operator [9,17]. in succinctly encoded trees in constant time, this yields a succinct index (which we improve with Thms. 1 and 3).

Lemma 2 (Fischer [9]). *Given an array $A[1, n]$ of totally ordered objects, there is a succinct index using $2n + o(n)$ bits and supporting RMQs in zero accesses to A and $\mathcal{O}(1)$ accesses to the index. This index can be built in $\mathcal{O}(n)$ time.*

2.3   Adaptive Sorting and Compression of Permutations

Sorting a permutation in the comparison model requires $\Theta(n \lg n)$ comparisons in the worst case over permutations of n elements. Yet, better results can be achieved for some parameterized classes of permutations. For a fixed permutation π, Knuth [18] considered *Runs* (contiguous ascending subsequences), counted by $|\mathsf{Runs}| = 1 + |\{i \ : \ 1 \le i < n, \pi_{i+1} < \pi_i\}|$; Levcopoulos and Petersson [19] introduced *Shuffled Up-Sequences* and its generalization *Shuffled Monotone Sequences*, respectively counted by $|\mathsf{SUS}| = \min\{k : \pi$ is covered by k increasing subsequences$\}$, and $|\mathsf{SMS}| = \min\{k : \pi$ is covered by k monotone subsequences$\}$. Barbay and Navarro [2] introduced strict variants of some of those concepts, namely *Strict Runs* and *Strict Shuffled Up-Sequences*, where sorted subsequences are composed of consecutive integers (e.g., $(2, 3, 4, 1, 5, 6, 7, 8)$ has two runs but three strict runs), counted by $|\mathsf{SRuns}|$ and $|\mathsf{SSUS}|$, respectively. For any of those five measures of disorder X, there is a variant of the merge-sort algorithm which sorts a permutation π, of size n and of measure of presortedness X, in time $\mathcal{O}(n(1 + \lg \mathsf{X}))$, which is within a constant factor of optimal in the worst case among instances of fixed size n and fixed values of X (this is not necessarily true for other measures of disorder).

As the merging cost induced by a subsequence is increasing with its length, the sorting time of a permutation can be improved by rebalancing the merging tree [2]. The complexity can then be expressed more precisely as a function of the *entropy* of the relative sizes of the sorted subsequences identified, where the entropy $\mathcal{H}(\mathsf{Seq})$ of a sequence $\mathsf{Seq} = \langle n_1, n_2, \ldots, n_r \rangle$ of r positive integers adding up to n is defined as $\mathcal{H}(\mathsf{Seq}) = \sum_{i=1}^{r} \frac{n_i}{n} \lg \frac{n}{n_i}$. This entropy satisfies $(r-1) \lg n \leq n\mathcal{H}(\mathsf{Seq}) \leq n \lg r$ by concavity of the logarithm, a formula which we will use later.

Barbay and Navarro [2] observed that each adaptive sorting algorithm in the comparison model also describes an encoding of the permutation π that it sorts, so that it can be used to compress permutations from specific classes to less than the information-theoretic lower bound of $n \lg n$ bits. Furthermore they used the similarity of the execution of the merge-sort algorithm with a Wavelet Tree [14], to support the application of $\pi()$ and its inverse $\pi^{-1}()$ in time logarithmic in the disorder of the permutation π (as measured by $|\mathsf{Runs}|, |\mathsf{SRuns}|, |\mathsf{SUS}|, |\mathsf{SSUS}|$ or $|\mathsf{SMS}|$) in the worst case. We summarize their technique in Lemma 3 below, in a way independent of the partition chosen for the permutation, and focusing only on the merging part of the sorting.

Lemma 3 (Barbay and Navarro [2]). *Given a partition of an array π of n totally ordered objects into $|\mathsf{Seq}|$ sorted subsequences of respective lengths $\mathsf{Seq} = \langle n_1, n_2, \ldots, n_{|\mathsf{Seq}|} \rangle$, these subsequences can be merged with $n(1 + \mathcal{H}(\mathsf{Seq}))$ comparisons on π and $\mathcal{O}(n(1 + \mathcal{H}(\mathsf{Seq})))$ total running time. This merging can be encoded using at most $(1 + \mathcal{H}(\mathsf{Seq}))(n + o(n)) + \mathcal{O}(|\mathsf{Seq}| \lg n)$ bits so that it supports the computation of $\pi(i)$ and $\pi^{-1}(i)$ in time $\mathcal{O}(1 + \lg |\mathsf{Seq}|)$ in the worst case $\forall i \in [1..n]$, and in time $\mathcal{O}(1 + \mathcal{H}(\mathsf{Seq}))$ on average when i is chosen uniformly at random in $[1..n]$.*

3   Compressed Succinct Indexes for Range Minima

We now explain how to improve on the result of Lemma 2 for permutations that are partially ordered. We consider only the case where the input A is a permutation of $[1..n]$: if this is not the case, we can sort the elements in A by rank, considering earlier occurrences of equal elements as smaller. Our first and simplest compressed data structure for RMQs uses an amount of space which is a function of $|\mathsf{SRuns}|$, the number of strict runs in π. Beside its simplicity, its interest resides in that it uses a total space within $o(n)$ bits on permutations where $|\mathsf{SRuns}| \in o(n)$, and introduces techniques which we will use in Thms. 2 and 3.

Theorem 1. *Given an array $A[1, n]$ of totally ordered objects, composed of $|\mathsf{SRuns}|$ strict runs, there is a compressed succinct index using $\lceil \lg \binom{n}{|\mathsf{SRuns}|} \rceil + 2|\mathsf{SRuns}| + o(n)$ bits which supports RMQs in zero accesses to A and $\mathcal{O}(1)$ accesses to the index.*

Proof. We mark the beginnings of the runs in A with a 1 in a bit-vector $B[1, n]$, and represent B with the compressed succinct data structure from Raman et al. [24], using $\lceil \lg \binom{n}{|\mathsf{SRuns}|} \rceil + o(n)$ bits. Further, we define A' as the (conceptual) array consisting of the heads of A's runs ($A'[i] = A[select_1(B, i)]$). We build the LRM-Tree from Lemma 2 on A' using $2|\mathsf{SRuns}|(1 + o(1))$ bits. To answer a query

$\text{RMQ}_A(i, j)$, we compute $x = rank_1(B, i)$ and $y = rank_1(B, j)$, then compute $m' = \text{RMQ}_{A'}(x, y)$ as the minimum of the heads of those runs that overlap the query interval, and map it back to its position in A by $m = select_1(B, m')$. Then if $m < i$, we return i as the final answer to $\text{RMQ}_A(i, j)$, otherwise we return m. The correctness of this algorithm follows from the fact that only i and the heads that are contained in the query interval can be the range minimum. Because the runs are strict, the former occurs if and only if the head of the run containing i is smaller than all other heads in the query range.   □

The same idea as in Thm. 1 applied to more general runs yields another compressed succinct index for RMQs, potentially smaller but this time requiring to compare two elements from the input to answer RMQs.

Theorem 2. *Given an array $A[1, n]$ of totally ordered objects, composed of $|\mathsf{Runs}|$ runs, there is a compressed succinct index using $2|\mathsf{Runs}| + \lceil \lg \binom{n}{|\mathsf{Runs}|} \rceil + o(n)$ bits and supporting RMQs in 1 comparison within A and $\mathcal{O}(1)$ accesses to the index.*

Proof. We build the same data structures as in Thm. 1, using $2|\mathsf{Runs}| + \lceil \lg \binom{n}{|\mathsf{Runs}|} \rceil + o(n)$ bits. To answer a query $\text{RMQ}_A(i, j)$, we compute $x = rank_1(B, i)$ and $y = rank_1(B, j)$. If $x = y$, return i. Otherwise, compute $m' = \text{RMQ}_{A'}(x + 1, y)$, and map it back to its position in A by $m = select_1(B, m')$. The final answer is i if $A[i] < A[m]$, and m otherwise.   □

To achieve a compressed succinct index which never accesses the array and whose space usage is a function of $|\mathsf{Runs}|$, we need more space and a more heavy machinery, as shown next. The main idea is that a permutation with few runs results in a compressible LRM-Tree, where many nodes have out-degree 1.

Theorem 3. *Given an array $A[1, n]$ of totally ordered objects, composed of $|\mathsf{Runs}|$ runs, there is a compressed succinct index using $2|\mathsf{Runs}| \lg n + o(n)$ bits, and supporting RMQs in zero accesses to A and $\mathcal{O}(1)$ accesses to the index.*

Proof. We build the LRM-Tree $\mathcal{T}_A$ from Sect. 2.1 directly on A, and then compress it with the tree representation of Jansson et al. [17].

To see that this results in the claimed space, let n_k denote the number of nodes in $\mathcal{T}_A$ with out-degree $k \geq 0$. Let $(i_1, j_1), \ldots, (i_{|\mathsf{Runs}|}, j_{|\mathsf{Runs}|})$ be an encoding of the runs in A as (start, end), and look at a pair (i_x, j_x). We have $\text{PSV}_A(k) = k - 1$ for all $k \in [i_x + 1..j_x]$, and so the nodes in $[i_x..j_x]$ form a path in $\mathcal{T}_A$, possibly interrupted by branches stemming from heads i_y of other runs $y > x$ with $\text{PSV}_A(i_y) \in [i_x..j_x - 1]$. Hence $n_0 = |\mathsf{Runs}|$, and $n_1 \geq n - |\mathsf{Runs}| - (|\mathsf{Runs}| - 1) > n - 2|\mathsf{Runs}|$, as in the worst case the values $\text{PSV}_A(i_y)$ for $i_y \in \{i_2, i_3, \ldots, i_{|\mathsf{Runs}|}\}$ are all different.

As an illustrative example, look again at the tree in Fig. 1. It has $n_0 = 9$ leaves, corresponding to the runs $\langle 15 \rangle$, $\langle 8, 13 \rangle$, $\langle 7, 11, 16 \rangle$, $\langle 1, 10 \rangle$, $\langle 9, 14 \rangle$, $\langle 2, 12 \rangle$, $\langle 3, 6 \rangle$, $\langle 5 \rangle$, and $\langle 4 \rangle$ in A. The first four runs have a PSV of $A[0] = -\infty$ for their corresponding head elements, the next two head-PSVs point to $A[7] = 1$, the next one to $A[11] = 2$, and the last two to $A[13] = 3$. Hence, the heads of the runs "destroy" exactly four of the $n - n_0 + 1$ potential degree-1 nodes in the tree, so $n_1 = n - n_0 - 4 + 1 = 16 - 9 - 3 = 4$.

Now $\mathcal{T}_A$, with degree-distribution $n_0, \ldots, n_{n-1}$, is compressed into $nH^*(\mathcal{T}_A) + O\left(\frac{n(\lg \lg n)^2}{\lg n}\right)$ bits [17], where

$$nH^*(\mathcal{T}_A) = \lg\left(\frac{1}{n}\binom{n}{n_0, n_1, \ldots, n_{n-1}}\right)$$

is the so-called *tree entropy* [17] of $\mathcal{T}_A$. This representation supports all navigational operations in $\mathcal{T}_A$ in constant time, and in particular those required for Lemma 2. A rough inequality yields a bound on the number of possible such LRM-Trees:

$$\binom{n}{n_0, n_1, \ldots, n_{n-1}} = \frac{n!}{n_0! n_1! \ldots n_{n-1}!} \leq \frac{n!}{n_1!} \leq \frac{n!}{(n - 2|\mathsf{Runs}|)!} \leq n^{2|\mathsf{Runs}|} \, ,$$

from which one easily bounds the space usage of the compressed succinct index:

$$nH^*(\mathcal{T}_A) \leq \lg\left(\frac{1}{n}n^{2|\mathsf{Runs}|}\right) = \lg\left(n^{2|\mathsf{Runs}|-1}\right) = (2|\mathsf{Runs}| - 1)\lg n < 2|\mathsf{Runs}|\lg n \, .$$

Adding the space required to index the structure of Jansson et al. [17] yields the claimed space bound. □

4   Sorting Permutations

Barbay and Navarro [2] showed how to use the decomposition of a permutation π into $|\mathsf{Runs}|$ ascending consecutive runs of respective lengths Runs to sort adaptively to their entropy $\mathcal{H}(\mathsf{Runs})$. Those runs entirely partition the LRM-Tree of π: one can easily draw the partition corresponding to the runs considered by Barbay and Navarro [2] by iteratively tagging the leftmost maximal untagged leaf-to-root path of the LRM-Tree. For instance, the permutation of Figure 1 has nine runs ($\langle 15 \rangle$, $\langle 8, 13 \rangle$, $\langle 7, 11, 16 \rangle$, $\langle 1, 10 \rangle$, $\langle 9, 14 \rangle$, $\langle 2, 12 \rangle$, $\langle 3, 6 \rangle$, $\langle 5 \rangle$, and $\langle 4 \rangle$), of respective sizes given by the vector $< 1, 2, 3, 2, 2, 2, 2, 1, 1 >$.

But *any* partition of the LRM-Tree into branches (such that the values traversed by the path are increasing) can be used to sort π, and a partition of smaller entropy yields a faster merging phase. To continue with the previous example, the nodes of the LRM-Tree of Figure 1 can be partitioned differently, so that the vector formed by the sizes of the increasing subsequences it describes has lower entropy. One such partition would be $\langle 15 \rangle$, $\langle 8, 13 \rangle$, $\langle 7, 11, 16 \rangle$, $\langle 1, 2, 3, 4 \rangle$, $\langle 10 \rangle$, $\langle 9, 14 \rangle$, $\langle 12 \rangle$, $\langle 6 \rangle$, and $\langle 5 \rangle$, of respective sizes given by the vector $< 1, 2, 3, 4, 1, 2, 1, 1, 1 >$.

Definition 3 (LRM-Partition). *An LRM-Partition P of an LRM-Tree $\mathcal{T}$ for an array A is a partition of the nodes of $\mathcal{T}$ into $|\mathsf{LRM}|$ down-paths, i.e. paths starting at some branching node of the tree, and ending at a leaf. The* entropy *of*

P is $\mathcal{H}(P) = \mathcal{H}(r_1, \ldots, r_{|\mathsf{LRM}|})$, where $r_1, \ldots, r_{|\mathsf{LRM}|}$ are the lengths of the down-paths in P. P is optimal if its entropy is minimal among all the LRM-partitions of $\mathcal{T}$. The entropy of this optimal partition is the LRM-entropy of the LRM-Tree $\mathcal{T}$ and, by extension, the LRM-entropy of the array A.

Note that since there are exactly $|\mathsf{Runs}|$ leaves in the LRM-Tree, there will always be $|\mathsf{Runs}|$ down-paths in the LRM-partition; hence $|\mathsf{LRM}| = |\mathsf{Runs}|$. We first define a particular LRM-partition and prove that its entropy is minimal. Then we show how it can be computed in linear time.

Definition 4 (Left-Most Spinal LRM-Partition). *Given an LRM-Tree $\mathcal{T}$, the* left-most spinal chord *of $\mathcal{T}$ is the leftmost path among the longest root-to-leaf paths in $\mathcal{T}$; and the* left-most spinal LRM-partition *is defined recursively as follows. Removing the left-most spinal chord of $\mathcal{T}$ leaves a forest of shallower trees, which are partitioned recursively. The left-most spinal partition is obtained by concatenating all resulting LRM-partitions in arbitrary order.* LRM *denotes the vector formed by the $|\mathsf{LRM}|$ lengths of the subsequences in the partition.*

For instance, the left-most spinal LRM-partition of the LRM-tree given in Figure 1 is quite easy to build: the first left-most spinal chord is $-\infty, 1, 2, 3, 6$, which removal leaves a forest of simple branches. The resulting partition is $\langle 15 \rangle$, $\langle 8, 13 \rangle$, $\langle 7, 11, 16 \rangle$, $\langle 1, 2, 3, 6 \rangle$, $\langle 10 \rangle$, $\langle 9, 14 \rangle$, $\langle 12 \rangle$, $\langle 5 \rangle$, and $\langle 4 \rangle$, of respective sizes given by the vector $< 1, 2, 3, 4, 1, 2, 1, 1, 1 >$.

The LRM-partition, by successively extracting increasing subsequences of maximal length, actually yields a partition of minimal entropy, as shown in the following lemma.

Lemma 4. *The entropy of the left-most spinal LRM-partition is minimal among all LRM-partitions.*

Proof. Given an LRM-Tree $\mathcal{T}$, consider the leftmost leaf L_0 among the leaves of maximal depth in $\mathcal{T}$. We prove that there is always an optimal LRM-partition which contains the down-path $(-\infty, L_0)$. Applying this property recursively in the trees produced by removing the nodes of $(-\infty, L_0)$ from $\mathcal{T}$ yields the optimality of the leftmost LRM-partition.

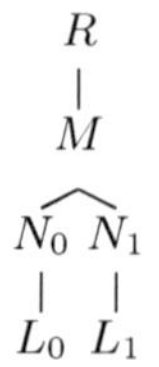

R
$|$
M
$\widehat{N_0 \ N_1}$
$| \quad |$
$L_0 \ L_1$

Fig. 2. Consider an arbitrary LRM-partition P and the down-path (N_0, L_0) in P finishing at L_0. If $N_0 \neq -\infty$ (that is, N_0 is not the root), then consider the parent M of N_0 and the down-path (R, L_1) which contains M and finishes at a leaf L_1. Call N_1 the child of M on the path to L_1.

Consider an arbitrary LRM-partition P and the nodes R, M, N_0, N_1 and L_1 as described in Figure 2. Call r the number of nodes in (R, M), d_0 the number of nodes in (N_0, L_0), and d_1 the number of nodes in (N_1, L_1). Note that $d_1 \leq d_0$

because L_0 is one of the deepest leaves. Thus the LRM-partition P has a down-path (N_0, L_0) of length d_0 and another (R, L_1) of length $r + d_1$. We build a new LRM-partition P' by switching some parts of the down-paths, so that one goes from R to L_0 and the other from N_1 to L_1, with new down-path lengths $r + d_0$ and d_1, respectively.

Let $\langle n_1, n_2, \ldots, n_{|\mathsf{LRM}|}\rangle$ be the down-path lengths in P, such that $\mathcal{H}(P) = \mathcal{H}(n_1, n_2, \ldots, n_{|\mathsf{LRM}|}) = n \lg n - \sum_{i=1}^{|\mathsf{LRM}|} n_i \lg n_i$. Without loss of generality (the entropy is invariant to the order of the parameters), assume that $n_1 = d_0$ and $n_2 = r + d_1$ are the down-paths we have considered: they are replaced in P' by down-paths of length $n'_1 = r + d_0$ and $n'_2 = d_1$. The variation in entropy is $[(r + d_1) \lg(r + d_1) + d_0 \lg d_0] - [(r + d_0) \lg(r + d_0) + d_1 \lg d_1]$, which can be rewritten as $f(d_1) - f(d_0)$ with $f(x) = (r + x) \lg(r + x) - x \lg x$. Since the function $f(x) = (r + x) \lg(r + x) - x \lg x$ has positive derivative and $d_1 \leq d_0$, the difference is non-positive (and strictly negative if $d_1 < d_0$, which would imply that P was not optimal). Iterating this argument until the path of the LRM-partition containing L_0 is rooted in $-\infty$ yields an LRM-partition of entropy no larger than that of the LRM-partition P, and one which contains the down-path $(-\infty, L_0)$.

Applying this argument to an optimal LRM-partition demonstrates that there is always an LRM-partition which contains the down-path $(-\infty, L_0)$. This, in turn, applied recursively to the subtrees obtained by removing the nodes from the path $(-\infty, L_0)$ from $\mathcal{T}$, shows the minimality of the entropy of the left-most spinal LRM-partition. □

While the definition of the left-most spinal LRM-partition is constructive, building this partition in linear time requires some sophistication, described in the following lemma:

Lemma 5. *Given an LRM-Tree $\mathcal{T}$, there is an algorithm which computes its left-most spinal LRM-partition in linear overall time (without accessing the original array).*

Proof. Given an LRM-Tree $\mathcal{T}$ (and potentially no access to the array from which it originated), we first set up an array D containing the *depths* of the nodes in $\mathcal{T}$, listed in preorder. We then index D for range maximum queries in linear time using Lemma 2. Since D contains only internal data, the number of accesses to it matters only to the running time of the algorithm (they are distinct from accesses to the array at the construction of $\mathcal{T}$). Now the deepest node in $\mathcal{T}$ can be found by a range maximum query over the whole array, supported in constant time. From this node, we follow the path to the root, and save the corresponding nodes as the first subsequence. This divides A into disconnected subsequences, which can be processed recursively using the same algorithm, as the nodes in any subtree of $\mathcal{T}$ form an interval in D. We do so until all elements in A have been assigned to a subsequence. Note that, in the recursive steps, the numbers in D are not anymore the depths of the corresponding nodes in the remaining subtrees. Yet, as all depths listed in D differ by the same offset from their depths in any connected subtree, this does not affect the result of the range maximum queries.   □

Note that the left-most spinal LRM-partition is not much more expensive to compute than the partition into ascending consecutive runs [2]: at most $2(n-1)$ comparisons between elements of the array for the LRM-partition instead of $n-1$ for the Runs-Partition. Note also that $\mathcal{H}(\mathsf{LRM}) \leq \mathcal{H}(\mathsf{Runs})$, since the partition of π into consecutive ascending runs is just one LRM-partition among many. The concept of LRM-partitions yields a new adaptive sorting algorithm:

Theorem 4. *Let π be a permutation of size n, and of LRM-Entropy $\mathcal{H}(\mathsf{LRM})$. The LRM-Sorting algorithm sorts π in a total of at most $n(3 + \mathcal{H}(\mathsf{LRM})) - 2$ comparisons between elements of π and in total running time of $\mathcal{O}(n(1 + \mathcal{H}(\mathsf{LRM})))$.*

Proof. Obtaining the left-most optimal LRM-partition P composed of runs of respective lengths LRM through Lemma 5 uses at most $2(n-1)$ comparisons between elements of π and $\mathcal{O}(n)$ total running time. Now sorting π is just a matter of applying Lemma 3: it merges the subsequences of P in $n(1 + \mathcal{H}(\mathsf{LRM}))$ additional comparisons between elements of π and $\mathcal{O}(|\mathsf{LRM}| \lg |\mathsf{LRM}|)$ additional internal operations. The sum of those complexities yields $n(3 + \mathcal{H}(\mathsf{LRM})) - 2$ data comparisons, and since $|\mathsf{LRM}| \lg |\mathsf{LRM}| < n\mathcal{H}(\mathsf{LRM}) + \lg |\mathsf{LRM}|$ by concavity of the logarithm, the total time complexity is in $\mathcal{O}(n(1 + \mathcal{H}(\mathsf{LRM})))$.   □

On instances where $\mathcal{H}(\mathsf{LRM}) = \mathcal{H}(\mathsf{Runs})$, LRM-Sorting can actually perform $n-1$ more comparisons than Runs-Sorting, due to the cost of the construction of the LRM-Tree. Yet, the entropy of the LRM-partition is never larger than the entropy of the Runs partition ($\mathcal{H}(\mathsf{LRM}) \leq \mathcal{H}(\mathsf{Runs})$), which ensures that LRM-sorting's asymptotical performance is never worse than Runs-sorting's performance [2]. Furthermore, LRM-Sorting is arbitrarily faster than Runs-Sorting on permutations with few consecutive inversions, as the lower entropy of the LRM-partition more than compensates for the additional cost of computing the LRM-Tree. For instance, for $n > 2$ odd and $\pi = 1, 3, 2, 5, 4, \ldots, 2i+1, 2i, \ldots, n, n-1$, $|\mathsf{Runs}| = |\mathsf{LRM}| = n/2$, $\mathsf{Runs} = \langle 2, \ldots, 2 \rangle$ and $\mathsf{LRM} = \langle n/2 + 1, 1, \ldots, 1 \rangle$, so that the entropy of LRM is arbitrarily smaller than the one of Runs.

When $\mathcal{H}(\mathsf{LRM})$ is much larger than $\mathcal{H}(\mathsf{SUS})$, the merging of the LRM-partition can actually require many more comparisons than the merging of the SUS partition produced by Levcopoulos and Petersson's algorithm [19]. For instance, for $n > 2$ even and $\pi = 1, n/2+1, 2, n/2+2, \ldots, n/2, n$, $|\mathsf{LRM}| = |\mathsf{Runs}| = n/2$ and $\mathcal{H}(\mathsf{LRM}) = \lg \frac{n}{2}$, whereas $|\mathsf{SUS}| = 2$ and $\mathcal{H}(\mathsf{SUS}) = \lg 2$.

Yet, the high cost of computing the SUS partition (up to $n(1 + \mathcal{H}(\mathsf{SUS}))$ additional comparisons within the array, as opposed to only $2(n-1)$ for the LRM-partition) means that on instances where $\mathcal{H}(\mathsf{LRM}) \in [\mathcal{H}(\mathsf{SUS}), 2\mathcal{H}(\mathsf{SUS}) - 1]$, LRM-Sorting actually performs *fewer* comparisons within the array than SUS-Sorting (if only potentially half, given that $\mathcal{H}(\mathsf{SUS}) \leq \mathcal{H}(\mathsf{LRM})$). Consider for instance, for $n > 2$ multiple of 3, $\pi = 1, 2, n, 3, 4, n-1, 5, 6, n-2, \ldots 2n/3 + 1$:

there LRM = SUS = $\langle 2n/3 + 1, 1, \ldots, 1 \rangle$, so that LRM and SUS have the same entropy, and LRM-sorting outperforms SUS-sorting. A similar reasoning applies to the comparison of the worst-case performances of LRM-Sorting and SMS-Sorting.

Another major advantage of LRM-Sorting over SUS and SMS sorting is that the optimal partition can be computed in linear time, whereas no such linear time algorithm is known to compute the partition of minimal entropy of π into Shuffled Up-Sequences or Shuffled Monotone Sequences; the notation $\mathcal{H}(\text{SUS})$ is defined only as the entropy of the partition of π produced by Levcopoulos and Petersson's algorithm [19], which only promises the smallest number of Shuffled Up-Sequences [2].

LRM-Sorting generally improves on both Runs-Sorting and SUS-Sorting in the number of comparisons performed within the input array. As mentioned in the Introduction, this is important in cases where the internal data structures used by the algorithm do fit in main memory, but not the input itself. Furthermore, we show in the next section that this difference in performance implies an even more meaningful difference in the size of the compressed data structures for permutations corresponding to those sorting algorithms.

5   Compressing Permutations

As shown by Barbay and Navarro [2], sorting opportunistically in the comparison model yields a compression scheme for permutations, and with some more work a compressed succinct data structure supporting the direct and inverse operators in time logarithmic on the disorder of the permutation. We show that the sorting algorithm of Thm. 4 corresponds to a compressed succinct data structure for permutations which supports the direct and reverse operators in time logarithmic on its LRM-Entropy (defined in the previous section), while often using less space than previous solutions. The essential component of our solution is a data structure for encoding an LRM-partition P. In order to apply Lemma 3, our data structure must efficiently support two operators:

- the operator $map(i)$ indicates, for each position $i \in [1..n]$ in the input permutation π, the corresponding subsequence s of P, and the relative position p of i in this subsequence;
- the operator $unmap(s, p)$ is the reverse of $map()$: given a subsequence $s \in [1..|\text{LRM}|]$ of P and a position $p \in [1..n_s]$ in s, it indicates the corresponding position i in π.

We obviously cannot afford to rewrite the numbers of π in the order described by the partition, which would use $n \lg n$ bits. A naive solution would be to encode this partition as a string S over alphabet $[1..|\text{LRM}|]$, using a succinct data structure supporting the *access*, *rank* and *select* operators on it. This solution is not suitable as it would require at the very least $n\mathcal{H}(\text{Runs})$ bits *only to encode the LRM-partition*, making this encoding worse than the |Runs| compressed succinct data structure [2]. We describe a more complex data structure which uses less space, and which supports the desired operators in constant time.

Lemma 6. *Let P be an LRM-partition consisting of* |LRM| *subsequences of respective lengths given by the vector* LRM, *summing to n. There is a succinct data structure using* $2|\text{LRM}|\lg n + \mathcal{O}(|\text{LRM}|) + o(n)$ *bits which supports the operators* map *and* unmap *on P in constant time (without accessing the original array).*

Proof. The main idea of the data structure is that the subsequences of an LRM-partition P for a permutation π are not as general as, say, the subsequences of a partition into |SUS| up-sequences. For each pair of subsequences (u, v), either the positions of u and v belong to disjoint intervals of π, or the values corresponding to u (resp. v) all fall between two values from v (resp. u).

As such, the subsequences in P can be organized into a forest of ordinal trees, where (1) the internal nodes of the trees correspond to the |LRM| subsequences of P, organized so that the node u is the parent of the node v if the positions of the subsequence corresponding to v are contained between two positions of the subsequence corresponding to u, (2) the children of a node are ordered in the same order as their corresponding subsequences in the permutation, and (3) the leaves of the trees correspond to the n positions in π, children of the internal node u corresponding to the subsequence they belong to.

For instance in Figure 3, the permutation $\pi = (4, 5, 9, 6, 8, 1, 3, 7, 2)$ has the LRM-partition $\langle 4, 5, 6, 8 \rangle, \langle 9 \rangle, \langle 1, 3, 7 \rangle, \langle 2 \rangle$, whose encoding can be visualized by the expression $(45(9)68)(137)(2)$ and encoded by the balanced parenthesis expression $(()()(())()())(()()()())$ (note that this is a forest, not a tree, hence the excess of '('s versus ')'s is going to zero several times inside the expression).

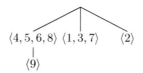

Fig. 3. Given a permutation $\pi = (4, 5, 9, 6, 8, 1, 3, 7, 2)$, its LRM-partition $\langle 4, 5, 6, 8 \rangle, \langle 9 \rangle, \langle 1, 3, 7 \rangle, \langle 2 \rangle$ can be visualized by the expression $(45(9)68)(137)(2)$ and encoded as a forest

Given a position $i \in [1..n]$ in π, the corresponding subsequence s of P is simply obtained by finding the parent of the i-th leaf, and returning its preorder rank among internal nodes. The relative position p of i in this subsequence is given by the number of its left siblings which are leaves. Conversely, given the rank $s \in [1..|\text{LRM}|]$ of a subsequence in P and a position $p \in [1..n_s]$ in this subsequence, the corresponding position i in π is computed by finding the s-th internal node in preorder, selecting its p-th child which is a leaf, and computing the preorder rank of this node among all the leaves of the tree.

We represent such a forest using the structure of Jansson et al. [17] by adding a fake root node to the forest. The only operation it does not support is counting the number of leaf siblings to the left of a node, and finding the p-th leaf child of a node. Jansson et al.'s structure [17] encodes a DFUDS representation [4] of the tree, where each node with d children is represented as d opening parentheses followed by a closing parenthesis: "$(\cdots ())$". Thus we set up an additional bitmap, of the same length and aligned to the parentheses string of Jansson et al.'s structure, where we mark with a one each opening parenthesis that corresponds to an internal node (the remaining parentheses, opening or closing, are set to zero). Then the operations are easily carried out using *rank* and *select* on this bitmap and the one from Jansson et al.'s structure.

Since the forest has n leaves and $|\mathsf{LRM}|$ internal nodes, Jansson et al.'s structure [17] takes space $H^* + o(n)$ bits, where $H^* = \lg \binom{n+|\mathsf{LRM}|}{n,n_1,\ldots,n_{n-1}} \le \lg \frac{(n+|\mathsf{LRM}|)!}{n!} \le \lg \left((n+|\mathsf{LRM}|)^{|\mathsf{LRM}|} \right) = |\mathsf{LRM}| \lg (n+|\mathsf{LRM}|) = |\mathsf{LRM}| \lg n + \mathcal{O}(|\mathsf{LRM}|)$. On the other hand, the bitmap that we added is of length $2(n+|\mathsf{LRM}|) \le 4n$ and has exactly $|\mathsf{LRM}|$ 1s, and thus a compressed representation [24] requires $|\mathsf{LRM}| \lg n + \mathcal{O}(|\mathsf{LRM}|) + o(n)$ additional bits. □

Given the data structure for LRM-partitions from Lemma 6, and applying the merging data structure from Lemma 3 immediately yields a compressed succinct data structure for permutations. Note that the index and the data are interwoven in a single data structure (i.e., this encoding is not a succinct index [1]), so we express the complexity of its operators as a single measure (as opposed to previous ones, for which we distinguished data and index complexity).

Theorem 5. *Let π be a permutation of size n, such that it has an optimal LRM-partition of size $|\mathsf{LRM}|$ and entropy $\mathcal{H}(\mathsf{LRM})$. There is a compressed succinct data structure using $(1 + \mathcal{H}(\mathsf{LRM}))(n + o(n)) + \mathcal{O}(|\mathsf{LRM}| \lg n)$ bits, supporting the computation of $\pi(i)$ and $\pi^{-1}(i)$ in time $\mathcal{O}(1 + \lg |\mathsf{LRM}|)$ in the worst case $\forall i \in [1..n]$, and in time $\mathcal{O}(1 + \mathcal{H}(\mathsf{LRM}))$ on average when i is chosen uniformly at random in $[1..n]$. It can be computed in at most $n(3 + \mathcal{H}(\mathsf{LRM})) - 2$ comparisons in π and total running time of $\mathcal{O}(n(1 + \mathcal{H}(\mathsf{LRM})))$.*

Proof. Lemma 6 yields a data structure for an optimal LRM-partition of π using $2|\mathsf{LRM}| \lg n + \mathcal{O}(|\mathsf{LRM}|) + o(n)$ bits, and supports the *map* and *unmap* operators in constant time. The merging data structure from Lemma 3 requires $(1 + \mathcal{H}(\mathsf{LRM}))(n + o(n)) + \mathcal{O}(|\mathsf{LRM}| \lg n)$ bits, and supports the operators $\pi()$ and $\pi^{-1}()$ in the time described, through the additional calls to the operators *map()* and *unmap()*. The latter space is asymptotically dominant. □

References

1. Barbay, J., He, M., Munro, J.I., Rao, S.S.: Succinct indexes for strings, binary relations, and multi-labeled trees. In: Proc. SODA, pp. 680–689. ACM/SIAM (2007)
2. Barbay, J., Navarro, G.: Compressed representations of permutations, and applications. In: Proc. STACS, pp. 111–122. IBFI Schloss Dagstuhl (2009)
3. Bender, M.A., Farach-Colton, M., Pemmasani, G., Skiena, S., Sumazin, P.: Lowest common ancestors in trees and directed acyclic graphs. J. Algorithms 57(2), 75–94 (2005)
4. Benoit, D., Demaine, E.D., Munro, J.I., Raman, R., Raman, V., Rao, S.S.: Representing trees of higher degree. Algorithmica 43(4), 275–292 (2005)

5. Brodal, G.S., Davoodi, P., Rao, S.S.: On space efficient two dimensional range minimum data structures. In: de Berg, M., Meyer, U. (eds.) ESA 2010. LNCS, vol. 6347, pp. 171–182. Springer, Heidelberg (2010)
6. Chen, K.-Y., Chao, K.-M.: On the range maximum-sum segment query problem. In: Fleischer, R., Trippen, G. (eds.) ISAAC 2004. LNCS, vol. 3341, pp. 294–305. Springer, Heidelberg (2004)
7. Crochemore, M., Iliopoulos, C.S., Kubica, M., Rahman, M.S., Walen, T.: Improved algorithms for the range next value problem and applications. In: Proc. STACS, pp. 205–216. IBFI Schloss Dagstuhl (2008)
8. Daskalakis, C., Karp, R.M., Mossel, E., Riesenfeld, S., Verbin, E.: Sorting and selection in posets. In: Proc. SODA, pp. 392–401. ACM/SIAM (2009)
9. Fischer, J.: Optimal succinctness for range minimum queries. In: López-Ortiz, A. (ed.) LATIN 2010. LNCS, vol. 6034, pp. 158–169. Springer, Heidelberg (2010)
10. Fischer, J., Heun, V., Stühler, H.M.: Practical entropy bounded schemes for $O(1)$-range minimum queries. In: Proc. DCC, pp. 272–281. IEEE Press, Los Alamitos (2008)
11. Fischer, J., Mäkinen, V., Navarro, G.: Faster entropy-bounded compressed suffix trees. Theor. Comput. Sci. 410(51), 5354–5364 (2009)
12. Gál, A., Miltersen, P.B.: The cell probe complexity of succinct data structures. Theor. Comput. Sci. 379(3), 405–417 (2007)
13. Golynski, A.: Optimal lower bounds for rank and select indexes. Theor. Comput. Sci. 387(3), 348–359 (2007)
14. Grossi, R., Gupta, A., Vitter, J.S.: High-order entropy-compressed text indexes. In: Proc. SODA, pp. 841–850. ACM/SIAM (2003)
15. Huffman, D.: A method for the construction of minimum-redundancy codes. Proceedings of the I.R.E. 40, 1090–1101 (1952)
16. Jacobson, G.: Space-efficient static trees and graphs. In: Proc. FOCS, pp. 549–554. IEEE Computer Society, Los Alamitos (1989)
17. Jansson, J., Sadakane, K., Sung, W.-K.: Ultra-succinct representation of ordered trees. In: Proc. SODA, pp. 575–584. ACM/SIAM (2007)
18. Knuth, D.E.: Art of Computer Programming, 2nd edn. Sorting and Searching, vol. 3. Addison-Wesley Professional, Reading (1998)
19. Levcopoulos, C., Petersson, O.: Sorting shuffled monotone sequences. Inf. Comput. 112(1), 37–50 (1994)
20. Mäkinen, V., Navarro, G.: Implicit compression boosting with applications to self-indexing. In: Ziviani, N., Baeza-Yates, R. (eds.) SPIRE 2007. LNCS, vol. 4726, pp. 229–241. Springer, Heidelberg (2007)
21. Munro, J.I., Raman, R., Raman, V., Rao, S.S.: Succinct representations of permutations. In: Baeten, J.C.M., Lenstra, J.K., Parrow, J., Woeginger, G.J. (eds.) ICALP 2003. LNCS, vol. 2719, pp. 345–356. Springer, Heidelberg (2003)
22. Navarro, G., Mäkinen, V.: Compressed full-text indexes. ACM Computing Surveys 39(1), Article No. 2 (2007)
23. Pătraşcu, M.: Succincter. In: Proc. FOCS, pp. 305–313. IEEE Computer Society, Los Alamitos (2008)
24. Raman, R., Raman, V., Rao, S.S.: Succinct indexable dictionaries with applications to encoding k-ary trees and multisets. ACM Transactions on Algorithms 3(4), Art. 43 (2007)
25. Sadakane, K., Grossi, R.: Squeezing succinct data structures into entropy bounds. In: Proc. SODA, pp. 1230–1239. ACM/SIAM (2006)
26. Sadakane, K., Navarro, G.: Fully-functional succinct trees. In: Proc. SODA, pp. 134–149. ACM/SIAM (2010)

Substring Range Reporting

Philip Bille and Inge Li Gørtz

Technical University of Denmark
{phbi,ilg}@imm.dtu.dk

Abstract. We revisit various string indexing problems with range reporting features, namely, position-restricted substring searching, indexing substrings with gaps, and indexing substrings with intervals. We
obtain the following main results.

- We give efficient reductions for each of the above problems to a new
  problem, which we call *substring range reporting*. Hence, we unify
  the previous work by showing that we may restrict our attention to
  a single problem rather than studying each of the above problems
  individually.
- We show how to solve substring range reporting with optimal query
  time and little space. Combined with our reductions this leads to
  significantly improved time-space trade-offs for the above problems.
  In particular, for each problem we obtain the first solutions with
  optimal time query and $O(n \log^{O(1)} n)$ space, where n is the length
  of the indexed string.

Our bounds for substring range reporting are based on a novel combination of suffix trees and range reporting data structures. The reductions
are simple and general and may apply to other combinations of string
indexing with range reporting.

1 Introduction

Given a string S of length n the *string indexing problem* is to preprocess S
into a compact representation that efficiently supports *substring queries*, that
is, given another string P of length m report all occurrences of substrings in S
that match P. Combining the classic suffix tree data structure [13] with perfect
hashing [12] leads to an optimal time-space trade-off for string indexing, i.e., an
$O(n)$ space representation that supports queries in $O(m + \text{occ})$ time, where occ
is the number of occurrences of P in S.

In recent years, several extensions of string indexing problems that add *range
reporting* features have been proposed. For instance, Mäkinen and Navarro proposed the *position-restricted substring searching problem* [17, 18]. Here, queries
take an additional range $[a, b]$ of positions in S and the goal is to report the
occurrences of P within $S[a, b]$. For such extensions of string indexing no optimal time-space trade-off is known. For instance, for position-restricted substring
searching one can either get $O(n \log^{\varepsilon} n)$ space (for any constant $\varepsilon > 0$) and
$O(m + \log \log n + \text{occ})$ query time or $O(n^{1+\varepsilon})$ space with $O(m + \text{occ})$ query

R. Giancarlo and G. Manzini (Eds.): CPM 2011, LNCS 6661, pp. 299–308, 2011.
© Springer-Verlag Berlin Heidelberg 2011

time [17, 18, 7]. Hence, removing the $\log \log n$ term in the query comes at the cost of significantly increasing the space.

In this paper, we revisit a number string indexing problems with range reporting features, namely *position-restricted substring searching, indexing substrings with gaps*, and *indexing substrings with intervals*. We achieve the following results.

- We give efficient reductions for each of the above problems to a new problem, which we call *substring range reporting*. Hence, we unify the previous work by showing that we may restrict our attention to a single problem rather than studying each of the above problems individually.
- We show how to solve substring range reporting with optimal query time and little space. Combined with our reductions this leads to significantly improved time-space trade-offs for all of the above problems. For instance, we show how to solve position-restricted substring searching in $O(n \log^{\varepsilon} n)$ space and $O(m + \text{occ})$ query time.

Our bounds for substring range reporting are based on a novel combination of suffix trees and range reporting data structures. The reductions are simple and general and may apply to other combinations of string indexing with range reporting.

1.1   Substring Range Reporting

Let S be a string where each position is associated with a integer value in the range $[0, u]$. The integer associated with position i in S is the *label* of position i, denoted $\text{label}(i)$, and we call S a *labeled string*. Given a labeled string S, the *substring range reporting problem* is to compactly represent S while supporting *substring range reporting queries*, that is, given a string P and a pair of integers a and b, $0 \leq a \leq b \leq u$, report all starting positions in S that match P and whose labels are in the range $[a, b]$.

We assume a standard unit-cost RAM model with word size w and a standard instruction set including arithmetic operations, bitwise boolean operations, and shifts. We assume that a label can be stored in a constant number of words and therefore $w = \Theta(\log u)$. The space complexity is the number of words used by the algorithm. All bounds mentioned in this paper are valid in this model of computation.

To solve substring range reporting a basic approach is to combine a suffix tree with a 2D range reporting data structure. A query for a pattern P and range $[a, b]$ consists of a search in the suffix tree and then a 2D range reporting query with $[a, b]$ and the lexicographic range of suffixes defined P. This is essentially the overall approach used in the known solutions for position-restricted substring searching [17, 18, 7, 8, 22, 4], which is a special case of substring range reporting (see the next section).

Depending on the choice of the 2D range reporting data structure this approach leads to different trade-offs. In particular, if we plug in the 2D range reporting data structure of Alstrup et al. [2], we get a solution with $O(n \log^\varepsilon n)$ space and $O(m + \log \log u + \text{occ})$ query time (see Mäkinen and Navarro [17,18]). The $\log \log u$ term in the query time is from the range reporting query. Alternatively, if we use a fast data structure for the range successor problem [7,22] to do the range reporting, we get optimal $O(m + \text{occ})$ query time but increase the space to at least $\Omega(n^{1+\varepsilon})$. Indeed, since any 2D range reporting data structure with $O(n \log^{O(1)} n)$ space must use $\Omega(\log \log u)$ query time [20], we cannot hope to avoid this blowup in space with this approach.

Our first main contribution is a new and simple technique that overcomes the inherent problem of the previous approach. We show the following result.

Theorem 1. *Let S be a labeled string of length n with labels in the range $[0, u]$. For any constants $\varepsilon, \delta > 0$, we can solve substring range reporting using $O(n(\log^\varepsilon n + \log \log u))$ space, $O(n(\log n + \log^\delta u))$ expected preprocessing time, and $O(m + \text{occ})$ query time, for a pattern string of length m.*

Compared to the previous results we achieve optimal query time with an additional $O(n \log \log u)$ term in the space. For the applications considered here, we have that $u = O(n)$ and therefore the space bound simplifies to $O(n(\log^\varepsilon n + \log \log u)) = O(n \log^\varepsilon n)$. Hence, in this case there is no asymptotic space overhead.

The key idea to obtain Theorem 1 is a new and simple combination of suffix trees with multiple range reporting data structures for both 1 and 2 dimensions. Our solution handles queries differently depending on the length of the input pattern such that the overall query is optimized accordingly.

Interestingly, the idea of using different query algorithms depending on the length of the pattern is closely related to the concept of *filtering search* introduced for the standard range reporting problem by Chazelle as early as 1986 [5]. Our new results show that this idea is also useful in combinatorial pattern matching.

1.2   Applications

Our second main contribution is to show that substring range reporting actually captures several other string indexing problems. In particular, we show how to reduce the following problems to substring range reporting.

- *Position-restricted substring searching:* Given a string S of length n, construct a data structure supporting the following query: Given a string P and query interval $[a, b]$, with $1 \le a \le b \le n$, return the positions of substrings in S matching P whose positions are in the interval $[a, b]$.
- *Indexing substrings with intervals:* Given a string S of length n, and a set of intervals $\pi = \{[s_1, f_1], [s_2, f_2], \ldots, [s_{|\pi|}, f_{|\pi|}]\}$ such that $s_i, f_i \in [1, n]$ and $s_i \le f_i$, for all $1 \le i \le |\pi|$, construct a data structure supporting the following query: Given a string P and query interval $[a, b]$, with $1 \le a \le b \le n$, return the positions of substrings in S matching P whose positions are in $[a, b]$ *and* in one of the intervals in π.

- *Indexing substrings with gaps:* Given a string S of length n and an integer d, the problem is to construct a data structure supporting the following query: Given two strings P_1 and P_2 return all positions of substrings in S matching $P_1 \circ \star^d \circ P_2$. Here $\circ$ denotes concatenation and $\star$ is a wildcard matching all characters in the alphabet.

Previous results. Let m be the length of P. Mäkinen and Navarro [17, 18] introduced the position-restricted substring searching problem. Their fastest solution uses $O(n \log^\varepsilon n)$ space, $O(n \log n)$ expected preprocessing time, and $O(m + \log \log n + \text{occ})$ query time. Crochemore et al. [7] proposed another solution using $O(n^{1+\varepsilon})$ space, $O(n^{1+\varepsilon})$ preprocessing time, and $O(m + \text{occ})$ query time (see also Section 1.1). Using techniques from range non-overlapping indexing [6] it is possible to improve these bounds for small alphabet sizes [21]. Several succinct versions of the problem have also been proposed [17, 18, 22, 4]. All of these have significantly worse query time since they require superconstant time per reported occurrence. Finally, Crochemore et al. [9] studied a restricted version of the problem with $a = 1$ or $b = n$.

For the indexing substrings with intervals problem, Crochemore et al. [7, 8] gave a solution with $O(n \log^2 n)$ space, $O(|\pi| + n \log^3 n)$ expected preprocessing time, and $O(m + \log \log n + \text{occ})$ query time. They also showed how to achieve $O(n^{1+\varepsilon})$ space, $O(n^{1+\varepsilon} + |\pi|)$ preprocessing time, and $O(m + \text{occ})$ query time. Several papers [3, 14, 16] have studied the property matching problem, which is similar to the indexing substrings with intervals problem, but where both start and end point of the match must be in the same interval.

Iliopoulos and Rahman [15] studied the problem of indexing substrings with gaps. They gave a solution using $O(n \log^\varepsilon n)$ space, $O(n \log n)$ expected preprocessing time, and $O(m + \log\log n + \text{occ})$ query time, where m is the length of the two input strings. Crochemore and Tischler recently proposed a variant of the problem [10].

Our results. We reduce position-restricted substring searching, indexing substrings with intervals, and indexing substrings with gaps to substring range reporting. Applying Theorem 1 with our new reductions, we get the following result.

Theorem 2. *Let S be a string of length n and let m be the length of the query. For any constant $\varepsilon > 0$, we can solve*

- (i) *Position-restricted substring searching using $O(n \log^\varepsilon n)$ space, $O(n \log n)$ expected preprocessing time, and $O(m + \text{occ})$ query time.*
- (ii) *Indexing substrings with intervals using $O(n \log^\varepsilon n)$ space, $O(|\pi| + n \log n)$ expected preprocessing time, and $O(m + \text{occ})$ query time.*
- (iii) *Indexing substrings with gaps using $O(n \log^\varepsilon n)$ space, $O(n \log n)$ expected preprocessing time, and $O(m + \text{occ})$ query time (m is the size of the two input strings).*

This improves the best known time-space trade-offs for all three problems, that all suffer from the trade-off inherent in 2D range reporting.

The reductions are simple and general and may apply to other combinations of string indexing with range reporting.

2   Basic Concepts

2.1   Strings and Suffix Trees

Throughout the section we will let S be a labeled string of length $|S| = n$ with labels in $[0, u]$. We denote the character at position i by $S[i]$ and the substrings from position i to j by $S[i, j]$. The substrings $S[1, j]$ and $S[i, n]$ are the *prefixes* and *suffixes* of S, respectively. The *reverse* of S is S^R. We denote the label of position i by $\text{label}_S(i)$. The *order* of suffix $S[i, n]$, denoted $\text{order}_S(i)$, is the lexicographic order of $S[i, n]$ among the suffixes of S.

The *suffix tree* for S, denoted T_S, is the compacted trie storing all suffixes of S [13]. The *depth* of a node v in T_S is the number of edges on the path from v to the root. Each of the edges in T_S is associated with some substring of S. The children of each node are sorted from left to right in increasing alphabetic order of the first character of the substring associated with the edge leading to them. The concatenation of substrings from the root to v is denoted $\text{str}_S(v)$. The *string depth* of v, denoted $\text{strdepth}_S(v)$, is the length of $\text{str}_S(v)$. The *locus* of a string P, denoted $\text{locus}_S(P)$, is the minimum depth node v such that P is a prefix of $\text{str}_S(v)$. If P is not a prefix of a substring in S we define $\text{locus}_S(P)$ to be $\perp$.

Each leaf ℓ in T_S uniquely corresponds to a suffix in S, namely, the suffix $\text{str}_S(\ell)$. Hence, we will use $\text{label}_S(\ell)$ and $\text{order}_S(\ell)$ to refer to the label and order of the corresponding suffix. For an internal node v we extend the notation such that

$$\text{label}_S(v) = \{\text{label}_S(\ell) \mid \ell \text{ is a descendant leaf of } v\}$$
$$\text{order}_S(v) = \{\text{order}_S(\ell) \mid \ell \text{ is a descendant leaf of } v\}.$$

Since children of a node are sorted, the left to right order of the leaves in T_S corresponds to the lexicographic order of the suffixes of S. Hence, for any node v, $\text{order}_S(v)$ is an interval. We denote the left and right endpoints of this interval by l_v and r_v. When the underlying string S is clear from the context we will often drop the subscript $_S$ for brevity.

The suffix tree for S uses $O(n)$ space and can be constructed in $O(\text{sort}(n))$ time, where $\text{sort}(n)$ is the time for sorting n values in the model of computation [11]. For our results we only need a comparison-based $O(n \log n)$ sorting algorithm. Let P be a string of length m. If $\text{locus}_S(P) = \perp$ then P does not occur as a substring in S. Otherwise, the substrings in S that match P are the suffixes in $\text{order}_S(\text{locus}_S(P))$. Hence, we can compute all occurrences of P in S by traversing the suffix tree from the root to $\text{locus}_S(P)$ and then report all suffixes stored in the subtree. Using perfect hashing [12] to represent the outgoing edges of each node in T_S we achieve an $O(n)$ solution to string indexing that supports queries in $O(m + \text{occ})$ time (here occ is the total number of occurrences of P in S).

2.2   Range Reporting

Let $X \subseteq \{0, \ldots, u\}^d$ be a set of points in a d-dimensional grid. The *range reporting problem* in d-dimensions is to compactly represent X while supporting *range reporting queries*, that is, given a rectangle $R = [a_1, b_1] \times \cdots \times [a_d, b_d]$ report all points in the set $R \cap X$. We use the following results for range reporting in 1 and 2 dimensions.

Lemma 1 (Alstrup et al. [1], Mortensen et al. [19]). *For a set of n points in $[0, u]$ and any constant $\gamma > 0$, we can solve 1D range reporting using $O(n)$ space, $O(n \log^\gamma u)$ expected preprocessing time and $O(1 + \text{occ})$ query time.*

Lemma 2 (Alstrup et al. [2]). *For a set of n points in $[0, u] \times [0, u]$ and any constant $\varepsilon > 0$, we can solve 2D range reporting using $O(n \log^\varepsilon n)$ space, $O(n \log n)$ expected preprocessing time, and $O(\log \log u + \text{occ})$ query time.*

3   Substring Range Reporting

We now show Theorem 1. Recall that S is a labeled string of length n with labels from $[0, u]$.

3.1   The Data Structure

Our substring range reporting data structure consists of the following components.

- The suffix tree T_S for S. For each node v in T_S we also store l_v and r_v. We partition T_S into a *top tree* and a number of *bottom trees*. The top tree consists of all nodes in T_S whose string depth is at most $\log \log u$ and all their children. The trees induced by the remaining nodes are the forest of bottom trees.
- A 2D range reporting data structure on the set of points $\{(\text{order}_S(i), \text{label}_S(i)) \mid i \in \{1, \ldots, n\}\}$.
- For each node v in the top tree, a 1D range reporting data structure on the set $\{\text{label}_S(i) \mid i \in \text{order}_S(v)\}$.

We analyze the space and preprocessing time for the data structure. We use the range reporting data structures from Lemmas 1 and 2. The space for the suffix tree is $O(n)$ and the space for the 2D range reporting data structure is $O(n \log^\varepsilon n)$, for any constant $\varepsilon > 0$. We bound the space for the (potentially $\Omega(n)$) 1D range reporting data structures stored for the top tree. Let V_d be the set of nodes in the top tree with depth d. Since the sets $\text{order}_S(v)$, $v \in V_d$, partition the set of descendant leaves of nodes in V_d, the total size of these sets is as most n. Hence, the total size of the 1D range reporting data structures for the nodes in V_d is therefore $O(n)$. Since there are at most $\log \log u + 1$ levels in the top tree, the space for all 1D range reporting data structures is $O(n \log \log u)$. Hence, the total space for the data structure is $O(n(\log^\varepsilon n + \log \log u))$.

We can construct the suffix tree in $O(\text{sort}(n))$ time and the 2D range reporting data structure in $O(n \log n)$ expected time. For any constant $\gamma > 0$, the expected preprocessing time for all 1D range reporting data structures is

$$O\left(\sum_{v \text{ in top tree}} |\text{order}_S(v)| \log^\gamma u\right) = O(n \log \log u \log^\gamma u) = O(n \log^{2\gamma} u).$$

Setting $\delta = 2\gamma$ we use $O(n(\log n + \log^\delta u))$ expected preprocessing time in total.

3.2   Substring Range Queries

Let P be a string of length m, and let a and b be a pair of integers, $0 \leq a \leq b \leq u$. To answer a substring range query we want to compute the set of starting positions for P whose labels are in $[a, b]$. First, we compute the node $v = \text{locus}_S(P)$. If $v = \bot$ then P is not a substring of S, and we return the empty set. Otherwise, we compute the set of descendant leaves of v with labels in $[a, b]$. There are two cases to consider.

   (i) If v is in the top tree we query the 1D range reporting data structure for v with the interval $[a, b]$.
  (ii) If v is in a bottom tree, we query the 2D range reporting data with the rectangle $[l_v, r_v] \times [a, b]$.

Given the points returned by the range reporting data structures, we output the corresponding starting positions of the corresponding suffixes. From the definition of the data structure it follows that these are precisely the occurrences of P within the range $[a, b]$. Next consider the time complexity. We find $\text{locus}_S(P)$ in $O(m)$ time (see Section 2). In case (i) we use $O(1 + \text{occ})$ time to compute the result by Lemma 1. Hence, the total time for a substring range query for case (i) is $O(m + \text{occ})$. In case (ii) we use $O(\log \log u + \text{occ})$ time to compute the result by Lemma 2. We have that $v = \text{locus}_S(P)$ is in a bottom tree and therefore $m \geq \text{strdepth}(\text{parent}(\text{locus}_S(v))) > \log \log u$. Hence, the total time to answer a substring range query in case (ii) is $O(m + \log \log u + \text{occ}) = O(m + \text{occ})$. Thus, in both cases we use $O(m + \text{occ})$ time.

Summing up, our solution uses $O(n(\log^\varepsilon n + \log \log u))$ space, $O(n(\log n + \log^\delta u))$ expected preprocessing time, and $O(m + \text{occ})$ query time. This completes the proof of Theorem 1.

4   Applications

In this section we show how to improve the results for the three problems position-restricted substring searching, indexing substrings with intervals, and indexing gapped substrings, using our data structure for substring range reporting. Let $\text{REPORT}_S(P, a, b)$ denote a substring range reporting query on string S with parameters P, a, and b.

4.1   Position-Restricted Substring Searching

We can reduce position-restricted substring searching to substring range reporting by setting $\text{label}(i) = i$ for all $i = 1, \dots, n$. To answer a query we return the result of the substring range query $\text{REPORT}_S(P, a, b)$. Since each label is equal to the position, it follows that the solution to the substring range reporting instance immediately gives a solution to position-restricted substring searching. Applying Theorem 1 with $u = n$, this proves Theorem 2(i).

4.2   Indexing Substrings with Intervals

We can reduce indexing substrings with intervals to substring range reporting by setting

$$\text{label}(i) = \begin{cases} i & \text{if } i \in \varphi \text{ for some } \varphi \in \pi, \\ 0 & \text{otherwise.} \end{cases}$$

To answer a query we return the result of the substring range reporting query $\text{REPORT}_S(P, a, b)$. Let I be the solution to the indexing substrings with intervals instance and let I' be the solution to the substring range reporting instance derived by the above reduction. Then $i \in I \Leftrightarrow i \in I'$.

To prove this assume $i \in I$. Then $i \in \varphi$ for some $\varphi \in \pi$ and $i \in [a, b]$. From $i \in \varphi$ and the definition of $\text{label}(i)$ it follows that $\text{label}(i) = i$. Thus, $\text{label}(i) = i \in [a, b]$ and thus $i \in I'$. Assume $i \in I'$. Then $\text{label}(i) \in [a, b]$. Since $a > 0$ also $\text{label}(i) > 0$, and it follows that $\text{label}(i) = i$. By the reduction this means that $i \in \varphi$ for some $\varphi \in \pi$. Since $i = \text{label}(i)$, we have $i \in [a, b]$ and therefore $i \in I$.

We can construct the labeling in $O(n + |\pi|)$ if the intervals are sorted by startpoint or endpoint. Otherwise additional time for sorting is needed. A similar approach is used in the solution by Crochemore et al. [7].

Applying Theorem 1 with $u = n$, this proves Theorem 2(ii).

4.3   Indexing Substrings with Gaps

We can reduce the indexing substrings with gaps problem to substring range reporting as follows. Construct the suffix tree of the reverse of S, i.e., the suffix tree T_{S^R} for S^R. For each node v in T_{S^R} also store l_v and r_v. Set

$$\text{label}_S(i) = \begin{cases} \text{order}_{S^R}(n - i + d + 2) & \text{for } i \geq d + 2, \\ 0 & \text{otherwise.} \end{cases}$$

To answer a query find the locus node v of P_1^R in T_{S^R}. Then use the substring range reporting data structure to return all positions of substrings in S matching P_2 whose labels are in the range $[l_v, r_v]$. For each position i returned by $\text{REPORT}_S(P_2, l_v, r_v)$, return $i - |P_1| - d$. See Fig. 1 for an example.

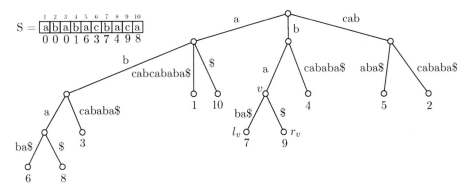

Fig. 1. A string S, the labeling for $d = 2$ (below the string), and the suffix tree of T_{S^R}. Given a query $P_1 = \text{ab}$ and $P_2 = \text{bac}$ we find $v = \text{locus}_{S^R}(\text{ba})$ (marked in the suffix tree). We have $l_v = 6$ and $r_v = 7$ from the left-to-right-order in the T_{S_R}. The substring range reporting query $\text{REPORT}_s(P_2, 6, 7)$ returns 7. Hence, we report the occurrence at position $7 - 2 - 2 = 3$.

Correctness of the reduction. We will now show that the reduction is correct. Let I be the solution to the indexing substrings with gaps instance and let I' be the solution to the substring range reporting instance derived by the above reduction. We will show $i \in I \Leftrightarrow i \in I'$. Let $m_i = |P_i|$ for $i = 1, 2$.

If $i \in I$ then there is an occurrence of P_1 at position i in S and an occurrence of P_2 at position $i' = i + m_1 + d$ in S. It follows directly, that there is an occurrence of P_1^R at position $i'' = n - (i + m_1) + 2$ in S^R. By definition,

$$\text{label}_S(i') = \text{label}_S(i + m_1 + d) = \text{order}_{S^R}(n - (i + m_1 + d) + d + 2) = \text{order}_{S^R}(i''),$$

where the second equality follows from the fact that $i + m_1 + d \geq d + 2$. Since there is an occurrence of P_1^R at position i'' in S^R, we have $\text{label}_S(i') = \text{order}_{S^R}(i'') \in \text{order}_{S^R}(\text{locus}_{S^R}(P_1^R))$. Thus, $\text{label}_S(i') \in [l_v, r_v]$, and since there is an occurrence of P_2 at position i' in S, we have $i' - m_1 - d = i \in I'$.

If $i \in I'$ then there is an occurrence of P_2 at position $i' = i + m_1 + d$ with $\text{label}(i')$ in the range $[l_v, r_v]$, where $v = \text{locus}_{S^R}(P_1^R)$. We need to show that this implies that there is an occurrence of P_1 at position i in S. By definition,

$$\text{label}_S(i') = \text{order}_{S^R}(n - i' + d + 2) = \text{order}_{S^R}(n - i - m_1 + 2).$$

Let $i'' = n - i - m_1 + 2$. Since $\text{order}_{S^R}(i'') = \text{label}_S(i') \in [l_v, r_v]$, there is an occurrence of P_1^R at position i'' in S^R. It follows directly, that there is an occurrence of P_1 at position $n - i'' - m_1 + 2 = n - (n - i - m_1 + 2) - m_1 + 2 = i$ in S. Therefore, $i \in I$.

Complexity. Construction of the suffix tree T_{S^R} takes time $O(n \log n)$ and the labeling can be constructed in time $O(n)$. Both use space $O(n)$. It takes $O(m_1)$ time to find the locus nodes of P_1^R in T_{S^R}. The substring range reporting query takes time $O(m_2 + \text{occ})$. Thus the total query time is $O(m + \text{occ})$.

Applying Theorem 1 with $u = n$, this completes the proof of Theorem 2(iii).

References

1. Alstrup, S., Brodal, G., Rauhe, T.: Optimal static range reporting in one dimension. In: Proc. 33rd STOC, pp. 476–482 (2001)
2. Alstrup, S.,Stølting Brodal, G., Rauhe, T.: New data structures for orthogonal range searching. In: Proc. 41st FOCS, pp. 198–207 (2000)
3. Amir, A., Chencinski, E., Iliopoulos, C.S., Kopelowitz, T., Zhang, H.: Property matching and weighted matching. Theoret. Comput. Sci. 395(2-3), 298–310 (2008)
4. Bose, P., He, M., Maheshwari, A., Morin, P.: Succinct orthogonal range search structures on a grid with applications to text indexing. In: Dehne, F., Gavrilova, M., Sack, J.-R., Tóth, C.D. (eds.) WADS 2009. LNCS, vol. 5664, pp. 98–109. Springer, Heidelberg (2009)
5. Chazelle, B.: Filtering search: A new approach to query-answering. SIAM J. Comput. 15(3), 703–724 (1986)
6. Cohen, H., Porat, E.: Range non-overlapping indexing. In: Dong, Y., Du, D.-Z., Ibarra, O. (eds.) ISAAC 2009. LNCS, vol. 5878, pp. 1044–1053. Springer, Heidelberg (2009)
7. Crochemore, M., Iliopoulos, C.S., Kubica, M., Rahman, M.S., Walen, T.: Improved algorithms for the range next value problem and applications. In: Proc. 25th STACS, pp. 205–216 (2008)
8. Crochemore, M., Iliopoulos, C.S., Kubica, M., Rahman, M.S., Walen, T.: Finding patterns in given intervals. Fundam. Inform. 101(3), 173–186 (2010)
9. Crochemore, M., Iliopoulos, C.S., Rahman, M.S.: Optimal prefix and suffix queries on texts. Inf. Process. Lett. 108(5), 320–325 (2008)
10. Crochemore, M., Tischler, G.: The gapped suffix array: A new index structure for fast approximate matching. In: Chavez, E., Lonardi, S. (eds.) SPIRE 2010. LNCS, vol. 6393, pp. 359–364. Springer, Heidelberg (2010)
11. Farach-Colton, M., Ferragina, P., Muthukrishnan, S.: On the sorting-complexity of suffix tree construction. J. ACM 47(6), 987–1011 (2000)
12. Fredman, M.L., Komlós, J., Szemerédi, E.: Storing a sparse table with $O(1)$ worst case access time. J. ACM 31, 538–544 (1984)
13. Gusfield, D.: Algorithms on strings, trees, and sequences: computer science and computational biology, Cambridge (1997)
14. Iliopoulos, C.S., Rahman, M.S.: Faster index for property matching. Inf. Process. Lett. 105(6), 218–223 (2008)
15. Iliopoulos, C.S., Rahman, M.S.: Indexing factors with gaps. Algorithmica 55(1), 60–70 (2009)
16. Juan, M., Liu, J., Wang, Y.: Errata for "Faster index for property matching". Inf. Process. Lett. 109(18), 1027–1029 (2009)
17. Mäkinen, V., Navarro, G.: Position-restricted substring searching. In: Correa, J.R., Hevia, A., Kiwi, M. (eds.) LATIN 2006. LNCS, vol. 3887, pp. 703–714. Springer, Heidelberg (2006)
18. Mäkinen, V., Navarro, G.: Rank and select revisited and extended. Theoret. Comput. Sci. 387(3), 332–347 (2007)
19. Mortensen, C.W., Pagh, R., Pătraşcu, M.: On dynamic range reporting in one dimension. In: Proc. 37th STOC, pp. 104–111 (2005)
20. Pătraşcu, M., Thorup, M.: Time-space trade-offs for predecessor search. In: Proc. 38th STOC, pp. 232–240 (2006)
21. Porat, E.: Personal communication (2011)
22. Yu, C.-C., Hon, W.-K., Wang, B.-F.: Improved data structures for the orthogonal range successor problem. Comput. Geometry 44(3), 148–159 (2011)

Faster Subsequence and Don't-Care Pattern Matching on Compressed Texts

Takanori Yamamoto[1], Hideo Bannai[1],
Shunsuke Inenaga[2], and Masayuki Takeda[1]

[1] Department of Informatics, Kyushu University
[2] Graduate School of Information Science and Electrical Engineering,
Kyushu University
744 Motooka, Nishiku, Fukuoka 819–0395, Japan
takanori.yamamoto@i.kyushu-u.ac.jp,
{bannai,takeda}@inf.kyushu-u.ac.jp,
inenaga@c.csce.kyushu-u.ac.jp

Abstract. Subsequence pattern matching problems on compressed text were first considered by Cégielski *et al.* (Window Subsequence Problems for Compressed Texts, Proc. CSR 2006, LNCS 3967, pp. 127–136), where the principal problem is: given a string T represented as a straight line program (SLP) $\mathcal{T}$ of size n, a string P of size m, compute the number of minimal subsequence occurrences of P in T. We present an $O(nm)$ time algorithm for solving all variations of the problem introduced by Cégielski *et al.*. This improves the previous best known algorithm of Tiskin (Towards approximate matching in compressed strings: Local subsequence recognition, Proc. CSR 2011), which runs in $O(nm \log m)$ time. We further show that our algorithms can be modified to solve a wider range of problems in the same $O(nm)$ time complexity, and present the first matching algorithms for patterns containing VLDC (variable length don't care) symbols, as well as for patterns containing FLDC (fixed length don't care) symbols, on SLP compressed texts.

1 Introduction

A *straight-line program* (*SLP*) [6] is a context free grammar in the Chomsky normal form that derives a single string. SLPs are a widely accepted abstract model of various text compression schemes, since texts compressed by any grammar-based compression algorithm (e.g. [12,8]) can be represented as SLPs, and those compressed by the LZ-family (e.g., [16,17]) can be quickly transformed to SLPs. An SLP of a string of size N can be as small as $O(\log N)$. SLPs are a promising representation of a given string, not only for reducing the storage size of the data, but for efficiently conducting various string processing operations [13,5]. Recently, *self indices* based on SLPs have also appeared [4].

Subsequence pattern matching [1] and its related problems have extensively been studied. Window subsequences are also known as serial episodes in data mining applications [10]. Now our interest is: Can we efficiently solve subsequence

R. Giancarlo and G. Manzini (Eds.): CPM 2011, LNCS 6661, pp. 309–322, 2011.

matching problems on compressed strings? When both text and pattern are given as SLPs, subsequence matching is NP-hard [9]. Therefore, in the sequel we only consider the case where the text is given as an SLP, while the pattern is given as an uncompressed string.

Subsequence problems on SLP-compressed texts were first considered in [3]. The principal problem considered is to compute the number of minimal subsequence occurrences of P in T. They presented $O(nm^2 \log m)$ time algorithms for solving the problems for an SLP of size n and subsequence pattern of length m. Later, an improved algorithm running in time $O(nm^{1.5})$ was presented by Tiskin [14]. Later, Tiskin improved the running time to $O(nm \log m)$ [15]. In this paper, we further reduce the time complexities to $O(nm)$.

The contribution of this paper is twofold. Firstly, we improve the algorithm for building the L and R arrays of [3], from $O(nm^2 \log m)$ to $O(nm)$, therefore reducing the overall time complexity of the algorithms for the subsequence pattern matching problems to $O(nm)$. Following the ideas of [3], we give a simpler presentation of these algorithms.

Secondly, we show that the algorithm can be extended to cope with patterns that contain *don't care* symbols, and give $O(nm)$-time matching algorithms for patterns containing VLDC (variable length don't care) symbols, as well as an $O(nm)$-time matching algorithm for patterns containing FLDC (fixed length don't care) symbols. There has been work on pattern matching for patterns containing FLDC symbols on a compressed representation of Sturmian words [2]. On the other hand, our algorithms can search *arbitrary* SLPs for patterns containing don't cares, and hence are applicable to more practical compressed texts.

2   Preliminaries

2.1   Strings

Let Σ be a finite *alphabet*. An element of Σ^* is called a *string*. The length of a string T is denoted by $|T|$. The empty string ε is a string of length 0, namely, $|\varepsilon| = 0$. For a string $T = XYZ$, X, Y and Z are called a *prefix*, *substring*, and *suffix* of T, respectively. The i-th character of a string T is denoted by $T[i]$ for $0 \le i \le |T| - 1$, and the substring of a string T that begins at position i and ends at position j is denoted by $T[i : j]$ for $0 \le i \le j \le |T| - 1$. For convenience, let $T[i : j] = \varepsilon$ if $j < i$.

A string P of length m is a *subsequence* of string T, if there exist indices $0 \le i_0 < \cdots < i_{m-1} \le |T| - 1$ such that $P[0] = T[i_0], \ldots, P[m - 1] = T[i_{m-1}]$. The pair (i_0, i_{m-1}) is called an *occurrence* of subsequence P in T. Let $Occ(T, P)$ denote the set of all occurrences of subsequence P in T. An occurrence $(u, v) \in Occ(T, P)$ is *minimal* if P is *not* a subsequence of $T[u + 1 : v]$ nor $T[u : v - 1]$. For strings X, Y, if an occurrence $(u, v) \in Occ(XY, P)$ satisfies $0 \le u < |X|$ and $|X| \le v < |XY|$, we say that this occurrence *crosses* X and Y.

2.2 Straight Line Programs

In this paper, we treat strings described in terms of *straight line programs* (*SLPs*). A straight line program $\mathcal{T}$ is a sequence of assignments such that $X_1 = expr_1, X_2 = expr_2, \ldots, X_n = expr_n$, where each X_i is a variable and each $expr_i$ is an expression, where $expr_i = a$ ($a \in \Sigma$), or $expr_i = X_\ell X_r$ ($\ell, r < i$).

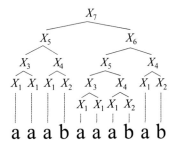

Denote by T the string derived from the last variable X_n of the program $\mathcal{T}$. The *size* of the program $\mathcal{T}$ is the number n of assignments in $\mathcal{T}$. Note that $|T| = O(2^n)$.

Let $val(X_i)$ represent the string derived from X_i. When it is not confusing, we identify a variable X_i with $val(X_i)$. Then, $|X_i|$ denotes the length of the string X_i derives. For assignment $X_i = X_\ell X_r$, if an occurrence (u,v) of subsequence P in $val(X_i)$ crosses $val(X_\ell)$ and $val(X_r)$, we say that (u,v) is a crossing subsequence occurrence of P in X_i.

Fig. 1. An example of an SLP $X_1 = \mathsf{a}$, $X_2 = \mathsf{b}$, $X_3 = X_1 X_1$, $X_4 = X_1 X_2$, $X_5 = X_3 X_4$, $X_6 = X_5 X_4$, $X_7 = X_5 X_6$, that derives string $\mathsf{aaabaaabab}$

3   Subsequence Matching Problems on Compressed Texts

This section is organized as follows: We first review an $O(nm)$ time algorithm for calculating tables Q^L and Q^R, which can determine whether a string P of length m is a subsequence of the string derived from an SLP $\mathcal{T}$ of size n (Subsequence Recognition). A brief description of the algorithm appears in [14], where it is noted that the algorithm "*has been known in folklore*", which was pointed out by Y. Lifshits. We then describe how to efficiently compute auxiliary tables L and R using Q^L and Q^R. Following the ideas in [3], we use L and R to give straightforward descriptions of $O(nm)$ time algorithms for solving the problem of finding all minimal subsequence occurrences of a pattern in a SLP-compressed text (Subsequence Matching), and its window-accumulated version (Window Subsequence Matching).

3.1 Subsequence Recognition

For $i = 1, \ldots, n, j = 0, \ldots, m$, let $Q^L(i,j)$ denote the length of the longest prefix of $P[j:m-1]$ which is a subsequence of X_i. We have that P is a subsequence of T, if and only if $Q^L(n,0) = m$.

Lemma 1 ([14]). *Given a pattern P of length m and an SLP $\mathcal{T}$ of size n representing text T, $Q^L(i,j)$ for $i = 1, \ldots, n, j = 0, \ldots, m$ can be calculated in $O(nm)$ time and space.*

Proof. $Q^L(i,j)$ can be defined recursively, as follows. For the base case, if $X_i = a$ for some $a \in \Sigma$, then

$$Q^L(i,j) = \begin{cases} 1 & \text{if } j < m \text{ and } P[j] = a, \\ 0 & \text{otherwise.} \end{cases} \tag{1}$$

If $X_i = X_\ell X_r$, then

$$Q^L(i,j) = Q^L(\ell,j) + Q^L(r,j') \tag{2}$$

where $j' = j + Q^L(\ell,j)$, because $P[j : j + Q^L(\ell,j) - 1]$ is the longest prefix of $P[j : m-1]$ that is a subsequence of X_ℓ, and the rest is the longest prefix of $P[j + Q^L(\ell,j) : m-1]$ that is a subsequence of X_r. Since each $Q^L(i,j)$ can be calculated in constant time, $Q^L(i,j)$ for $i = 1, \ldots, n, j = 0, \ldots, m$ can be calculated in $O(nm)$ time and space.   □

Thus we can test whether a pattern P is a subsequence of an SLP $\mathcal{T}$ in $O(nm)$ time.

We similarly define $Q^R(i,j)$ as the length of the longest suffix of $P[0 : m-j-1]$ that is a subsequence of X_i, which can also be calculated in $O(nm)$ time and space.

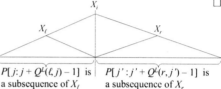

Fig. 2. Lemma 1. $Q^L(i,j) = Q^L(\ell,j) + Q^L(r,j')$ where $j' = j + Q^L(\ell,j)$. $Q^L(\ell,j)$ is the length of the prefix of $P[j : m-1]$ which is a subsequence of X_ℓ, and $Q^L(r,j')$ is the length of the prefix of the rest of it.

3.2 Subsequence Matching

Auxiliary Tables. We next define $L(i,j)$ and $R(i,j)$ that are central to the algorithm presented in [3]. We define $L(i,j)$ as the length of the shortest prefix of X_i, for which $P[j : m-1]$ is a subsequence. When there is no such prefix of X_i, $L(i,j)$ is defined as ∞. We similarly define $R(i,j)$ as the length of the shortest suffix of X_i, for which $P[0 : m-j-1]$ is a subsequence. When there is no such suffix of X_i, $R(i,j)$ is defined as ∞. Only these values for L (resp. R) corresponding to suffixes (resp. prefixes) of P are required in the algorithms which follow. However, the algorithm presented in [3] required the values for L and R corresponding to *all substrings* of P to compute these values, therefore making the running time of the algorithm $O(nm^2 \log m)$. We improve their algorithm by showing that we can calculate $L(i,j)$ (resp. $R(i,j)$) using only values corresponding to suffixes (resp. prefixes) of P with support from $Q^L(i,j)$ (resp. $Q^R(i,j)$), and reduce the running time to $O(nm)$.

Lemma 2. *Given a pattern P of length m, an SLP $\mathcal{T}$ of size n representing text T, and $Q^L(i,j)$ (resp. $Q^R(i,j)$) for $i = 1, \ldots, n, j = 0, \ldots, m$, $L(i,j)$ (resp. $R(i,j)$) for all $i = 1, \ldots, n, j = 0, \ldots, m$ can be calculated in $O(nm)$ time and space.*

Proof. We shall only describe how to calculate $L(i, j)$ using $Q^L(i, j)$, since the case for $R(i, j)$ and $Q^R(i, j)$ is essentially the same. $L(i, j)$ can be defined recursively as follows: For the base case, if $X_i = a$ for some $a \in \Sigma$, then

$$L(i, j) = \begin{cases} 0 & \text{if } j = m, \\ 1 & \text{if } j = m - 1 \text{ and } P[j : m - 1] = a, \\ \infty & \text{otherwise.} \end{cases}$$

If $X_i = X_\ell X_r$, then

$$L(i, j) = \begin{cases} L(\ell, j) & \text{if } L(\ell, j) \neq \infty, \\ |X_\ell| + L(r, j') & \text{if } L(\ell, j) = \infty, \end{cases}$$

where $j' = j + Q^L(\ell, j)$. This is because: When $L(\ell, j) \neq \infty$, $P[j : m - 1]$ is a subsequence of X_ℓ, and $L(\ell, j)$ is the length of the shortest prefix of X_ℓ for which $P[j : m - 1]$ is a subsequence. Since X_ℓ is a prefix of X_i, the length of the shortest prefix of $X_i = X_\ell X_r$ for which $P[j : m - 1]$ is a subsequence is clearly equal to $L(\ell, j)$. When $L(\ell, j) = \infty$, $P[j : m - 1]$ is not a subsequence of X_ℓ. This implies that the value of $L(i, j)$ is at least $|X_\ell|$. The exact value of $L(i, j)$ can be efficiently computed from $Q^L(\ell, j)$, as follows. Since $L(i, j) - |X_\ell|$ equals to the shortest prefix of X_r for which $P[j' : m - 1]$ is a subsequence, we have $L(i, j) - |X_\ell| = L(r, j')$ where $j' = j + Q^L(\ell, j)$.

Therefore, given the Q^L table, each $L(i, j)$ can be computed in constant time. Hence $L(i, j)$ for all $i = 1, \ldots, n, j = 0, \ldots, m$ can be computed in $O(nm)$ time and space. □

Counting Minimal Occurrences. For text T represented by an SLP $\mathcal{T}$ of size n, we show how to calculate the number of minimal occurrences of subsequence P of length m in T in time $O(nm)$, using $L(i, j)$ and $R(i, j)$. Let M_i denote the number of minimal occurrences of P in $val(X_i)$. Since $val(X_n) = T$, the desired output is the value of M_n.

Our algorithm is based essentially on the same ideas as described in [3]. However, we note that they did not provide a rigorous proof of correctness, and the pseudo-code shown in their paper seems to contain some errors. Below, we give a simple presentation of the algorithm and a proof of correctness.

For any variable $X_i = X_\ell X_r$, let $C(\ell, r)$ denote the number of minimal occurrences of P in X_i that cross X_ℓ and X_r.

Lemma 3. *Given a pattern P of length m, an SLP $\mathcal{T}$ of size n, and $C(\ell, r)$ for all variables of form $X_i = X_\ell X_r$, the values M_i for $i = 1, \ldots, n$ can be calculated in $O(n)$ time.*

Proof. M_i is recursively computable as follows. For the base case, if $X_i = a$ for some $a \in \Sigma$, then $M_i = 0$ if $P \neq a$ and $M_i = 1$ if $P = a$. If $X_i = X_\ell X_r$, then $M_i = M_\ell + M_r + C(\ell, r)$. Hence we can compute M_i for all $i = 1, \ldots, n$ recursively, in total of $O(n)$ time. □

What remains is how to calculate $C(\ell, r)$ for all variables of type $X_i = X_\ell X_r$. For $k = 0, \ldots, m$, consider the following pairs (u_k, v_k) where u_k is the beginning position in X_i, of the shortest suffix of X_l for which $P[0 : m - 1 - k]$ is a subsequence (or $-\infty$ if such a suffix does not exist), and v_k is the ending position in X_i, of the shortest prefix of X_r for which $P[m - k : m - 1]$ is a subsequence (or ∞ if such a prefix does not exist), i.e., $u_k = |X_\ell| - R(\ell, k)$ and $v_k = |X_\ell| + L(r, m - k) - 1$ (see also Fig. 3 (Left)). Clearly u_k and v_k are monotonically non-decreasing, that is, $u_{k-1} \le u_k < |X_\ell| = u_m$, and $v_0 = |X_\ell| - 1 < v_k \le v_{k+1}$ for $k = 1, \ldots, m - 1$. When both $0 \le u_k < |X_\ell|$ and $|X_\ell| \le v_k < |X_i|$ hold, then (u_k, v_k) is a crossing subsequence occurrence of P in X_i. Note that neither (u_0, v_0) nor (u_m, v_m) are crossing occurrences. Let $Occ^{SS}(\ell, r) = \{(u_k, v_k) \mid k = 1, \ldots, m-1\}$. It is easy to see that every minimal crossing subsequence occurrence of P in X_i must be an element of $Occ^{SS}(\ell, r)$, and it remains to identify them.

Lemma 4. $(u_k, v_k) \in Occ^{SS}(\ell, r)$ *is a minimal occurrence if and only if* $\nexists k' \in \{0, \ldots, m\}$ *s.t.* $(u_k, v_k) \ne (u_{k'}, v_{k'})$, *and* $u_k \le u_{k'}$ *and* $v_{k'} \le v_k$.

Proof. ($\Longrightarrow$) If for some $k' \in \{0, \ldots, m\}$ s.t. $(u_{k'}, v_{k'}) \ne (u_k, v_k)$ we have $u_k \le u_{k'}$ and $v_{k'} \le v_k$, then (u_k, v_k) cannot be a minimal occurrence by definition.

($\Longleftarrow$) We show the contraposition. Assume (u_k, v_k) is not a minimal occurrence. If $u_k = -\infty$ (or resp. $v_k = \infty$), then $u_k \le u_0 = -\infty$ (resp. $v_m \le v_k = \infty$) and from the monotonicity of u_ks and v_ks, we can choose $k' = 0$ (resp. $k' = m$). If $u_k \ne -\infty$ and $v_k \ne \infty$, there exist some occurrence $(u, v) \ne (u_k, v_k)$ s.t. $u_k \le u$ and $v \le v_k$. If (u, v) is a crossing occurrence, then a minimal occurrence $(u_{k'}, v_{k'})$ can be chosen from $Occ^{SS}(\ell, r)$ s.t. $u \le u_{k'}$ and $v_{k'} \le v$. If it is not, then $v \le |X_l| - 1$ or $u \ge |X_l|$, and we can choose (u_0, v_0) or (u_m, v_m). $\square$

Lemma 5. *Consider* $(u_k, v_k) \in Occ^{SS}(\ell, r)$, *and let* $K = \{k' \mid (u_k, v_k) = (u_{k'}, v_{k'}), \ k' = 1, \ldots, m - 1\}$, $k_s = \min K$ *and* $k_e = \max K$. *Then,* (u_k, v_k) *is minimal if and only if* $u_{k_s - 1} < u_k$ *and* $v_k < v_{k_e + 1}$.

Proof. From the monotonicity of u_k and v_k, and from Lemma 4, we have that (u_k, v_k) is minimal if and only if

$$\nexists k' \in \{0, \ldots, m\} \text{ s.t. } (u_k, v_k) \ne (u_{k'}, v_{k'}), \ (u_k \le u_{k'}) \wedge (v_{k'} \le v_k)$$
$$\Longleftrightarrow \forall k' \in \{0, \ldots, m\} \text{ s.t. } (u_{k'}, v_{k'}) \ne (u_k, v_k), \ (u_{k'} < u_k) \vee (v_k < v_{k'})$$
$$\Longleftrightarrow ((u_{k_s - 1} < u_k) \vee (v_k < v_{k_s - 1})) \wedge ((u_{k_e + 1} < u_k) \vee (v_k < v_{k_e + 1}))$$
$$\Longleftrightarrow (u_{k_s - 1} < u_k) \wedge (v_k < v_{k_e + 1}). \qquad \square$$

Lemma 6. *Given a pattern P of length m, an SLP T of size n, and $L(i, j)$, $R(i, j)$ for $i = 1, \ldots, n, j = 0, \ldots, m$, $C(\ell, r)$ for all variables of form $X_i = X_\ell X_r$, can be computed in total of $O(nm)$ time.*

Proof. A pseudo-code of our algorithm which computes $C(\ell, r)$ is shown in Algorithm 1 (see also Fig. 3 (Right)). The time complexity is clearly $O(m)$ for each $X_i = X_\ell X_r$, and hence $O(nm)$ in total. The correctness is due to Lemma 5. $\square$

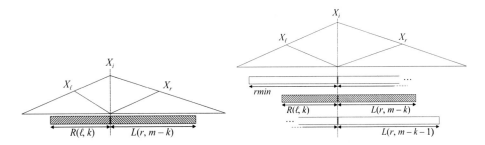

Fig. 3. (Left) If $R(\ell, k) \neq \infty$ and $L(r, m - k) \neq \infty$, there is a crossing subsequence occurrence of P. $P[0 : k - 1]$ is a subsequence of $X_\ell[|X_\ell| - R(\ell, k) : |X_\ell| - 1]$, and $P[k : m - 1]$ is a subsequence of $X_r[0 : L(r, m - k) - 1]$. (Right) Illustration of Algorithm 1. When $rmin > R(\ell, k)$ and $L(r, m - k) < L(r, m - k - 1)$, then $(|X_\ell| - R(\ell, k), |X_\ell| + L(r, m - k) - 1)$ is a crossing minimal occurrence. We then update $rmin \leftarrow R(\ell, k)$ to find the next crossing minimal occurrence.

Algorithm 1. Counting Minimal Crossing Subsequence Occurrences.

Input: SLP variable $X_i = X_\ell X_r$, pattern P, auxiliary tables L, R.
Output: The number of minimal crossing subsequence occurrences $C(\ell, r)$.
1 $C \leftarrow 0$; $rmin \leftarrow R(\ell, 0)$;
2 **for** $k \leftarrow 1$ **to** $m - 1$ **do**
3   **if** $rmin > R(\ell, k)$ **and** $L(r, m - k) < L(r, m - k - 1)$ **then**
4           $C \leftarrow C + 1$; $rmin \leftarrow R(\ell, k)$;

5 **return** C ;

Finally, we obtain the main result of this section.

Theorem 1. *Given a pattern P of length m and an SLP $\mathcal{T}$ of size n representing text T, the number of minimal subsequence occurrences of P in T can be calculated in $O(nm)$ time.*

Window Subsequence Matching. Cégielski *et al.* [3] introduced several window-accumulated variants of subsequence pattern matching on compressed texts. The principal problem is: Given an SLP $\mathcal{T}$ generating text T, a pattern P, and non-negative integer w, count the number of minimal subsequence occurrences (u, v) of P in T such that $v - u + 1 \leq w$.

Our algorithm for counting minimal occurrences can readily be extended to this window-accumulated variant. See Algorithm 1. By simply adding "$R(\ell, k) + L(r, m - k) \leq w$" in the **if**-condition of line 3, we can solve the problem in the same complexity $O(nm)$. We remark that the other variants considered in [3] can also be solved in the same complexity. Details are omitted due to lack of space.

4   Don't-Care Pattern Matching Problems on Compressed Texts

In this section we show that the ideas of Section 3 can be extended to solve pattern matching problems for patterns with fixed length don't care (FLDC) and variable length don't care (VLDC) symbols, in the same complexity $O(nm)$.

4.1   FLDC Pattern Matching on Compressed Texts

We can find *substrings* of X_i matching P, the same way as counting minimal subsequence occurrences. If a subsequence P of T occurs in (i_0, i_{m-1}) and $i_{m-1} - i_0 + 1 = m$, obviously the substring $T[i_0 : i_{m-1}]$, is equal to P.

The above idea can be extended to a pattern matching problem where the pattern includes *fixed length don't care (FLDC)* symbols. Let the symbol '∘' denote a don't care character that can match an arbitrary character in Σ. We call $P \in (\Sigma \cup \{\circ\})^*$ an *FLDC pattern*. An FLDC pattern P of length m occurs in string T at position i_0, if $T[i_0 + i] = P[i]$ or $P[i] = \circ$ for all $0 \le i \le m - 1$.

To count the occurrences of an FLDC pattern P using our window subsequence matching algorithms, we only need to count minimal subsequence occurrences of P that fit in a window of size $|P|$ with the exception that ∘ can match any single character. We can do this by simply modifying the base cases of $Q^L(i, j)$ and $L(i, j)$ as follows: If $X_i = a$ for some $a \in \Sigma$, then

$$Q^L(i,j) = \begin{cases} 1 & \text{if } j < m \text{ and } (P[j] = a \text{ or } P[j] = \circ), \\ 0 & \text{otherwise.} \end{cases}$$

$$L(i,j) = \begin{cases} 0 & \text{if } j = m, \\ 1 & \text{if } j = m - 1 \text{ and } (P[j : m - 1] = a \text{ or } P[j : m - 1] = \circ), \\ \infty & \text{otherwise.} \end{cases}$$

The base cases of $Q^R(i, j)$ and $R(i, j)$ should be modified similarly as well.

4.2   VLDC Pattern Matching on Compressed Texts

Let the symbol '⋆' denote a variable-length don't care character that can match an arbitrary string in Σ^*. We call $P \in (\Sigma \cup \{\star\})^*$ a *variable-length don't care (VLDC)* pattern. In the sequel, we only consider VLDC patterns that start and end with ⋆, and the ⋆'s do not occur consecutively. Consider any VLDC pattern $P = \star s_1 \star s_2 \star \cdots \star s_{m'} \star$, where each $s_j \in \Sigma^+$. The length of P is $m = \sum_{j=1}^{m'} |s_j|$. Each s_j is called the j-th *segment* of P. VLDC pattern P is said to *match* a string $T \in \Sigma^*$ if there exist indices $0 \le i_0 < i_0 + |s_1| \le i_1 < i_1 + |s_2| \le \cdots < i_{m'-1} + |s_{m'}| \le |T| - 1$ such that $s_1 = T[i_0 : i_0 + |s_1| - 1], \ldots, s_{m'} = T[i_{m'-1} : i_{m'-1} + |s_{m'}| - 1]$. The pair $(i_0, i_{m'-1} + |s_{m'}| - 1)$ is called an *occurrence* of VLDC pattern P in T. An occurrence (u, v) of VLDC pattern P in T is *minimal* if neither $(u+1, v)$ nor $(u, v-1)$ is an occurrence of P in T. Note that if each segment is a single character, then the above notion is equivalent to that of subsequences.

In what follows, we present how to compute minimal occurrences of a VLDC pattern in an SLP-compressed text. We will extend the notion of the auxiliary tables L, R, Q^L, and Q^R to cope with VLDC pattern matching. In so doing, we firstly introduce some new notion.

For any $X_i = X_\ell X_r$ and s_j, let

$$Occ^\ddagger(X_i, s_j) =$$

$$\left\{ k \;\middle|\; \begin{array}{l} X_\ell[|X_\ell| - k : |X_\ell| - 1] = s_j[0 : k-1], \\ X_r[0 : |s_j| - k - 1] = s_j[k : |s_j| - 1], k = 1, \ldots, \min\{|s_j| - 1, |X_\ell|\} \end{array} \right\}.$$

Namely, values in $Occ^\ddagger(X_i, s_j)$ correspond to lengths of overlap with X_ℓ, for all crossing *substring* occurrences of s_j in X_i. We can compute $Occ^\ddagger(X_i, s_j)$ for all $i = 1, \ldots, n$, $j = 1, \ldots, m'$ in total of $O(nm)$ time and space, as follows: Let h be the length of the longest segment of P. We decompress the prefix and suffix of length h of each variable X_i, i.e., we compute strings $A_i = X_i[|X_i| - \min\{h, |X_i|\} : |X_i| - 1]$ and $B_i = X_i[0 : \min\{h, |X_i|\} - 1]$. This can be done in total of $O(nm)$ time and space. Let $X_i = X_\ell X_r$. We can then compute $Occ^\ddagger(X_i, s_j)$ in $O(|s_j|)$-time by using any standard linear-time pattern matching algorithm (e.g. [7]) for text $A_\ell B_r$ and pattern s_j. Moreover, $Occ^\ddagger(X_i, s_j)$ forms a single arithmetic progression [11], and can thus be represented as the first element, the last element, and the number of elements, which require only $O(1)$ space. Overall it takes $O(nm)$ time and space to compute $Occ^\ddagger(X_i, s_j)$ for all $i = 1, \ldots, n$, $j = 1, \ldots, m'$.

Let $LCP(X_i, s_j, k)$ denote the length of the longest common prefix of X_i and $s_j[k : |s_j| - 1]$. We can also compute $LCP(X_i, s_j, k)$ in $O(nm)$ time and space for all $i = 1, \ldots, n$, $j = 1, \ldots, m'$, $k = 0, \ldots, |s_j|$, by the following recursion: For the base case, if $X_i = a$ for some $a \in \Sigma$, then $LCP(X_i, s_j, k) = 0$ if $X_i \neq s_j[k]$, and $LCP(X_i, s_j, k) = 1$ if $X_i = s_j[k]$. If $X_i = X_\ell X_r$, then

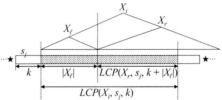

Fig. 4. Illustration of the recursion for $LCP(X_i, s_j, k)$. If $LCP(\ell, j, k) = |X_\ell|$, then $LCP(i, j, k) = |X_\ell| + LCP(r, j, k + |X_\ell|)$.

$$LCP(X_i, s_j, k) = \begin{cases} |X_\ell| + LCP(X_r, s_j, k + |X_\ell|) & \text{if } LCP(X_\ell, s_j, k) = |X_\ell|, \\ LCP(X_\ell, s_j, k) & \text{otherwise.} \end{cases}$$

Let $LCS(X_i, s_j, k)$ denote the length of the longest common suffix of X_i and $s_j[0 : |s_j| - k - 1]$. $LCS(X_i, s_j, k)$ can also be computed similarly in $O(nm)$ time and space.

For any VLDC pattern $P = \star s_1 \star s_2 \star \cdots \star s_{m'} \star$, we define a sub-pattern $segsub^L(P, j, k, q)$ of P, for $j = 1, \ldots, m' + 1$, $k = 0, \ldots, |s_j| - 1$, $q = 0, \ldots, m' - j + 1$, as follows:

$$segsub^L(P, j, k, q) = \begin{cases} \varepsilon & \text{if } q = 0 \text{ or } j > m', \\ \star s_j \star \cdots \star s_{j+q-1} \star & \text{if } q > 0, j \leq m', k = 0, \\ s_j[k : |s_j| - 1] \star \cdots \star s_{j+q-1} \star & \text{if } q > 0, j \leq m', k > 0. \end{cases}$$

Let $Q^L(i, j, k)$ denote the maximum number of segments in the sub-patterns $segsub^L(P, j, k, q)$ that match $val(X_i)$, i.e.,

$$Q^L(i, j, k) = \max\{q \mid segsub^L(P, j, k, q) \text{ matches } val(X_i)\}.$$

Also, we define $L(i, j, k)$ as the length of the shortest prefix of $val(X_i)$ that matches the sub-pattern giving $Q^L(i, j, k)$, i.e.,

$$L(i, j, k) = \min\{p \mid segsub^L(P, j, k, Q^L(i, j, k)) \text{ matches } X_i[0 : p - 1]\}$$

We define $Q^R(i, j, k)$ and $R(i, j, k)$ similarly, but be careful that $segsub^R(P, j, k, q)$ for $j = 0, \ldots, m', k = 0, \ldots, |s_j| - 1, q = 0, \ldots, j$ is defined as follows:

$$segsub^R(P, j, k, q) = \begin{cases} \varepsilon & \text{if } q = 0 \text{ or } j = 0, \\ \star s_{j-q+1} \star \cdots \star s_j \star & \text{if } q > 0, j > 0, k = 0, \\ \star s_{j-q+1} \star \cdots \star s_j[0 : |s_j| - k - 1] & \text{if } q > 0, j > 0, k > 0. \end{cases}$$

Lemma 7. *Given an SLP T and VLDC pattern $P = \star s_1 \star \cdots \star s_{m'} \star$, $Q^L(i, j, k)$ (resp. $Q^R(i, j, k)$) and $L(i, j, k)$ (resp. $R(i, j, k)$) can be also computed in $O(nm)$ time and space for all $i = 1, \ldots, n$, $j = 1, \ldots, m' + 1$ (resp. $j = 0, \ldots, m'$) and $k = 0, \ldots, |s_j| - 1$.*

Proof. $Q^L(i, j, k)$ and $L(i, j, k)$ can be defined recursively as follows. For the base case, $X_i = a$, $(a \in \Sigma)$, then

$$Q^L(i, j, k) = \begin{cases} 1 & \text{if } 1 \leq j \leq m' \text{ and } k = |s_j| - 1 \text{ and } s_j[|s_j| - 1] = a, \\ 0 & \text{otherwise.} \end{cases}$$

$$L(i, j, k) = \begin{cases} 1 & \text{if } Q^L(i, j, k) > 0, \\ 0 & \text{otherwise.} \end{cases}$$

If $X_i = X_\ell X_r$, $|s_j| - k > |X_\ell|$ and $k > 0$, then

$$Q^L(i, j, k) = \begin{cases} Q^L(r, j, k + |X_\ell|) & \text{if } LCP(X_\ell, s_j, k) = |X_\ell|, \\ 0 & \text{if } LCP(X_\ell, s_j, k) < |X_\ell|. \end{cases} \tag{3}$$

$$L(i, j, k) = \begin{cases} |X_\ell| + L(r, j, k + |X_\ell|) & \text{if } Q^L(i, j, k) > 0, \\ 0 & \text{if } Q^L(i, j, k) = 0. \end{cases} \tag{4}$$

(See also Fig 5 (Left).)

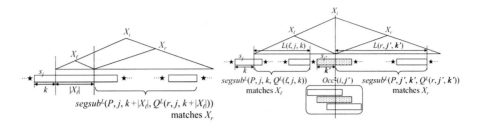

Fig. 5. (Left) Illustration of Equations (3) and (4) of Lemma 7. If $|s_j| - k > |X_\ell|$ and $LCP(i, j, k) = |X_\ell|$, then $Q^L(i, j, k) = Q^L(r, j, k + |X_\ell|)$ and $L(i, j, k) = |X_\ell| + L(r, j, k + |X_\ell|)$. (Right) Illustration of Equation (5) of Lemma 7. j' and k' can be computed in $O(1)$ time. Then, $Q^L(i, j, k)$ and $L(i, j, k)$ can be also computed in $O(1)$ time. Since $s_{j'}$ and $s_{j'-1}$ cannot overlap, k' must satisfy $k' + L(\ell, j, k) \leq |X_\ell|$.

If $X_i = X_\ell X_r$ and, $|s_j| - k \leq |X_\ell|$ or $k = 0$, then let $j' = j + Q^L(\ell, j, k)$ and $k' = \max\{x \mid x \in Occ^\ddagger(X_i, s_{j'}) \cup \{0\}, x + L(\ell, j, k) \leq |X_\ell|\}$. $Q^L(i, j, k)$ and $L(i, j, k)$ can be computed as follows: If $k = 0$ or $Q^L(\ell, j, k) > 0$, then $Q^L(i, j, k) = Q^L(\ell, j, k) + Q^L(r, j', k')$ and

$$L(i, j, k) = \begin{cases} |X_\ell| + L(r, j', k') & \text{if } Q^L(r, j', k') > 0, \\ L(\ell, j, k) & \text{if } Q^L(r, j', k') = 0. \end{cases} \quad (5)$$

(See also Fig 5 (Right).)

Otherwise ($k > 0$ and $Q^L(\ell, j, k) = 0$), $Q^L(i, j, k) = 0$ and $L(i, j, k) = 0$.

j' and k' can be computed in $O(1)$ time if $Q^L(\ell, j, k)$, $L(\ell, j, k)$ and $Occ^\ddagger(X_i, s_{j'})$ are already computed, and $Occ^\ddagger$ is represented as an arithmetic progression. Hence $Q^L(i, j, k)$ and $L(i, j, k)$ for all $i = 1, \ldots, n$, $j = 1, \ldots, m'$, and $k = 0, \ldots, |s_j| - 1$ can be computed in $O(nm)$ time and space. $Q^R(i, j, k)$ and $R(i, j, k)$ can be computed similarly using $LCS(X_i, s_j, k)$. □

An occurrence (u, v) of VLDC pattern P is a crossing occurrence in $X_i = X_\ell X_r$ if $0 \leq u < |X_\ell|$ and $|X_\ell| \leq v < |X_i|$. Let M_i and $C(\ell, r)$ denote the number of minimal occurrences and the number of minimal crossing occurrences of VLDC pattern P in $X_i = X_\ell X_r$, respectively.

Lemma 8. *Given a VLDC pattern P of length m, an SLP $\mathcal{T}$ of size n, and $C(\ell, r)$ for all variables of form $X_i = X_\ell X_r$, the values M_i for $i = 1, \ldots, n$ can be calculated in $O(n)$ time.*

Proof. M_i can be defined recursively as follows. For the base case ($X_i = a \in \Sigma$), if $P = \star a \star$ then $M_i = 1$, otherwise $M_i = 0$. For the case $X_i = X_\ell X_r$, $M_i = M_\ell + M_r + C(\ell, r)$. Thus M_i can be computed for all $i = 1, \ldots, n$, in $O(n)$ total time and space, if $C(\ell, r)$ for all variables of form $X_i = X_\ell X_r$ are already computed.

In what follows we describe how to compute $C(\ell, r)$ for each $X_i = X_\ell X_r$ in $O(m)$ time. Algorithm 2 shows a pseudo-code of our algorithm to compute $C(\ell, r)$. For convenience, for $i = 1, \ldots, n, j = 0, \ldots, m'$ and $k \in Occ^{\ddagger}(X_i, s_j) \cup \{0\}$, let

$$
\mathbf{L}(i, j, k) = \begin{cases} 0 & \text{if } j = 0, \\ L(i, j, k) & \text{if } j > 0 \text{ and } Q^L(i, j, k) = m' - j + 1, \\ \infty & \text{otherwise.} \end{cases}
$$

$$
\mathbf{R}(i, j, k) = \begin{cases} 0 & \text{if } j = 0, \\ R(i, j, k) & \text{if } j > 0 \text{ and } Q^R(i, j, k) = j, \\ \infty & \text{otherwise.} \end{cases}
$$

Note that conceptually, the tables L and R for subsequences correspond to $\mathbf{L}$ and $\mathbf{R}$ defined above, and when $segsub^L(P, j, k, m' - j + 1)$ does not match X_i, then $\mathbf{L}(i, j, k) = \infty$, and when $segsub^R(P, j, k, j)$ does not match X_i, then $\mathbf{R}(i, j, k) = \infty$. Hence we can compute the number of crossing VLDC pattern occurrences in a similar way to the case of subsequence patterns.

Care is taken for possible crossing occurrences when a segment is crossing X_i. For any j and $k > 0$, only occurrences $(|X_\ell| - \mathbf{R}(\ell, j, |s_j| - k), |X_\ell| + \mathbf{L}(r, j, k) - 1)$ for which $k \in Occ^{\ddagger}(X_i, s_j)$ can be crossing occurrences of P in X_i (see also Fig. 6 (Left)). For $j = 2, \ldots, m'$ and $k = 0$, occurrences $(|X_\ell| - \mathbf{R}(\ell, j - 1, 0), |X_\ell| + \mathbf{L}(r, j, 0) - 1)$ can be crossing occurrences of P in X_i (see also Fig. 6 (Right)). By checking these possible crossing occurrences in decreasing order of j and k, we can compute the number of crossing occurrences as described in Algorithm 2. Since the number of candidates is $d = \Sigma_{j=1}^{m'} |Occ^{\ddagger}(X_i, s_j)| + m' + 1 = O(m)$, we can compute all the crossing occurrences in a total of $O(nm)$ time and space. □

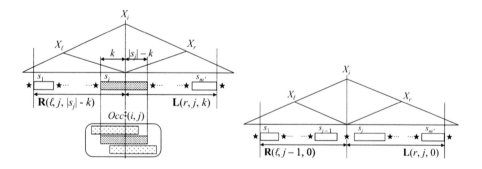

Fig. 6. Illustration of Algorithm 2. (Left) If $k \in Occ^{\ddagger}(X_i, s_j)$, $\mathbf{R}(\ell, j, |s_j| - k) \neq \infty$ and $\mathbf{L}(r, j, k) \neq \infty$, then $(|X_\ell| - \mathbf{R}(\ell, j, |s_j| - k), |X_\ell| + \mathbf{L}(r, j, k) - 1)$ is a candidate of a crossing occurrence. (Right) If $k = 0$, $\mathbf{R}(\ell, j - 1, 0) \neq \infty$ and $\mathbf{L}(r, j, 0) \neq \infty$, then $(|X_\ell| - \mathbf{R}(\ell, j - 1, 0), |X_\ell| + \mathbf{L}(r, j, 0) - 1)$ is a candidate of a crossing occurrence.

Algorithm 2. Counting Minimal Crossing VLDC Occurrences.

Input: SLP variable $X_i = X_\ell X_r$, pattern P, auxiliary tables $L(i, j, k)$, $R(i, j, k)$.
Output: The number of minimal crossing VLDC occurrences $C(\ell, r)$.

1   $d \leftarrow 0$; $(R[0], L[0]) \leftarrow (\mathbf{R}(\ell, m', 0), 0)$;
2   **for** $j \leftarrow m'$ **to** 1 **do**
3   **forall the** $k \in Occ^{\ddagger}(X_i, s_j)$ *in descending order* **do**
4        $d \leftarrow d + 1$; $(R[d], L[d]) \leftarrow (\mathbf{R}(\ell, j, |s_j| - k), \mathbf{L}(r, j, k))$;
5      $d \leftarrow d + 1$; $(R[d], L[d]) \leftarrow (\mathbf{R}(\ell, j - 1, 0), \mathbf{L}(r, j, 0))$;
6   $C \leftarrow 0$; $rmin \leftarrow R[0]$;
7   **for** $d' \leftarrow 1$ **to** $d - 1$ **do**
8   **if** $rmin > R[d']$ *and* $L[d'] < L[d' + 1]$ **then**
9        $C \leftarrow C + 1$; $rmin \leftarrow R[d']$;
10 **return** C ;

Consequently, we obtain the main result of this section:

Theorem 2. *Given a VLDC pattern P of length m and an SLP $\mathcal{T}$ of size n representing text T, the number of minimal occurrences of P in T can be calculated in $O(nm)$ time.*

Window VLDC Pattern Matching. This algorithm for VLDC patterns can be also extended to window-accumulated problems by adding the condition "$R[d'] + L[d'] \leq w$".

5   Conclusion

All algorithms we presented in this paper run in $O(nm)$ time and space. A natural open problem is if this can be reduced further. Other open problems are mixing variable and fixed length don't care symbols, and constraining the minimum and maximum lengths of strings that variable-length don't care symbols can match.

Acknowledgments

This work was supported by KAKENHI 22680014.

References

1. Baeza-Yates, R.A.: Searching subsequences. Theoretical Computer Science 78(2), 363–376 (1991)
2. Baturo, P., Rytter, W.: Compressed string-matching in standard sturmian words. Theoretical Computer Science 410(30–32), 2804–2810 (2009)
3. Cégielski, P., Guessarian, I., Lifshits, Y., Matiyasevich, Y.: Window subsequence problems for compressed texts. In: Grigoriev, D., Harrison, J., Hirsch, E.A. (eds.) CSR 2006. LNCS, vol. 3967, pp. 127–136. Springer, Heidelberg (2006)

4. Claude, F., Navarro, G.: Self-indexed text compression using straight-line programs. In: Královič, R., Niwiński, D. (eds.) MFCS 2009. LNCS, vol. 5734, pp. 235–246. Springer, Heidelberg (2009)
5. Hermelin, D., Landau, G.M., Landau, S., Weimann, O.: A unified algorithm for accelerating edit-distance computation via text-compression. In: Proc. STACS 2009, pp. 529–540 (2009)
6. Karpinski, M., Rytter, W., Shinohara, A.: An efficient pattern-matching algorithm for strings with short descriptions. Nordic Journal of Computing 4, 172–186 (1997)
7. Knuth, D.E., Morris, J.H., Pratt, V.R.: Fast pattern matching in strings. SIAM J. Comput. 6(2), 323–350 (1977)
8. Larsson, N.J., Moffat, A.: Offline dictionary-based compression. In: Proc. Data Compression Conference 1999, pp. 296–305. IEEE Computer Society Press, Los Alamitos (1999)
9. Lifshits, Y., Lohrey, M.: Querying and embedding compressed texts. In: Královič, R., Urzyczyn, P. (eds.) MFCS 2006. LNCS, vol. 4162, pp. 681–692. Springer, Heidelberg (2006)
10. Mannila, H., Toivonen, H., Verkamo, A.I.: Discovery of frequent episodes in event sequences. Data Mining and Knowledge Discovery 1(3), 259–289 (1997)
11. Miyazaki, M., Shinohara, A., Takeda, M.: An improved pattern matching algorithm for strings in terms of straight-line programs. In: CPM 1997. LNCS, vol. 1264, pp. 1–11. Springer, Heidelberg (1997)
12. Nevill-Manning, C.G., Witten, I.H., Maulsby, D.L.: Compression by induction of hierarchical grammars. In: Data Compression Conference 1994, pp. 244–253. IEEE Computer Society Press, Los Alamitos (1994)
13. Rytter, W.: Grammar compression, LZ-encodings, and string algorithms with implicit input. In: Díaz, J., Karhumäki, J., Lepistö, A., Sannella, D. (eds.) ICALP 2004. LNCS, vol. 3142, pp. 15–27. Springer, Heidelberg (2004)
14. Tiskin, A.: Faster subsequence recognition in compressed strings. J. Math. Sci. 158(5), 759–769 (2009)
15. Tiskin, A.: Towards approximate matching in compressed strings: Local subsequence recognition. In: Proc. CSR 2011 (to appear, 2011)
16. Ziv, J., Lempel, A.: A universal algorithm for sequential data compression. IEEE Transactions on Information Theory IT-23(3), 337–349 (1977)
17. Ziv, J., Lempel, A.: Compression of individual sequences via variable-length coding. IEEE Transactions on Information Theory 24(5), 530–536 (1978)

A Combinatorial Model of Phyllotaxis Perturbations in *Arabidopsis thaliana*

Yassin Refahi[1], Etienne Farcot[1,*], Yann Guédon[1], Fabrice Besnard[2],
Teva Vernoux[2], and Christophe Godin[1]

[1] CIRAD/INRA/INRIA, Virtual Plants INRIA team, UMR AGAP,
TA A-108/02, 34398 Montpellier Cedex 5, France
[2] RDP, ENS/CNRS/INRA/Univ. Lyon, 46,
allée d'Italie 69364 LYON cedex 07, France
`etienne.farcot@inria.fr`

Abstract. Phyllotaxis is the geometric arrangement of organs in plants, and is known to be highly regular. However, experimental data (from *Arabidopsis thaliana*) show that this regularity is in fact subject to specific patterns of permutations. In this paper we introduce a model for these patterns, as well as algorithms designed to identify these patterns in noisy experimental data. These algorithms thus incorporate a denoising step which is based on Gaussian-like distributions for circular data for which a common dispersion parameter has been previously estimated. The application of the proposed algorithms allows us to confirm the plausibility of the proposed model, and to characterize the patterns observed in a specific mutant. The algorithms are available in the OpenAlea software platform for plant modelling [10].

1 Introduction

Vascular plants produce new organs at the tip of the stem in a highly organised fashion. This patterning process occurs in small groups of stem cells, the so-called shoot apical meristem (SAM), and generates regular patterns called phyllotaxis [6]. The phyllotaxis of the model plant *Arabidopsis thaliana* follows a spiral, where single organs are initiated successively at an approximately constant divergence angle from the previous organ. The most frequent angle found in nature is the golden angle, close to 137.5°, and leads to the so-called Fibonacci phyllotaxis.

The geometric regularity of this phenomenon has impelled scientists to use mathematical approaches since early studies, two centuries ago. However a complete understanding of the biological processes that drive phyllotaxis is still far from complete. Most models are mechanistic, and allow for an explanation of the occurrence of a limited number of theoretical divergence angles (including 137.5°), as well as constrained transitions between successive angles in a given plant, see e.g. [1,3]. One leading principle of these models is based on the SAM functioning, where the appearance of new organs – called *primordia* – is supposed

---

* Corresponding author.

R. Giancarlo and G. Manzini (Eds.): CPM 2011, LNCS 6661, pp. 323–335, 2011.
© Springer-Verlag Berlin Heidelberg 2011

to be precluded both in the center of the SAM and in the vicinity of previously formed primordia. This is explained in terms of an inhibitory field surrounding existing primordia.

In this paper, we are interested in the variations of angles between consecutive organs in real plants. These angles may be subject to noise and perturbations. Only few studies have been devoted to this problem. For instance, statistical tests have been proposed to distinguish between random and regular phyllotactic patterns, or combinations thereof [4,5]. Perturbations in the phyllotactic patterns have also been observed in a study about transitions between different phyllotactic modes in real plants [2]. It was suggested that these perturbations might result from permutations in the order of appearance of organs along the phyllotactic spiral.

In this paper, we build up further on this initial idea. We first considered both reference (wild-type) plants with spiral phyllotaxis (model plant *Arabidopsis thaliana*) and mutant plants that were markedly perturbed in their phyllotaxy. We developed a combinatorial model for the type of perturbations observed in spiral phyllotaxis. Uncertainty is taken into account by assuming that each measured angle can correspond to several theoretical angles among those predicted by the model. Algorithms are proposed to detect such patterns in sequences of angles, and generate all candidate sequences from noisy data. For a given theoretical angle, the corresponding observed angles are modeled by a Gaussian-like distribution for circular data. For each candidate theoretical angle, the posterior probability of the measured angle is computed and compared to a threshold. This allows to reduce the set of candidate sequences.

2 Model Formulation

The exploratory analysis of our measured angles highlighted two characteristics of the measured divergence angle sequences. For a given plant, let α denote the canonical divergence angle:

- The measured divergence angles covered almost all the possible values (between 0 and 360°) with highest frequencies around the canonical Fibonacci angle of 137.5°. At least four classes of divergence angles were apparent but they were not unambiguously separated.
- Short segments (i.e. sub-sequences) of non-canonical divergence angles were identified along measured sequences and were more frequent in the mutant. They seemed to follow constrained patterns, or motifs.

In particular, a motif corresponding approximately to $(2\alpha, -\alpha, 2\alpha)$ was frequently observed in wild-type and even more often in mutants (see Figure 1). This motif, which was already observed in [2], can be simply explained by a permutation of two consecutive organs on the stem,without changing their angular positions. This led us to hypothesize that the segments of non-canonical angles could be explained by permutations involving 2 or even 3 organs (the most realistic numbers given the structure of the SAM). Let us now formulate this idea in more precise terms.

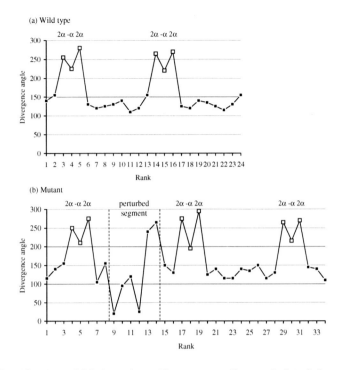

Fig. 1. Identification of M-shaped motifs corresponding to isolated 2-permutations. The perturbed segments cannot be easily explained on the mutant individual.

An ideal sequence would simply be a repetition of the canonical angle, of the form $(\alpha, \alpha, ..., \alpha)$. Since we assume that permutations occur, all terms in a sequence $S = (\mu_1, ..., \mu_\ell)$ of divergence angle will in fact verify $\mu_j \in \alpha\mathbb{Z}^* = \{i\alpha \mid i \in \mathbb{Z}, i \neq 0\}$. We define the corresponding absolute angles as follows:

$$v_0 = 0, \quad v_i = \sum_{j=1}^{i} \mu_j, \quad V(S) = (v_0, ..., v_\ell), \quad i \in \{0, ..., \ell\}. \tag{1}$$

From V we define a series containing the order of appearance of organs if the first is 0. We name it *order index series* of S, denoted $U(S)$, or simply $U = (u_0, u_1, ..., u_\ell)$ when S is clear from the context:

$$u_i = \frac{1}{\alpha}(v_i - v_J), \quad 0 \leq i \leq \ell \quad \text{where} \quad J = \arg\min_{j \in \{0 \cdots \ell\}} v_j. \tag{2}$$

From the definition it is clear that if $J > 0$ then $v_J < 0$ since $v_0 = 0$. This may occur when the sequence starts with permuted angles, a fact related to the left truncation of observed sequence with respect to complete sequences. If the sequence S follows a spiral phyllotaxis we have

$$u_i = i, \quad \forall i \in \{0, ..., \ell\}.$$

We call the sequence S n-admissible if a finite number of permutations, applied to disjoint blocks of at most n successive organs, results in an ordered sequence.

Definition 1. *A sequence* $S = (\mu_1, ..., \mu_\ell) \in (\alpha\mathbb{Z}^*)^\ell$ *is* n-admissible, *for some* $n \in \{1, ..., \ell\}$, *if and only if its associated order index series* U *satisfies:*
$\forall i \in \{1, 2, ..., \ell\}$, $u_i \neq i \Rightarrow \exists j, k \in \{0, 1, ..., \ell\}$, $j \leq i \leq k$, $k - j + 1 \leq n$, $(u_j, ..., u_k)$ *is a permutation of* $(j, ..., k)$, *i.e. their underlying sets are equal:* $\{u_j, ..., u_k\} = \{j, ..., k\}$.
Such a $(u_j, ..., u_k)$, *of length in* $\{1, ..., n\}$, *is called a shuffled block.*

Property 1. If S is n-admissible, then U is a permutation of $(0, ..., \ell)$. In general, the converse holds only for a certain $n \in \{2, ..., \ell\}$.
U is a permutation of $(0, ..., \ell) \iff \exists n \in \{1, ..., \ell\}$ s.t. S is n-admissible.

Example 1. The sequence $S = (-\alpha, 2\alpha, 3\alpha, -\alpha, -\alpha, 3\alpha)$ is 3-admissible, but not 2-admissible. Indeed, its absolute angles are $V = (0, -\alpha, \alpha, 4\alpha, 3\alpha, 2\alpha, 5\alpha)$. Hence $v_J = v_1 = -\alpha$, and $U = (1, 0, 2, 5, 4, 3, 6)$. Then, $(u_0, u_1) = (1, 0)$ and $(u_3, u_4, u_5) = (5, 4, 3)$ are shuffled blocks of length at most 3, and suffice to reconstruct the canonical sequence $(0, 1, 2, 3, 4, 5, 6)$.

For n-admissible sequences, the μ_i only belong to a finite subset of $\alpha\mathbb{Z}$:

Property 2. The divergence angles of an n-admissible sequence, take values in:

$$D_n = \{i\alpha \,|\, (1 - n) \leq i \leq (2n - 1), \ i \neq 0\}.$$

Proof. Let μ_i be a divergence angle in an n-admissible sequence, , we know from Eq. (1)-(2) that $\mu_i = (u_i - u_{i-1})\alpha$. In other words, note that, up to the multiplicative constant α, S is the first-order differenced sequence of U. There are four possible cases for u_{i-1} and u_i:

1. Neither u_{i-1} nor u_i are in any shuffled block, so $u_{i-1} = (i - 1)$, $u_i = i$ and $\mu_i = \alpha$.
2. u_{i-1} is in a shuffled block but u_i is not in a shuffled block, so $u_i = i$, and $u_{i-1} \in \{(i - n), ..., (i - 2)\}$ then $\mu_i \in \{2\alpha, ..., n\alpha\}$.
3. u_{i-1} is not in a shuffled block but u_i is in a shuffled block, so $u_{i-1} = (i - 1)$ and $u_i \in \{i + 1, ..., (i + n - 1)\}$ then $\mu_i \in \{2\alpha, ..., n\alpha\}$.
4. Both u_{i-1} and u_i are in a shuffled block.
   - u_{i-1} and u_i are in the same shuffled block so $u_i, u_{i-1} \in \{j, ..., (n+j-1)\}$, for some $j < i - 1$. Hence $\mu_i \in \{(1 - n)\alpha, ..., (n - 1)\alpha\} \setminus \{0\}$.
   - u_{i-1} and u_i are in two different but chained shuffled blocks so $u_{i-1} \in \{(i-n), ..., (i-2)\}$, $u_i \in \{i+1, ..., (i+n-1)\}$ then $\mu_i \in \{3\alpha, ..., (2n-1)\alpha\}$.

   $\square$

In general, the concatenation of two n-admissible sequences is not n-admissible. However this can be true after translating only the first angle of the second sequence. As we show now after two preliminary observations.

Proposition 1. *Let* $S = (\mu_1, ..., \mu_\ell)$ *be* n-admissible *and* $V(S)$ *and* $U(S)$ *be the sequence of absolute angles and the order index series respectively. Then* $J < n$ *and* $0 \leq -v_J < n\alpha$, *where* J *is defined as in (1).*

Proof. First we prove that $J < n$. By construction of order index series we know that $u_J = (v_J - v_J)/\alpha = 0$.

If $J = 0$ then $J < n$. We suppose that $J \neq 0$, therefore $u_J \neq J$ and by Definition 1, $\exists j, k \in \{0, 1, ..., \ell\}$, $j \leq J \leq k$, $k - j + 1 \leq n$, $u_J = 0 \in \{u_j...u_k\} = \{j...k\}$. Thus $j = 0$, and $k - j + 1 = k + 1 \leq n$. Hence $J \leq k < n$.

Now we prove $0 \leq -v_J < n\alpha$. Since $u_0 = -v_J/\alpha \geq 0$ by definition, this amounts to $0 \leq u_0 < n$, in which only the second part remains to be proved. It holds obviously for $u_0 = 0$. Otherwise, u_0 is part of a shuffled block $\{u_j...u_k\} = \{j...k\}$, where $j = 0$, and $k < n$, whence $u_0 < n$.   □

The order index series of two concatenated sequences does not always begin with the order index series of the first sequence. However, this is true if the first sequence is long enough.

Proposition 2. *Let $S = (\mu_1, ..., \mu_i)$, $P = (\mu_{i+1}, ..., \mu_{i+k})$. If $i \geq n$, then $U(S)$ is subsequence of $U(S \cdot P)$ where $S \cdot P = (\mu_1, ..., \mu_{i+k})$ denotes the concatenation of S and P.*

Proof. $V(S)$ is subsequence of $V(S \cdot P)$. Therefore $U(S)$ is a subsequence of $U(S \cdot P)$ iff $v_J = v_{J'}$ where J and J' are defined as in Eq. (2) for S and $S \cdot P$ respectively. Since $J < n, J' < n$ from Proposition 1, the minimal element of V and V' appears among their first $n - 1$ elements, which they share if $i \geq n$.   □

Proposition 3. *Let $S = (\mu_1, ..., \mu_i)$, $i \geq n$, and $P = (\mu_{i+1}, ..., \mu_{i+k})$. Let $U(S) = (u_0, ..., u_i)$ and $U(P) = (u'_0, u'_1, ..., u'_k)$ be the order index series of S and P respectively. Suppose that S is n-admissible.*
Then, the concatenated sequence $S \cdot P = (\mu_1, ..., \mu_{i+k})$ is n-admissible iff

$$P|_{u_i} \doteq \left(\mu_{i+1} + (u_i - i)\alpha, \; \mu_{i+2}, ..., \mu_{i+k} \right) \quad \text{is } n\text{-admissible and } u'_0 = 0. \quad (3)$$

Proof. We use again the identity $\mu_i = \alpha(u_i - u_{i-1})$.

Let $U = U(S \cdot P) = (u_0, ..., u_{i+k})$ denote the order index series $S \cdot P$. Since from Proposition 2 we know that $U(S)$ is a subsequence of $U(S \cdot P)$, we can easily show that $U(S \cdot P) = U(S) \cdot (u'_0 + i, u'_1 + i, ..., u'_k + i)$. From Property 1 we know that $\{u_0, ...u_i\} = \{0, ..., i\}$. Since S is n-admissible and $u'_0 = 0$ then it is clear that $S \cdot P$ is n-admissible if and only if $(u_{i+1}, ..., u_{i+k})$ is a permutation of $(i + 1, ..., i + k)$ that can be decomposed into disjoint shuffled blocks of length $\leq n$, or equivalently for $(u_{i+1} - i, ..., u_{i+k} - i)$ and $(1, ..., k)$. In other words, $S \cdot P$ is n-admissible iff $(u_{i+1} - i, ..., u_{i+k} - i)$ is the order index series of an n-admissible sequence. From the initial remark, the divergence angle sequence leading to this order index series can be written as

$$\alpha \left(u_{i+1} - i, (u_{i+2} - i) - (u_{i+1} - i), ..., (u_{i+k} - i) - (u_{i+k-1} - i) \right)$$
$$= \alpha \left(u_{i+1} - i, u_{i+2} - u_{i+1}, ..., u_{i+k} - u_{i+k-1} \right),$$

where the multiplication by α is applied to each component. Then, the same remark again shows that this sequence is exactly $P|_{u_i}$.   □

It will be useful in the last section to scan sequences backwards. One shall then rely on reversibility of the n-admissible property:

Property 3. Let $S = (\mu_1, ..., \mu_\ell)$ be a sequence of divergence angles. S in n-admissible iff the reversed sequence $S' = (\mu_\ell, ..., \mu_1)$ is n-admissible.

Proof. Let $U(S) = (u_0, u_1, ..., u_\ell)$ be the order index series of S, we know that $S = ((u_1 - u_0)\alpha, ..., (u_\ell - u_{\ell-1})\alpha)$ and $S' = ((u_\ell - u_{\ell-1})\alpha, ..., (u_1 - u_0)\alpha)$. Moreover, $V(S)$ and $V(S')$ obviously have the same minimum, say v_J. Then,

$$U(S') = (-v_J, (u_\ell - u_{\ell-1}) - v_J, (u_\ell - u_{\ell-2}) - v_J, ..., (u_\ell - u_0) - v_J)$$
$$= u_\ell - v_J - (U(S))',$$

where $(U(S))' = (u_\ell, ..., u_0)$ is the reversed order index sequence of S. It is clear that the latter can be decomposed into shuffled blocks of length $\le n$ iff $U(S)$ itself can. Since $U(S')$ is seen above to be a translation of this reversed sequence it also shares this property. $\square$

Property 2 defines the theoretical angles that may occur in an n-admissible sequence, but the measured angles are never exactly in D_n, and could correspond to two or more of these theoretical angles. This may lead to several n-admissible sequences. This will later be stored as a suffix tree.

Definition 2. *A labelled tree $T = (V, E, L)$, where $L : V \to D_n$, is called an n-admissible tree if all leaves have a common depth $\ell \in \mathbb{N}$, and every path from the root to a leaf is labelled by an n-admissible sequence.*

Let Γ be a mapping that for each measured angle proposes candidate theoretical angles among those in D_n

$$\Gamma : [0, 360°) \longrightarrow 2^{D_n}$$
$$x_i \longmapsto C_i = \{\mu_{i1}, \mu_{i2}, , ..., \mu_{ik}\} \subset D_n \tag{4}$$

We also consider a function $\omega : [0, 360°) \times D_n \to [0, 1]$ that returns a confidence level $\omega(x_i, \mu_q)$ – typically a probability – for each (x_i, μ_q).

3   Detecting n-Admissibility in Noisy Sequences

3.1   Problems

Given the Γ function above, a set of measured angles will generate a possibly high number of candidate sequences.

Problem 1. Let $x = (x_1, ..., x_\ell) \in [0, 360°)^\ell$ be measured angles, and $C = \prod_{i=1}^\ell C_i \subset D_n^\ell$ where $C_i = \Gamma(x_i)$. The task is to find all n-admissible $S = (\mu_1, \mu_2, ..., \mu_\ell)$ in C.

In order to deal with this problem we first need to know whether a given sequence of divergence angles is n-admissible.

Problem 2. Given a sequence S of divergence angles, the task is to determine whether S is n-admissible.

The following is a straightforward observation which is a special case of lemma 10.3 in [8]. It will be used to recognise n-admissible sequences.

Lemma 1. *Let U be a permutation of $\{1, ..., \ell\}$. Then for all $1 \leq i < j \leq \ell$, $\{u_j, ..., u_k\} = \{j, ..., k\}$ iff $\min\{u_j, ..., u_k\} = j$ and $\max\{u_j, ..., u_k\} = k$.*

3.2 Algorithms

First we propose an algorithm to solve Problem 2. In order to use Lemma 1, it first checks whether $\{u_0, ..., u_\ell\} \neq \{0, 1, ..., \ell\}$ in linear time, using a Parikh mapping [8]. Then, it determines whether an input sequence is n-admissible by a single scan of the sequence from left to right. The time complexity of the algorithm is $O(\ell)$ where ℓ is length of the input sequence.

---

n-admissible algorithm:
input: n, S # *S: a sequence of length ℓ a priori composed of theoretical divergence angles*
output: Boolean (true or false)
Begin
if S_0 not in D_n:
    return False # *since $(\mu_{i+1} + (u_i - i)\alpha$ in $P|_{u_i}$ in proposition 3 could be not in D_n*
Construct the order index series $U(S)$
if $\{u_0, ..., u_\ell\} \neq \{0, 1, ..., \ell\}$:   return False # *using Parikh mapping*
i:=0
while $i \leq \ell$:
    if $u_i \neq i$:   lo:=u_i; up:=u_i; j:=i+1
        while true:
            if $j > \ell$:   return false
            if j - i > n - 1: return false
            lo:=min(u_j, lo); up:=max(u_j, up)
            if (lo = i) & (up = j): i:=j+1 ; **break** # *$(u_i, ..., u_j)$ shuffled block*
            j:=j+1
    else: i:=i+1
return true # *the order index series $U(S)$ can also be returned if needed*
End

Remark 1. The notion of shuffled block can be seen as a special case of interval [8], Ch. 10. However, because it is much more specific, existing interval extraction algorithms would return invalid subsequences, whence the need for a new algorithm as above.

Now we can deal with Problem 1. A naive algorithm would construct all candidate sequences $S = \mu_1, \mu_2, ..., \mu_\ell$, and then apply the **n-admissible algorithm**. The number of candidate sequences equals $\prod_{i=1}^{\ell} |C_i|$, where the $|C_i|$ is the cardinality of C_i. Since C_i are typically not singletons, $|C|$ increases exponentially with ℓ. Therefore we propose a lookahead algorithm to explore the search space avoiding non necessary paths, as sketched below. The source code is available for more details [10].

---

n-admissible tree algorithm:
input: Γ, n, $(x_1, ..., x_\ell)$ # *a sequence of measured divergence angles of length ℓ*
output: n-admissible tree
# *with nodes labelled by both divergence angles μ_i and order indices u_i*
Begin
$T := \{root\}$; $\mu(root) := 0$; $u(root) := 0$
find:= false
while True:
    nLeaves:= *nonterminal_leaves*(T) # *leaves of depth $< \ell$*
    if nLeaves is empty: return T
    for *leaf* in nLeaves:
        $d :=$ depth($leaf$) ; $m :=$ min(n, $\ell - d$);
        for $k \in \{1, ..., m\}$:
            for $P \in \Gamma(x_{d+1}) \times ... \times \Gamma(x_{d+k})$:
                Compute $P|_{u(leaf)}$ # *cf. (3), Proposition 3*
                if n-admissible($P|_{u(leaf)}$): # *also returns μ and u for nodes in P*
                    append all nodes on P to *leaf*
End

Thanks to the use of Proposition 3, n-admissibility can be tested on subsequences only. The time complexity of the n-admissible tree algorithm increases exponentially with the lookahead limit n. When the returned tree contains only one n-admissible sequence (as was generally the case in practice), it is more precisely $O(l \times (k)^n)$ where $k = max(|C_i|)$.

Proposition 4. *Let us call $\mathcal{A}_n(C)$ the set of n-admissible sequences in C, and $\pi(T)$ the set of all (labels of) paths in T, from the root to the leaves. Then $\pi(T) = \mathcal{A}_n(C)$, i.e. the tree built in the* **n-admissible tree** *algorithm contains exactly the n-admissible sequences in C.*

Proof. The inclusion $\pi(T) \subset \mathcal{A}_n(C)$ is clear, since in the 2nd for loop only paths which are n-admissible can be added, thanks to Proposition 3.
To see that the converse holds, it suffices to remark that given an n-admissible sequence $(\mu_1, ..., \mu_\ell)$, one of the n subsequences

$$(\mu_1, ..., \mu_{\ell-1}), \quad (\mu_1, ..., \mu_{\ell-2}) \quad \cdots \quad (\mu_1, ..., \mu_{\ell-n})$$

must be n-admissible as well, as follows from the definition. Because all these subsequences are tested in the for loop, there cannot be an n-admissible sequence that is not detected by the algorithm, and thus $\pi(T) \supset \mathcal{A}_n(C)$. $\square$

Further Pruning.
In the case where $\mathcal{A}_n(C)$ is large, one may use the weights $\omega(x_i, \mu_j)$ to sort the sequences according to a confidence level. Actually, to a given path with labels $\mu_1, ..., \mu_\ell$ in the computed n-admissible tree, one may naturally assign the weight $\prod_j \omega(x_j, \mu_j)$. Then, all paths in the tree can be ordered according to their weight.

These weights can also be used to limit the size of the constructed tree, by ruling out all candidate paths whose weight is below a certain threshold. Since the weight of each node is lower than 1, the weight of a path can only be lower than that of any of its subpaths. Hence, it is possible in the `for` loop of the algorithm to prune not only the non admissible paths, but also those having a weight below a threshold. This is how we have actually implemented the algorithm, using posterior probabilities for weights, and adding a threshold as an input to the algorithm, as explained in the next section.

4   Results

4.1   Assignment of Measured Angles to Theoretical Angles

We have used the proposed algorithms to analyse the sequences of measured angles. The Γ function was parametrised using a statistical model. In a first step, a hidden Markov chain was estimated on the basis of the pooled wild-type + mutant measured divergence angle sequences in order to estimate an angle measurement uncertainty parameter; see more details in [9]. In this hidden Markov chain, the states of the non-observable Markov chain represents "theoretical" divergence angles while the von Mises observation distributions attached to each state of the non-observable Markov chain represents measurement uncertainty. The von Mises distribution [7], also known as the circular Gaussian distribution, is a univariate Gaussian-like periodic distribution for a variable $x \in [0, 360°)$. Let $g(x; \mu_q, \kappa)$ denote the probability density function of the von Mises distribution, with parameters μ_q (mean direction) and κ (concentration parameter). The main output of this first step of analysis was the estimated common concentration parameter (inverse variance) κ. This parameter corresponds to a standard deviation of approximately $\sigma = 18°$ (for our set of measured angles). Using this standard deviation, the Γ mapping (4) for our data is defined as follows:

$$\Gamma(x_i) = \{\mu_q \in D_n \mid \mu_q - \rho\sigma \leq x_i \leq \mu_q + \rho\sigma\}$$

The intervals defined by parameters ρ and σ correspond typically to a cumulative probability of 0.9975 with respect to the angle distribution centered at μ_q. For each possible theoretical angle of index q, the posterior probability

$$\omega(x_i, \gamma) = \frac{g(x_i; \mu_q, \kappa)}{\sum_{\mu_r \in D_n} g(x_i; \mu_r, \kappa)}$$

was calculated and compared with a predefined threshold (typical value 0.05) to decide whether this angle should be kept or rejected for the labelling of the

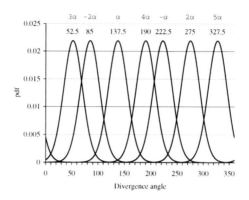

Fig. 2. The estimated von Mises distributions used for the Γ mapping

sequence in the `n-admissible tree` algorithm. The implicit underlying hypothesis was that the theoretical angles were *a priori* equally probable.

4.2   Interpretation on Our Dataset

We applied the modeling approach to our data set, see Figure 3 for an example where the predicted divergence angle sequence is in close agreement with the measured divergence angle sequence. For some sequences however, there was no n-admissible sequence in C as defined in Problem 1. This was often due to either some error in the measurement of divergence angles, or to too large deviations between an observed angle and any predicted angle corresponding to a valid prediction. A single measured angle x_i is non-explained, if no theoretical angle in $\Gamma(x_i)$ leads to a non-empty output of the `n-admissible tree` algorithm. Due

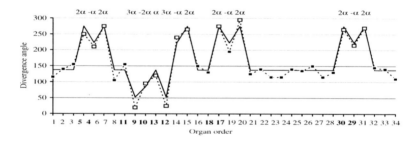

Fig. 3. Mutant individual: prediction of the of divergence angle sequence (continuous line) and labelling of the divergence angles within the permuted segments

to the dependencies induced by the permutation patterns, the angle at which the algorithm fails may in fact result from an isolated error, earlier in the sequence. Hence, all shuffled blocks preceding a non-explained angle should be marked as not valid. To achieve this goal, we define splitting points. The notion of splitting point can be viewed as a deterministic analogue of a regeneration point for a stochastic process. A regeneration point is a time instant at which the future of the process depends only of its state at that instant and is thus independent of its past before that instant. The process is thus reborn at a regeneration point.

Definition 3. *Let S be a sequence of divergence angles and $U(S)$ be the order index series of S, u_i is a splitting point iff $u_i = i$ and u_i is not in a shuffled block.*

Using the notion of splitting point, we implemented a procedure which was applied after the `n-admissible tree` algorithm. It consisted in a backtracking starting at the non-explained angle and progressing towards a splitting point. This allowed us to automatically invalidate blocks of angles preceding a value at which the algorithm failed.

To refine this analysis, we used reversibility (Property 3), and applied the whole procedure to both measured sequences and their reverse. Then, the intersection of angles invalidated on a sequence and its reverse, was often reduced to a single angle. Moreover these angles were likely due to measurement errors, typically the omission of one angle in the series, leading to an isolated 2α in a sequence of canonical angles α, see Figure 4. An expert investigation of these automatically detected subsequences enabled us to find with increased accuracy those angles which were not explained by our model. The proposed modeling

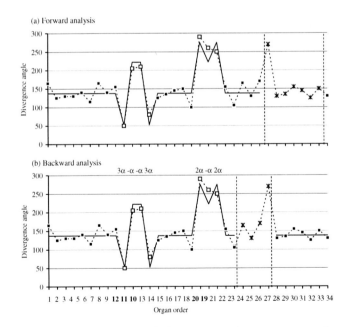

Fig. 4. Analysis of the sequence in both directions to detect segments that are invalid with respect to the permutation assumption. The invalid segments are delimited by dashed lines. The continuous line corresponds to the predicted divergence angles.

approach allows to explain a very large proportion of the non-canonical divergence angles despite the relatively high standard deviation (approx. 18°) of the estimated von Mises distributions. This indicates that the proposed model correctly describes the phyllotactic patterns of *Arabidopsis thaliana*. Wild-type plants were characterized by relatively frequent occurrences of 2-permutations generally isolated while mutants were characterized by the frequent occurrences of both 2- and 3-permutations whose succession generates highly complex motifs, see Figure 3 for an example. A summary of the results is shown in Table 1, with more precise counts of patterns in Table 2.

Table 1. Summary of the permutation patterns observed in both wild type and mutant plants

	Wild-type plant	Mutant
# sequences/# organs	82/2405	89/ 2815
% of non-canonical angles	15%	37%
% of unexplained angles	2%	5%
# individuals, Lucas phyllotaxis	2	2
# 2-permutations	123	297
# 3-permutations	3	53

The term Lucas phyllotaxis refers to a spiral phyllotaxis were the canonical divergence angle α is 99.5°. Although rarer than the Fibonacci spiral ($\alpha = 137.5°$), it is known to occur in nature, and was able to explain two wild-type and two mutant sequences, for which 137.5° failed.

Table 2. Permuted segments up to length 5. These segments are delimited by two splitting points. The divergence angle sequence is the first-order differenced organ sequence. By convention, the origin of the organ sequence is 0 (not indicated).

	organ order	divergence angles	wild-type	mutant
2-permutation	2 1 3	2 -1 2	90	193
	3 2 1 4	3 -1 -1 3	1	11
3-permutation	3 1 2 4	3 -2 1 2	1	9
	2 3 1 4	2 1 -2 3		13
total			2	33
2 2-permutations 2 1 4 3 5		2 -1 3 -1 2	16	32

References

1. Adler, I., Barabé, D., Jean, R.V.: A History of the Study of Phyllotaxis. Annals of Botany 80(3), 231–244 (1997)
2. Couder, Y.: Initial transitions, order and disorder in phyllotactic patterns: the ontogeny of Helianthus annuus. A case study. Acta Societatis Botanicorum Poloniae 67, 129–150 (1998)

3. Douady, S., Couder, Y.: Phyllotaxis as a self organizing process: I,II,III. Journal of Theoretical Biology 178, 255–312 (1996)
4. Jeune, B., Barabé, J.: Statistical Recognition of Random and Regular Phyllotactic Patterns. Annals of Botany 94, 913–917 (2004)
5. Jeune, B., Barabé, J.: A stochastic approach to phyllotactic pattern analysis. Journal of Theoretical Biology 238, 52–59 (2006)
6. Kuhlemeier, C.: Phyllotaxis. Trends in Plant Science 12, 143–150 (2007)
7. Mardia, K.V., Jupp, P.E.: Directional Statistics. John Wiley & Sons, Chichester (2000)
8. Parida, L.: Pattern Discovery in Bioinformatics: Theory & Algorithms. CRC Mathematical Computational Biology Series. Chapman & Hall, Sydney (2008)
9. Refahi, Y., Guédon, Y., Besnard, F., Farcot, E., Godin, C., Vernoux, T.: Analyzing perturbations in phyllotaxis of Arabidopsis thaliana. In: 6th International Workshop on Functional-Structural Plant Models, Davis, California, pp. 170–172 (2010)
10. http://openalea.gforge.inria.fr/doc/vplants/phyllotaxis_analysis/doc/_build/html/contents.html

Tractability and Approximability
of Maximal Strip Recovery

Laurent Bulteau[1], Guillaume Fertin[1], Minghui Jiang[2], and Irena Rusu[1]

[1] Laboratoire d'Informatique de Nantes-Atlantique (LINA), UMR CNRS 6241
Université de Nantes, 2 rue de la Houssinière, 44322 Nantes Cedex 3, France
{Laurent.Bulteau,Guillaume.Fertin,Irena.Rusu}@univ-nantes.fr
[2] Department of Computer Science, Utah State University, Logan, UT 84322, USA
mjiang@cc.usu.edu

Abstract. An essential task in comparative genomics is usually to decompose two or more genomes into synteny blocks, that is, segments of
chromosomes with similar contents. In this paper, we study the MAXI-
MAL STRIP RECOVERY problem (MSR) [Zheng et al. 07], which aims at
finding an optimal decomposition of a set of genomes into synteny blocks,
amidst possible noise and ambiguities. We present a panel of new or improved FPT and approximation algorithms for the MSR problem and its
variants. Our main results include the first FPT algorithm for the variant δ-gap-MSR-d, an FPT algorithm for CMSR-d and δ-gap-CMSR-d
running in time $O(2.360^k \operatorname{poly}(nd))$, where k is the number of markers or
genes considered as erroneous, and a $(d + 1.5)$-approximation algorithm
for CMSR-d and δ-gap-CMSR-d.

1 Introduction

An essential task in comparative genomics is usually to decompose two or more
genomes into synteny blocks, that is, segments of chromosomes with similar
contents. This task is non-trivial when the genomic maps contain noise and
ambiguities, which need to be removed before we can give a precise synteny
block decomposition. This is the objective of the MAXIMAL STRIP RECOVERY
problem (MSR) [10]: to delete a set of markers (genes) from the genomic maps
until the remaining markers can be partitioned into a set of strips (synteny
blocks) of maximum total length.

We review some definitions. A genome consists of one or more chromosomes;
each chromosome is a sequence of genes. Correspondingly, a *genomic map* consists of one or more sequences of gene markers. Each marker is a signed integer
representing a gene: the absolute value of the integer represents the family of the
gene; the sign of the integer represents the orientation. A marker has *duplicates*
if it is contained more than once in some genomic map, possibly in different orientations. A *strip* of $d \geq 2$ genomic maps is a sequence of *at least two* markers
appearing consecutively in each map, such that the order of the markers and the
orientation of each marker are either both preserved or both reversed. The *re-
versed opposite* of a sequence $s = \langle x_1, \ldots, x_h \rangle$ is $-s = \langle -x_h, \ldots, -x_1 \rangle$. The MSR
problem on d input maps is the following maximization problem MSR-d [2,10]:

R. Giancarlo and G. Manzini (Eds.): CPM 2011, LNCS 6661, pp. 336–349, 2011.

PROBLEM MSR-d

INPUT: d genomic maps $G_1, \ldots, G_d$ each containing n markers without duplicates.

SOLUTION: d subsequences $G'_1, \ldots, G'_d$ of $G_1, \ldots, G_d$ respectively, each containing the same ℓ markers, such that all the markers in $G'_1, \ldots, G'_d$ can be partitioned into strips.

PARAMETER: the number ℓ of selected markers.

The maximization problem MSR-d that maximizes the parameter ℓ, the number of selected markers, has a *complement* minimization problem called CMSR-d [9,8] that minimizes the parameter $k = n - \ell$, the number of deleted markers. For genomic maps of close species with few errors, k can be much smaller than ℓ, thus approximation and FPT algorithms are sometimes more relevant for CMSR than for MSR. We refer to Figure 1 for an example.

Given d subsequences $G'_1, \ldots, G'_d$ of d genomic maps $G_1, \ldots, G_d$, respectively, the *gap* between two consecutive markers a and b of G'_i is the number of markers appearing between a and b in G_i, a and b excluded. The *gap* of a strip s is the maximum gap between any two consecutive markers of s in any map G'_i. The deleted markers between markers of a strip correspond to noise and ambiguities, which occur infrequently. A synteny block is a segment of chromosomes that remain undisrupted by genome rearrangements during evolution. Consecutive elements of a synteny block can only be separated in a dataset due to noise and ambiguities. Thus a strip having a large gap is unlikely to correspond to a synteny block; see [3] for an empirical analysis. This leads to the following gap-constrained variant of MSR-d [1]:

PROBLEM δ-gap-MSR-d

INPUT: d genomic maps $G_1, \ldots, G_d$ each containing n markers without duplicates.

SOLUTION: d subsequences $G'_1, \ldots, G'_d$ of $G_1, \ldots, G_d$ respectively, each containing the same ℓ markers, such that all the markers in $G'_1, \ldots, G'_d$ can be partitioned into strips, and such that each strip has gap at most δ.

PARAMETER: the number ℓ of selected markers.

No doubt that MSR-d is a more elegant problem from a theoretical perspective, but δ-gap-MSR-d could be more relevant in biological applications. The gap-constrained variant of CMSR-d, denoted δ-gap-CMSR-d, can be similarly defined. Similarly to MSR-d and CMSR-d, the parameter for δ-gap-MSR-d is ℓ,

$$
\begin{array}{llllllllll}
G_1 = & 1 & 5 & -3 & 2 & 6 & 4 & 8 & 7 \\
G_2 = & 1 & 5 & -3 & -8 & 7 & -6 & 2 & 4 \\
G_3 = & -8 & 2 & 7 & -6 & -4 & 3 & -5 & -1
\end{array}
\qquad
\begin{array}{llllll}
G'_1 = & 1 & 5 & -3 & 6 & 8 \\
G'_2 = & 1 & 5 & -3 & -8 & -6 \\
G'_3 = & -8 & -6 & 3 & -5 & -1
\end{array}
$$

Fig. 1. Three genomic maps G_1, G_2, G_3, and an optimal solution G'_1, G'_2, G'_3 for MSR-3. The markers $2, 4, 7$ are deleted; the markers $1, 3, 5, 6, 8$ are selected in two strips $\langle 1, 5, -3 \rangle$ and $\langle 6, 8 \rangle$ of G'_1, G'_2, G'_3. The gap of the strip $\langle 1, 5, -3 \rangle$ is 0; the gap of the strip $\langle 6, 8 \rangle$ is 2, since there are 2 markers between -8 and -6 in G_3.

and the parameter for δ-gap-CMSR-d is k. In most cases, δ and d are assumed to be constants, although our FPT algorithm in Theorem 3 does not depend on this assumption and can take δ and d as parameters besides ℓ. There is no known direct reduction from δ-gap-MSR-d to MSR-d or vice versa. Although the gap constraint appears to be an additional burden that the algorithm has to take care of, it also limits the set of candidate strips and their intersection pattern, especially when δ is small, which may make the problem easier to handle.

For the four variants of the maximal strip recovery problem, MSR-d, CMSR-d, δ-gap-MSR-d, and δ-gap-CMSR-d, several hardness results have been obtained [2,9,6,1,7,8], and a variety of algorithms have been developed, including heuristics [10], approximation algorithms [2,1,5], and FPT algorithms [9,5]. For example, it is known that MSR-d admits a $2d$-approximation algorithm for any $d \geq 2$ [2,8], and that δ-gap-MSR-d admits a $2d$-approximation algorithm for any $d \geq 2$ and $\delta \geq 1$ and a 1.8-approximation algorithm for $d = 2$ and $\delta = 1$ [1]. Refer also to [11,5] for some very recent development on the CMSR problem parallel to our work. The following two theorems summarize some basic hardness results regarding these problems:

Theorem 1. [6,1,8] *MSR-d, CMSR-d, δ-gap-MSR-d, and δ-gap-CMSR-d are APX-hard for any $d \geq 2$ and $\delta \geq 2$, even if all markers appear in positive orientation in all genomic maps; 1-gap-MSR-d and 1-gap-CMSR-d are NP-hard for any $d \geq 2$.*

Theorem 2. [7] *MSR-d is W[1]-hard for any $d \geq 4$, even if all markers appear in positive orientation in all genomic maps.*

In this paper, we present a panel of new or improved FPT and approximation algorithms. Our positive results, together with some previous results, are summarized in Table 1. Due to space constraints, we present only three main results in this extended abstract. These results are (i) an FPT algorithm for δ-gap-MSR-d running in time $2^{O(d\delta\ell)}n$, (ii) an FPT algorithm for CMSR-d and δ-gap-CMSR-d, running in time $O(2.360^k\mathsf{poly}(nd))$, and (iii) a $(d+1.5)$-approximation algorithm for CMSR-d and δ-gap-CMSR-d.

Preliminaries. Given a set of genomic maps (either the set of d original maps given as input or some set of reduced maps during the execution of a recursive algorithm), if a maximal sequence of markers form a strip in these maps, then these markers are either all selected or all deleted in *any* optimal solution. This is because any solution that includes only a subset of the markers in a strip can be extended to a better solution that includes all markers in that strip. Hence, these markers can be treated as an atomic unit, and called a *super-marker*, whose *size* is the number of markers it contains. Note that the size of a super-marker is always at least 2. A marker that does not belong to any super-marker is a *single-marker*. We use the term *single-super-marker* to refer to either a single-marker or a super-marker. A common step of our algorithms is to partition the markers

Table 1. Positive results for variants of MSR

Problem	Best FPT algorithm (running time)
δ-gap-MSR-d	$O(2^t t d\delta^2 + n d\delta)$ [Theorem 3, Section 2] with $t = \ell(1 + \frac{3}{2}d\delta)$
CMSR-d	$O(2.360^k \mathsf{poly}(nd))$ [Theorem 4, Section 3]
δ-gap-CMSR-d ($\delta \geq 2$)	$O(2.360^k \mathsf{poly}(nd))$ [Theorem 4, Section 3]
1-gap-CMSR-d	$O(2^k \mathsf{poly}(nd))$ [See full version]

Problem	Best approximation ratio
MSR-d	$2d$ [2,8]
δ-gap-MSR-d ($\delta \geq 4$)	$2d$ [1]
1-gap-MSR-d ($d \geq 3$)	$0.75d + 0.75 + \epsilon$ [See full version]
1-gap-MSR-2	1.8 [1]
2-gap-MSR-d	$1.5d + \epsilon$ [See full version]
3-gap-MSR-d	$1.5d + 0.75 + \epsilon$ [See full version]
CMSR-d ($d \geq 3$)	$d + 1.5$ [Theorem 5, Section 4]
CMSR-2	3 [5]
δ-gap-CMSR-d	$d + 1.5$ [Theorem 5, Section 4]
1-gap-CMSR-2	2.778 [See full version]

into single-super-markers. If a set of genomic maps contains only super-markers, then we have a straightforward decomposition into strips, without deleting any marker.

For two markers or two single-super-markers u and v, denote by $\mathsf{gap}(u,v)$ the set of markers that appear between u and v in at least one of the maps; clearly $\mathsf{gap}(u,v) = \mathsf{gap}(v,u)$. We call v a *candidate successor* (resp. *candidate predecessor*) of u, and write $v \succ u$ (resp. $v \prec u$), if the following two conditions are satisfied: (1) $\langle u, v \rangle$ (resp. $\langle v, u \rangle$) is a strip (satisfying the δ-gap constraint, if necessary) in the reduced maps with all markers in $\mathsf{gap}(u,v)$ deleted, (2) $\langle u, x, v \rangle$ (resp. $\langle v, x, u \rangle$) cannot be a strip for any $x \in \mathsf{gap}(u,v)$. The relation $\prec$ always refers to two markers of the original map, even if we temporarily work with reduced maps. Note that v is a candidate successor of u if and only if u is a candidate predecessor of v. In the example of Figure 1, we have $6 \prec 8$, and $\mathsf{gap}(6,8) = \{2,4,7\}$. The following lemma gives some basic properties of the function gap:

Lemma 1. (a) *Let u, v, w be three markers or single-super-markers. If u and v are two candidate successors of w with $u \neq v$, then $u \in \mathsf{gap}(w,v)$ and $v \in \mathsf{gap}(w,u)$.* (b) *Let u and v be two single-super-markers. If $u \prec v$ or $u \succ v$, then $\mathsf{gap}(u,v) \neq \emptyset$.*

2   An FPT Algorithm for δ-gap-MSR-d

In this section, we present the first FPT algorithm for δ-gap-MSR-d with the parameter ℓ. Recall that without the gap constraint, MSR-d with the parameter ℓ is W[1]-hard for any $d \geq 4$. In sharp contrast to the W[1]-hardness of MSR-d, we obtain a somewhat surprising result that δ-gap-MSR-d is in FPT, where ℓ is the parameter, and δ and d are constants. In fact, our FPT algorithm for δ-gap-MSR-d works even if d and δ are not constants: δ-gap-MSR-d is in FPT even with three combined parameters d, δ and ℓ.

Theorem 3. *Algorithm 1 finds an optimal solution for δ-gap-MSR-d for any $d \geq 2$ and $\delta \geq 1$, in time $O(2^t t d \delta^2 + n d \delta)$, where $t = \ell(1 + \frac{3}{2} d\delta)$.*

Algorithm 1. FPT algorithm for δ-gap-MSR-d

1: Gather all pairs of markers (u, v) such that $u \prec v$. Such pairs are called *candidate pairs*.
2: For each marker u, create a boolean variable x_u.
3: For each candidate pair (u, v), create a conjunctive boolean formula $f_{u,v} = x_u \wedge x_v \wedge \neg x_{g_1} \wedge \ldots \wedge \neg x_{g_s}$, where $g_1, \ldots, g_s$ are the markers in $\mathsf{gap}(u, v)$.
4: Delete the variables that do not appear in any formula or appear only in negative form in the formulas.
5: Enumerate all possible assignments to the remaining variables to find an optimal assignment that maximizes the number of variables appearing in positive form in at least one satisfied formula. Delete all markers whose variables are not assigned true values.
6: Return the resulting genomic maps.

Our algorithm is based on a simple idea: create a boolean variable for each marker (where true means the marker is selected in a solution, false that it is unselected), then test all possible assignments to find an optimal solution. To reduce the time complexity of this brute-force approach, we add a pruning step (line 4) to delete certain variables whose markers cannot appear in any optimal solution. The remaining variables form a "kernel" on which we can find an optimal solution in FPT time.

Given an optimal solution, which selects ℓ markers, we call a marker *active* if it appears within distance at most δ from a selected marker in some map. Then each map contains at most $\ell\delta + \frac{\ell}{2}\delta$ unselected active markers: at most δ after each selected marker, and at most δ before the first marker of each strip (note that the number of strips of this optimal solution is at most $\ell/2$). The total number of active markers is at most $\ell + d(\ell\delta + \frac{\ell}{2}\delta) = \ell(1 + \frac{3}{2}d\delta)$.

The pruning step in line 4 depends on the crucial observation that a non-active marker can never appear in positive form. Suppose for contradiction that a non-active marker u appears in a candidate pair with some marker v. Then u is at distance at most $\delta + 1$ from v in each map. Since u, as a non-active marker, must be at distance at least $\delta + 1$ from the selected markers in all maps, no selected markers can appear between u and v in any map, thus we can extend the optimal solution by selecting both u and v, a contradtiction.

Note that in line 4 the variables appearing at least once in positive form are never deleted, hence no formula becomes empty after deleting the variables that appear only in negative form. After line 4, the number of remaining variables is at most the number of active markers, which is at most $t = \ell(1 + \frac{3}{2}d\delta)$. Correspondingly, the number of formulas is at most $t(\delta+1)$, because any candidate pair consists of an active marker and one of the $\delta + 1$ markers immediately following it in the first map. Each formula contains at most $d\delta + 2$ variables.

The time complexity of line 1 is $O(nd\delta)$. In lines 2 and 3, the variables can be created in time $O(n)$, and the formulas can be created in time $O(t(\delta + 1)(d\delta + 2)) = O(td\delta^2)$. Similarly, line 4 can be executed in time $O(n + td\delta^2)$. Finally, line 5 can be executed in time $O(2^t t(\delta + 1)(d\delta + 2)) = O(2^t td\delta^2)$, so the overall time complexity is $O(2^t td\delta^2 + nd\delta)$.

3   An FPT Algorithm for CMSR-d and δ-gap-CMSR-d

In this section, we design an FPT algorithm for CMSR-d and δ-gap-CMSR-d, where the parameter is k, the number of deleted markers in the optimal solution.

Since super-markers are already strips in the input genomic maps, one may naturally be tempted to come up with the following algorithm. First, find all super-markers, and add them to the solution. Then, delete a subset of single-markers until all markers in the resulting maps can be partitioned into strips. The correctness of this algorithm on finding an exact solution, however, depends on the assumption that in some optimal solution no super-marker needs to be deleted, which is false as can be seen in the following counter-example:

$$
\begin{array}{ccccccc}
G_1 = & 4 & 1 & 2 & 3 & 5 & 6 & 7 \\
G_2 = & 6 & -3 & -2 & -1 & 7 & 4 & 5
\end{array}
$$

Here $\langle 1, 2, 3 \rangle$ forms a super-marker, but the optimal solution deletes $\langle 1, 2, 3 \rangle$ and selects $\langle 4, 5 \rangle$ and $\langle 6, 7 \rangle$ instead. An easy generalization of this counter-example shows that any super-marker of size strictly less than $2d$ is not guaranteed to be always selected in some optimal solution.

We observe that an FPT algorithm for CMSR-d and δ-gap-CMSR-d can be easily obtained using the bounded search tree method. In any feasible solution for the two problems, a single-marker x must be either deleted or selected. If x is selected, then at least one of its neighbors must be deleted. Since x has at most $2d$ neighbors (at most two in each map), this leads to a very simple algorithm running in time $O((2d + 1)^k \mathsf{poly}(nd))$. Parallel to our work, Jiang et al. [5] presented an FPT algorithm running in time $O(3^k \mathsf{poly}(nd))$. We next describe a carefully tuned FPT algorithm running in time $O(2.360^k \mathsf{poly}(nd))$. For convenience, we consider the decision problem associated with CMSR-d and δ-gap-CMSR-d, for which the parameter k is part of the input.

Theorem 4. *Algorithm 2 finds an exact solution for the decision problems associated with CMSR-d and δ-gap-CMSR-d, for any $\delta \geq 1$ and $d \geq 2$, in time $O(c^k \mathsf{poly}(nd))$, where $c < 2.360$ is the unique real root of the equation $2c^{-1} + 2c^{-3} = 1$.*

It is interesting to note that although the two problems MSR-d and δ-gap-MSR-d have very different complexities when parameterized by ℓ, their complements CMSR-d and δ-gap-CMSR-d are both tractable when parameterized by k.

The efficiency of Algorithm 2 is made possible by several optimizations justified by the following lemmas. These lemmas are all based on very simple observations. Note that although we consider the decision problem for simplicity, Algorithm 2 can be adapted to directly return the actual solution, instead of "true", when the input instance indeed has a solution of size k. Recall that the relation $\prec$ in lines 14-15 is defined for markers in the original maps — it remains unchanged through recursive calls, and can be precomputed.

Lemma 2. *Let x be a single-marker and w a super-marker. If x is selected in an optimal solution, and w is a candidate successor or predecessor of x with exactly one marker in $\mathsf{gap}(x, w)$, then there is an optimal solution where the marker in $\mathsf{gap}(x, w)$ is deleted.*

Algorithm 2. FPT algorithm for δ-gap-CMSR-d and CMSR-d

Input: d genomic maps $G_1, \ldots, G_d$ each containing n markers without duplicates, and two parameters $k \in \mathbb{N}$, $\delta \in \mathbb{N} \cup \{\infty\}$

1: **return** recurse($G_1, \ldots, G_d, k, \delta$, false)

Function recurse($G_1, \ldots, G_d, k, \delta$, skip_step_2b): boolean

1: **if** $k < 0$ **then**
2:     **return** false
3: Partition the markers into single-super-markers.
4: **if** there exists at least one single-marker in G_1 **then**
5:     $x \leftarrow$ the left-most single-marker in G_1
6: **else**
7:     **return** true
8: $s \leftarrow$ the first single-super-marker following x in G_1
9:   // 1: Assume x is deleted in the optimal solution
10: Create $G'_1, \ldots, G'_d$ by removing x from $G_1, \ldots, G_d$.
11: **if** recurse($G'_1, \ldots, G'_d, k - 1, \delta$, false) **then**
12:     **return** true
13:   // 2: Assume x is part of a strip in the optimal solution
14: $Y \leftarrow \{$ single-super-marker $y \mid x \prec y \}$ // the set of candidate successors
15: $Z \leftarrow \{$ super-marker $z \mid z \prec x \}$ // the set of candidate predecessors
16: **if** $\exists w_0 \in Y \cup Z$ a super-marker s.t. (x, w_0) satisfies the conditions of Lemma 2 **then**
17:     Create $G'_1, \ldots, G'_d$ by removing all markers in $\mathsf{gap}(x, w_0)$ from $G_1, \ldots, G_d$.
18:     **return** recurse($G'_1, \ldots, G'_d, k - 1, \delta$, false)
19: **if** $\exists s_0$ a single-marker s.t. (x, s_0) satisfies the conditions of Lemma 3 **then**
20:     Create $G'_1, \ldots, G'_d$ by removing s_0 from $G_1, \ldots, G_d$.
21:     **return** recurse($G'_1, \ldots, G'_d, k - 1, \delta$, false)
22:   // 2.a: Assume x is not at the end of its strip
23: **if** $Y \neq \emptyset$ **then**
24:     **if** recurse_2a($Y, x, G_1, \ldots, G_d, k, \delta$) **then**
25:         **return** true
26:   // 2.b: Assume x is at the end of its strip
27: **if** $Z \neq \emptyset$ **and** skip_step_2b=false **then**
28:     **if** recurse_2b($Z, x, s, G_1, \ldots, G_d, k, \delta$) **then**
29:         **return** true
30: **return** false

Algorithm 2 (continued)
___
Function recurse_2a$(Y, x, G_1, \ldots, G_d, k, \delta)$: boolean

1: **if** $\exists y_0 \in Y$ s.t. y_0 satisfies the conditions of Lemma 4 **then**
2:    **if** $\delta \in \mathbb{N}$ **and** y_0 is a single-marker **then**
3:       Replace y_0 by the unspecified marker $[y_0 \mid Y]$.
4:       $Y_0 \leftarrow \{y_0\}$
5: **else**
6:    $Y_0 \leftarrow Y$
7: **for all** $y \in Y_0$ **do**
8:    Create $G'_1, \ldots, G'_d$ by removing all markers in $\mathsf{gap}(x, y)$ from $G_1, \ldots, G_d$.
9:    **if** recurse$(G'_1, \ldots, G'_d, k - |\mathsf{gap}(x, y)|, \delta, \text{false})$ **then**
10:       **return** true
11: **return** false

Function recurse_2b$(Z, x, s, G_1, \ldots, G_d, k, \delta)$: boolean

1: **if** $\exists z_0 \in Z$ s.t. z_0 satisfies the conditions of Lemma 5 **then**
2:    $Z_0 \leftarrow \{z_0\}$
3: **else**
4:    $Z_0 \leftarrow Z$
5: **for all** $z \in Z_0$ **do**
6:    **if** z ends with an unspecified marker $[y_0 \mid Y]$ **and** $\exists y_1 \in Y$ s.t. $y_1 \prec x$ **then**
7:       Replace the unspecified marker $[y_0 \mid Y]$ by y_1.
8:    Create $G'_1, \ldots, G'_d$ by removing all markers in $\mathsf{gap}(x, z)$ from $G_1, \ldots, G_d$.
9:    skip_next_step_2b $\leftarrow s$ exists **and** s is a single-marker **and** $s \notin \mathsf{gap}(x, z)$
10:    **if** recurse$(G'_1, \ldots, G'_d, k - |\mathsf{gap}(x, z)|, \delta, \text{skip\_next\_step\_2b})$ **then**
11:       **return** true
12: **return** false
___

Lemma 3. *Let x be a single-marker and s a single-super-marker. If s appears in $\mathsf{gap}(x, w)$ for each w that is a candidate successor or predecessor of x, then s itself cannot be a candidate successor or predecessor of x, and any solution selecting x deletes s.*

Lemma 4. *(In this lemma we assume there is no gap constraint.) Let x be a single-marker and y a candidate successor of x such that all markers in $\mathsf{gap}(x, y)$ are single-markers and candidate successors of x. If x is part of some strip in an optimal solution, but not at the end of this strip, then there is an optimal solution where $\langle x, y \rangle$ is part of some strip.*

Lemma 5. *Let x be the first single-marker in G'_1. Let z be a candidate predecessor of x such that all markers in $\mathsf{gap}(x, z)$ are size-2 super-markers and candidate predecessors of x. If x appears at the end of a strip in an optimal solution, then there is an optimal solution where $\langle z, x \rangle$ is at the end of some strip.*

In addition to these four optimizations, we also use a "delayed commitment" optimization which is the equivalent of Lemma 4 when we need to observe a gap constraint. We consider the case where x is part, but not at the end, of some strip in the optimal solution, and where y is a single-marker and a candidate successor of x such that all markers in $\mathsf{gap}(x, y)$ are single-markers and candidate successors of x. In this case we delete all markers in $\mathsf{gap}(x, y)$ to make $\langle x, y \rangle$ a strip, but keep the possibility of replacing y by any marker $y_1 \in \mathsf{gap}(x, y)$, should necessity arise. We denote this unspecified marker by $[y \mid \mathsf{gap}(x, y)]$.

To prove the correctness of Algorithm 2, we need the following easy lemma from [10]:

Lemma 6. [10, Proposition 2] *We can decompose the strips of any optimal solution in such a way that (1) each strip contains at most 3 single-super-markers and (2) each strip containing 3 single-super-markers starts and ends with a single-marker.*

Let OPT be any optimal solution, and let us decompose the strips of OPT as in the above lemma. We show by induction that the solution found by Algorithm 2 has the same size as OPT. Let x be the left-most single-marker in G_1, then exactly one of the following three cases is true:

    1: x is deleted in OPT,
    2.a: There exists a single-super-marker y such that $\langle x, y \rangle$ is part of a strip in OPT,
    2.b: There exists a super-marker z such that $\langle z, x \rangle$ is a strip in OPT.

Note that in case 2.b, z cannot be a single-marker since it is to the left of x in G_1. By our choice of x, case 2.a can be split into the following two subcases:

    2.a.i: There exists a single-super-marker y such that $\langle x, y \rangle$ is a strip in OPT,
    2.a.ii: There exists a single-super-marker y and a single-marker y' such that $\langle x, y, y' \rangle$ is a strip in OPT.

Refer to Algorithm 2. In case 1, a solution is found in lines 9–12 of the function recurse. In case 2, i.e. in the case where x is part of an optimal solution, if either Lemma 2 or Lemma 3 can be applied, then again a solution is found. Otherwise, we are in case 2.a or 2.b.

Suppose we are in case 2.a. If $y \in Y_0$, then the function recurse_2a tests a branch in which $\langle x, y \rangle$ becomes part of some strip. Otherwise, there exists some $y_0 \in Y$ satisfying the conditions of Lemma 4. If there is no gap constraint, y is replaced by y_0, which does not change the size of the solution. If there is a gap constraint, y is replaced by the unspecified marker $u = [y_0 \mid Y]$, and we look further in case 2.a.i or 2.a.ii.

In case 2.a.i, we can replace y by y_0 since $\mathsf{gap}(x, y_0)$ has no more markers than $\mathsf{gap}(x, y)$. In case 2.a.ii, we can replace y by any y_1 such that $x \prec y_1 \prec y'$, since $\mathsf{gap}(x, y) \cup \{y\} \cup \mathsf{gap}(y, y')$ is the same set as $\mathsf{gap}(x, y_1) \cup \{y_1\} \cup \mathsf{gap}(y_1, y')$. This is what happens in case 2.b of a subsequent recursive call in which y' becomes the left-most single-marker in G_1.

Suppose we are in case 2.b. If $z \in Z_0$, then the function recurse_2b tests a branch in which $\langle z, x \rangle$ becomes a strip. Otherwise, Lemma 5 can be applied, which leaves the size of the optimal solution unchanged. In line 9 of recurse_2b, if s becomes the left-most single-marker in G_1 in the next recursive call of recurse, it cannot be at the end of a strip because x is already at the end of a strip.

Algorithm 3. $(d + 1.5)$-approximation for δ-gap-CMSR-d and CMSR-d

1: $X \leftarrow \{$ triples of markers $(z, x, y) \mid z \prec y$ and $\mathsf{gap}(z, y) = \{x\} \}$
2: Partition the markers into single-super-markers.
3: **for all** $(z, x, y) \in X$ **do**
4:     **if** x, y and z are not deleted **and** y or z is a single-marker **then**
5:         Delete x.
6:         Re-create all super-markers.
7: Delete all remaining single-markers.
8: Return the resulting genomic maps.

This completes the correctness proof. An anonymous reviewer of an earlier version of this paper commented that perhaps some further properties of the optimal solution, besides those already described in our lemmas, might be used to improve the time complexity further. This may be true, but we believe that such improvement would require significantly different ideas.

4 An Approximation Algorithm for CMSR-d and δ-gap-CMSR-d

In this section, we present a $(d + 1.5)$-approximation algorithm for the two minimization problems CMSR-d and δ-gap-CMSR-d. Recall that $2d$-approximation algorithms [2,8,1] were known for the two maximization problems MSR-d and δ-gap-MSR-d.

Theorem 5. *Algorithm 3 finds a $(d + 1.5)$-approximation for* CMSR-d *and δ-gap-CMSR-d for any $d \geq 2$ and $\delta \geq 1$.*

Let k be the number of deleted markers in an optimal solution. Then the number of single-markers in the input maps is at most $(2d + 1)k$ because each single-marker is either deleted or adjacent to a deleted marker. This immediately yields a $(2d + 1)$-approximation algorithm: simply delete all single-markers.

We refer to Figure 2 for a tight example for the $(2d + 1)$-approximation algorithm. Observe that after one single-marker is deleted, many other single-markers may be merged into strips. Algorithm 3 first identifies (line 1) all triples of markers (z, x, y) such that z and y can be merged into a strip $\langle z, y \rangle$ after x is deleted, then successively deletes (lines 2–6) "cost-efficient" single-markers x that can reduce at least one other single-marker y or z, and finally removes (line 7) the remaining single-markers.

Lemma 7. *For each triple (z, x, y) in the set X in Algorithm 3, at least one of the three markers x, y, z must be deleted in any feasible solution.*

Proof. We prove the lemma by contradiction. Suppose that all three markers x, y, z are selected in a solution. Assume wlog that the sequence $\langle z, x, y \rangle$ appears in some map. Then x must be in the same strip as z or y. Assume wlog that

$$
\begin{aligned}
G_1 &= z_d y_d \quad \cdots \quad z_3 y_3 \quad z_2 y_2 \quad z_1 \, x \, y_1 \\
G_2 &= z_1 y_1 \quad z_2 \, x \, y_2 \quad z_3 y_3 \quad \cdots \quad z_d y_d \\
G_3 &= z_1 y_1 \quad z_2 y_2 \quad z_3 \, x \, y_3 \quad \cdots \quad z_d y_d \\
\cdots &\quad \cdots \quad \cdots \quad \cdots \quad \cdots \quad \cdots \\
G_d &= z_1 y_1 \quad z_2 y_2 \quad z_3 y_3 \quad \cdots \quad z_d \, x \, y_d
\end{aligned}
$$

Fig. 2. A tight example for the $(2d+1)$-approximation algorithm. The optimal solution deletes one single-marker x instead of all $2d + 1$ single-markers.

$\langle z, x \rangle$ is part of some strip. Then $z \prec x$. Recall that $z \prec y$. Thus x and y are both candidate successors of z. By Lemma 1a, we have $y \in \mathsf{gap}(z, x)$, thus y must be deleted: a contradiction. $\qquad \square$

We next prove the approximation ratio of Algorithm 3. Let O be the set of deleted markers in an optimal solution; $|O| = k$. For each marker $x \notin O$, we define two sets $\Gamma_{succ}(x)$ and $\Gamma_{pred}(x)$ as follows. If x is followed by a marker y in a strip of O, $\Gamma_{succ}(x) = \mathsf{gap}(x, y)$; otherwise x is the last marker of its strip, $\Gamma_{succ}(x) = \emptyset$. If x is preceded by a marker z in a strip of O, $\Gamma_{pred}(x) = \mathsf{gap}(z, x)$; otherwise x is the first marker of its strip, $\Gamma_{pred}(x) = \emptyset$. Then, for each marker $x \notin O$, define $\gamma(x) = |\Gamma_{succ}(x)| + |\Gamma_{pred}(x)|$, and for each marker $x \in O$, define $\gamma(x) = 0$.

Refer to Algorithm 3. Let D be the set of markers deleted in line 5, let S be the set of single-markers that are merged into super-markers in line 6, and let R be the set of markers deleted in line 7. Let $R_1 = \{r \in R \mid \gamma(r) = 1\}$ and $R_2 = \{r \in R \mid \gamma(r) \geq 2\}$. Note that if x is a single-marker at the beginning of the algorithm, then $\gamma(x) = 0$ if and only if $x \in O$. Thus we have a partition $R = (R \cap O) \cup R_1 \cup R_2$. Also note that each marker $x \in O$ is counted by γ at most twice in each map: at most once in some $\Gamma_{pred}(y)$, and at most once in some $\Gamma_{succ}(z)$. Thus we have the following inequality:

$$
\sum_{x \text{ single-marker}} \gamma(x) \leq 2dk. \tag{1}
$$

Each marker $x \in D$ has a corresponding triple $(z, x, y) \in X$, where z or y is a single-marker. After x is deleted in line 5, z and y are merged into the same super-marker in line 6. Thus we have the following inequality:

$$
|D| \leq |S|. \tag{2}
$$

For each marker $x \in D - O$, let $\phi(x)$ be an arbitrary marker in the non-empty set $\{z, x, y\} \cap O$ (see Lemma 7). Obviously $\phi(x) \neq x$, thus $\phi(x) \in O - D$. We show that at most two markers in $D - O$ can have the same image by ϕ. Suppose that $\phi(x_1) = \phi(x_2) = \phi$ for two different markers $x_1, x_2 \in D - O$, where x_1 is deleted before x_2 in Algorithm 3. Then the marker ϕ is merged into a super-marker after x_1 is deleted, and again merged into a larger super-marker after x_2 is deleted. Since a marker has at most two neighbors in a super-marker, ϕ is necessarily a single-marker before x_1 is deleted, so it belongs to S, indeed

$S \cap O$. Moreover, after x_2 is deleted and ϕ is merged into a larger super-marker, ϕ cannot be adjacent to any other single-marker, say x_3. Therefore

$$|D - O| \leq |O - D| + |S \cap O|. \tag{3}$$

Let u be a marker such that $\gamma(u) = 1$. Then u belongs to some strip in the optimal solution, and it has a neighbor $v = \psi(u)$ in the same strip such that $\mathsf{gap}(u, v)$ contains only one marker, say x. Note that $u, v \notin O$ and $x \in O$. We claim that if u is a single-marker at the beginning of the algorithm, then either $u \in D \cup S$ or $v \in D$. This claim is clearly true if u or v is deleted by the algorithm in line 5. Otherwise, with $(v, x, u) \in X$ or $(u, x, v) \in X$, either x is not deleted because u is merged into a super-marker, or x is deleted: in both cases $u \in S$. This proves the claim. So for each $u \in R_1$, we have $v \in D$, indeed $v \in D - O$. Note that there can be at most two markers u_1 and u_2 with the same image v by ψ: the two neighbors of v in some strip in the optimal solution. Thus we have $|R_1| \leq 2|D - O|$. Moreover, if there are two markers u_1 and u_2 with the same image v, then $\gamma(v) \geq 2$. Therefore

$$|R_1| \leq \sum_{v \in D - O} \gamma(v). \tag{4}$$

Combining inequalities (1), (2), (3), and (4), the calculation in the following shows that the number of deleted markers, $|D| + |R|$, is at most $(d + 1.5)k$. Thus Algorithm 3 indeed finds a $(d + 1.5)$-approximation for δ-gap-CMSR-d and CMSR-d.

$$
\begin{aligned}
2dk &\geq \sum_{x \text{ single-marker}} \gamma(x) \quad \text{by (1)} \\
&= \sum_{x \in D - O} \gamma(x) + \sum_{x \in S - O} \gamma(x) + \sum_{x \in R_1} \gamma(x) + \sum_{x \in R_2} \gamma(x) \\
&\geq \sum_{x \in D - O} \gamma(x) + |S - O| + |R_1| + 2|R_2| \\
&\geq |S - O| + 2|R_1| + 2|R_2| \quad \text{by (4)}.
\end{aligned}
$$

$$
\begin{aligned}
|D| + |R| &= |D| + |R_1| + |R_2| + |R \cap O| \\
&\leq |D| + dk - \tfrac{1}{2}|S - O| + |R \cap O| \\
&= |D| + dk - \tfrac{1}{2}(|S| - |S \cap O|) + |R \cap O| \\
&\leq |D| + dk - \tfrac{1}{2}|D| + \tfrac{1}{2}|S \cap O| + |R \cap O| \quad \text{by (2)} \\
&= \tfrac{1}{2}(|D| + |S \cap O|) + |R \cap O| + dk \\
&= \tfrac{1}{2}(|D \cap O| + |D - O| + |S \cap O|) + |R \cap O| + dk \\
&\leq \tfrac{1}{2}(|D \cap O| + (|O - D| + |S \cap O|) + |S \cap O|) + |R \cap O| + dk \quad \text{by (3)} \\
&= \tfrac{1}{2}|O| + (|S \cap O| + |R \cap O|) + dk \\
&\leq \tfrac{1}{2}k + k + dk \\
&= \left(d + \tfrac{3}{2}\right) k.
\end{aligned}
$$

After our initial submission of this paper for publication, we learned from an anonymous reviewer that Jiang et al. [5] have very recently designed a 3-approximation algorithm for CMSR-2 based on a similar greedy approach. Their algorithm does not work for the gap-constrained variant, although it seems that the algorithm might be extended to a $(d+1)$-approximation for CMSR-d for all $d \geq 2$. Our solution gives uniform results on both variants.

$$
\begin{aligned}
G_1 &= z_d y_d & \cdots & & z_3 y_3 & z_2 y_2 & z_1 \, vu \, y_1 \\
G_2 &= z_1 y_1 & z_2 \, uv \, y_2 & z_3 y_3 & \cdots & & z_d y_d \\
G_3 &= z_1 y_1 & z_2 y_2 & z_3 \, uv \, y_3 & \cdots & & z_d y_d \\
\cdots & & \cdots & \cdots & \cdots & \cdots & \cdots \\
G_d &= z_1 y_1 & z_2 y_2 & z_3 y_3 & \cdots & & z_d \, uv \, y_d
\end{aligned}
$$

$$
\begin{aligned}
G_1 &= z_d y_d & \cdots & & z_3 y_3 & z_2 y_2 & z_1 \, -v-u \, y_1 \\
G_2 &= z_1 y_1 & z_2 \, uv \, y_2 & z_3 y_3 & \cdots & & z_d y_d \\
G_3 &= z_1 y_1 & z_2 y_2 & z_3 \, uv \, y_3 & \cdots & & z_d y_d \\
\cdots & & \cdots & \cdots & \cdots & \cdots & \cdots \\
G_d &= z_1 y_1 & z_2 y_2 & z_3 y_3 & \cdots & & z_d \, uv \, y_d
\end{aligned}
$$

Fig. 3. Upper: an almost-tight example for the $(d+1.5)$-approximation algorithm showing that its approximation ratio cannot be better than $d+1$; the optimal solution deletes the two single-markers u and v instead of all $2d+2$ single-markers. Lower: an example showing that no algorithm deleting only single-markers can achieve an approximation ratio better than d; the optimal solution deletes one super-marker $\langle u, v \rangle$ instead of $2d$ single-markers z_i and y_i, $1 \leq i \leq d$.

We refer to Figure 3 for two examples: the first example gives a lower bound of $d+1$ on the approximation ratio of Algorithm 3; the second example gives a lower bound of d on the approximation ratio of *any* algorithm for δ-gap-CMSR-d and CMSR-d that deletes only single-markers. Note that both our Algorithm 3 and the algorithm in [5] delete only single-markers.

Compared to the approximation upper bound of $2d$ [2,1,8] for the two maximization problems MSR-d and δ-gap-MSR-d, which almost matches (at least asymptotically) the current best lower bound of $\Omega(d/\log d)$ [8], our upper bound of $d+1.5$ for the two minimization problems CMSR-d and δ-gap-CMSR-d is still far away from the constant lower bound in [8]. It is an intriguing question whether CMSR-d and δ-gap-CMSR-d admit approximation algorithms with constant ratios independent of d.

References

1. Bulteau, L., Fertin, G., Rusu, I.: Maximal strip recovery problem with gaps: Hardness and approximation algorithms. In: Dong, Y., Du, D.-Z., Ibarra, O. (eds.) ISAAC 2009. LNCS, vol. 5878, pp. 710–719. Springer, Heidelberg (2009)
2. Chen, Z., Fu, B., Jiang, M., Zhu, B.: On recovering syntenic blocks from comparative maps. Journal of Combinatorial Optimization 18, 307–318 (2009)

3. Choi, V., Zheng, C., Zhu, Q., Sankoff, D.: Algorithms for the extraction of synteny blocks from comparative maps. In: Giancarlo, R., Hannenhalli, S. (eds.) WABI 2007. LNCS (LNBI), vol. 4645, pp. 277–288. Springer, Heidelberg (2007)
4. Halldórsson, M.M.: Approximating discrete collections via local improvements. In: Proceedings of the 6th Annual ACM-SIAM Symposium on Discrete Algorithms (SODA 1995), pp. 160–169 (1995)
5. Jiang, H., Li, Z., Lin, G., Wang, L., Zhu, B.: Exact and approximation algorithms for the complementary maximal strip recovery problem. In: Journal of Combinatorial Optimization (to appear), doi:10.1007/s10878-010-9366-y
6. Jiang, M.: Inapproximability of maximal strip recovery. In: Dong, Y., Du, D.-Z., Ibarra, O. (eds.) ISAAC 2009. LNCS, vol. 5878, pp. 616–625. Springer, Heidelberg (2009)
7. Jiang, M.: On the parameterized complexity of some optimization problems related to multiple-interval graphs. In: Amir, A., Parida, L. (eds.) CPM 2010. LNCS, vol. 6129, pp. 125–137. Springer, Heidelberg (2010)
8. Jiang, M.: Inapproximability of maximal strip recovery: II. In: Lee, D.-T., Chen, D.Z., Ying, S. (eds.) FAW 2010. LNCS, vol. 6213, pp. 53–64. Springer, Heidelberg (2010)
9. Wang, L., Zhu, B.: On the tractability of maximal strip recovery. In: Chen, J., Cooper, S.B. (eds.) TAMC 2009. LNCS, vol. 5532, pp. 400–409. Springer, Heidelberg (2009)
10. Zheng, C., Zhu, Q., Sankoff, D.: Removing noise and ambiguities from comparative maps in rearrangement analysis. IEEE/ACM Transactions on Computational Biology and Bioinformatics 4, 515–522 (2007)
11. Zhu, B.: Efficient exact and approximate algorithms for the complement of maximal strip recovery. In: Chen, B. (ed.) AAIM 2010. LNCS, vol. 6124, pp. 325–333. Springer, Heidelberg (2010)

Efficient Seeds Computation Revisited[*]

Michalis Christou[1], Maxime Crochemore[1,3], Costas S. Iliopoulos[1,4],
Marcin Kubica[2], Solon P. Pissis[1], Jakub Radoszewski[2,**],
Wojciech Rytter[2,5,***], Bartosz Szreder[2], and Tomasz Waleń[2]

[1] Dept. of Informatics, King's College London, London WC2R 2LS, UK
{michalis.christou,maxime.crochemore,csi,solon.pissis}@dcs.kcl.ac.uk
[2] Dept. of Mathematics, Computer Science and Mechanics,
University of Warsaw, Warsaw, Poland
{kubica,jrad,rytter,szreder,walen}@mimuw.edu.pl
[3] Université Paris-Est, France
[4] Digital Ecosystems & Business Intelligence Institute,
Curtin University of Technology, Perth WA 6845, Australia
[5] Dept. of Math. and Informatics,
Copernicus University, Toruń, Poland

Abstract. The notion of the cover is a generalization of a period of a
string, and there are linear time algorithms for finding the shortest cover.
The seed is a more complicated generalization of periodicity, it is a cover
of a superstring of a given string, and the shortest seed problem is of
much higher algorithmic difficulty. The problem is not well understood,
no linear time algorithm is known. In the paper we give linear time algorithms for some of its versions — computing shortest left-seed array,
longest left-seed array and checking for seeds of a given length. The algorithm for the last problem is used to compute the seed array of a string
(i.e., the shortest seeds for all the prefixes of the string) in $O(n^2)$ time.
We describe also a simpler alternative algorithm computing efficiently
the shortest seeds. As a by-product we obtain an $O(n \log (n/m))$ time
algorithm checking if the shortest seed has length at least m and finding
the corresponding seed. We also correct some important details missing
in the previously known shortest-seed algorithm (Iliopoulos et al., 1996).

1 Introduction

The notion of periodicity in strings is widely used in many fields, such as combinatorics on words, pattern matching, data compression and automata theory
(see [13,14]). It is of paramount importance in several applications, not to talk
about its theoretical aspects. The concept of quasiperiodicity is a generalization of the notion of periodicity, and was defined by Apostolico and Ehrenfeucht
in [2]. In a periodic repetition the occurrences of the period do not overlap. In
contrast, the quasiperiods of a quasiperiodic string may overlap.

[*] The authors thank an anonymous referee for proposing several insightful remarks.
[**] The author is supported by grant no. N206 568540 of the National Science Centre.
[***] The author is supported by grant no. N206 566740 of the National Science Centre.

R. Giancarlo and G. Manzini (Eds.): CPM 2011, LNCS 6661, pp. 350–363, 2011.
© Springer-Verlag Berlin Heidelberg 2011

We consider *words* (*strings*) over a finite alphabet Σ, $u \in \Sigma^*$; the empty word is denoted by ε; the positions in u are numbered from 1 to $|u|$. By Σ^n we denote the set of words of length n. By u^R we denote the reverse of the string u. For $u = u_1 u_2 \ldots u_n$, let us denote by $u[i \mathinner{\ldotp\ldotp} j]$ a *factor* of u equal to $u_i \ldots u_j$ (in particular $u[i] = u[i \mathinner{\ldotp\ldotp} i]$). Words $u[1 \mathinner{\ldotp\ldotp} i]$ are called *prefixes* of u, and words $u[i \mathinner{\ldotp\ldotp} n]$ are called *suffixes* of u. Words that are both prefixes and suffixes of u are called *borders* of u. By $\mathsf{border}(u)$ we denote the length of the longest border of u that is shorter than u. We say that a positive integer p is the (shortest) *period* of a word $u = u_1 \ldots u_n$ (notation: $p = \mathsf{per}(u)$) if p is the smallest positive number, such that $u_i = u_{i+p}$, for $i = 1, \ldots, n - p$. It is a known fact [6,8] that, for any string u, $\mathsf{per}(u) + \mathsf{border}(u) = |u|$.

We say that a string s *covers* the string u if every letter of u is contained in some occurrence of s as a factor of u. Then s is called a *cover* of u. We say that a string s is: a *seed* of u if s is a factor of u and u is a factor of some string w covered by s; a *left seed* of u if s is both a prefix and a seed of u; a *right seed* of u if s is both a suffix and a seed of u (equivalently, s^R is a left seed of u^R). Seeds were first defined and studied by Iliopoulos, Moore and Park [11], who gave an $O(n \log n)$ time algorithm computing all the seeds of a given string $u \in \Sigma^n$, in particular, the shortest seed of u.

By $\mathsf{cover}(u)$, $\mathsf{seed}(u)$, $\mathsf{lseed}(u)$ and $\mathsf{rseed}(u)$ we denote the length of the shortest: cover, seed, left seed and right seed of u, respectively. By $\mathsf{covermax}(u)$ and $\mathsf{lseedmax}(u)$ we denote the length of the longest cover and the longest left seed of u that is shorter than u, or 0 if none.

For a string $u \in \Sigma^n$, we define its: *period array* $\mathsf{P}[1 \mathinner{\ldotp\ldotp} n]$, *border array* $\mathsf{B}[1 \mathinner{\ldotp\ldotp} n]$, *suffix period array* $\mathsf{P}'[1 \mathinner{\ldotp\ldotp} n]$, *cover array* $\mathsf{C}[1 \mathinner{\ldotp\ldotp} n]$, *longest cover array* $\mathsf{C}^M[1 \mathinner{\ldotp\ldotp} n]$, *seed array* $\mathsf{Seed}[1 \mathinner{\ldotp\ldotp} n]$, *left-seed array* $\mathsf{LSeed}[1 \mathinner{\ldotp\ldotp} n]$, and *longest left-seed array* $\mathsf{LSeed}^M[1 \mathinner{\ldotp\ldotp} n]$ as follows:

$$\begin{aligned}
\mathsf{P}[i] &= \mathsf{per}(u[1 \mathinner{\ldotp\ldotp} i]), & \mathsf{B}[i] &= \mathsf{border}(u[1 \mathinner{\ldotp\ldotp} i]), \\
\mathsf{P}'[i] &= \mathsf{per}(u[i \mathinner{\ldotp\ldotp} n]), & \mathsf{C}[i] &= \mathsf{cover}(u[1 \mathinner{\ldotp\ldotp} i]), \\
\mathsf{C}^M[i] &= \mathsf{covermax}(u[1 \mathinner{\ldotp\ldotp} i]), & \mathsf{Seed}[i] &= \mathsf{seed}(u[1 \mathinner{\ldotp\ldotp} i]), \\
\mathsf{LSeed}[i] &= \mathsf{lseed}(u[1 \mathinner{\ldotp\ldotp} i]), & \mathsf{LSeed}^M[i] &= \mathsf{lseedmax}(u[1 \mathinner{\ldotp\ldotp} i]).
\end{aligned}$$

Table 1. An example string together with its periodic and quasiperiodic arrays. Note that the left-seed array and the seed array are non-decreasing.

i	1	2	3	4	5	6	7	8	9	10	11	12	13	14	15	16
$u[i]$	a	b	a	a	b	a	a	a	b	b	a	a	b	a	a	b
$\mathsf{P}[i]$	1	2	2	3	3	3	3	7	7	10	10	11	11	11	11	11
$\mathsf{B}[i]$	0	0	1	1	2	3	4	1	2	0	1	1	2	3	4	5
$\mathsf{C}[i]$	1	2	3	4	5	3	4	8	9	10	11	12	13	14	15	16
$\mathsf{C}^M[i]$	0	0	0	0	0	3	4	0	0	0	0	0	0	0	0	0
$\mathsf{LSeed}[i]$	1	2	2	3	3	3	3	4	4	10	10	11	11	11	11	11
$\mathsf{LSeed}^M[i]$	0	0	2	3	4	5	6	7	8	0	10	11	12	13	14	15
$\mathsf{Seed}[i]$	1	2	2	3	3	3	3	4	4	8	8	8	8	8	8	11

The border array, suffix border array and period array can be computed in $O(n)$ time [6,8]. Apostolico and Breslauer [1,4] gave an on-line $O(n)$ time algorithm computing the cover array $\mathsf{C}[1..n]$ of a string. Li and Smyth [12] provided an algorithm, having the same characteristics, for computing the longest cover array $\mathsf{C}^M[1..n]$ of a given string. Note that the array C^M enables computing all covers of all prefixes of the string, same property holds for the border array B. Unfortunately, the LSeed^M array does not share this property.

Table 1 shows the above defined arrays for $u = $ abaabaaabbaabaab. For example, for the prefix $u[1..13]$ the period equals 11, the border is ab, the cover is abaabaaabbaab, the left seed is abaabaaabba, the longest left seed is abaabaaabbaa, and the seed is baabaaab.

We list here several useful (though obvious) properties of covers and seeds.

Observation 1

(a) A cover of a cover of u is also a cover of u.
(b) A cover of a left (right) seed of u is also a left (right) seed of u.
(c) A cover of a seed of u is also a seed of u.
(d) If u is a factor of v then $\mathsf{seed}(u) \leq \mathsf{seed}(v)$.
(e) If u is a prefix of v then $\mathsf{lseed}(u) \leq \mathsf{lseed}(v)$.
(f) If s and s' are two covers of a string u, $|s'| < |s|$, then s' is a cover of s.
(g) If s is the shortest cover or the shortest left seed or the shortest seed of a string u then $\mathsf{per}(s) > |s|/2$.

For a set X of positive integers, let us define the *maxgap* of X as:

$$\mathsf{maxgap}(X) = \max\{b - a : a, b \text{ are consecutive numbers in } X\} \text{ or } 0 \text{ if } |X| \leq 1.$$

For example $\mathsf{maxgap}(\{1, 3, 8, 13, 17\}) = 5$.

For a factor v of u, let us define $Occ(v, u)$ as the set of starting positions of all occurrences of v in u. By $\mathit{first}(v)$ and $\mathit{last}(v)$ we denote $\min Occ(v, u)$ and $\max Occ(v, u)$ respectively. For the sake of simplicity, we will abuse the notation, and denote $\mathsf{maxgap}(v) = \mathsf{maxgap}(Occ(v, u))$.

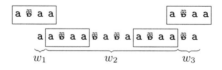

Fig. 1. The word $s = $ abaa is a border seed of $u = $ aabaababaabaaba

Assume s is a factor of u. Let us decompose the word u into $w_1 w_2 w_3$, where w_2 is the longest factor of u for which s is a border, i.e., $w_2 = u[\mathit{first}(s)..(\mathit{last}(s) + |s| - 1)]$. Then we say that s is a *border seed* of u if s is a seed of $w_1 \cdot s \cdot w_3$, see Fig. 1. The following fact is a corollary of Lemma 4, proved in Section 2.

Fact 2. *Let s be a factor of $u \in \Sigma^*$. The word s is a border seed of u if and only if $|s| \geq \max(P[first(s) + |s| - 1], \ P'[last(s)])$.*

Notions of maxgaps and border seeds provide a useful characterization of seeds.

Observation 3. *Let s be a factor of $u \in \Sigma^*$. The word s is a seed of u if and only if $|s| \geq maxgap(s)$ and s is a border seed of u.*

Several new and efficient algorithms related to seeds in strings are presented in this paper. Linear time algorithms computing left-seed array and longest left-seed array are given in Section 2. In Section 3 we show a linear time algorithm finding seed-of-a-given-length and apply it to computing the seed array of a string in $O(n^2)$ time. Finally, in Section 4 we describe an alternative simple $O(n \log n)$ time computation of the shortest seed, from which we obtain an $O(n \log (n/m))$ time algorithm checking if the shortest seed has length at least m (described in Section 5).

2   Computing Left-Seed Arrays

In this section we show two $O(n)$ time algorithms for computing the left-seed array and an $O(n)$ time algorithm for computing the longest left-seed array of a given string $u \in \Sigma^n$. We start by a simple characterization of the length of the shortest left seed of the whole string u — see Lemma 5. In its proof we utilize the following auxiliary lemma which shows a correspondence between the shortest left seed of u and shortest covers of all prefixes of u.

Lemma 4. *Let s be a prefix of u, and let j be the length of the longest prefix of u covered by s. Then s is a left seed of u if and only if $j \geq per(u)$.*

In particular, the shortest left seed s of u is the shortest cover of the corresponding prefix $u[1 \mathinner{.\,.} j]$.

Proof. ($\Rightarrow$) If s is a left seed of u then there exists a prefix p of s of length at least $n - j$ which is a suffix of u (see Fig. 2). We use here the fact, that $u[1 \mathinner{.\,.} j]$ is the *longest* prefix of u covered by s. Hence, p is a border of u, and consequently $border(u) \geq |p| \geq n - j$. Thus we obtain the desired inequality $j \geq per(u)$.

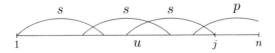

Fig. 2. Illustration of part ($\Rightarrow$) of Lemma 4

($\Leftarrow$) The inequality $j \geq per(u)$ implies that $v = u[1 \mathinner{.\,.} j]$ is a left seed of u (see Fig. 3). Hence, by Observation 1b, the word s, which is a cover of v, is also a left seed of u.

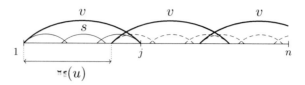

Fig. 3. Illustration of part ($\Leftarrow$) of Lemma 4

Finally, the "in particular" part is a consequence of Observation 1, parts b and f.   □

Lemma 5. *Let $u \in \Sigma^n$ and let $C[1 .. n]$ be its cover array. Then:*

$$lseed(u) = \min\{C[j] \ : \ j \geq per(u)\}. \tag{1}$$

Proof. By Lemma 4, the length of the shortest left seed of u can be found among the values $C[per(u)], \ldots, C[n]$. And conversely, for each of the values $C[j]$ for $per(u) \leq j \leq n$, there exists a left seed of u of length $C[j]$. Thus $lseed(u)$ equals the minimum of these values, which yields the formula (1).   □

Clearly, the formula (1) provides an $O(n)$ time algorithm for computing the shortest left seed of the whole string u. We show that, employing some algorithmic techniques, one can use this formula to compute shortest left seeds for all prefixes of u, i.e., computing the left-seed array of u, also in $O(n)$ time.

Theorem 1. *For $u \in \Sigma^n$, its left-seed array can be computed in $O(n)$ time.*

Proof. Applying (1) to all prefixes of u, we obtain:

$$\mathsf{LSeed}[i] = \min\{C[j] \ : \ P[i] \leq j \leq i\}. \tag{2}$$

Recall that both the period array $P[1 .. n]$ and the cover array $C[1 .. n]$ of u can be computed in $O(n)$ time [1,4,6,8].

The minimum in the formula (2) could be computed by data structures for Range-Minimum-Queries [9,15], however in this particular case we can apply a much simpler algorithm. Note that $P[i-1] \leq P[i]$, therefore the intervals of the form $[P[i], i]$ behave like a sliding window, i.e., both their endpoints are non-decreasing. We use a bidirectional queue Q which stores left-minimal elements in the current interval $[P[i], i]$ (w.r.t. the value $C[j]$). In other words, elements of Q are increasing and if Q during the step i contains an element j then $j \in [P[i], i]$ and $C[j] < C[j']$ for all $j < j' \leq i$. We obtain an $O(n)$ time algorithm ComputeLeftSeedArray.   □

ALGORITHM ComputeLeftSeedArray(u)

1: P$[1 \dots n]$:= period array of u; C$[1 \dots n]$:= cover array of u;
2: Q := *emptyBidirectionalQueue*;
3: **for** $i := 1$ **to** n **do**
4:    **while** (**not** *empty*(Q)) **and** (*front*(Q) < P$[i]$) **do** *popFront*(Q);
5:    **while** (**not** *empty*(Q)) **and** (C$[back(Q)] \geq$ C$[i]$) **do** *popBack*(Q);
6:    *pushBack*(Q, i);
7:    LSeed$[i]$:= C$[front(Q)]$;
8:    { Q stores left-minimal elements of the interval $[$P$[i], i]$ }
9: **return** LSeed$[1 \dots n]$;

Now we proceed to an alternative algorithm computing the left-seed array, which also utilizes the criterion from Lemma 4. We start with an auxiliary algorithm ComputeR-Array. It computes an array R$[1 \dots n]$ which stores, as R$[i]$, the length of the longest prefix of u for which $u[1 \dots i]$ is the shortest cover, 0 if none.

ALGORITHM ComputeR-Array(u)

1: C$[1 \dots n]$:= cover array of u;
2: **for** $i := 1$ **to** n **do** R$[i] := 0$;
3: **for** $i := 1$ **to** n **do** R$[$C$[i]] := i$;
4: **return** R$[1 \dots n]$;

The algorithm Alternative-ComputeLeftSeedArray computes the array LSeed from left to right. The current value of LSeed$[i]$ is stored in the variable ls, note that this value never decreases (by Observation 1e). Equivalently, for each i we have LSeed$[i-1] \leq$ LSeed$[i] \leq i$.

The particular value of LSeed$[i]$ is obtained using the necessary and sufficient condition from Lemma 4: LSeed$[i] = ls$ if ls is the smallest number such that $|w| \geq$ per($u[1 \dots i]$) = P$[i]$, where w is the longest prefix of $u[1 \dots i]$ that is covered by $u[1 \dots ls]$. We slightly modify this condition, substituting w with the longest prefix w' of the very word u that is covered by $u[1 \dots ls]$. Thus we obtain the condition R$[ls] \geq$ P$[i]$ utilized in the pseudocode below.

ALGORITHM Alternative-ComputeLeftSeedArray(u)

1: P$[1 \dots n]$:= period array of u; R$[1 \dots n]$:= ComputeR-Array(u);
2: LSeed$[0] := 0$; $ls := 0$;
3: **for** $i := 1$ **to** n **do**
4:    { An invariant of the loop: $ls =$ LSeed$[i-1]$. }
5:    **while** R$[ls] <$ P$[i]$ **do** $ls := ls + 1$;
6:    LSeed$[i] := ls$;
7: **return** LSeed$[1 \dots n]$;

Theorem 2. *Algorithm Alternative-ComputeLeftSeedArray runs in linear time.*

Proof. Recall that the arrays $P[1 . . n]$ and $C[1 . . n]$ can be computed in linear time [1,4,6,8]. The array $R[1 . . n]$ is obviously also computed in linear time.

It suffices to prove that the total number of steps of the while-loop in the algorithm Alternative-ComputeLeftSeedArray is linear in terms of n. In each step of the loop, the value of ls increases by one; this variable never decreases and it cannot exceed n. Hence, the while-loop performs at most n steps and the whole algorithm runs in $O(n)$ time. □

Concluding this section, we describe a linear-time algorithm computing the longest left-seed array, $\mathsf{LSeed}^M[1 . . n]$, of the string $u \in \Sigma^n$. The following lemma gives a simple characterization of the length of the longest left seed of the whole string u.

Lemma 6. *Let $u \in \Sigma^n$. If $\mathsf{per}(u) < n$ then $\mathsf{lseedmax}(u) = n - 1$, otherwise $\mathsf{lseedmax}(u) = 0$.*

Proof. First consider the case $\mathsf{per}(u) = n$. We show that $\mathsf{lseed}(u) = n$, consequently $\mathsf{lseedmax}(u)$ equals 0. Assume to the contrary that $\mathsf{lseed}(u) < n$. Then, a non-empty prefix of the minimal left seed of u, say w, is a suffix of u (consider the occurrence of the left seed that covers $u[n]$). Hence, $n - |w|$ is a period of u, a contradiction.

Assume now that $\mathsf{per}(u) < n$. Then u is a prefix of the word $u[1 . . \mathsf{per}(u)] \cdot u[1 . . n - 1]$ which is covered by $u[1 . . n - 1]$. Therefore $u[1 . . n - 1]$ is a left seed of u, $\mathsf{lseedmax}(u) \geq n - 1$, consequently $\mathsf{lseedmax}(u) = n - 1$. □

Using Lemma 6 we obtain $\mathsf{LSeed}^M[i] = i - 1$ or $\mathsf{LSeed}^M[i] = 0$ for every i, depending on whether $P[i] < i$ or not. We obtain the following result.

Theorem 3. *Longest left-seed array of $u \in \Sigma^n$ can be computed in $O(n)$ time.*

3   Computing Seeds of Given Length and Seed Array

In this section we show an $O(n^2)$ time algorithm computing the seed array $\mathsf{Seed}[1 . . n]$ of a given string $u \in \Sigma^n$, note that a trivial approach — computing the shortest seed for every prefix of u — yields $O(n^2 \log n)$ time complexity. In our solution we utilize a subroutine: testing whether u has a seed of a given length k. The following theorem shows that this test can be performed in $O(n)$ time.

Theorem 4. *It can be checked whether a given string $u \in \Sigma^n$ has a seed of a given length k in $O(n)$ time.*

Proof. Assume we have already computed in $O(n)$ time the suffix array SUF and the LCP array of longest common prefixes, see [6]. In the algorithm we start by dividing all factors of u of length k into groups corresponding to equal words. Every such group can be described as a maximal interval $[i . . j]$ in the suffix array SUF, such that each of the values $\mathsf{LCP}[i + 1], \mathsf{LCP}[i + 2], \ldots, \mathsf{LCP}[j]$ is at least k. The collection of such intervals can be constructed in $O(n)$ time by a single traversal of the LCP and SUF arrays (lines 1–9 of Algorithm SeedsOfA-GivenLength). Moreover, using Bucket Sort, we can transform this representation into a collection of lists, each of which describes the set $Occ(v, u)$ for some factor v of u, $v \in \Sigma^k$ (lines 10–11 of the algorithm). This can be done in linear time, provided that we use the same set of buckets in each sorting and initialize them just once.

Now we process each of the lists separately, checking the conditions from Observation 3: in lines 14–18 of the algorithm we check the "maxgap" condition, and in line 19 the "border seed" condition, employing Fact 2.

Thus, having computed the arrays SUF and LCP, and the period arrays $P[1 . . n]$ and $P'[1 . . n]$ of u, we can find all seeds of u of length k in $O(n)$ total time.   □

ALGORITHM SeedsOfAGivenLength(u, k)

  1: $P[1 . . n] :=$ period array of u; $P'[1 . . n] :=$ suffix period array of u;
  2: $\mathsf{SUF}[1 . . n] :=$ suffix array of u; $\mathsf{LCP}[1 . . n] :=$ lcp array of u;
  3: $Lists := emptyList$;
  4: $j := 1$;
  5: **while** $j \leq n$ **do**
  6:     $List := \{\mathsf{SUF}[j]\}$;
  7:     **while** $j < n$ **and** $\mathsf{LCP}[j + 1] \geq k$ **do**
  8:        $j := j + 1$; $List := append(List, \mathsf{SUF}[j])$;
  9:     $j := j + 1$; $Lists := append(Lists, List)$;
10: **for all** $List$ **in** $Lists$ **do**
11:     $BucketSort(List)$; { using the same set of buckets }
12: **for all** $List$ **in** $Lists$ **do**
13:     $first := prev := n$; $last := 1$; $covers :=$ **true**;
14:     **for all** i **in** $List$ **do**
15:        $first := \min(first, i)$; $last := \max(last, i)$;
16:        **if** $i > prev + k$ **then**
17:           $covers :=$ **false**;
18:        $prev := i$;
19:     **if** $covers$ **and** $(k \geq \max(P[first + k - 1], P'[last]))$ **then**
20:        **print** "$u[first . . (first + k - 1)]$ is a seed of u";

We compute the elements of the seed array $\mathsf{Seed}[1 . . n]$ from left to right, i.e., in the order of increasing lengths of prefixes of u. Note that $\mathsf{Seed}[i + 1] \geq \mathsf{Seed}[i]$ for any $1 \leq i \leq n - 1$, this is due to Observation 1d. If $\mathsf{Seed}[i + 1] > \mathsf{Seed}[i]$ then we increase the current length of the seed by one letter at a time, in total at most $n - 1$ such operations are performed. Each time we query for existence of a seed of a given length using the algorithm from Theorem 4. Thus we obtain $O(n^2)$ time complexity.

Theorem 5. *The seed array of a string $u \in \Sigma^n$ can be computed in $O(n^2)$ time.*

4 Alternative Algorithm for Shortest Seeds

In this section we present a new approach to shortest seeds computation based on very simple independent processing of disjoint chains in the suffix tree. It simplifies the computation of shortest seeds considerably.

Our algorithm is also based on a slightly modified version of Observation 3, formulated below as Lemma 7, which allows to relax the definition of maxgaps. We discuss an algorithmically easier version of maxgaps, called prefix maxgaps, and show that it can substitute maxgap values when looking for the shortest seed.

We start by analyzing the "border seed" condition. We introduce somewhat more abstract representation of sets of factors of u, called *prefix families*, and show how to find in them the shortest border seeds of u. Afterwards the key algorithm for computing prefix maxgaps is presented. Finally, both techniques are utilized to compute the shortest seed.

Let us fix the input string $u \in \Sigma^n$. For $v \in \Sigma^*$, by $PREF(v)$ we denote the set of all prefixes of v and by $PREF(v, k)$ we denote $PREF(v) \cap \Sigma^k \Sigma^*$ (*limited prefix subset*).

Let $\mathcal{F}$ be a family of limited prefix subsets of some factors of u, we call $\mathcal{F}$ a *prefix family*. Every element $PREF(v, k) \in \mathcal{F}$ can be represented in a canonical form, by a tuple of integers: $(first(v), last(v), k, |v|)$. Such a representation requires only constant space per element. By $\mathsf{bseed}(u, \mathcal{F})$ we denote the shortest border seed of u contained in some element of $\mathcal{F}$.

Example 1. Let $u = \mathsf{aabaababaabaaba}$ be the example word from Fig. 1. Let:

$$\mathcal{F} = \{PREF(\mathsf{abaab}, 4),\ PREF(\mathsf{babaa}, 4)\} = \{(2, 10, 4, 5),\ (6, 6, 4, 5)\}.$$

Note that $\bigcup \mathcal{F} = \{\mathsf{abaa}, \mathsf{abaab}, \mathsf{baba}, \mathsf{babaa}\}$. Then $\mathsf{bseed}(u, \mathcal{F}) = \mathsf{abaa}$.

The proof of the following fact is present implicitly in [11] (type-A and type-B seeds).

Theorem 6. *Let $u \in \Sigma^n$ and let $\mathcal{F}$ be a prefix family given in a canonical form. Then $\mathsf{bseed}(u, \mathcal{F})$ can be computed in linear time.*

Alternative proof of Theorem 6.   There is an alternative algorithm for computing $\mathsf{bseed}(u, \mathcal{F})$, based on a special version of Find-Union data structure. Recall that $\mathsf{B}[1 \mathbin{.\,.} n]$ is the border-array of u. Denote by $FirstGE(\mathcal{I}, c)$ (*first-greater-equal*) a query:

$$FirstGE(\mathcal{I}, c) = \min\{i \ : \ i \in \mathcal{I},\ \mathsf{B}[i] \geq c\},$$

where $\mathcal{I}$ is a subinterval of $[1 \mathbin{.\,.} n]$. We assume that $\min \emptyset = +\infty$. A sequence of linear number of such queries, sorted according to non-decreasing values of c, can be easily answered in linear time, using an interval version of Find-Union data structure, see [7,10]. The following algorithm applies the condition for border seed from Fact 2 to every element of $\mathcal{F}$, with $\mathsf{P}[first(s) + |s| - 1]$ substituted by $first(s) + |s| - 1 - \mathsf{B}[first(s) + |s| - 1]$. We omit the details.   □

ALGORITHM ComputeBorderSeed(u, $\mathcal{F}$)

1: $bseed := +\infty$;
2: **for all** $(first(v), last(v), k, |v|)$ **in** $\mathcal{F}$, in non-decreasing order of $first(v)$ **do**
3:     $k := \max(\mathsf{P}'[last(v)], k)$;
4:     $\mathcal{I} := [first(v) + k - 1, \ first(v) + |v| - 1]$;
5:     $pos := FirstGE(\mathcal{I}, \ first(v) - 1)$;
6:     $bseed := \min(bseed, \ pos - first(v) + 1)$;
7: **return** $bseed$;

Computation of the shortest seeds via prefix maxgaps. Let $T(u)$ be the suffix tree of u, recall that it can be constructed in $O(n)$ time [6,8]. By $Nodes(u)$ we denote the set of factors of u corresponding to explicit nodes of $T(u)$, for simplicity we identify the nodes with the strings they represent. For $v \in Nodes(u)$, the set $Occ(v, u)$ corresponds to leaf list of the node v (i.e., the set of values of leaves in the subtree rooted at v), denoted as $LL(v)$. Note that $first(v) = \min LL(v)$ and $last(v) = \max LL(v)$, and such values can be computed for all $v \in Nodes(u)$ in $O(n)$ time. For $v \in Nodes(u)$, we define the *prefix maxgap* of v as:

$$\Delta(v) \ = \ \max\{\mathsf{maxgap}(w) \ : \ w \in PREF(v)\}.$$

Equivalently, $\Delta(v)$ is the maximum of maxgap values on the path from v to the root of $T(u)$. We introduce an auxiliary problem:

Prefix Maxgap Problem:
    given a word $u \in \Sigma^n$, compute $\Delta(v)$ for all $v \in Nodes(u)$.

The following lemma (an alternative formulation of Observation 3) shows that prefix maxgaps can be used instead of maxgaps in searching for seeds. This is important since computation of prefix maxgaps $\Delta(v)$ is simple, in comparison with $\mathsf{maxgap}(v)$ — this is due to the fact that the $\Delta(v)$ values on each path down the suffix tree $T(u)$ are non-decreasing. Efficient computation of $\mathsf{maxgap}(v)$ requires using augmented height-balanced trees [5] or other rather sophisticated techniques [3]. The shortest-seed algorithm in [11] also computes prefix maxgaps instead of maxgaps, however this observation is missing in [11].

Lemma 7. *Let s be a factor of $u \in \Sigma^*$ and let w be the shortest element of $Nodes(u)$ such that $s \in PREF(w)$. The word s is a seed of u if and only if $|s| \geq \Delta(w)$ and s is a border seed of u.*

Proof. If s corresponds to an element of $Nodes(u)$, then $s = w$. Otherwise, s corresponds to an implicit node in an edge in the suffix tree, and w is the lower end of the edge. Note that in both cases we have $\Delta(w) \geq \mathsf{maxgap}(w) = \mathsf{maxgap}(s)$. By Observation 3, this implies part ($\Leftarrow$) of the conclusion. As for the part ($\Rightarrow$), it suffices to show that $|s| \geq \Delta(w)$.

Assume, to the contrary, that $|s| < \Delta(w)$. Let $v \in PREF(w) \cap Nodes(u)$ be the word for which $\mathsf{maxgap}(v) = \Delta(w)$, and let a, b be consecutive elements of the set $Occ(v, u)$ for which $a + \mathsf{maxgap}(v) = b$.

Let us note that no occurrence of s starts at any of the positions $a+1, \ldots, b-1$. Moreover, none of the suffixes of the form $u[i \mathinner{.\,.} n]$, for $a + 1 \leq i \leq b - 1$, is a prefix of s. Indeed, v is a prefix of s of length at most $n - b + 1$, and such an occurrence of s (or its prefix) would imply an extra occurrence of v. Note that at most $|s| \leq b - a - 1$ first positions in the interval $[a, b]$ can be covered by an occurrence of s in u (at position a or earlier) or by a suffix of s which is a prefix of u. Hence, position $b - 1$ is not covered by s at all, a contradiction.   □

By Lemma 7, to complete the shortest seed algorithm it suffices to solve the Prefix Maxgap Problem (this is further clarified in the ComputeShortestSeed algorithm below). For this, we consider the following problem. By $SORT(X)$ we denote the sorted sequence of elements of $X \subseteq \{1, 2, \ldots, n\}$.

Chain Prefix Maxgap Problem
Input: a family of disjoint sets $X_1, X_2, \ldots, X_k \subseteq \{1, 2, \ldots, n\}$
       together with $SORT(X_1 \cup X_2 \cup \ldots \cup X_k)$.
       The size of the input is $m = \sum |X_i|$.
Output: the numbers $\Delta_i = \max_{j \leq i} \mathsf{maxgap}(X_j \cup X_{j+1} \cup \ldots \cup X_k)$.

Theorem 7. *The Chain Prefix Maxgap Problem can be solved in $O(m)$ time using an auxiliary array of size n.*

Proof. Initially we have the list $L = SORT(X_1 \cup X_2 \ldots \cup X_k)$. Let *pred* and *suc* denote the predecessor and successor of an element of L. The elements of L store a Boolean flag *marked*, initially set to false. In the algorithm we use an auxiliary array $pos[1 \mathinner{.\,.} n]$ such that $pos[i]$ is a pointer to the element of value i in L, if there is no such element then the value of $pos[i]$ can be arbitrary. Obviously the algorithm takes $O(m)$ time.   □

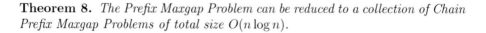

ALGORITHM ChainPrefixMaxgap(L)

1: $\Delta_1 := \mathsf{maxgap}(L)$; { naive computation }
2: **for** $j := 2$ **to** k **do**
3:     $\Delta_j := \Delta_{j-1}$;
4:     **for all** i in X_{j-1} **do** $marked(pos[i]) := \mathbf{true}$;
5:     **for all** i in X_{j-1} **do**
6:         $p := pred(pos[i])$; $q := suc(pos[i])$;
7:         **if** $(p \neq \mathbf{nil})$ **and** $(q \neq \mathbf{nil})$ **and** (**not** $marked(p)$) **and** (**not** $marked(q)$) **then**
8:             $\Delta_j := \max(\Delta_j, \; value(q) - value(p))$;
9:         $delete(L, pos[i])$;

Theorem 8. *The Prefix Maxgap Problem can be reduced to a collection of Chain Prefix Maxgap Problems of total size $O(n \log n)$.*

Proof. We solve a more abstract version of the Prefix Maxgap Problem. We are given an arbitrary tree T with n leaves annotated with distinct integers from the interval $[1, n]$, and we need to compute the values $\Delta(v)$ for all $v \in Nodes(T)$, defined as follows: $\mathsf{maxgap}(v) = \mathsf{maxgap}(LL(v))$, where $LL(v)$ is the leaf list of v, and $\Delta(v)$ is the maximum of the values maxgap on the path from v to the root of T. We start by sorting $LL(root(T))$, which can be done in $O(n)$ time. Throughout the algorithm we store a global auxiliary array $pos[1 .. n]$, required in the ChainPrefixMaxgap algorithm.

Let us find a *heaviest path* P in T, i.e., a path from the root down to a leaf, such that all *hanging* subtrees are of size at most $|T|/2$ each. The values of $\Delta(v)$ for $v \in P$ can all be computed in $O(n)$ time, using a reduction to the Chain Prefix Maxgap Problem (see Fig. 4).

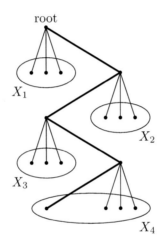

Fig. 4. A tree with an example heaviest path P (in bold). The values $\Delta(v)$ for $v \in P$ can be computed using a reduction to the Chain Prefix Maxgap Problem with the sets X_1 through X_4.

Then we perform the computation recursively for the hanging subtrees, previously sorting $LL(T')$ for each hanging subtree T'. Such sorting operations can be performed in $O(n)$ total time for all hanging subtrees.

At each level of recursion we need a linear amount of time, and the depth of recursion is logarithmic. Hence, the total size of invoked Chain Prefix Maxgap Problems is $O(n \log n)$.   □

Now we proceed to the shortest seed computation. In the algorithm we consider all factors of u, dividing them into groups corresponding to elements of $Nodes(u)$. Let $w \in Nodes(u)$ and let v be its parent. Let $s \in PREF(w)$ be a word containing v as a proper prefix, i.e., $s \in PREF(w, |v| + 1)$. By Lemma 7, the word s is a seed of u if and only if $|s| \geq \Delta(w)$ and s is a border seed of u.

Using the previously described reductions (Theorems 6–8), we obtain the following algorithm:

ALGORITHM ComputeShortestSeed(u)

1: Construct the suffix tree $T(u)$ for the input string u;
2: Solve the Prefix Maxgap Problem for $T(u)$ using the ChainPrefixMaxgap
3:     algorithm — in $O(n \log n)$ total time (Theorems 7 and 8);
4: $\mathcal{F} := \{\ PREF(w,\ \max(|v| + 1, \Delta(w)))\ :\ (v, w)$ is an edge in $T(u)\ \}$;
5: **return** bseed($u, \mathcal{F}$); { Theorem 6 }

Observe that the *workhorse* of the algorithm is the chain version of the Prefix Maxgap Problem, which has a fairly simple linear time solution. The main problem is of a structural nature, we have a collection of very simple problems each computable in linear time but the total size is not linear. This identifies the bottleneck of the algorithm from the complexity point of view.

5   Long Seeds

Note that the most time-expensive part of the ComputeShortestSeed algorithm is the computation of prefix maxgaps, all the remaining operations are performed in $O(n)$ time. Using this observation we can show a more efficient algorithm computing the shortest seed provided that its length m is sufficiently large. For example if $m = \Theta(n)$ then we obtain an $O(n)$ time algorithm for the shortest seed.

Theorem 9. *One can check if the shortest seed of a given string u has length at least m in $O(n \log (n/m))$ time, where $n = |u|$. If so, a corresponding seed can be reported within the same time complexity.*

Proof. We show how to modify the ComputeShortestSeed algorithm. Denote by s the shortest seed of u, $|s| = m$.

By Observation 1g, the longest overlap between consecutive occurrences of s in u is at most $\frac{m}{2}$, therefore the number of occurrences of s in u is at most $\frac{2n}{m}$. Hence, searching for the shortest seed of length at least m, it suffices to consider nodes v of the suffix tree $T(u)$ for which: $|v| \geq m$ and $|LL(v)| \leq \frac{2n}{m}$.

Thus, we are only interested in prefix maxgaps for nodes in several subtrees of $T(u)$, each of which contains $O(n/m)$ nodes. Thanks to the small size of each subtree, the algorithm ComputeShortestSeed finds all such prefix maxgaps in $O(n \log (n/m))$ time. Please note that using this algorithm for each node we obtain a prefix maxgap only in its subtree (not necessarily in the whole tree), however Lemma 7 can be simply adjusted to such a modified definition of prefix maxgaps.   □

References

1. Apostolico, A., Breslauer, D.: Of periods, quasiperiods, repetitions and covers. In: Structures in Logic and Computer Science, pp. 236–248 (1997)
2. Apostolico, A., Ehrenfeucht, A.: Efficient detection of quasiperiodicities in strings. Theor. Comput. Sci. 119(2), 247–265 (1993)
3. Berkman, O., Iliopoulos, C.S., Park, K.: The subtree max gap problem with application to parallel string covering. Inf. Comput. 123(1), 127–137 (1995)
4. Breslauer, D.: An on-line string superprimitivity test. Inf. Process. Lett. 44(6), 345–347 (1992)
5. Brodal, G.S., Pedersen, C.N.S.: Finding maximal quasiperiodicities in strings. In: Giancarlo, R., Sankoff, D. (eds.) CPM 2000. LNCS, vol. 1848, pp. 397–411. Springer, Heidelberg (2000)
6. Crochemore, M., Hancart, C., Lecroq, T.: Algorithms on Strings. Cambridge University Press, Cambridge (2007)
7. Crochemore, M., Iliopoulos, C., Kubica, M., Radoszewski, J., Rytter, W., Waleń, T.: Extracting powers and periods in a string from its runs structure. In: Chavez, E., Lonardi, S. (eds.) SPIRE 2010. LNCS, vol. 6393, pp. 258–269. Springer, Heidelberg (2010)
8. Crochemore, M., Rytter, W.: Jewels of Stringology. World Scientific, Singapore (2003)
9. Fischer, J., Heun, V.: A new succinct representation of RMQ-information and improvements in the enhanced suffix array. In: Chen, B., Paterson, M., Zhang, G. (eds.) ESCAPE 2007. LNCS, vol. 4614, pp. 459–470. Springer, Heidelberg (2007)
10. Gabow, H.N., Tarjan, R.E.: A linear-time algorithm for a special case of disjoint set union. In: Proceedings of the 15th Annual ACM Symposium on Theory of Computing (STOC), pp. 246–251 (1983)
11. Iliopoulos, C.S., Moore, D.W.G., Park, K.: Covering a string. Algorithmica 16(3), 288–297 (1996)
12. Li, Y., Smyth, W.F.: Computing the cover array in linear time. Algorithmica 32(1), 95–106 (2002)
13. Lothaire, M. (ed.): Algebraic Combinatorics on Words. Cambridge University Press, Cambridge (2001)
14. Lothaire, M. (ed.): Applied Combinatorics on Words. Cambridge University Press, Cambridge (2005)
15. Sadakane, K.: Succinct data structures for flexible text retrieval systems. J. Discrete Algorithms 5(1), 12–22 (2007)

Efficient Matching of Biological Sequences Allowing for Non-overlapping Inversions[*]

Domenico Cantone, Salvatore Cristofaro, and Simone Faro

Università degli Studi di Catania, Dipartimento di Matematica e Informatica
Viale Andrea Doria 6, I-95125, Catania, Italy
{cantone,cristofaro,faro} @dmi.unict.it

Abstract. Inversions are a class of chromosomal mutations, widely regarded as one of the major mechanisms for reorganizing the genome.

In this paper we present a new algorithm for the approximate string matching problem allowing for non-overlapping inversions which runs in $\mathcal{O}(nm)$ worst-case time and $\mathcal{O}(m^2)$-space, for a character sequence of size n and pattern of size m. This improves upon a previous $\mathcal{O}(nm^2)$-time algorithm.

1 Introduction

Retrieving information and teasing out the meaning of biological sequences are central problems in modern biology. Generally, basic biological information is stored in strings of nucleic acids (DNA, RNA) or amino acids (proteins). Aligning sequences helps in revealing their shared characteristics, while matching sequences can infer useful information from them. With the availability of large amounts of DNA data, matching of nucleotide sequences has become an important application and there is an increasing demand for fast computer methods for data analysis and retrieval.

Approximate string matching is a fundamental problem in text processing. It consists in finding approximate matches of a pattern in a text. The precision of a match is measured in terms of the sum of the costs of the edit operations necessary to convert the string into an exact match.

Most classical models, as for instance the Levenshtein or Damerau edit distance, assume that changes between strings occur only locally (for an in-depth survey on approximate string matching, see [7]). However, evidence shows that large scale changes, like duplications, translocations, and inversions, are common events in genetic evolution [4]. For instance, chromosomal inversions are rearrangements in which a segment of a chromosome is reversed end to end. Notice that inversions do not involve any loss of genetic information, but simply rearrange the linear gene sequence.

In this paper we are interested in the approximate string matching problem allowing for non-overlapping inversions. Much work has been made for the closely

---

[*] This work has been partly supported by G.N.C.S., Istituto Nazionale di Alta Matematica "Francesco Severi".

R. Giancarlo and G. Manzini (Eds.): CPM 2011, LNCS 6661, pp. 364–375, 2011.

related sequence alignment problem with inversions. Although the latter problem does not have a known polynomial algorithm in its full generality, when restricted to non-overlapping inversions it admits polynomial solutions. A first solution was proposed by Schöniger and Waterman [8]. Their algorithm, based on dynamic programming, runs in $\mathcal{O}(n^2m^2)$-time and $\mathcal{O}(n^2m^2)$-space on input sequences of length n and m. Later, Gao $et\ al.$ [2] developed a space-efficient variant which requires only $\mathcal{O}(nm)$-space (and still $\mathcal{O}(n^2m^2)$-time). More recently, Vellozo $et\ al.$ [9] proposed a $\mathcal{O}(nm^2)$-time and $\mathcal{O}(nm)$-space algorithm, within the more general framework of an edit graph.

Although proposed for the sequence alignment problem, the algorithm by Vellozo $et\ al.$ could also be adapted to the approximate string matching problem with non-overlapping inversions, yielding a $\mathcal{O}(nm^3)$-time and $\mathcal{O}(m^2)$-space solution to the latter problem. A more efficient solution, which runs in $\mathcal{O}(nm^2)$-time and $\mathcal{O}(m^2)$-space, was presented by Cantone $et\ al.$ [1]. They actually addressed a slightly more general problem, allowing also for translocations of equal length adjacent factors besides non-overlapping inversions. A very recent algorithm by Grabowski $et\ al.$ [5] solves the same matching problem, i.e., when translocations and non-overlapping inversions are allowed, in $\mathcal{O}(nm^2)$-time and $\mathcal{O}(m)$-space, obtaining better performances in practical cases.

In this paper we present an algorithm for the approximate string matching problem with non-overlapping inversions which runs in $\mathcal{O}(nm)$ worst-case time and $\mathcal{O}(m^2)$-space.

The paper is organized as follows. In Section 2 we provide the basic terminology and definitions. Next, in Section 3 we present a general $\mathcal{O}(nm^2)$-time and $\mathcal{O}(m^2)$-space algorithm for the approximate matching problem with non-overlapping inversions, based on the dynamic programming approach. Such algorithm will then be refined in Section 4, yielding a $\mathcal{O}(nm)$-time and $\mathcal{O}(m^2)$-space algorithm which constitutes the main result of the paper. Finally we draw our conclusions in Section 5.

2   Basic Notions and Properties

A string p of length $m \geq 0$ is represented as a finite array $p[0 .. m-1]$. In such a case we also write $|p| = m$. In particular, for $m = 0$ we obtain the empty string, denoted by ε. The concatenation of strings p and q is denoted as $p.q$ or, more simply, as pq. We denote with p^R the reversal of p, i.e., string p written in reverse order. Notice that $|p| = |p^R|$ and $(p^R)^R = p$. Moreover, for any two strings p and q, we have that $(p.q)^R = (q^R.p^R)$.

Given a nonempty string p and an integer i, we denote by $p[i]$ the $(i+1)$st symbol of p from left to right, if $0 \leq i < |p|$, otherwise we consider $p[i]$ as undefined.[1] Likewise, we denote with $p[i .. j]$ the substring of p contained between the $(i+1)$st and the $(j+1)$st symbol of p (both inclusive), for $0 \leq i \leq j < |p|$. Moreover, we put $p_j = p[0 .. j]$, for $0 \leq j < |p|$.

---

[1] When $p[i]$ is undefined, the condition $p[i] = c$, for any character symbol c, will be regarded as false, whereas the condition $p[i] \neq c$ will be regarded as true.

We say that p is a *prefix* (resp., *suffix*) of q, and write $p \sqsubseteq q$ (resp., $p \sqsupseteq q$), if there is a string s such that $q = p.s$ (resp., $q = s.p$). A string p is a *border* of q if both $p \sqsubseteq q$ and $p \sqsupseteq q$ hold. The set of the borders of p is denoted by $borders(p)$.

For a set S of strings, we denote by $\|S\|$ the collection of the lengths of the strings belonging to S, i.e, $\|S\| = \{|p| : p \in S\}$.

For strings p and q, we denote by $\langle p, q \rangle$ the set of all suffixes s of p such that s^R is a suffix of q, i.e., $\langle p, q \rangle = \{s : s \sqsupseteq p \text{ and } s^R \sqsupseteq q\}$.

The following lemma states useful properties of the set of borders of two strings.

Lemma 1. *For all strings p, q, v, w, and z, and every alphabet symbol c, the following facts hold:*

(a) *if $v, w \in \langle p, q \rangle$, then either $v \in borders(w)$ or $w \in borders(v)$;*
(b) *if $v, w \in \langle p, q \rangle$ and $|v| \geq |w|$, then $w \in borders(v)$;*
(c) *if $v \in \langle p, q \rangle$ and $w \in borders(v)$, then $w \in \langle p, q \rangle$;*
(d) *if z is the longest string belonging to $\langle p, q \rangle$, then $\langle p, q \rangle = borders(z)$;*
(e) *$\langle p, q.c \rangle = \{c.s : s \in \langle p, q \rangle \text{ and } c.s \sqsupseteq p\} \cup \{\varepsilon\}$;*
(f) *$\|\langle p, q.c \rangle\| = \{\ell + 1 : \ell \in \|\langle p, q \rangle\| \text{ and } p[|p| - 1 - \ell] = c\} \cup \{0\}$.*

Proof. First of all we notice that (b) and (f) are immediate consequences of (a) and (e), respectively; similarly, (d) follows plainly from (b) and (c). Thus, we only need to prove (a), (c), and (e).

We begin with (a). Let $v, w \in \langle p, q \rangle$. By the very definition of $\langle p, q \rangle$ we have

$$v \sqsupseteq p, \quad v^R \sqsupseteq q, \quad w \sqsupseteq p, \quad \text{and} \quad w^R \sqsupseteq q.$$

Without loss of generality, let us assume that $|v| \leq |w|$. Then, from $v \sqsupseteq p$ and $w \sqsupseteq p$ we have $v \sqsupseteq w$; likewise, from $v^R \sqsupseteq q$ and $w^R \sqsupseteq q$ we have $v^R \sqsupseteq w^R$. The latter implies $v \sqsubseteq w$, which, together the previously established relation $v \sqsupseteq w$, yields $v \in borders(w)$, proving (a).

Concerning (c), let $v \in \langle p, q \rangle$ and $w \in borders(v)$. Then we have $v \sqsupseteq p$ and $w \sqsupseteq v$, so that $w \sqsupseteq p$. Likewise, we have $v^R \sqsupseteq q$ and $w \sqsubseteq v$. The latter is equivalent to $w^R \sqsupseteq v^R$, so that $w^R \sqsupseteq q$. From $w \sqsupseteq p$ and $w^R \sqsupseteq q$ it follows that $w \in \langle p, q \rangle$, proving (c).

Finally, we turn to the proof of (e). Let $v \in \langle p, q.c \rangle$, where c is a character. Then $v \sqsupseteq p$ and $v^R \sqsupseteq q.c$. If $v \neq \varepsilon$, then $v = c.s$, for a string s such that $s \sqsubseteq q^R$. But then $s \sqsupseteq (q^R)^R = q$, which, together with $s \sqsupseteq p$, implies

$$\langle p, q.c \rangle \subseteq \{c.s : s \in \langle p, q \rangle \text{ and } c.s \sqsupseteq p\} \cup \{\varepsilon\}.$$

To show the converse inclusion, we observe preliminarily that $\varepsilon \in \langle p, q.c \rangle$. Let $s \in \langle p, q \rangle$ such that $c.s \sqsupseteq p$. Then $s^R \sqsupseteq q$, which implies $(c.s)^R = s^R.c \sqsupseteq q.c$. The latter, together with $c.s \sqsupseteq p$, implies $c.s \in \langle p, q.c \rangle$. Thus

$$\{c.s : s \in \langle p, q \rangle \text{ and } c.s \sqsupseteq p\} \cup \{\varepsilon\} \subseteq \langle p, q.c \rangle,$$

which together with the previously established inclusion proves (e). $\qquad\square$

Given two strings p and q of the same length m, an *inverted decomposition* of p and q is a sequence $(\ell_1, \ell_2, \ldots, \ell_k)$ of lengths such that:

(a) $1 \leq \ell_i \leq m$, for $1 \leq i \leq k$;

(b) $\sum_{i=1}^{k} \ell_i = m$;

(c) $p[L_j \mathrel{..} L_{j+1}] = (q[L_j \mathrel{..} L_{j+1}])^{\mathrm{R}}$, for $0 \leq j < k$, and where $L_j = \sum_{i=1}^{j} \ell_i$ (so that $L_0 = 0$).

When p and q admit an inverted decomposition, we write $p \bowtie q$.

Observe that an inverted decomposition $(\ell_1, \ell_2, \ldots, \ell_k)$ of p and q induces a sequence of strings $(s_1, s_2, \ldots, s_k)$ such that $s_1 s_2 \cdots s_k = p$ and $s_1^{\mathrm{R}} s_2^{\mathrm{R}} \cdots s_k^{\mathrm{R}} = q$, and conversely. Thus, we plainly have that $p \bowtie q$ iff $q \bowtie p$. Additionally, the following property can be easily proved.

Lemma 2. *For all strings p and q, we have that $p \bowtie q$ holds iff (exactly) one of the following two conditions holds:*

(a) $p = q = \varepsilon$, or

(b) $p = v.z$ and $q = w.z^{\mathrm{R}}$, for a string $z \neq \varepsilon$ and strings v and w such that $v \bowtie w$. □

Given a text t of length n, a pattern p of length m is said to *match with non-overlapping inversions* (or to have an *occurrence with non-overlapping inversions*) at location i of t if $p \bowtie t[i \mathrel{..} i + m - 1]$, i.e., if there exists an inverted decomposition of p and $t[i \mathrel{..} i + m - 1]$.

The *approximate matching problem with non-overlapping inversions* is to find all locations i in a given text t at which a given pattern p matches with non-overlapping inversions.

For the sake of simplicity, in the rest of the paper we will refer to non-overlapping inversions simply as *inversions*, since this will generate no confusion.

3   A General Dynamic Programming Approach

In this section we present a general dynamic programming algorithm for the pattern matching problem with inversions. Our algorithm, which will be named DPInversionMatcher, is characterized by a $\mathcal{O}(nm^2)$-time and a $\mathcal{O}(m^2)$-space complexity, where m and n are the length of the pattern and text, respectively. In the next section we will then show how it can be refined so as to improve its time complexity to $\mathcal{O}(nm)$.

As above, let t be a text of length n and p a pattern of length m. The algorithm DPInversionMatcher solves the matching problem with inversions by computing the occurrences of all prefixes of the pattern in continuously increasing prefixes of the text using a dynamic programming approach. That is, during its $(i+1)$st iteration, for $i = 0, 1, \ldots, n-m$, our algorithm establishes whether $p_j \bowtie t[i \mathrel{..} i+j]$, for each $j = 0, 1, \ldots, m - 1$, exploiting information gathered during previous iterations.

To begin with, we denote by $M(j, i)$ the set of all integral values k, with $0 \le k \le j$, such that the prefix p_k has an occurrence with inversions at location i of the text, or more formally

$$M(j, i) = \begin{cases} \{0 \le k \le j : p_k \bowtie t[i .. i+k]\} \cup \{-1\} & \text{if } i \ge 0 \text{ and } j \ge 0 \\ \{-1\} & \text{otherwise,} \end{cases}$$

for $-m \le i \le n - m$ and $0 \le j < m$.

Then notice that $p_j \bowtie t[i .. i+j]$ iff $j \in M(j, i)$ and hence $p \bowtie t[i .. i+m-1]$ iff $m - 1 \in M(m-1, i)$. Thus the matching problem with inversions can be solved by computing the sets $M(m-1, i)$, for increasing values of i.

We also define $R(j, i)$ as the set of the lengths of all strings s such that $s \sqsupseteq p_j$ and $s^R \sqsupseteq t_{i+j}$, or more formally

$$R(j, i) = \begin{cases} \|\langle p_j, t_{i+j} \rangle\| & \text{if } 0 \le j < m \text{ and } 0 \le i \le m - n \\ \{0\} & \text{otherwise.} \end{cases}$$

By Lemma 2, we obtain the following recursive relation

$$M(j, i) = \begin{cases} M(j-1, i) \cup \{j\} & \text{if } j - \ell \in M(j-1, i), \text{ for some } \ell \in R(j, i) \\ M(j-1, i) & \text{otherwise,} \end{cases}$$

where $0 \le j < m$ and $-m \le i \le n - m$, which allows to reduce the computation of the set $M(j, i)$ to that of the sets $M(j-1, i)$ and $R(j, i)$.

Likewise, the sets $R(j, i)$ can be computed by the recursive relation

$$R(j, i) = \{0\} \cup \{\ell + 1 : \ell \in R(j, i-1) \text{ and } p[j - \ell] = t[i + j]\},$$

with $0 \le j < m$ and $0 < i \le n - m$, which follows from Lemma 1(f).

The above considerations translate directly into the algorithm DPInversion-Matcher in Fig. 1. Sets $R(j, i)$ are maintained by an array R of dimension m; more precisely, just after iteration i of the for-loop at line 3, we have that $\mathsf{R}[j] = R(j, i)$. Similarly, sets $M(j, i)$ are maintained by a single set(-variable) M, which is initialized to $\{-1\}$ at the beginning of iteration i of the for-loop at line 3 (this corresponds to the set $M(-1, i)$). Then, during the execution of the subsequent for-loop at line 5, the set M is expanded so as to take in sequence the relevant elements $M(0, i), M(1, i), \ldots, M(m-1, i)$; more precisely, just after the execution of iteration j of the for-loop at line 5 we have that $\mathsf{M} = M(j, i)$.

The set M can be implemented as a linear array $\mathcal{A}$ of length $m + 1$ of Boolean values such that

$$\mathcal{A}[j] = \begin{cases} \text{true,} & \text{if } j - 1 \in \mathsf{M} \\ \text{false,} & \text{otherwise,} \end{cases}$$

for $0 \le j \le m$. Likewise, each set $\mathsf{R}[j]$ can be implemented as a linked list (or possibly as an array of length $m + 1$ of Boolean values, as well). Then it follows easily that the algorithm DPInversionMatcher in Fig. 1 has a $\mathcal{O}(m^2)$-space complexity and a $\mathcal{O}(nm^2)$-time complexity. Indeed, the computation of the set $\mathsf{R}[j]$ at line 6 and the conditional test at line 7 require $\mathcal{O}(j)$-time, for $0 \le j < m$.

```
DPInversionMatcher(p, m, t, n)
  1.   for j := 0 to m − 1 do
  2.     R[j] := {0}
  3.   for i := −m + 1 to n − m do
  4.     M := {−1}
  5.     for j = max(−i, 0) to m − 1 do
  6.       R[j] := {0} ∪ {ℓ + 1 : ℓ ∈ R[j] \ {j + 1} AND p[j − ℓ] = t[i + j]}
  7.       if (∃ℓ ∈ R[j] : j − ℓ ∈ M) then
  8.         M := M ∪ {j}
  9.       if (m − 1 ∈ M) then
 10.         output(i)
```

Fig. 1. The algorithm DPInversionMatcher for the matching problem with inversions

4   The Algorithm InversionSampling

In this section we present a refinement of the algorithm DPInversionMatcher presented before. The new algorithm, named InversionSampling, achieves a $\mathcal{O}(nm)$ worst-case time complexity and, as before, requires $\mathcal{O}(m^2)$ additional space.

The main idea upon which the new algorithm is based is that we do not need to maintain explicitly the whole set $R(j, i)$ to evaluate the conditional test at line 7. In particular we show that by efficiently computing the values in the set $R(j, i)$, each conditional test at line 7 can be performed in amortized $\mathcal{O}(1)$-time.

Specifically, as will be proved in Lemma 5 below, during each iteration of the algorithm DPInversionMatcher, just before the execution of the conditional test at line 7, the following condition holds

$$\begin{aligned}\text{either } &\{\ell \in R(j, i) : j - \ell \in M(j - 1, i)\} = \emptyset \quad \text{(when the test is false)}, \\ \text{or } &\{\ell \in R(j, i) : j - \ell \in M(j - 1, i)\} = R(j, i) \setminus \{0\}.\end{aligned} \tag{1}$$

Thus it follows that, for each $0 < j < m$ and $-m \leq i \leq n-m$, if $\max(R(j, i)) \in M(j - 1, i)$ then $\{\ell \in R(j, i) : j - \ell \in M(j - 1, i)\} = R(j, i) \setminus \{0\}$.

Since just before the execution of the conditional test at line 7 of the algorithm DPInversionMatcher we have that $j \notin M(j-1, i)$, the condition '$(\exists \ell \in \mathsf{R}[j] : j - \ell \in \mathsf{M})$' at line 7 can be replaced by the condition '$j - e(\mathsf{R}[j]) \in \mathsf{M}$', for any function $e(\cdot)$ such that $e(\mathsf{R}[j]) \in (\mathsf{R}[j] \setminus \{0\}) \cup \{\max(\mathsf{R}[j])\}$ holds, without affecting the correctness of the algorithm. In particular, we choose $e(\cdot) \equiv \max(\cdot)$ and describe an efficient way to compute the value $\max(R(j, i))$, which allows to reduce the time complexity of the searching-phase of the algorithm to $\mathcal{O}(nm)$.

Recalling that $R(j, i) = \|\langle p_j, t_{i+j} \rangle\|$, it turns out that the maximum of the set $R(j, i)$, for $-m < i \leq n - m$ and $0 \leq j < m$, can be computed from the maximum of the set $R(j, i-1)$, without any need to compute explicitly the whole set $R(j, i)$. This can be done by using the following relation:

$$\max(\|\langle p_j, t_{i+j} \rangle\|) = \max\{\ell + 1 : \ell \in \langle p_j, t_{i+j-1} \rangle \text{ and } p[j - \ell] = t[i + j]\}, \tag{2}$$

```
procMPT(p, m)
 1.  for k := 0 do m − 1 do
 2.     i := 0
 3.     j := W[k, k] := −1
 4.     while (i < (m − k)) do
 5.        while ((j > −1) and
               (p[i + k] ≠ p[j + k])) do
 6.           j := W[k, k + j]
 7.        i := i + 1
 8.        j := j + 1
 9.        W[k, k + i] := j
10.  return(W)
```

```
InversionSampling(p, m, t, n)
 1.  W := procMPT(p, m)
 2.  for j := 0 to m − 1 do
 3.     K[j] := 0
 4.  for i = −m + 1 to n − m do
 5.     M = {−1}
 6.     for j := max(−i, 0) to m − 1 do
 7.        while ((K[j] > 0)
                and (p[j − K[j]] ≠ t[i + j])) do
 8.           K[j] := W[j + 1 − K[j], j + 1]
 9.        if p[j − K[j]] = t[i + j] then
10.           K[j] := K[j] + 1
11.        if (j − K[j] ∈ M) then
12.           M := M ∪ {j}
13.     if (m − 1) ∈ M then
14.        output(i)
```

Fig. 2. (On the left) the procedure for computing the table W, and (on the right) the variant InversionSampling of the algorithm DPInversionMatcher

which will be proved in Lemma 6 below.

Let $\|\langle p_j, t_{i+j-1}\rangle\|$ be the set $\{\ell_1, \ell_2, \ldots, \ell_k\}$, with $\ell_i > \ell_{i+1}$, for all $0 < i < k$, and $\ell_k = 0$. For the computation of the set $\max(\|\langle p_j, t_{i+j}\rangle\|)$ we start from the value $\ell_1 = \max(\|\langle p_j, t_{i+j-1}\rangle\|)$, and examine in sequence the items $\ell_1, \ell_2, \ldots, \ell_k$ until we find a value ℓ_i such that $p[j - \ell_i] = t[i + m − 1]$ or we reach $\ell_k = 0$. If ℓ is the value obtained by such scanning process, we check whether $p[j - \ell] = t[i + j]$ and, in this case, we conclude that $\max(\|\langle p_j, t_{i+j}\rangle\|) = \ell + 1$; otherwise we conclude that $\max(\|\langle p_j, t_{i+j}\rangle\|) = 0$.

The above procedure requires to know in advance the set $\|\langle p_j, t_{i+j-1}\rangle\|$. To this purpose let us put

$$\pi(p_j, h) = \begin{cases} \max(\|borders(p[h\mathinner{..}j]) \setminus \{p[h\mathinner{..}j]\}\|) & \text{if } 0 \leq h \leq j \\ -1 & \text{otherwise.} \end{cases}$$

For $i = 1, \ldots, k$, let us also put $v_i = p[j + 1 − \ell_i \mathinner{..} j]$. Then, since v_{i+1} is a border of v_i (by Lemma 1(b)), we have that $\ell_{i+1} = \pi(p_j, j + 1 − \ell_i)$, for $0 < i < k$.

Such values can be precomputed and collected into a table W of dimensions $(m + 1) \times (m + 1)$, where $W[0, 0] = −1$ and $W[h, k] = \pi(p_{k-1}, h)$, for $0 < k \leq m$ and $0 \leq h \leq k$ (the values of the remaining entries of W are not relevant).

Table W can be computed in $\mathcal{O}(m^2)$-time and space by means of the procedure procMPT in Fig. 2, which is a generalization of the procedure used by the Morris-Pratt algorithm [6] for computing the length of the longest proper border of $s[0\mathinner{..}j]$, for a given string s with $0 \leq j < |s|$ (see also [3], where this function is called the *prefix function* of the pattern).

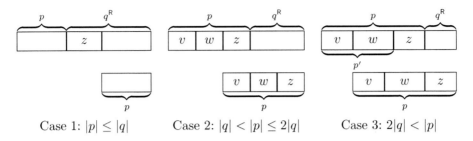

Case 1: $|p| \leq |q|$ Case 2: $|q| < |p| \leq 2|q|$ Case 3: $2|q| < |p|$

Fig. 3. The three cases considered in Lemma 3

The resulting algorithm, InversionSampling, is presented in Fig. 2. Notice that the part of the code from line 7 up to line 10 implements the assignment $K[j] := \max(R(j,i))$.

4.1   Correctness Issues

In this section we prove the validity of (1) and (2), upon which the correctness of the algorithm InversionSampling is based. In particular, they will be direct consequences of Lemmas 5 and 6, respectively.

We first state and prove two useful properties related to the suffixes of inverted strings, which will be used in our main results.

Lemma 3. *Let p and q be strings such that $p \sqsupseteq p.q^R$. Then there exist two strings q_1 and q_2 such that (a) $q = q_1.q_2$ and (b) $p.q^R = q_1^R.q_2^R.p$.* $\square$

Proof. Let $p \sqsupseteq p.q^R$. To begin with, notice that if $|q| = 0$, the lemma follows trivially. So, let us suppose that $|q| > 0$ and assume inductively that the lemma holds for any pair p', q' of strings such that $|p'| < |p|$ and $p' \sqsupseteq p'.q'^R$.

We distinguish the following three cases (see Fig. 3 for a pictorial illustration).

Case 1: $|p| \leq |q|$. From $p \sqsupseteq p.q^R$ and $|p| \leq |q|$, it follows that $p \sqsupseteq q^R$, so that $q^R = z.p$, for some string z. Putting $q_1 = p^R$ and $q_2 = z^R$, we have then $q_1.q_2 = p^R.z^R = (z.p)^R = (q^R)^R = q$, and $p.q^R = p.z.p = (p^R)^R.(z^R)^R.p = q_1^R.q_2^R.p$ and therefore (a) and (b) are both satisfied in the present case.

Case 2: $|q| < |p| \leq 2|q|$. Let z be the suffix of p such that $|z| = |p| - |q| \leq |q|$. Observe that $2|z| \leq |z| + |q| = |p|$, so that $|z| \leq \lfloor |p|/2 \rfloor$. Therefore p can be decomposed as $p = v.w.z$, with $|v| = |z|$ and $|w.z| = |q|$. But since $p \sqsupseteq p.q^R$, we have $v = z$ and $q^R = w.z$. If we put $q_1 = z^R$ and $q_2 = w^R$, so that $z = q_1^R$ and $w = q_2^R$, we have $q = (q^R)^R = (w.z)^R = (q_2^R.q_1^R)^R = (q_1^R)^R.(q_2^R)^R = q_1.q_2$ and $p.q^R = (z.w.z).(w.z) = (z.w).(z.w.z) = z.w.p = q_1^R.q_2^R.p$, proving (a) and (b) in the present case.

Case 3: $2|q| < |p|$. Let v and z be, respectively, the prefix and the suffix of p such that $|v| = |z| = |q|$. Plainly, $z = q^R$, as $p \sqsupseteq p.q^R$. In addition, since $|v| + |z| = 2|q| < |p|$, it follows that $p = v.w.z$, for a nonempty string w. Let us put $p' = v.w$, so that $p = p'.z$. Observe that $|p'| = |p| - |q| < |p|$, since

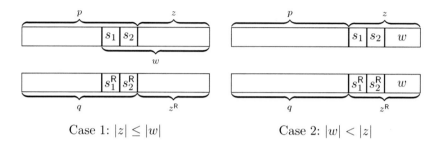

Case 1: $|z| \leq |w|$ Case 2: $|w| < |z|$

Fig. 4. The two cases considered in Lemma 4

$|q| > 0$. We have also $p' \sqsupseteq p'.z$, so by induction $z^{\mathsf{R}} = q_1.q_2$ (i.e., $q = q_1.q_2$) and $p'.z = q_1^{\mathsf{R}}.q_2^{\mathsf{R}}.p'$, for some strings q_1 and q_2. Hence, $p.q^{\mathsf{R}} = (v.w.z).q^{\mathsf{R}} = (p'.z).q^{\mathsf{R}} = (q_1^{\mathsf{R}}.q_2^{\mathsf{R}}.p').q^{\mathsf{R}} = q_1^{\mathsf{R}}.q_2^{\mathsf{R}}.(p'.q^{\mathsf{R}}) = q_1^{\mathsf{R}}.q_2^{\mathsf{R}}.(p'.z) = q_1^{\mathsf{R}}.q_2^{\mathsf{R}}.p$, so that (a) and (b) hold in this last case too, completing the proof of the lemma. □

Lemma 4. *Let p and q be strings of the same length. Then we have $p.z \bowtie q.z^{\mathsf{R}}$ if and only if $p \bowtie q$, for every string z.*

Proof. To begin with, notice that if $p \bowtie q$ then, plainly, $p.z \bowtie q.z^{\mathsf{R}}$. Thus, it is enough to prove the converse implication, namely that $p.z \bowtie q.z^{\mathsf{R}}$ implies $p \bowtie q$. So, let p, q, and z be nonempty strings such that $p.z \bowtie q.z^{\mathsf{R}}$ and assume inductively that the lemma is true for all triplets p', q', z', with $|z'| < |z|$, such that $p'.z' \bowtie q'.z'^{\mathsf{R}}$. By Lemma 2, there are strings u, v, and $w \neq \varepsilon$ such that $u \bowtie v$, $p.z = u.w$, and $q.z^{\mathsf{R}} = v.w^{\mathsf{R}}$.

We consider first the case in which $|z| \leq |w|$ (this is illustrated in Fig. 4, on the left). Since $z \sqsupseteq u.w$ and $z^{\mathsf{R}} \sqsupseteq v.w^{\mathsf{R}}$, we have that $z \sqsupseteq w$ and $z^{\mathsf{R}} \sqsupseteq w^{\mathsf{R}}$. Let s be the string such that $w = s.z$. Then, $z^{\mathsf{R}} \sqsupseteq w^{\mathsf{R}} = (s.z)^{\mathsf{R}} = z^{\mathsf{R}}.s^{\mathsf{R}}$ and hence, by Lemma 3, there are strings s_1 and s_2 such that $s_1.s_2 = s$ (which implies that $w = s_1.s_2.z$) and $z^{\mathsf{R}}.s^{\mathsf{R}} = s_1^{\mathsf{R}}.s_2^{\mathsf{R}}.z^{\mathsf{R}}$, i.e., $w^{\mathsf{R}} = s_1^{\mathsf{R}}.s_2^{\mathsf{R}}.z^{\mathsf{R}}$. Therefore, we have that $p.z = u.s_1.s_2.z$ and $q.z^{\mathsf{R}} = v.s_1^{\mathsf{R}}.s_2^{\mathsf{R}}.z^{\mathsf{R}}$. These equalities imply, respectively, that $p = u.s_1.s_2$ and $q = v.s_1^{\mathsf{R}}.s_2^{\mathsf{R}}$, and hence, as $u \bowtie v$, by a double application of Lemma 2, we get $p \bowtie q$.

Let us consider next the case in which $|w| < |z|$ (this is illustrated in Fig. 4, on the right). Since $w \sqsupseteq p.z$ and $w^{\mathsf{R}} \sqsupseteq q.z^{\mathsf{R}}$, in this case we have that $w \sqsupseteq z$ and $w^{\mathsf{R}} \sqsupseteq z^{\mathsf{R}}$. Let s be the string such that $z = s.w$. Then, $w^{\mathsf{R}} \sqsupseteq z^{\mathsf{R}} = (s.w)^{\mathsf{R}} = w^{\mathsf{R}}.s^{\mathsf{R}}$, and hence, by Lemma 3, there are strings s_1 and s_2 such that $s_1.s_2 = s$ (which implies that $z = s_1.s_2.w$) and $w^{\mathsf{R}}.s^{\mathsf{R}} = s_1^{\mathsf{R}}.s_2^{\mathsf{R}}.w^{\mathsf{R}}$, i.e., $z^{\mathsf{R}} = s_1^{\mathsf{R}}.s_2^{\mathsf{R}}.w^{\mathsf{R}}$. Therefore, we have that $p.z = p.s_1.s_2.w$ and $q.z^{\mathsf{R}} = q.s_1^{\mathsf{R}}.s_2^{\mathsf{R}}.w^{\mathsf{R}}$, which imply, respectively, that $u.w = p.s_1.s_2.w$ and $v.w^{\mathsf{R}} = q.s_1^{\mathsf{R}}.s_2^{\mathsf{R}}.w^{\mathsf{R}}$, so that $u = p.s_1.s_2$ and $v = q.s_1^{\mathsf{R}}.s_2^{\mathsf{R}}$. Since $|s_2| < |z|$, by induction we deduce $p.s_1 \bowtie q.s_1^{\mathsf{R}}$ from $p.s_1.s_2 = u \bowtie v = q.s_1^{\mathsf{R}}.s_2^{\mathsf{R}}$. Likewise, since $|s_1| < |z|$, again by induction we deduce $p \bowtie q$ from $p.s_1 \bowtie q.s_1^{\mathsf{R}}$. Thus, $p \bowtie q$ holds even when $|w| < |z|$, concluding the proof of the lemma. □

Correctness of (1) is a direct consequence of the following lemma.

Lemma 5. *Let $D(j, i) = \{\ell \in R(j, i) : j - \ell \in M(j - 1, i)\}$, for $0 < j < m$ and $-m \leq i \leq n - m$. Then we have either $D(j, i) = \emptyset$ or $D(j, i) = R(j, i) \setminus \{0\}$.*

Proof. First of all, note that if $D(j, i) \neq \emptyset$ then $j \in M(j, i)$, i.e., $p_j \bowtie t[i .. i + j]$, so that by Lemma 4, we must have $p_{j-k} \bowtie t[i .. i + j - k]$ for all $k \in R(j, i) \setminus \{0\}$. But $p_{j-k} \bowtie t[i .. i + j - k]$, with $k \neq 0$, implies that $j - k \in M(j - 1, i)$, and thus $R(j, i) \setminus \{0\} \subseteq D(j, i)$. The converse implication, i.e., $D(j, i) \subseteq R(j, i) \setminus \{0\}$, holds trivially, since $j \notin M(j - 1, i)$.   □

Finally, relation (2), which allows to compute the maximum of the set $R(j, i)$ from the maximum of the set $R(j, i - 1)$, is established in the following lemma.

Lemma 6. *Given two strings p and q, with $|p| = m$, and a character c, we have*

$$\max(\|\langle p, q.c \rangle\|) = \max\{|v| + 1 : v \in \langle p, q \rangle \text{ and } p[m - 1 - |v|] = c\}.$$

Proof. Let z be the longest string belonging to $\langle p, q \rangle$, so that $|z| = \max(\|\langle p, q \rangle\|)$, and let $v_1, v_2, \ldots, v_k$ be the borders of z, ordered by their decreasing lengths. Observe that if $v, w \in \langle p, q \rangle$, then $v \sqsupseteq p$ and $w \sqsupseteq p$, so that if v and w have the same length they must coincide. Hence, the set $\langle p, q \rangle$ cannot contain any two distinct strings of the same length. It also follows that the longest string belonging to $\langle p, q \rangle$ is well (and uniquely) defined. Also, note that a string z cannot have two distinct borders of the same length. Thus we have

$$|v_1| > |v_2| > \cdots > |v_k|,$$

with $v_1 = z$ and $v_k = \varepsilon$. Then, from Lemma 1(d) it follows that $\langle p, q \rangle = \{v_1, v_2, \ldots, v_k\}$ which, by Lemma 1(d), yields

$$\|\langle p, q.c \rangle\| = \{|v| + 1 : v \in \{v_1, \ldots, v_k\} \text{ and } p[m - 1 - |v|] = c\} \cup \{0\},$$

completing the proof of the lemma.   □

4.2   Worst-Case Time Analysis

We show now that the worst-case time complexity $T(n, m)$ of the algorithm InversionSampling reported in Fig. 2 is $\mathcal{O}(nm)$, for an input text t of length n and pattern p of length m.

To begin with, we observe that the preprocessing phase of the algorithm requires $\mathcal{O}(m^2)$-time (and space), due to the computation of the table W and the initialization at line 2. Next we evaluate the complexity of the searching phase, namely of the for-loop at line 4. Let us denote by A the set of pairs $\{-m + 1, \ldots, n - m\} \times \{0 \ldots, m - 1\}$. For each pair $(i, j) \in A$, we let C1(i, j) be the number of times that the while-loop at line 7 is executed during iteration i of the for-loop at line 4, and we let $K(i, j)$ be the value contained in K$[j]$ just after the termination of such iteration; in addition, we put C2$(i, j) = 1$, if the assignment instruction at line 10 is executed during iteration i, otherwise we put $\texttt{C2}(i, j) = 0$. Plainly, we have that

$$T(n, m) = \mathcal{O}\left(\sum_{i=-m+1}^{n-m} \sum_{j=0}^{m-1} (\texttt{C1}(i, j) + 1) \right), \tag{3}$$

and therefore it is enough to prove that the double summation in (3) is asymptotically bounded above by the product nm.

Since $\texttt{C2}(i, j) \leq 1$ for each $(i, j) \in A$, we have that, for $0 \leq j < m$,

$$\sum_{i=-m+1}^{n-m} \texttt{C2}(i, j) \leq n. \tag{4}$$

On the other hand, we have also that

$$K(i + 1, j) - \texttt{C2}(i + 1, j) \leq K(i, j) - \texttt{C1}(i + 1, j), \tag{5}$$

for all $(i, j) \in A$ such that $i < n - m$. Indeed, during iteration i, the value contained in $\mathsf{K}[j]$ just after the execution of the while-loop at line 7 (i.e., $K(i + 1, j) - \texttt{C2}(i + 1, j)$) can never exceed the value contained in $\mathsf{K}[j]$ just before this execution minus the number of times that the while-loop iterates (i.e., $K(i, j) - \texttt{C1}(i + 1, j)$), since $\mathsf{K}[j]$ is decremented at least by one unit during each iteration of the while-loop at line 7. Thus it follows that

$$0 \leq K(h, j) \leq \sum_{i=-m+1}^{h} \texttt{C2}(i, j) - \sum_{i=-m+1}^{h} \texttt{C1}(i, j), \tag{6}$$

for all $(h, j) \in A$, as can be verified by induction on h, using (5). From (6) it follows that

$$\sum_{j=0}^{m-1} \sum_{i=-m+1}^{n-m} (\texttt{C1}(i, j) + 1) \leq \sum_{j=0}^{m-1} \sum_{i=-m+1}^{n-m} (\texttt{C2}(i, j) + 1),$$

and thus, using (4), we finally obtain that

$$\sum_{j=0}^{m-1} \sum_{i=-m+1}^{n-m} (\texttt{C1}(i, j) + 1) \leq (n + 1)m,$$

which in turn, by (3), yields $T(n, m) = \mathcal{O}(nm)$.

5   Conclusions and Future Work

In this paper we have presented an algorithm to solve the pattern matching problem under a string distance which allows inversions of non-overlapping factors. The algorithm, named InversionSampling, has worst case $\mathcal{O}(nm)$-time and $\mathcal{O}(m^2)$-space complexity, where m and n are the length of the pattern and the length of the text, respectively. We are currently working on an efficient variant of the present algorithm with a linear average time complexity.

Acknowledgements

We thank the anonymous reviewers for their helpful comments.

References

1. Cantone, D., Faro, S., Giaquinta, E.: Approximate string matching allowing for inversions and translocations. In: Holub, J., Zdárek, J. (eds.) Proceedings of the Prague Stringology Conference 2010, pp. 37–51. Czech Technical University, Prague, Czech Republic (2010)
2. Chen, Z., Gao, Y., Lin, G., Niewiadomski, R., Wang, Y., Wu, J.: A space-efficient algorithm for sequence alignment with inversions and reversals. Theor. Comput. Sci. 325(3), 361–372 (2004)
3. Cormen, T.H., Leiserson, C.E., Rivest, R.L., Stein, C.: Introduction to Algorithms, 2nd edn. MIT Press, Cambridge (2001)
4. Devos, K.M., Atkinson, M.D., Chinoy, C.N., Francis, H.A., Harcourt, R.L., Koebner, R.M.D., Liu, C.J., Masoj, P., Xie, D.X., Gale, M.D.: Chromosomal rearrangements in the rye genome relative to that of wheat. TAG Theoretical and Applied Genetics 85, 673–680 (1993)
5. Grabowski, S., Faro, S., Giaquinta, E.: String matching with inversions and translocations in linear average time (most of the time). Information Processing Letters 111(11), 516–520 (2011)
6. Morris Jr, J.H., Pratt, V.R.: A linear pattern-matching algorithm. Report 40. University of California, Berkeley (1970)
7. Navarro, G.: A guided tour to approximate string matching. ACM Comput. Surv. 33(1), 31–88 (2001)
8. Schniger, M., Waterman, M.: A local algorithm for DNA sequence alignment with inversions. Bulletin of Mathematical Biology 54, 521–536 (1992)
9. Vellozo, A.F., Alves, C.E.R., Pereira do Lago, A.: Alignment with non-overlapping inversions in $\mathcal{O}(n^3)$-time. In: WABI, pp. 186–196 (2006)

A Coarse-to-Fine Approach to Computing the k-Best Viterbi Paths

Jesper Nielsen

Bioinformatics Research Centre, Aarhus University,
C.F. Møllers Alle 8, DK-8000 Aarhus C, Denmark
jn@birc.au.dk

Abstract. The Hidden Markov Model (HMM) is a probabilistic model used widely in the fields of Bioinformatics and Speech Recognition. Efficient algorithms for solving the most common problems are well known, yet they all have a running time that is quadratic in the number of hidden states, which can be problematic for models with very large state spaces. The Viterbi algorithm is used to find the maximum likelihood hidden state sequence, and it has earlier been shown that a coarse-to-fine modification can significantly speed up this algorithm on some models. We propose combining work on a k-best version of Viterbi algorithm with the coarse-to-fine framework. This algorithm may be used to approximate the total likelihood of the model, or to evaluate the goodness of the Viterbi path on very large models.

Keywords: coarse-to-fine; k-best; Viterbi; Hidden Markov Models.

1 Introduction

A Hidden Markov Model (HMM) [16] is a probabilistic model, in which there is a series of hidden states evolving through time, each state depending only on the previous state. At each time-step the current hidden state emits an observable symbol, with the hidden state determining the probability of a given observable symbol being emitted. HMMs are used widely in many fields, particularly Bioinformatics [1, 5–7, 9, 13, 14, 18, 19, 21], and Speech Recognition [4, 15, 20].

One of the reasons for the success of the HMM framework is probably the existence of simple and efficient algorithms for solving the most common problems associated with HMMs. The Viterbi algorithm computes the maximum likelihood sequence of hidden states given the model and observed symbols; the forward algorithm computes the total likelihood of the observed symbols given the model; the backward algorithm, used together with the forward algorithm, can give the total likelihood of a given hidden state at a given point in time; and this again can be used by the Baum-Welch algorithm to learn the probabilities in the model from data.

The Viterbi, forward, and backward algorithms are all similar and all have an execution time that is linear in the product of the number of time-steps and the number of possible transitions between hidden states. For most use cases this is

R. Giancarlo and G. Manzini (Eds.): CPM 2011, LNCS 6661, pp. 376–387, 2011.
© Springer-Verlag Berlin Heidelberg 2011

good enough, but if the transition matrix is dense the number of transitions is quadratic in the number of states, and if the number of states is large this can be a problem.

To achieve fast execution of the Viterbi algorithm on HMMs with a large number of hidden states a coarse-to-fine framework has been proposed and used successfully [2, 10, 17]. The idea is to approximate the desired HMM by a series of coarse HMMs with increasingly fewer states, each of the coarse states corresponding to several states in the previous, finer, HMM. Next, the Viterbi algorithm is used to find the maximally likely path through the coarse HMM and replace all states in this path with the states represented by it in the finer HMM. This is repeated until you find a path containing only states from the original HMM. If the coarse HMMs are constructed correctly this will be the exact maximum likelihood path in your original HMM.

Another extension to the Viterbi algorithm is a k-best version [11]. Instead of only finding the single most likely path the k most likely ones are found. In this paper we propose to combine the coarse-to-fine technique with the k-best Viterbi algorithm, giving a k-best Viterbi algorithm that is fast on very large HMMs.

A coarse-to-fine k-best Viterbi algorithm has also been proposed in [3], but that article does, strictly speaking, not use HMMs, and the way they use coarse-to-fine means that they only get an approximate solution.

2   Methods

We will use a notation similar to that in [16]. Let the set of N distinct hidden states be denoted by $S = \{S_1, S_2, ..., S_N\}$, and let $Q = q_1 q_2 \cdots q_T$ be the sequence of T actual hidden states. Such a sequence of states we will also call a *path*. Similarly let $V = \{V_1, V_2, ..., V_M\}$ be the set of M distinct observable symbols, and $O = O_1 O_2 \cdots O_T$ the sequence of T actual observations. Formally, an HMM is a three-tuple $\lambda = (A, B, \pi)$, where $A = \{a_{ij}\}$, $a_{ij} = \mathbb{P}(q_t = j \,|\, q_{t-1} = i)$, with $i, j \in S$, is the transition matrix, $B = \{b_j(o)\}$, $b_j(o) = \mathbb{P}(O_t = o \,|\, q_t = j)$, with $j \in S$ and $o \in V$, is the distribution of observable symbols, and finally $\pi_i = \mathbb{P}(q_1 = i)$, for $i \in S$, is the initial distribution vector.

The probability of a given sequence of hidden states Q and observed symbols O is then:

$$\begin{aligned}
\mathbb{P}(O, Q \,|\, \lambda) &= \mathbb{P}(O \,|\, Q, \lambda)\,\mathbb{P}(Q \,|\, \lambda) \\
&= \left(\prod_{t=1}^{T} \mathbb{P}(O_t \,|\, q_t, \lambda) \right) \left(\mathbb{P}(q_1 \,|\, \lambda) \prod_{t=2}^{T} \mathbb{P}(q_t \,|\, q_{t-1}, \lambda) \right) \\
&= \left(\prod_{t=1}^{T} b_{q_t}(O_t) \right) \left(\pi_{q_1} \prod_{t=2}^{T} a_{q_{t-1} q_t} \right) .
\end{aligned}$$

In most real-world scenarios we would not know the path of actual hidden states Q. We are going to assume that only the observations O and the model parameters λ are known, in this article.

2.1   The Viterbi Algorithm

The classical way to estimate Q is to find the maximum likelihood Q using the Viterbi algorithm [16]. We want to maximize

$$\mathbb{P}(Q \mid O, \lambda) \propto \mathbb{P}(Q, O \mid \lambda) \ .$$

This is done by defining $\delta_t(i)$ to be the likelihood of the most likely sequence of states from time 1 to time t and ending in state i

$$\delta_t(i) = \max_{q_1, q_2, \ldots, q_{t-1}} \{\mathbb{P}(q_1 q_2 \cdots q_{t-1}, q_t = i, O_1 O_2 \cdots O_t \mid \lambda)\},$$

which can be computed efficiently using dynamic programming

$$\delta_t(i) = \begin{cases} \pi_i b_i(O_1) & \text{if } t = 1 \\ \max\limits_{j \in S} \{\delta_{t-1}(j) a_{ji}\} b_i(O_t) & \text{otherwise} \end{cases} \ .$$

The above technically only gives rise to the likelihood of the path, but the actual path can be found by backtracking which entries gave rise to the result of each max operation.

The above algorithm has an execution time of $O(N^2 T)$. This is fast enough for many practical purposes, but due to the N^2 term the algorithm may be inadequate if N is big.

2.2   Coarse-to-Fine

In the case of large N, a coarse-to-fine approach may be used [17]. Let $\mathcal{T}$ be a tree with hidden states S as leaves. Let $\mathcal{R}(\mathcal{T})$ be the root of $\mathcal{T}$. Finally, let $c(i)$ be the set of immediate children of node i, where i is any node in $\mathcal{T}$. If i is a leaf in $\mathcal{T}$, that is if $i \in S$, we set $c(i) = \{i\}$ for mathematical convenience.

The nodes in $\mathcal{T}$ are going to be the hidden states in a new HMM, so we also need to define the probabilities in this new HMM. The probabilities for an internal node in the tree is simply going to be the maximum over the probabilities for all the children

$$a_{ij}^{\mathcal{T}} = \begin{cases} a_{ij} & \text{if } i \text{ and } j \text{ are leaves} \\ \max\limits_{i' \in c(i), j' \in c(j)} \{a_{i'j'}^{\mathcal{T}}\} & \text{otherwise} \end{cases},$$

$$b_j^{\mathcal{T}}(k) = \begin{cases} b_j(k) & \text{if } j \text{ is a leaf} \\ \max\limits_{j' \in c(j)} \{b_{j'}^{\mathcal{T}}(k)\} & \text{otherwise} \end{cases},$$

and

$$\pi_i^{\mathcal{T}} = \begin{cases} \pi_i & \text{if } i \text{ is a leaf} \\ \max\limits_{i' \in c(i)} \{\pi_{i'}^{\mathcal{T}}\} & \text{otherwise} \end{cases} \ .$$

Strictly speaking $\lambda^{\mathcal{T}} = (A^{\mathcal{T}}, B^{\mathcal{T}}, \pi^{\mathcal{T}})$ is not an HMM, because $A^{\mathcal{T}}$, $B^{\mathcal{T}}$, and $\pi^{\mathcal{T}}$ no longer define probabilities, since they do not necessarily sum to one. Also, it turns out that we do not actually need to compute the exact max, but that any upper bound will work, though a tighter bound should give a better execution time. We need to modify the Viterbi algorithm, such that it allows a different set of states for each time-step, thus let $viterbi(O, S^{1n}, ..., S^{Tn}, \lambda^T)$ compute $Q^n = q_1^n q_2^n \cdots q_T^n$, the most likely sequence of states, emitting the observed symbols, constrained to $q_t^n \in S^{tn}$. If that Viterbi algorithm finds a path that only contains states that are leaves in $\mathcal{T}$, we will call it a *true solution*, since it is also a solution in the original HMM. Otherwise it is an *estimate*. The algorithm proposed by [17] is to start by setting $S^{t1} = c(\mathcal{R}(\mathcal{T}))$, repeatedly use the above Viterbi algorithm to find the most likely path Q^n and replace all states on that path by their children $S^{t(n+1)} = (S^{tn} \setminus \{q_t^n\}) \cup c(q_t^n)$, until a true solution is found. During the execution, the algorithm can visit several states that are not associated with the final true solution, but [17] shows that once a true solution is found, it will also be the maximally likely path Q in the original HMM. This runs the Viterbi algorithm several times, but with a very small state space, and may therefore be faster than the original Viterbi algorithm on the full state space. The speed depends very much upon finding the true solution Q in few iterations, and not spending time visiting states unrelated to Q. How well this succeeds depends on the concrete model, and how $\mathcal{T}$ is built.

2.3 k-Best

Our work is based on the work of Huang and Chiang. In [11] they suggest four different algorithms for computing k-best Viterbi paths, numbered zero through three. The first algorithm is too inefficient to warrant our attention, and the second algorithm is an optimization that is not relevant to this work. We are going to use their algorithms two and three, and refer them as HC2 and HC3 respectively. Define $\delta_t^k(i)$ to be the likelihood of the kth most likely path from time 1 to time t ending in state i, thus $\delta_t^1(i) = \delta_t(i)$. In the original Viterbi algorithm, we create a table of $\delta_t(i)$ for all (t, i) combinations. The HC2 algorithm simply extends this to storing a list of length k, instead of a single entry in this table. The observation for HC3 is that the majority of the cells will not be involved in all of the k most likely paths, so we will delay the computation of $\delta_t^k(i)$ until $\delta_t^{k-1}(i)$ has actually been used in a path.

To explain how HC3 works in the framework of this article, define $h_t(i)$ to be a heap associated with state i at time t. $h_t(i)$ contains values of the form $\delta_{t-1}^l(j)a_{ji}$, and is used to determine where to get the solution for the next $\delta_t^k(i)$ from. j refers to the source state the solution is from, and l indicates the rank of the solution from j. Obviously, the first column $t = 1$ does not have a previous column to get solutions from. For the first column the only path ending at a given state is the path containing only that state. Therefore we set $\delta_1^1(i) = \pi_i b_i(O_1)$ and do not use heaps for that column. For the remaining columns the heaps are initially built from $\{\delta_{t-1}^1(j)a_{ji}\}_{j \in S}$, remembering $\delta_{t-1}^1(j)$ can be found from $\delta_{t-1}(j)$.

To compute $\delta_t^k(i)$, define two functions: $processTop(t, i)$ which is a utility function that processes the top of the heap $h_t(i)$, and $getSolution(t, i, k)$ which actually finds and returns $\delta_t^k(i)$. $processTop(t, i)$ simply pops the top $\delta_{t-1}^l(j)a_{ji}$ from $h_t(i)$, and computes the next $\delta_t^{k+1}(i) = \delta_{t-1}^l(j)a_{ji}b_i(O_t)$. If one more solution $\delta_{t-1}^{l+1}(j)$ from the source state j exists, it is pushed on to $h_t(i)$ to replaced the item that was just popped. Such a solution will not exist if there simply are not enough paths from time 1 ending in state i, at time t. For example $\delta_1^2(i)$ does not exist, and neither does $\delta_2^k(i)$ for k larger than the total number of states N.

getSolution(t, i, k) starts by checking if $\delta_t^k(i)$ has already been computed. If it has, then we simply return it. Otherwise it repeatedly calls $processTop$ until $\delta_t^{k-1}(i)$ has been computed. At this point $\delta_t^k(i)$ can be computed by peeking at the top of the heap. We do not call $processTop$, as that would pop the value off the heap and require us to push a new value. This new value might not actually be needed, and by deferring the computation of it we can save a significant amount of work.

To actually get the k best paths for the entire HMM greedily consume and replace the best solution from the last column in the dynamic programming table, corresponding to $t = T$. This can be done efficiently, using a heap similarly to above: Build the heap from $\{\delta_T^1(i)\}_{i \in S}$. Next, repeatedly pop the best solution $\delta_T^l(i)$ off the stack, return it to the user, and insert the next candidate solution $\delta_T^{l+1}(i)$, from the source state i, if such a solution exists. Since you do not need to know k before the algorithm is run, but can keep pulling new solutions until any arbitrary condition is satisfied, this is an on-line algorithm.

2.4   Coarse-to-Fine k-Best

The contribution of this paper is to combine the above k-best algorithms with the coarse-to-fine framework. For the HC2 algorithm this is relatively straightforward. The algorithm is the same as the coarse-to-fine Viterbi algorithm, except we find the k best paths in each iteration, using the HC2 algorithm, and split all states involved in any of these. We refer to this as the C2FHC2 algorithm. One may note that since the states near the root of the tree represent many leaves they may have a significant fraction of the k best paths passing through them. We have experimented with a version that counts the number of possible paths passing through a state and only split the minimal amount of nodes, necessary to find k paths in the final iteration. It performed significantly worse than the naive approach, and therefore it is the naive version that is presented in this paper. There may exist a better strategy for deciding how many nodes to split in each iteration.

Extending the HC3 algorithm to C2FHC3 is more involved. We propose keeping the same basic structure as the HC3 algorithm, but updating $processTop$ and getSolution to also handle the splitting of nodes into its children. Care is needed with the value returned by getSolution as we need to know whether it is an estimate or a true solution. Furthermore we introduce the concept of a level of an estimate which is used in a heuristic to help make sure that the splitting of states is distributed evenly over all the time-steps, so that any local optima are discovered to be such, as early as possible. Define the level of an estimate to be the depth in $\mathcal{T}$ of the highest non-simple state on the solution path. We are going to produce estimates of increasing levels, so we will at some point reach the leaves of $\mathcal{T}$, and thus have a true solution. Intuitively we start by asking *getSolution* to give us a solution at level 0. Next, repeatedly poll for solutions that are at one level deeper than the solution previously returned, thus pushing the path toward the leaves, until a true solution is returned.

Previously we used the k parameter in the *getSolution*(t, i, k) function and $\delta_t^k(i)$ to indicate the rank of a solution. We now allow these to take the value of $est(m)$, to denote an estimate at level m. Thus *getSolution*$(t, i, est(m))$ requests the computation of an estimate at level m, while *getSolution*(t, i, k) still requests the computation of the kth best true solution. If an estimate at level m is requested *getSolution* is allowed to return a solution at a deeper level or the first true solution $\delta_t^1(i)$.

processTop still starts by popping the most promising value $\delta_{t-1}^l(j)a_{ji}^{\mathcal{T}}$ from $h_t(i)$. If this is an estimate at some level $l = est(m)$, we want to improve this estimate. If the source state j is already deeper than the level m, or j cannot be split, because it is a simple state, we obtain this better estimate by calling *getSolution*$(t - 1, j, est(m + 1))$ and getting an estimate at a higher level from j. If, instead, j is a candidate for splitting, a better solution can be obtained by doing that. If l is not an estimate we can use it to compute the next true solution for this state $\delta_t^{k+1}(i) = \delta_{t-1}^l(j)a_{ji}^{\mathcal{T}}b_i^{\mathcal{T}}(O_t)$, pushing the next solution $\delta_{t-1}^{l+1}(j)a_{ji}^{\mathcal{T}}$ on to $h_t(i)$, if it exists.

Finally, *getSolution*(t, i, k) also needs to be updated. First it checks whether an acceptable solution has already been computed, remembering that $\delta_t^1(i)$ is an acceptable solution if k is an estimate. If no such solution is found *processTop* is called until one is found, possibly on the heap.

The above does leave out the details of the base cases. *processTop* refers to the previous time-step, which is still not well-defined for the first column. As in the original HC3 algorithm we do not use any heap for the first column, set $\delta_1^1(i) = \pi_i^{\mathcal{T}} b_i^{\mathcal{T}}(O_1)$, and initialize the remaining heaps from the first estimate of the preceding states $\{getSolution(t - 1, j, est(0))\}_{j \in c(\mathcal{R}(\mathcal{T}))}$.

2.5   Building $\mathcal{T}$

The tree $\mathcal{T}$ is irrelevant for the correctness of the result, but it can have a very large impact on the execution time of the algorithm. There are many different ways to build $\mathcal{T}$, but for our experiments we built it bottom-up, based on a cost function $K(\mathcal{T})$. Start out with each state $q \in S$ being a small tree containing only itself as root. Now build the tree by repeatedly joining the two trees giving the cheapest result according to the cost function K until only one tree remains. The motivation for the cost function K is to minimize the expected number of states visited. To define the cost function we first find the *a priori* probability of the hidden states $\mathbb{P}(q)$ from the stationary distribution of the transition matrix. From this we can also find the probability of a subtree as the sum of the probabilities of the children $\mathbb{P}(\mathcal{T}) = \sum_{\mathcal{T}' \in c(\mathcal{R}(\mathcal{T}))} \mathbb{P}(\mathcal{T}')$, and the probability of an observable symbol $o \in V$ as $\mathbb{P}(o) = \sum_{q \in S} \mathbb{P}(o \,|\, q) \, \mathbb{P}(q)$. Furthermore define

$$R(o \,|\, \mathcal{T}) = \begin{cases} \mathbb{P}(o \,|\, q) & \text{if } \mathcal{T} \text{ is a single state } q \\ \max_{\mathcal{T}' \in c(\mathcal{R}(\mathcal{T}))} \{ R(o \,|\, \mathcal{T}') \} & \text{otherwise} \end{cases},$$

and $R(\mathcal{T} \,|\, o) = \frac{R(o \,|\, \mathcal{T}) \mathbb{P}(\mathcal{T})}{\mathbb{P}(o)}$. Using this we also define

$$K(\mathcal{T} \,|\, o) = \begin{cases} R(\mathcal{T} \,|\, o) & \text{if } \mathcal{T} \text{ is a single state } q \\ R(\mathcal{T} \,|\, o) \left(1 + \sum_{\mathcal{T}' \in c(\mathcal{R}(\mathcal{T}))} K(\mathcal{T}' \,|\, o) \right) & \text{otherwise} \end{cases},$$

and finally $K(\mathcal{T}) = \sum_{o \in V} K(\mathcal{T} \,|\, o) \mathbb{P}(o)$. If we cache $R(o \,|\, \mathcal{T})$ and $K(\mathcal{T} \,|\, o)$ for the children of $\mathcal{T}$, we can compute $K(\mathcal{T})$ in time $O(M)$, and the entire tree can be built in $O(N^2 M)$, using a quad-tree [8], if we ignore the time it takes to find the stationary distribution of the transition matrix. In our implementation that distribution is approximated by multiplying the transition matrix to the initial probability vector 25 times, which is $O(N^2)$.

3   Results

The model parameters λ are important for the tree $\mathcal{T}$ and the running time of the algorithm. Therefore we have experimented with four different ways to generate them. The first is to set the emission probabilities to the uniform distribution, while the transition probabilities have been randomly drawn. Thus the emissions are ignored by the HMM and the most likely path is determined only by the transition matrix. Similarly we have used an HMM where the transition probabilities are the uniform distribution, while the emission probabilities are randomly chosen. This gives an HMM where the most likely path is determined only by the observed sequence. The third parameter set was built randomly based on a tree, with states clustered closely in the tree also resembling each other, to give a random HMM that is guaranteed to have some structure the clustering algorithm can exploit. Finally, we have used a completely random HMM, where both the transition and emission probabilities were randomly drawn.

Only the sum of the likelihood of the found Viterbi paths are computed and timed, thus no backtracking is performed. The shown values include both the time to build $\mathcal{T}$ and the time to run the algorithm, but not the time to read input from disk. All experiments were run on three different HMMs and the lines in the plot follows the averages of them. The experiments were run on a MacPro with two Intel quad-core Xeon processors running at 2.26GHz and with 8GB of main memory. All the benchmarks were run on using our own implementations of the algorithms.

In the experiments shown in Fig. 1 we have benchmarked the execution time of the algorithm against the number of states N. The execution time of neither the HC2 nor the HC3 algorithms changes significantly between different HMMs, which was expected. The HC3 method is good for small state spaces, while the

Number of hidden states versus execution time

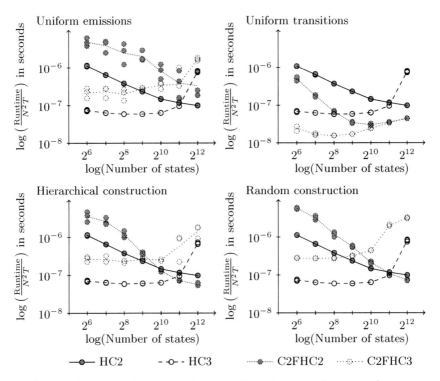

Fig. 1. Results from experiments testing execution time as a function of state space size. $k = 1000$, $M = 10$ and $T = 100$. Plotted are the running time divided by the number of states squared and the sequence length. Each experiment was repeated on three different models, with the line showing the average.

HC2 method performs better for larger state spaces. The models without emission probabilities are generally worst-cases for the coarse-to-fine methods and the models without transitions are best-cases. Without transitions it is trivial to find the most likely path and for those models profiling show that the time to build $\mathcal{T}$ dominates for $N > 2^7$. Without that time included the coarse-to-fine methods can be several orders of magnitude faster than the non-coarse-to-fine methods. Note how well the C2FHC3 performs for those models, although it generally is slow. The hierarchical and random models show the methods under more realistic conditions, and we see that all the coarse-to-fine methods perform badly when the state space is small, but that C2FHC2 can be somewhat faster than the competing algorithms when N is sufficiently large.

In Fig. 2 we show results from benchmarks of the impact of the k parameter. What we see from these graphs is that the speed of HC2 versus HC3 depends very much on the k parameter. HC3 might take a long time to build the heaps, but getting the next solution is extremely cheap, once they are built. HC2 is more sensitive to a large k.

Number of solutions k versus execution time

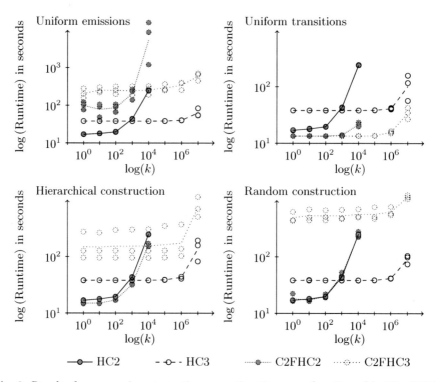

Fig. 2. Results from experiments testing execution time as a function of k. $N = 2000$, $M = 10$ and $T = 100$. HC2 and C2FHC2 were stopped early due to excessive memory usage. Each experiment was repeated on three different models, with the line showing the average.

We have also applied our methods to the technique presented in [21]. The topic of that paper is to jointly estimate genetic crossover and gene conversion rates, which they do using a hill-climbing method to find the maximum likelihood parameter set, with the likelihood of a given set of parameters computed using a number of HMMs. In the article the forward method [16] is used to compute the exact $\mathbb{P}(Q \mid \lambda)$, but since we are only interested in the shape of the fitness landscape, and not the actual values, the result from a k-best Viterbi algorithm might be a good approximation. We ran our experiment on data with 40 sequences of length 35 generated by the *ms* program [12]. The method of Yin et al. generates several different HMMs of increasing complexity, with the biggest having a number of states that is cubic in the number of input sequences. The generated HMMs have a structure such that we can build $\mathcal{T}$ in time $O(N)$, using a domain-specific algorithm, and that the transition probabilities can be computed on-demand in time $O(1)$. However, the structure also allows the forward algorithm, that normally runs in time $O(N^2T)$, to be computed in time

Table 1. Running times for algorithms using an HMM to estimate the likelihood of data given parameters on genetic crossover and gene conversion rates. Many different HMMs are generated, but in the worst case the parameters are $N = 65640$, $M = 2, T = 35$, and we set $k = 1000$. The method of Yin et al. is based on a modified forward algorithm, using domain-specific knowledge, running in time $O(NT)$.

Method	Time
HC2	34669 s
HC3	8858 s
C2FHC2	831 s
C2FHC3	55.3s
Forward	137 s
Yin et al.	15.2s

$O(NT)$. For the timing of the $O(NT)$ forward algorithm the implementation of Yin et al. was used. For comparison we have also implemented a straightforward $O(N^2T)$ forward algorithm.

The results are shown in Table 1. We may notice that C2FHC3 is surprisingly fast compared to the previous experiments. The generated HMMs have only 16 distinct transition probabilities, and since there are many thousands of states the most relevant of our previous experiments would likely be the one with uniform transition probabilities. However, in that experiment our methods were limited by the time it took to construct the tree $\mathcal{T}$, and that is less of an issue in this application.

4   Conclusion

We have shown how the coarse-to-fine heuristic can be combined with a k-best Viterbi algorithm in a way that can achieve a significant speed-up for HMMs with large state spaces in cases where the model has a suitable structure.

Remembering that the total likelihood of the data is the sum of the likelihood of all paths through all hidden states this may be used to approximate the forward algorithm for HMMs where the state space is so large that it is infeasible to run the traditional algorithm. Alternatively it may be used in its own right to find the k-best paths or as an extension to the Viterbi algorithm that will also give an informal sense of the variance and reliability of the result.

Building the tree and predicting whether this approach will work well for a given model remains an unsolved problem although the algorithm for building trees suggested in this paper seems to work well in general. The goal of this method is primarily to be faster than quadratic in the number hidden states. However building the transition matrix is quadratic in the number of hidden states, so this method might be especially suitable to models that have enough structure that the tree and all probabilities can be computed efficiently on demand, without storing them.

Source code can be downloaded from `www.birc.dk/~jn/c2flib`.

Acknowledgments. Part of this research was carried out at the University of California, Berkeley. We thank Yun S. Song and Michael I. Jordan for suggesting the problem and Junming Yin for discussions on it.

References

1. Albrechtsen, A., Sand Korneliussen, T., Moltke, I., van Overseem Hansen, T., Nielsen, F.C., Nielsen, R.: Relatedness mapping and tracts of relatedness for genome-wide data in the presence of linkage disequilibrium. Genetic Epidemiology 33(3), 266–274 (2009)
2. Charniak, E.: A maximum-entropy-inspired parser. In: Proceedings of the 1st North American Chapter of the Association for Computational Linguistics conference, pp. 132–139. Morgan Kaufmann Inc., San Francisco (2000)
3. Charniak, E., Johnson, M.: Coarse-to-fine n -best parsing and MaxEnt discriminative reranking. In: Proceedings of the 43rd Annual Meeting on Association for Computational Linguistics - ACL 2005, June 1, pp. 173–180 (2005)
4. Chong, J., Yi, Y., Faria, A., Satish, N., Keutzer, K.: Data-parallel large vocabulary continuous speech recognition on graphics processors. In: Proceedings of the 1st Annual Workshop on Emerging Applications and Many Core Architecture (EAMA), pp. 23–35. sn (2008)
5. Drinnenberg, I., Weinberg, D., Xie, K., Mower, J., Wolfe, K., Fink, G., Bartel, D.: RNAi in Budding Yeast. Science 326(5952), 544 (2009)
6. Du, J., Rozowsky, J., Korbel, J., Zhang, Z., Royce, T., Schultz, M., Snyder, M.: A Supervised Hidden Markov Model Framework for Efficiently Segmenting Tiling Array Data in Transcriptional and ChIP-chip Experiments: Systematically Incorporating Validated Biological Knowledge. Bioinformatics (2008)
7. Dutheil, J.Y., Ganapathy, G., Hobolth, A., Mailund, T., Uyenoyama, M.K., Schierup, M.H.: Ancestral Population Genomics: The Coalescent Hidden Markov Model Approach. Genetics 183, 259–274 (2009)
8. Finkel, R., Bentley, J.: Quad trees a data structure for retrieval on composite keys. Acta informatica 4(1), 1–9 (1974)
9. Fridlyand, J., Snijders, A., Pinkel, D., Albertson, D., Jain, A.: Hidden Markov models approach to the analysis of array CGH data. Journal of Multivariate Analysis 90(1), 132–153 (2004)
10. Goodman, J.: Global thresholding and multiple-pass parsing. In: Proceedings of the Second Conference on Empirical Methods in Natural Language Processing, pp. 11–25 (1997)
11. Huang, L., Chiang, D.: Better k-best parsing. In: Proc. of IWPT, pp. 53–64 (2005)
12. Hudson, R.: Generating samples under a Wright Fisher neutral model of genetic variation. Bioinformatics 18(2), 337 (2002)
13. Karplus, K., Barrett, C., Cline, M., Diekhans, M., Grate, L., Hughey, R.: Predicting protein structure using only sequence information. Proteins Suppl. 3, 121–125 (1999)
14. Knapp, K., Chen, Y.P.P.: An evaluation of contemporary hidden Markov model genefinders with a predicted exon taxonomy. Nucleic acids research 35(1), 317–324 (2007)
15. Kupiec, J.: Robust part-of-speech tagging using a hidden Markov model. Computer Speech & Language 6(3), 225–242 (1992)

16. Rabiner, L.: A tutorial on hidden Markov models and selected applications in speech recognition (1990)
17. Raphael, C.: Coarse-to-fine dynamic programming. IEEE Transactions on Pattern Analysis and Machine Intelligence 23(12), 1379–1390 (2001)
18. Senf, A., Chen, X.W.: Identification of genes involved in the same pathways using a Hidden Markov Model-based approach. Bioinformatics (Oxford, England) 25(22), 2945–2954 (2009)
19. Wang, K., Li, M., Hadley, D., Liu, R., Glessner, J., Grant, S.F.A., Hakonarson, H., Bucan, M.: PennCNV: an integrated hidden Markov model designed for high-resolution copy number variation detection in whole-genome SNP genotyping data. Genome research 17(11), 1665–1674 (2007)
20. Willett, D., Neukirchen, C., Rigoll, G.: Efficient search with posterior probability estimates in HMM-based speech recognition. In: Proceedings of the 1998 IEEE International Conference on Acoustics, Speech and Signal Processing, ICASSP 1998 (Cat. No.98CH36181), vol. 2, pp. 821–824. IEEE, Los Alamitos (1998)
21. Yin, J., Jordan, M.I., Song, Y.S.: Joint estimation of gene conversion rates and mean conversion tract lengths from population SNP data. Bioinformatics (Oxford, England) 25(12), i231–i239 (2009)

Finding Approximate and Constrained Motifs in Graphs

Riccardo Dondi[1], Guillaume Fertin[2], and Stéphane Vialette[3]

[1] Dipartimento di Scienze dei Linguaggi, della Comunicazione e degli Studi Culturali
Università degli Studi di Bergamo, Via Donizetti 3, 24129 Bergamo - Italy
riccardo.dondi@unimib.it
[2] Laboratoire d'Informatique de Nantes-Atlantique (LINA), UMR CNRS 6241
Université de Nantes, 2 rue de la Houssinière, 44322 Nantes Cedex 3 - France
guillaume.fertin@univ-nantes.fr
[3] IGM-LabInfo, CNRS UMR 8049, Université Paris-Est,
5 Bd Descartes 77454 Marne-la-Vallée, France
vialette@univ-mlv.fr

Abstract. One of the emerging topics in the analysis of biological networks is the inference of motifs inside a network. In the context of metabolic network analysis, a recent approach introduced in [14], represents the network as a vertex-colored graph, while a motif $\mathcal{M}$ is represented as a multiset of colors. An occurrence of a motif $\mathcal{M}$ in a vertex-colored graph G is a connected induced subgraph of G whose vertex set is colored exactly as $\mathcal{M}$. We investigate three different variants of the initial problem. The first two variants, MIN-ADD and MIN-SUBSTITUTE, deal with approximate occurrences of a motif in the graph, while the third variant, CONSTRAINED GRAPH MOTIF (or CGM for short), constrains the motif to contain a given set of vertices. We investigate the classical and parameterized complexity of the three problems. We show that MIN-ADD and MIN-SUBSTITUTE are NP-hard, even when $\mathcal{M}$ is a set, and the graph is a tree of degree bounded by 4 in which each color appears at most twice. Moreover, we show that MIN-SUBSTITUTE is in FPT when parameterized by the size of $\mathcal{M}$. Finally, we consider the parameterized complexity of the CGM problem, and we give a fixed-parameter algorithm for graphs of bounded treewidth, while we show that the problem is $W[2]$-hard, even if the input graph has diameter 2.

1 Introduction

The problem of analyzing biological networks such as protein-protein interaction networks and metabolic networks has become increasingly relevant in Computational Biology (see for example [5, 12, 13, 17–19]). While the classical approach is based on graph-theoretical topology of the motif, a recent approach introduced in [14] aims at discovering functional motifs that do not rely on the conservation of the topology, but that are simply connected components of the network. This approach has been formalized as a graph problem (named GRAPH MOTIF), in

R. Giancarlo and G. Manzini (Eds.): CPM 2011, LNCS 6661, pp. 388–401, 2011.

which given a vertex-colored graph $G = (V, E)$ and a multiset $\mathcal{M}$ of colors, the goal is to find a subset $V' \subseteq V$ which is connected and whose vertex set is colored exactly as $\mathcal{M}$.

The GRAPH MOTIF problem has been widely studied, and some variants have been introduced. The original problem is known to be NP-complete [14], even if the input graph is a tree with maximum degree 3 and the motif is a set [10], and if the input graph is a bipartite graph with maximum degree 4 and the motif is built over two colors only [10]. It is easy to see that GRAPH MOTIF admits a polynomial time algorithm when the input graph is a tree and each color occurs at most twice in the input tree. The GRAPH MOTIF problem is known to be in FPT, when parameterized by the size of the motif [4, 10, 11], while it is W[1]-hard when parameterized by the number of distinct colors in the motif, even in the case the input graph is a tree [10]. Recently, the kernelization complexity of the problem has also been considered [1].

Different variants of the GRAPH MOTIF problem have been introduced. Such variants either modify the requirement of connectedness [7], or look for approximate occurrences of the motif, where some colors are allowed to be inserted or deleted in an occurrence of the motif [5, 8, 11]. Following this direction, we consider three variants of the GRAPH MOTIF problem. In the first two variants, we relax the constraint that each color of $\mathcal{M}$ must appear in an occurrence of the motif, and we allow for the adding (MIN-ADD) or the substitution (MIN-SUBSTITUTE) of some colors. These two problems are motivated by the fact that, due to experimental errors, there may not exist an exact occurrence of the motif $\mathcal{M}$ in the graph G. In the third variant, CONSTRAINED GRAPH MOTIF (or CGM, for short), we strengthen the requirement of connectedness, constraining some vertices of the input graph to be part of an occurrence of a motif $\mathcal{M}$. This is motivated by the fact that, due to a previous knowledge on the structure of the network, we may require some of the vertices to be contained in any occurrence of $\mathcal{M}$.

The rest of the paper is organized as follows. In Section 2, we give some preliminary definitions and we formally define the problems. In Section 3, we show that MIN-SUBSTITUTE and MIN-ADD are NP-hard, even when $\mathcal{M}$ is a set, the input graph is a tree T of degree bounded by 4 and each color has at most two occurrences in T. Notice that under the same hypotheses, the GRAPH MOTIF problem admits a polynomial time algorithm. In Section 4, we give an FPT algorithm for MIN-SUBSTITUTE. In Section 5, we discuss the parameterized complexity of the CGM problem, when the parameter is the number of colors not belonging to mandatory vertices ; in Section 5.1, we show that CGM is fixed-parameter tractable for graphs of bounded treewidth, while in Section 5.2 we show that CGM is $W[2]$-hard, even if the diameter of the input graph is bounded by 2. Some of the proofs are omitted due to space constraints.

2   Preliminaries

In this section, we recall basic notations used in the rest of the paper. Given a graph $G = (V, E)$ and $V' \subseteq V$, we denote by $G[V']$ the subgraph of G induced by V', that is $G[V'] = (V', E')$ and $\{u, v\} \in E'$ iff $u, v \in V'$ and $\{u, v\} \in E$. Given a vertex $v \in V$, we denote by $N(v)$ the set of vertices in G adjacent to v. We recall that a graph is cubic when each vertex has degree 3.

Let G be a connected graph, where every vertex $u \in V(G)$ is assigned a color $c(u)$ from a set $\mathcal{C}$ of colors. For any subset V' of V, let $C(V')$ be the multiset of colors assigned to the vertices in V'. Let $\mathcal{M}$ be a multiset of colors, whose colors are taken from the set $\mathcal{C}$. Given a colored graph G and a subset of vertices $V' \subseteq V(G)$, $C(V')$ is said to *match* a multiset of colors $\mathcal{M}$ if $C(V')$ is equal to $\mathcal{M}$. In this case, by abuse of notation, we say that V' matches $\mathcal{M}$. Given a subset of vertices $V' \subseteq V(G)$ such that V' matches $\mathcal{M}$ and $G[V']$ is connected, then V' is called an *occurrence* of $\mathcal{M}$ in G. A motif $\mathcal{M}$ is said *colorful* when $\mathcal{M}$ is a set of colors (rather than a multiset).

In this paper, we consider three variants of the GRAPH MOTIF problem. For two of them, MIN-ADD and MIN-SUBSTITUTE, we look for a vertex set V' of $G = (V, E)$, such that $G[V']$ is connected and $C(V')$ is not necessarily equal to $\mathcal{M}$. Furthermore, we consider a constrained variant of the GRAPH MOTIF problem, CGM, where the input consists of a vertex colored graph and a set of mandatory vertices that must belong to any occurrence of motif $\mathcal{M}$.

Let us introduce the first two variants of GRAPH MOTIF problem.

MIN-ADD (decision version)
Input : A multiset of colors $\mathcal{M}$, a vertex-colored graph $G = (V, E)$, an integer p.
Question : Is there a subset $V' \subseteq V$, such that $G[V']$ is connected, $C(V') \supseteq \mathcal{M}$ and $|C(V') \setminus \mathcal{M}| \leq p$?

MIN-SUBSTITUTE (decision version)
Input : A multiset of colors $\mathcal{M}$, a vertex-colored graph $G = (V, E)$, an integer p.
Question : Is there a subset $V' \subseteq V$, such that $G[V']$ is connected and $C(V')$ can be obtained with at most p substitutions from $\mathcal{M}$?

Notice that, in case $p = 0$, both MIN-ADD and MIN-SUBSTITUTE are equivalent to the GRAPH MOTIF problem. As a consequence, MIN-ADD and MIN-SUBSTITUTE are both NP-hard when the motif is colorful, the input graph consists of a tree T and each color has at most 3 occurrences in T [10]. Furthermore, MIN-ADD (resp. MIN-SUBSTITUTE) cannot be approximated within any approximation factor, and does not admit any fixed-parameter tractable algorithm, when the parameter is the number of added colors (resp. the number of substitutions). Notice that MIN-ADD is in FPT, when parameterized by $|\mathcal{M}|$. Indeed, in [11], a variant of GRAPH MOTIF, called Multiset Graph Motif With Gaps (MGMG), is considered: given an input graph G and a motif $\mathcal{M}$, we look for an occurrence of $\mathcal{M}$ that is allowed to contain gaps. Note that this is precisely MIN-ADD, where the gaps represent colors to be added to $\mathcal{M}$. As in [11] it is shown that MGMG is in FPT when parameterized by $|\mathcal{M}|$, we can conclude that the MIN-ADD problem is in FPT. Furthermore, in case the motif is colorful, a fixed-parameter algorithm for MIN-ADD has been given in [5].

Let us now consider a different variant of the GRAPH MOTIF problem, called Constrained Graph Motif (CGM).

Constrained Graph Motif (CGM)
Input : A multiset of colors $\mathcal{M}$, a vertex-colored graph $G = (V, E)$, a set of mandatory vertices $V_M \subseteq V$.
Question : Is there a subset $V' \subseteq V$, such that $G[V']$ is connected, $C(V') = \mathcal{M}$ and $V_M \subseteq V'$?

Given an instance of CGM, define the *optional occurrences* C_o as $C_o = \mathcal{M} \setminus C(V_M)$.

The CGM problem is NP-complete, since the GRAPH MOTIF problem is NP-complete [10, 14]. It is easy to see that CGM is fixed-parameter tractable, when the parameter is the size of the motif. Indeed, recall that GRAPH MOTIF is fixed-parameter tractable. By recoloring the graph, assigning a unique color to each vertex in V_M, and by modifying accordingly $\mathcal{M}$, we can conclude that each occurrence of $\mathcal{M}$ in G must include all the vertices in V_M.

In Section 5, we investigate the parameterized complexity of the CGM problem, when the parameter is the number of optional occurrences. Notice that the Minimum (Unweighted) Steiner Tree problem is a restriction of the CGM problem, where the non mandatory vertices in the Steiner Tree problem correspond to optional occurrences in CGM. As the Minimum (Unweighted) Steiner Tree problem is W[2]-hard when parameterized by the number of non mandatory vertices [6], it follows that the CGM problem is W[2]-hard when parameterized by the number of optional occurrences.

In Section 5.1, we will consider the case where the input graph has bounded treewidth and we will use a tree decomposition of the graph. Let us recall the definition of tree decomposition of a graph [9, 15]. Given a graph $G = (V, E)$, a tree decomposition of G is a pair $\langle \{X_i : i \in I\}, T \rangle$, such that each X_i is called a *bag*, and T is a tree having as vertices the elements of I and such that:

1. $\cup_{i \in I} X_i = V$;
2. for each edge $\{u, v\} \in E$, there is a bag X_i with $u, v \in X_i$;
3. for each i, j, k in V, if j is on the path from i to k in G, then $X_i \cap X_k \subseteq X_j$.

The width of $\langle \{X_i : i \in I\}, T \rangle$ is equal to $\max\{|X_i| : i \in I\} - 1$ and the treewidth of a graph G is equal to the minimum δ such that G has a tree decomposition of width δ. A tree decomposition $\langle \{X_i, i \in \{1, \ldots, p\}\}, T \rangle$ of a graph G is *nice* (see [15]) when, given a vertex i of the tree decomposition, i has at most two children and the following conditions hold:

1. if i has two children j and k, then $X_i = X_j = X_k$;
2. if i has exactly one child j, then one of the following conditions holds:
   (a) $|X_i| = |X_j| + 1$, and then $X_j \subset X_i$; or
   (b) $|X_i| = |X_j| - 1$, and then $X_j \supset X_i$.

In the rest of the paper, in order to extend some results from the case when $\mathcal{M}$ is colorful to the general case, we use the recoloring technique introduced in [4], based on the color-coding technique [3]. The recoloring technique starts from a general motif $\mathcal{M}$ and computes a colorful motif C, recoloring accordingly the vertices of the input graph G. Let V' be an occurrence of $\mathcal{M}$ in the graph G, then V' achieves a *colorful recoloring* if $C(V')$ is colorful after the recoloring of $\mathcal{M}$ and G. In [4], the following result was shown:

Lemma 1 (Betzler et al. [4]). *Given a motif $\mathcal{M}$, the number of trials to achieve a colorful recoloring of $\mathcal{M}$ with an error probability of ε is $|\ln(\varepsilon)| \cdot O(e^{|\mathcal{M}|})$.*

3   NP-Hardness of **Min-Substitute** and **Min-Add**

In this section, we show that MIN-SUBSTITUTE and MIN-ADD are NP-hard, even if the input graph is a tree, the motif is colorful and each color has at most two occurrences in the input tree. Recall that, under the same hypotheses, the GRAPH MOTIF problem admits a polynomial time algorithm.

Theorem 1. *The* MIN-SUBSTITUTE *problem is NP-hard, even when the input graph is a tree of maximum degree 4, each color occurs at most twice in the input graph and the motif is colorful.*

Proof. We give a reduction from the Minimum Vertex-Cover on Cubic Graphs problem (MIN-VCC). Let $G = (V, E)$ be a cubic graph with $V = \{v_1, v_2, \ldots, v_n\}$, the MIN-VCC problem asks for a subset $V' \subseteq V$ of size at most p, such that for each $\{v_i, v_j\} \in E$ at least one of v_i, v_j is in V'. MIN-VCC is known to be NP-hard [2]. Starting from G, we construct an instance of the MIN-SUBSTITUTE problem which consists of a tree T and a set of colors $\mathcal{M}$. For any vertex $v_i \in V$, let $e_{i,j}$, $1 \leq j \leq 3$, be its 3 incident edges, ordered arbitrarily. The tree $T = (V_T, E_T)$ is defined as follows (see Figure 1):

- $V_T = \{l_i, a_{i,1}, a_{i,2} : 1 \leq i \leq n\} \cup \{s_i : 1 \leq i \leq p\} \cup \{t_i : 1 \leq i \leq n+1\} \cup \{e_{i,j} : 1 \leq i \leq n \wedge 1 \leq j \leq 3\}$;
- $E_T = \{\{l_i, l_{i+1}\} : 1 \leq i < n\} \cup \{\{s_i, s_{i+1}\} : 1 \leq i < p\} \cup \{\{t_i, t_{i+1}\} : 1 \leq i < n+1\} \cup \{\{l_n, t_1\}\} \cup \{\{t_{n+1}, s_1\}\} \cup \{\{l_i, a_{i,1}\}, \{a_{i,1}, a_{i,2}\} : 1 \leq i \leq n\} \cup \{\{a_{i,2}, e_{i,j}\} : 1 \leq i \leq n \wedge 1 \leq j \leq 3\}$.

Clearly, this construction gives us a tree of maximum degree 4. Let us describe the colors assigned to each vertex of $V(G)$. Each vertex l_i, $1 \leq i \leq n$, is assigned a unique color $c(l_i)$, each vertex s_i, $1 \leq i \leq p$, is assigned a unique color $c(s_i)$, and each each vertex t_i, $1 \leq i \leq n+1$, is assigned a unique color $c(t_i)$. The two vertices $a_{i,1}$, $a_{i,2}$, $1 \leq i \leq n$, are assigned the same color $c(v_i)$. Finally, each vertex $e_{i,x}$ in V_T, $1 \leq i \leq n$ and $1 \leq x \leq 3$, associated to an edge $e_{i,j} = \{v_i, v_j\}$ in E, is assigned color $c(e_{i,j})$. Each color occurs at most twice in T, as each color $c(e_{i,j})$ is associated to two vertices of T, while each color $c(v_i)$ is associated to

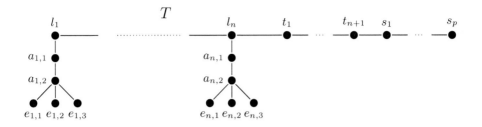

Fig. 1. Illustration of the reduction from MIN-VCC to MIN-SUBSTITUTE

vertices $a_{i,1}$, $a_{i,2}$. $\mathcal{M}$ is a set of colors defined as follows: $\mathcal{M} = \{c(l_i) : 1 \leq i \leq n\} \cup \{c(s_i) : 1 \leq i \leq p\} \cup \{c(v_i) : 1 \leq i \leq n\} \cup \{c(e_{i,j}) : e_{i,j} \in E\}$. Notice that no occurrence of a color $c(t_i)$, $1 \leq i \leq n+1$, belongs to $\mathcal{M}$.

Starting from a vertex cover $V' \subseteq V$ of G of size at most p, a solution $V_{T'}$ of MIN-SUBSTITUTE, that substitutes p colors from $\mathcal{M}$, is obtained as follows. Given an edge $e_{i,j} = \{v_i, v_j\}$, define $e_{i,j}^{min} = \min\{i, j\}$. The vertex set $V_{T'}$ defined as follows:
$V_{T'} = \{l_i, a_{i,1} : 1 \leq i \leq n\} \cup \{t_i : 1 \leq i \leq p - |V'|\} \cup \{a_{i,2} : v_i \in V'\} \cup \{e_{i,x} : c(e_{i,x}) = c(e_{i,j}) \wedge i = e_{i,j}^{min}\}$.

By construction and since V' is a vertex cover, $V_{T'}$ induces a subtree of T. It is easy to see that, given $C(V_{T'}) = \mathcal{M}'$, $\mathcal{M}'$ can be obtained from $\mathcal{M}$ by p substitutions.

Let us consider now a solution $V_{T'}$ of MIN-SUBSTITUTE, where $C(V_{T'}) = \mathcal{M}'$, $|\mathcal{M}'| = |\mathcal{M}|$, and $\mathcal{M}'$ can be obtained from $\mathcal{M}$ with at most p substitutions. First, we show that $V_{T'}$ does not contain a vertex of the set $\{s_i : 1 \leq i \leq p\}$. Indeed, assume that a vertex s_i is part of $V_{T'}$; by construction the set of vertices $\{t_j : 1 \leq j \leq n+1\}$ must belong to $V_{T'}$, and since $\mathcal{M}$ does not contain occurrences of any color $c(t_j)$, $1 \leq j \leq n+1$, it follows that $\mathcal{M}'$ requires at least $n+1$ substitutions. Notice that $n+1 > p$, as each vertex cover V' of G has size at most n. Hence, we can assume that $V_{T'}$ does not contain any vertex in the set $\{s_i : 1 \leq i \leq p\}$. It follows that all the colors $c(s_i)$, $1 \leq i \leq p$, in $\mathcal{M}$ must be substituted, and, since by hypothesis $\mathcal{M}'$ can be obtained from $\mathcal{M}$ with at most p substitutions, it follows that only the colors $c(s_i)$, $1 \leq i \leq p$, are substituted. Hence $\{l_i, a_{i,1} : 1 \leq i \leq n\} \subseteq V_{T'}$ and $\mathcal{M}' \supseteq \{c(e_{i,j}) : e_{i,j} \in E\}$. Since $T[V_{T'}]$ must be connected, it follows that each vertex colored $c(e_{i,j})$ must be connected to some vertex $a_{i,2} \in V_{T'}$ colored by $c(v_i)$. Define $V' = \{v_i : a_{i,2} \in V_{T'}\}$; then V' is a cover of G of size at most p, which completes the proof. $\square$

Theorem 2. *The MIN-ADD problem is NP-hard, even when the input graph is a tree of maximum degree 4, each color occurs at most twice in the input graph and the motif is colorful.*

Proof. (Sketch) The result follows from a reduction from MIN-VCC similar to that of Theorem 1. Given an instance of MIN-VCC, an instance $(T, \mathcal{M})$ of MIN-ADD is constructed as follows. $T = (V_T, E_T)$ is defined as follows (see Fig. 2):

- $V_T = \{l_i, a_{i,1}, a_{i,2} : 1 \leq i \leq n\} \cup \{e_{i,j} : 1 \leq i \leq n \land 1 \leq j \leq 3\}$;
- $E_T = \{\{l_i, l_{i+1}\} : 1 \leq i < n\} \cup \{\{l_i, a_{i,1}\}, \{a_{i,1}, a_{i,2}\} : 1 \leq i \leq n\} \cup \{\{a_{i,2}, e_{i,j}\} : 1 \leq i \leq n \land 1 \leq j \leq 3\}$.

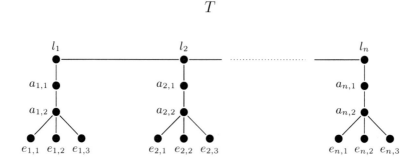

Fig. 2. Illustration of the reduction from MIN-VCC to MIN-ADD

Each vertex l_i, $1 \leq i \leq n$, is assigned a unique color $c(l_i)$, $1 \leq i \leq n$. The two vertices $a_{i,1}$, $a_{i,2}$, $1 \leq i \leq n$, are assigned the same color $c(v_i)$. Finally, each vertex $e_{i,x}$ in V_T, $1 \leq i \leq n$ and $1 \leq x \leq 3$, associated to an edge $e_{i,j} = \{v_i, v_j\}$ in E, is assigned color $c(e_{i,j})$. $\mathcal{M}$ is a set of colors defined as follows: $\mathcal{M} = \{c(l_i) : 1 \leq i \leq n\} \cup \{c(v_i) : 1 \leq i \leq n\} \cup \{c(e_{i,j}) : e_{i,j} \in E\}$.

It can be proved that starting from a vertex cover $V' \subseteq V$ of G, we can compute in polynomial time a solution $V_{T'}$ of MIN-ADD such that $C(V_{T'}) \supseteq \mathcal{M}$ and $|C(V_{T'})| \leq |\mathcal{M}| + |V'|$. Conversely, starting from a solution $V_{T'}$ of MIN-ADD such that $C(V_{T'}) \supseteq \mathcal{M}$ and $|C(V_{T'})| \leq |\mathcal{M}| + p$, we can compute a vertex cover V' of G such that $|V'| \leq p$.   □

4   Parameterized Complexity of **Min-Substitute**

In this section, we discuss the parameterized complexity of MIN-SUBSTITUTE, when parameterized by $|\mathcal{M}|$. We recall that MIN-SUBSTITUTE is not in FPT, as discussed in Section 2, when parameterized by the the size of the solution (i.e., the number of substituted colors).

Let us first consider the case where the motif $\mathcal{M}$ is colorful (i.e., $\mathcal{M}$ is a set). The algorithm is based on dynamic programming. Let $(G = (V, E), \mathcal{M})$ be an instance of MIN-SUBSTITUTE. Instead of computing directly a solution for MIN-SUBSTITUTE, we compute a solution for a slightly different problem, where we visit the vertices of a connected component of G, allowing to visit some vertices more than once. Let v be a vertex of the input graph G, let $C \subseteq \mathcal{M}$ be a subset of colors, let k be the number of vertices of the solution of MIN-SUBSTITUTE we are looking for, and define $S[v, C, k]$ as the minimum value z required by a visit of a connected set V_T of vertices of G such that:

1. $v \in V_T$;
2. exactly k visits of vertices in V_T are done;
3. $C(V_T)$ matches q colors of C, where $z = k - q$.

Notice that a vertex of V_T may be visited more than once, while the overall number of visits must be k. Now, let us define the dynamic programming recurrence to compute $S[v, C, k]$.

$$S[v, C, k] = \min_{C' \subseteq C, u \in N(v), k_1 + k_2 = k} \left\{ S[v, C', k_1] + S[u, C \setminus C', k_2] \right\}. \qquad (1)$$

For the base cases: $S[u, C', 1] = 0$, when $c(u) \in C'$, for each $C' \subseteq C$ and $u \in V$, and $S[u, C', 1] = 1$ when $c(u) \notin C'$. Now, let us prove the correctness of Recurrence (1).

Lemma 2. *Let $(G, \mathcal{M})$ be an instance of* MIN-SUBSTITUTE, *let v be a vertex of G, and let C be a subset of $\mathcal{M}$. There is a visit of a connected vertex set V_T of G, such that $v \in V_T$, the vertices of V_T are visited k times, and $C(V_T)$ matches q colors of C, iff there exists an entry $S[v, C, k] = z$, where $z = k - q$.*

An optimal solution for the MIN-SUBSTITUTE problem can be found as follows. We look for the minimal value z in the entries $S[v, \mathcal{M}, |\mathcal{M}|]$, with $v \in V$. Notice that this value may be associated to a visit of a connected vertex set V_T, where some of the vertices may be visited repeatedly. Each repeated visit of a vertex represents a color to be substituted, since $\mathcal{M}$ is colorful. It follows that we can compute a feasible solution for MIN-SUBSTITUTE by replacing these repeated visits by some connected components adjacent to V_T without increasing the number of substitutions.

The time complexity of the algorithm is $O^*(3^{|\mathcal{M}|})$, as we have to consider all possible subsets $C \subseteq \mathcal{M}$ and for each subset C we have to consider all possibile bipartitions of C. Indeed, there are $O(3^{|\mathcal{M}|})$ possible bipartitions of all possible subsets C of $\mathcal{M}$. In order to extend the results to a multiset, we apply the recoloring technique described in [4]. Combining Lemma 2 with Lemma 1, and we get that MIN-SUBSTITUTE, parameterized by $|\mathcal{M}|$, can be solved in time $O^*((3e)^{O(|\mathcal{M}|)})$.

5   Parameterized Complexity of CGM

In this section, we consider the parameterized complexity of CGM, where the parameter is the number k of optional occurrences C_o, that is $k = |C_o|$, where $C_o = \mathcal{M} \setminus C(V_M)$. First, in Section 5.1, we show that CGM is fixed-parameter tractable, when the input graph is of bounded treewidth ; then, in Section 5.2, we prove that CGM is W[2]-hard, even when the input graph is of diameter 2.

5.1   An FPT Algorithm for Graphs of Bounded Treewidth

Here, we describe a fixed-parameter algorithm for CGM for graphs of bounded treewidth. Let $(G = (V, E), \mathcal{M}, V_M)$ be an instance of CGM, and let us first consider the case where the motif $\mathcal{M}$ is colorful.

Denote by k the number of optional occurrences and by δ the treewidth of graph G. The algorithm is based on a nice tree decomposition of G (see Section 2 for the definition of nice tree decomposition of a graph). We also consider a slightly more general problem, where instead of requiring that an occurrence of a motif consists of a single connected component, we may have an occurrence consisting of at most $\delta + 1$ connected components, where the different connected components are induced by a partition of a bag X_i of the nice tree decomposition. Given a vertex i of the tree decomposition of G, denote by $T[i]$ the subtree of the nice tree decomposition rooted at i and let $V(T[i]) = \{u \in X_j : j \in T[i]\}$.

Now, consider a set X_i, $1 \le i \le p$, of the nice tree decomposition $\langle\{X_i, i \in \{1, \ldots, p\}\}, T\rangle$. From the definition of treewidth, it follows that $|X_i| \le \delta + 1$. Now, let us define a mapping function f_i associated to the vertices of X_i, as follows.

Definition 1. *Let X_i be a bag of the nice tree decomposition of G. A mapping function f_i from X_i to $\{0, 1, \ldots, \delta + 1\}$ is feasible when*

1. *$f_i(v) \ne 0$ for each mandatory vertex v in X_i;*
2. *for each pair of vertices $u, v \in X_i$ such that $c(u) = c(v)$, then $f_i(u) = 0$ or $f_i(v) = 0$;*
3. *define $X_i^l = \{v \in X_i : f(v_i) = l\}$, $l \in \{1, \ldots, \delta + 1\}$, and $X_i' = \cup_l X_i^l$, then X_i^l is a maximal connected component of $G[X_i']$.*

A feasible mapping f_i represents a partition of a subset $X_i' \subseteq X_i$ in at most $\delta + 1$ connected components, where $f_i(v) = p \ne 0$ implies that v belongs to the p-th connected component, while $f_i(v) = 0$ implies that v does not belong to X_i'.

Definition 2. *Let W be a set of vertices of $V(T[i])$, consisting of the connected components $W_1, W_2, \ldots, W_z$. Let f_i be a feasible mapping of X_i in $\{0, 1, \ldots \delta + 1\}$, then W is mapped (or partitioned) according to f_i if:*

1. *for each p, $1 \le p \le z$, $W_p \cap X_i \ne \emptyset$, and there exists exactly one l, $1 \le l \le \delta + 1$, such that $W_p \cap X_i = X_i^l$*
2. *for each l, $1 \le l \le \delta + 1$, such that $X_i^l \ne \emptyset$, there exists exactly one p, $1 \le p \le z$, such that $X_i^l = W_p \cap X_i$.*

Notice that by Definition 2, if a vertex u of W is not in X_i, then there exists a vertex v in $W \cap X_i$ such that v and u are in the same connected component W_x of W, v is assigned some label $l \ne 0$, and all the vertices of $W_x \cap X_i$ are assigned the same label l.

Given two sets X_i and X_j of a nice tree decomposition, and two feasible mappings $f_i : X_i \to \{0, \ldots, \delta + 1\}$ and $f_j : X_j \to \{0, \ldots, \delta + 1\}$, then f_i and f_j are *consistent* if, for each $v \in X_i \cap X_j$, $f_i(v) = f_j(v)$.

Let i be a vertex of the nice tree decomposition, with exactly one child j, such that $|X_i| = |X_j| + 1$ and $X_j \subset X_i$, with $v \in X_i \setminus X_j$. Then, a feasible mapping f_i is an *extension* of a feasible mapping f_j, when either:

1. $f_i(v) = 0$; or
2. $f_i(v) = l$, $l \in \{1, \ldots, \delta + 1\}$, $f_i(u) \neq l$ for each $u \in X_i \cap X_j$, and f_i, f_j are consistent; or
3. there exists a value $l \in \{1, \ldots, \delta + 1\}$ such that
   (a) $f_i(v) = l$;
   (b) if $f_j(z) = 0$, then $f_i(z) = 0$, for $z \in X_i \cap X_j$;
   (c) if $f_i(z) \neq f_j(z)$, for $z \in X_i$ and $f_j(z) \neq 0$, then $f_i(z) = l$.

Given a feasible mapping f_i of X_i in $\{0, 1, \ldots \delta + 1\}$, define $c(X_i, f_i) = \{c \in C_o : \exists v \in X_i,\ c(v) = c \wedge f_i(v) \neq 0\}$.

Let us define the value $S[i, f_i, C']$, where i is a vertex of the nice tree decomposition of G, f_i is a feasible mapping function of the set X_i in $\{0, 1, \ldots \delta + 1\}$ and $C' \subseteq C_o$ be a subset of the set of optional occurrences. $S[i, f_i, C'] = 1$ when there exists a set W of vertices in the nice tree decomposition rooted at i, such that the vertices of W can be partitioned according to f_i, each mandatory vertex of $T[i]$ is in W, and the set of optional occurrences of $c(W)$ is C' ; else $S[i, f_i, C'] = 0$. Next, we describe how to compute $S[i, f_i, C']$ by dynamic programming, depending on the three different cases of a nice tree decomposition.

Case 1) Assume that vertex i has two children j and k (recall that $X_i = X_j = X_k$), then

$$S[i, f_i, C'] = \bigvee_{f_j, f_k, C_j, C_k} S[j, f_j, C_j] \wedge S[k, f_k, C_k],$$

where f_i, f_j, f_k are all feasible and consistent, $C' = (C_j \cup C_k)$ and $c(X_i, f_i) = C_j \cap C_k$.

Case 2) Assume that i has exactly one child j, such that $X_i = X_j \cup \{v\}$, then

$$S[i, f_i, C'] = \bigvee_{f_j, C_j} S[j, f_j, C_j],$$

where f_i and f_j are feasible, f_i is an extension of f_j, $C' = C_j \cup \{c(v)\}$ and $c(v) \notin C_j$, when $f_i(v) \neq 0$ and $v \notin V_M$, and $C' = C_j$ when $f_i(v) = 0$ or $v \in V_M$.

Case 3) Assume that X_i has exactly one child X_j, such that $X_i = X_j \setminus \{v\}$, then

$$S[i, f_i, C'] = \bigvee_{f_j} S[j, f_j, C'],$$

where f_i and f_j are feasible and consistent, and there is a vertex $z \in X_i \cap X_j$, such that $f_j(z) = f_j(v)$, with $v \in X_j \setminus X_i$, when $f_j(v) \neq 0$.

For the base cases (when X_i is a leaf of the nice tree decomposition), define $S[i, f_i, C'] = 1$ when there is a partition of the vertices of X_i according to the feasible function f_i, and $c(X_i, f_i) = C'$; else $S[i, f_i, C'] = 0$.

First, we prove the correctness of the above recurrences, then we discuss the time complexity of the algorithm.

Lemma 3. *Let f_i be a feasible mapping function of X_i, and let W be a set of vertices in $V(T[i])$, such that W contains all the mandatory vertices in $V(T[i])$, W can be mapped according to f_i and C' is the set of optional occurrences in $c(W)$. Then $S[i, f_i, C'] = 1$.*

Lemma 4. *Let $S[i, f_i, C'] = 1$ for a feasible mapping function f_i of X_i in $\{0, 1, \ldots, \delta+1\}$, then there exists a set W of vertices in $V(T[i])$ such that the set of optional occurrences in $c(W)$ is C', W contains all the mandatory vertices in $V(T[i])$ and the vertices of W can be mapped according to f_i.*

Theorem 3 shows how the values $S[i, f_i, C']$ are used to compute the existence of a feasible solution for CGM.

Theorem 3. *Let $(G = (V, E), \mathcal{M}, V_M)$ be an instance of the CGM problem. Then there is a solution W of CGM over instance of $(G, \mathcal{M}, V_M)$ iff there is a vertex i of the nice tree decomposition and a feasible function f_i that maps X_i in $\{0, x\}$, with $x \in \{1, \ldots, \delta+1\}$, such that $S[i, f_i, C_o] = 1$ and such $V_M \subseteq V(T[i])$.*

Proof. Assume that there is a vertex i of the nice tree decomposition and a feasible function f_i that maps X_i in $\{0, x\}$, with $x \in \{1, \ldots, \delta + 1\}$, such that $S[i, f_i, C_o] = 1$ and all the mandatory vertices of G are in $T[i]$. By Lemma 4, it follows that there is a set of vertices W in $V(T[i])$ that contains all the mandatory vertices of G, such that the set of optional occurrences in $c(W)$ is C_o and such that the vertices of W can be mapped according to f_i. Furthermore, notice that W consists of a single connected component. Hence W is a solution of CGM.

Consider the case where there is a solution W of CGM over instance $(G = (V, E), \mathcal{M}, V_M)$. Consider a vertex i of the tree decomposition of G such that all the vertices of W are contained in $V(T[i])$. By Lemma 3, it follows that $S[i, f_i, C_o] = 1$ for some feasible function f_i that maps X_i in $\{0, x\}$, with $x \in \{1, \ldots, \delta + 1\}$. □

Now, we discuss the time complexity of the above algorithm. Denote by n the size of V. Given a vertex i and the associated set X_i of the nice tree decomposition, the number of possible mapping functions of X_i into $\{0, \ldots, \delta+1\}$ is $O(\delta^\delta)$. The number of possible subsets C' is $O(2^k)$. Since the number of vertices of a nice tree decomposition is $O(n)$, it follows that we have $O(\delta^\delta n 2^k)$ entries $S[i, f_i, C]$. Given a mapping function f_i of X_i into $\{0, \ldots, \delta + 1\}$, computing an entry $S[i, f_i, C]$, given the entries of the children (or the child) of i, requires time at most $O(\delta^{2\delta} 2^{2k})$ (notice that the worst case occurs when i has two children). Hence the total time complexity is $O(\delta^{3\delta} n 2^{3k})$.

When a motif is a multiset of colors, we apply the recoloring technique presented in Lemma 1. As a consequence, CGM can be solved with error probability ε in time $O(|\ln(\varepsilon)| \delta^{3\delta} n 2^{4.4427k})$, for graphs of treewidth δ.

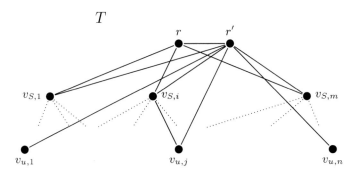

Fig. 3. Illustration of the reduction from MIN-SC to CGM ; notice that element $u_j \in S_i$

5.2 Hardness of Parameterization

The CGM problem parameterized by the number of optional occurrences is $W[2]$-hard, as stated in Section 2. Here we strengthen the result, showing that the problem is $W[2]$-hard even when the input graph is of diameter 2.

Theorem 4. *The* CGM *problem, parameterized by the number of optional occurrences, is* $W[2]$*-hard, even when the input graph is of diameter* 2.

Proof. (*Sketch*) We give a parameterized preserving reduction from Minimum Set Cover (MIN-SC). Given a universe $U = \{u_1, \ldots, u_n\}$ and a collection of sets $\mathcal{S} = \{S_1, \ldots, S_m\}$ over U, CGM asks for a collection $\mathcal{S}'$ of at most k sets of $\mathcal{S}$, such that $\bigcup_{S'_i \in \mathcal{S}'} S'_i = U$. MIN-SC is known to be $W[2]$-hard [16]. Let $(U, \mathcal{S})$ be an instance of MIN-SC, we define a corresponding instance $(G = (V, E), \mathcal{M}, V_M)$ of the CGM problem (see Fig. 3). The graph G of diameter 2 is defined as follows:

- $V = \{r\} \cup \{r'\} \cup \{v_{S,i} : 1 \leq i \leq m\} \cup \{v_{u,j} : 1 \leq j \leq n\}$;

- $E = \{\{r, r'\}\} \cup \{\{r, v_{S,i}\} : 1 \leq i \leq m\} \cup \{\{r', v_{S,i}\} : 1 \leq i \leq m\} \cup \{\{v_{S,i}, v_{u,j}\} : 1 \leq i \leq m \wedge u_j \in S_i\} \cup \{\{r', v_{u,j}\} : 1 \leq j \leq n\}$.

Vertex r and vertex r' are both colored by $c(r)$, vertex $v_{S,i}$ is colored by $c(S)$, $1 \leq i \leq m$, and vertex $v_{u,j}$ is colored by $c(u_j)$, $1 \leq j \leq n$. The motif $\mathcal{M}$ is a multiset containing one occurrence of color $c(r)$, one occurrence of each color $c(u_j)$, $1 \leq j \leq n$, and k occurrences of color $c(S)$. Finally, $V_M = V \setminus (\{v_{S,i} : 1 \leq i \leq m\} \cup \{r'\})$.

Then, it is possible to show that, given a solution of MIN-SC of size at most k, we can compute in polynomial time a solution of CGM over instance $(G = (V, E), \mathcal{M}, V_M)$. Similarly, it is possible to show that given an occurrence V_T of motif $\mathcal{M}$ in G, we can compute in polynomial time a solution of MIN-SC of size at most k. By construction, a solution of CGM over instance $(G = (V, E), \mathcal{M}, V_M)$ contains exactly k optional occurrences. Hence the reduction is parameter preserving, thus implying that CGM is $W[2]$-hard. □

References

1. Ambalath, A.M., Balasundaram, R., Rao H., C., Koppula, V., Misra, N., Philip, G., Ramanujan, M.S.: On the Kernelization Complexity of Colorful Motifs. In: Raman, V., Saurabh, S. (eds.) IPEC 2010. LNCS, vol. 6478, pp. 14–25. Springer, Heidelberg (2010)
2. Alimonti, P., Kann, V.: Some APX-Completeness Results for Cubic Graphs. Theor. Comput. Sci. 237(1-2), 123–134 (2000)
3. Alon, N., Yuster, R., Zwick, U.: Color Coding. Journal of the ACM 42(4), 844–856 (1995)
4. Betzler, N., Fellows, M.R., Komusiewicz, C., Niedermeier, R.: Parameterized Algorithms and Hardness Results for Some Graph Motif Problems. In: Ferragina, P., Landau, G.M. (eds.) CPM 2008. LNCS, vol. 5029, pp. 31–43. Springer, Heidelberg (2008)
5. Bruckner, S., Hüffner, F., Karp, R.M., Shamir, R., Sharan, R.: Topology-Free Querying of Protein Interaction Networks. In: Batzoglou, S. (ed.) RECOMB 2009. LNCS, vol. 5541, pp. 74–89. Springer, Heidelberg (2009)
6. Cesati, M.: Compendium of parameterized problems, http://bravo.ce.uniroma2.it/home/cesati/research/compendium.pdf
7. Dondi, R., Fertin, G., Vialette, S.: Weak Pattern Matching in Colored Graphs: Minimizing the Number of Connected Components. In: Italiano, G.F., Moggi, E., Laura, L. (eds.) ICTCS 2007, pp. 27–38. World Scientific, Singapore (2007)
8. Dondi, R., Fertin, G., Vialette, S.: Maximum Motif Problem in Vertex-Colored Graphs. In: Kucherov, G., Ukkonen, E. (eds.) CPM 2009 Lille. LNCS, vol. 5577, pp. 221–235. Springer, Heidelberg (2009)
9. Downey, R., Fellows, M.: Parameterized Complexity. Springer, Heidelberg (1999)
10. Fellows, M., Fertin, G., Hermelin, D., Vialette, S.: Sharp Tractability Borderlines for Finding Connected Motifs in Vertex-Colored Graphs. In: Arge, L., Cachin, C., Jurdziński, T., Tarlecki, A. (eds.) ICALP 2007. LNCS, vol. 4596, pp. 340–351. Springer, Heidelberg (2007)
11. Guillemot, S., Sikora, F.: Finding and Counting Vertex-Colored Subtrees. In: Hliněný, P., Kučera, A. (eds.) MFCS 2010. LNCS, vol. 6281, pp. 405–416. Springer, Heidelberg (2010)
12. Kelley, B.P., Sharan, R., Karp, R.M., Sittler, T., Root, D.E., Stockwell, B.R., Ideker, T.: Conserved Pathways within Bacteria and Yeast as Revealed by Global Protein Network Alignment. Proc. Nat. Acad. Sci. 100(20), 11394–11399 (2003)
13. Koyutürk, M., Grama, A., Szpankowski, W.: Pairwise Local Alignment of Protein Interaction Networks Guided by Models of Evolution. In: Miyano, S., Mesirov, J., Kasif, S., Istrail, S., Pevzner, P.A., Waterman, M. (eds.) RECOMB 2005. LNCS (LNBI), vol. 3500, pp. 48–65. Springer, Heidelberg (2005)
14. Lacroix, V., Fernandes, C.G., Sagot, M.F.: Motif Search in Graphs: Application to Metabolic Networks. IEEE/ACM Transactions on Computational Biology and Bioinformatics (TCBB) 3(4), 360–368 (2006)
15. Niedermeier, R.: Invitation to Fixed-Parameter Algorithms. Oxford University Press, Oxford (2006)
16. Paz, A., Moran, S.: Non Deterministic Polynomial Optimization Problems and Their Approximations. Theor. Comput. Sci. 15, 251–277 (1981)
17. Scott, J., Ideker, T., Karp, R.M., Sharan, R.: Efficient Algorithms for Detecting Signaling Pathways in Protein Interaction Networks. Journal of Computational Biology 13, 133–144 (2006)

18. Sharan, R., Ideker, T., Kelley, B., Shamir, R., Karp, R.M.: Identification of Protein Complexes by Comparative Analysis of Yeast and Bacterial Protein Interaction Data. In: Bourne, P.E., Gusfield, D. (eds.) RECOMB 2004, pp. 282–289. ACM Press, New York (2004)
19. Sharan, R., Suthram, S., Kelley, R., Kuhn, T., McCuine, S., Uetz, P., Sittler, K.R.M., Ideker, T.: Conserved Patterns of Protein Interaction in Multiple Species. Proc. Nat. Acad. Sci. 102(6), 1974–1979 (2005)

Improved MAX SNP-Hard Results for Finding an Edit Distance between Unordered Trees*

Kouichi Hirata[1], Yoshiyuki Yamamoto[2], and Tetsuji Kuboyama[3]

[1] Department of Artificial Intelligence
[2] Graduate School of Computer Science and Systems Engineering
Kyushu Institute of Technology
Kawazu 680-4, Iizuka 820-8502, Japan
{hirata,j673025y}@ai.kyutech.ac.jp
[3] Computer Center, Gakushuin University
Mejiro 1-5-1, Toshima, Tokyo 171-8588, Japan
kuboyama@gakushuin.ac.jp

Abstract. Zhang and Jiang (1994) have shown that the problem of *finding an edit distance between unordered trees* is MAX SNP-hard. In this paper, we show that this problem is MAX SNP-hard, even if (1) the height of trees is 2, (2) the degree of trees is 2, (3) the height of trees is 3 under a unit cost, and (4) the degree of trees is 2 under a unit cost.

1 Introduction

It is one of the important tasks for data mining from tree-structured data such as HTML and XML data for web mining or DNA and glycan data for bioinformatics to formulate such data as *rooted labeled trees* (*trees*, for short) and then compare them based on a distance measure between trees. The most famous distance measure between trees is an *edit distance* [1,7,9]. The edit distance is formulated as the minimum cost to transform from a tree to another tree by applying *edit operations* of a *substitution*, a *deletion*, and an *insertion* to trees.

After the algorithm to compute the edit distance for *ordered* trees has been proposed [7], the time complexity of it has been improved as $O(n^3)$ time [2] so far, where n is the maximum number of nodes in given two trees. On the other hand, the problem of finding an edit distance for *unordered* trees is intractable, that is, NP-hard [10] and MAX SNP-hard [8]. Here, the NP-hardness [10] has been shown in the case that *trees are binary under a unit cost*, where a unit cost is a cost assigning every edit operation to 1. Also the NP-hardness holds in the case that *the height of trees is 2 under a unit cost* (*cf.*, [11]).

On the other hand, the MAX SNP-hardness [8] has been shown just in the case that *the height of trees is 7 and the degrees of trees are unbounded under an indel unit cost*, not a unit cost. Here, an indel unit cost is a cost assigning

---

* This work is partially supported by Grand-in-Aid for Scientific Research 20500126, 20240014, 21500145 and 22240010 from the Ministry of Education, Culture, Sports, Science and Technology, Japan.

R. Giancarlo and G. Manzini (Eds.): CPM 2011, LNCS 6661, pp. 402–415, 2011.

a substitution to 2 and both a deletion and an insertion to 1. In particular, we cannot apply this proof to showing that the problem of finding an edit distance *under a unit cost* between unordered trees is MAX SNP-hard.

In this paper, as the improved MAX SNP-hard results, we show that the problem of finding an edit distance between unordered trees is MAX SNP-hard, even if (1) the height of trees is 2, (2) the degree of trees is 2, (3) the height of trees is 3 under a unit cost, and (4) the degree of trees is 2 under a unit cost. As the corollaries, we also show that the problem of finding the *largest common embedded tree* (called the *largest common sub-tree* in [8]) between unordered trees is MAX SNP-hard, even if (1) the height of trees is 2 and (2) the degree of trees is 2.

2 Edit Distance and Mapping

A *tree* is a connected graph without cycles. For a tree $T = (V, E)$, we denote V and E by $V(T)$ and $E(T)$, respectively. We sometimes denote $v \in V(T)$ by $v \in T$. A *rooted tree* is a tree with one node r chosen as its *root*.

For each node v in a rooted tree with the root r, let $UP_r(v)$ be the unique path from v to r. The *parent* of $v(\neq r)$ is its adjacent node on $UP_r(v)$ and the *ancestors* of $v(\neq r)$ are the nodes on $UP_r(v) - \{v\}$. We denote $u \leq v$ if v is an ancestor of u or $u = v$. The parent and the ancestors of the root r are undefined. We say that u is a *child* of v if v is the parent of u, and u is a *descendant* of v if v is an ancestor of u. A *leaf* is a node having no children.

Let T be a rooted tree and $v \in T$. The *degree* of v is the number of the children of v, and the *degree* $d(T)$ of T is the maximum degree for every node in T. The *height* of v is the number of edges in $UP_r(v)$, and the *height* $h(T)$ of T is the maximum height for every node in T.

We say that a rooted tree is *ordered* if a left-to-right order among siblings is given; *Unordered* otherwise. Also we say that a tree is *labeled* if each node is assigned a symbol from a fixed finite alphabet Σ, where we denote the label of a node v by $l(v)$, and sometimes identify v with $l(v)$. In this paper, we call a rooted unordered labeled tree a *tree*, simply.

Let T be a tree. Then, we call the following three operations *edit operations*. Also see Figure 1.

1. *Substitution*: Change the label of the node v in T (from l_1 to l_2).
2. *Deletion*: Delete a non-root node v in T (labeled by l_1) with a parent v' (labeled by l'), making the children of v become the children of v'. The children are inserted in the place of v as a subset of the children of v'.
3. *Insertion*: The complement of deletion. Insert a node v (labeled by l_2) as a child of v' (labeled by l') in T making v the parent of a subset of the children of v'.

For a special *blank* symbol $\varepsilon \notin \Sigma$, let $\Sigma_\varepsilon = \Sigma \cup \{\varepsilon\}$. Then, we represent each edit operation by $l_1 \mapsto l_2$, where $(l_1, l_2) \in (\Sigma_\varepsilon \times \Sigma_\varepsilon - \{(\varepsilon, \varepsilon)\})$. The operation is a substitution if $l_1 \neq \varepsilon$ and $l_2 \neq \varepsilon$, a deletion if $l_2 = \varepsilon$, and an insertion if $l_1 = \varepsilon$.

Substitution $l_1 \mapsto l_2$

Deletion $l_1 \mapsto \varepsilon$

Insertion $\varepsilon \mapsto l_2$

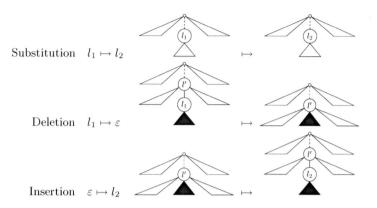

Fig. 1. Edit operations for trees

We define a *cost* $\gamma : (\Sigma_\varepsilon \times \Sigma_\varepsilon - \{(\varepsilon,\varepsilon)\}) \mapsto \mathbf{R}$ on pairs of labels. We constrain a cost γ to be a *metric*, that is, $\gamma(l_1,l_2) \geq 0$, $\gamma(l_1,l_1) = 0$, $\gamma(l_1,l_2) = \gamma(l_2,l_1)$ and $\gamma(l_1,l_3) \leq \gamma(l_1,l_2) + \gamma(l_2,l_3)$. In particular, we use the following two costs:

1. A *unit cost* μ [10,11]: $\mu(l_1,l_2) = 1$ if $l_1 \neq l_2$ and $l_1,l_2 \in \Sigma_\varepsilon$.
2. An *indel unit cost* ι_μ [3,8]: For $l_1,l_2 \in \Sigma$, $\iota_\mu(l_1,\varepsilon) = \iota_\mu(\varepsilon,l_2) = 1$, and $\iota_\mu(l_1,l_2) = 2$ if $l_1 \neq l_2$.

For a cost γ, we define the *cost* of an edit operation by setting $\gamma(l_1 \mapsto l_2) = \gamma(l_1,l_2)$. The *cost* of a sequence $S = s_1,\ldots,s_k$ of edit operations is given by $\gamma(S) = \sum_{i=1}^{k} \gamma(s_i)$. Then, an *edit distance* $\tau(T_1,T_2)$ between trees T_1 and T_2 are defined as $\min\{\gamma(S) \mid S$ is a sequence of edit operations transforming T_1 to $T_2\}$.

For trees T_1 and T_2, we say that the triple (M,T_1,T_2) is a *mapping* between T_1 and T_2 if $M \subseteq V(T_1) \times V(T_2)$ and every pair (v_1,w_1) and (v_2,w_2) in M satisfies the following conditions.

1. $v_1 = v_2$ iff $w_1 = w_2$ (one-to-one condition).
2. $v_1 \leq v_2$ iff $w_1 \leq w_2$ (ancestor condition).

We will use M instead of (M,T_1,T_2) when there is no confusion.

Let M be a mapping between T_1 and T_2. Also let I_1 (*resp.*, I_2) be the set of nodes in T_1 (*resp.*, T_2) but not in M. Then, the *cost* $\gamma(M)$ of M is given as $\sum_{(v,w)\in M} \gamma(l(v),l(w)) + \sum_{v\in I_1} \gamma(l(v),\varepsilon) + \sum_{w\in I_2} \gamma(\varepsilon,l(w))$. Tai [7] has shown that $\tau(T_1,T_2) = \min\{\gamma(M) \mid M$ is a mapping between T_1 and $T_2\}$.

Finally, we introduce the *least common embedded tree* [8]. Let T_1 and T_2 be two trees. A mapping f between T_1 and T_2 is an *embedding* of T_1 into T_2 if $l(u) = l(f(u))$. We say that T_1 is an *embedded tree* of T_2 if there exists an embedding of T_1 into T_2. A *common embedded tree* of T_1 and T_2 is any embedded tree of both T_1 and T_2, and the *largest common embedded tree* of T_1 and T_2 is a common embedded tree of T_1 and T_2 whose size is largest.

3   MAX SNP-Hard Results

Suppose that Π_1 and Π_2 are two optimization problems. Then, we say that Π_1 *L-reduces* to Π_2 [5] if there exist polynomial-time algorithms f, g and constants $\alpha, \beta > 0$ satisfying the following statements for an instance I of Π_1:

1. $opt(f(I)) \leq \alpha \cdot opt(I)$.
2. For a solution of $f(I)$ with weight s_2, the algorithm g produces in polynomial time a solution of I with weight s_1 such that $|s_1 - opt(I)| \leq \beta \cdot |s_2 - opt(f(I))|$.

If Π_1 *L-reduces* to Π_2 and Π_2 can be approximated in polynomial time within a factor of $1 + \varepsilon$, then Π_1 can be approximated within the factor $1 + \alpha\beta\varepsilon$. If Π_2 has a polynomial time approximation scheme (PTAS), then so does Π_1 [5].

A problem is MAX SNP-*hard* if every problem in MAX SNP can be *L*-reduced to it. Since the composition of two *L*-reductions is also an *L*-reduction, a problem is MAX SNP-hard if a MAX SNP-hard problem can be *L*-reduced to it. It is known that if any MAX SNP-hard problem has a PTAS, then P=NP. Hence, it is very unlikely for a MAX SNP-hard problem to have a PTAS [5].

In the remainder of this section, we use the *L*-reduction from the following MAX SNP-hard problem MAX 3SC-3:

MAXIMUM BOUNDED COVERING BY 3-SETS (MAX 3SC-3) [4]
INSTANCE: A finite set S and a collection $\mathcal{C}$ of 3-elements subset of S, where every element of S occurs at most three of the subsets in $\mathcal{C}$.
SOLUTION: Find the largest covering $\mathcal{C}' \subseteq \mathcal{C}$ of S, where a *covering* is a collection of mutually disjoint sets.

Throughout of this paper, for an instance of the MAX 3SC-3 problem, let $S = \{s_1, \ldots, s_m\}$ and $\mathcal{C} = \{C_1, \ldots, C_n\}$. Also let $C_i = \{s_{i1}, s_{i2}, s_{i3}\}$, where $s_{ij} \in S$. Furthermore, let $\mathcal{C}^*$ be the largest covering of $\mathcal{C}$.

In this paper, we investigate the following two optimization problems, where $h = \max\{h(T_1), h(T_2)\}$ and $d = \max\{d(T_1), d(T_2)\}$. If both h and d are bounded by constants c_1 and c_2, then the maximum number of leaves is also bounded by $c_1^{c_2}$, which implies that the problem of $\text{UTED}(c_1, c_2, \gamma)$ is tractable [6]. Hence, we pay our attention that either h or d is unbounded, denoted by $*$.

UNORDERED TREE EDIT DISTANCE WITH (h, d, γ) $(\text{UTED}(h, d, \gamma))$
INSTANCE: Two unordered trees T_1 and T_2 and a cost γ.
SOLUTION: Find an edit distance between T_1 and T_2 under a cost γ.

LARGEST COMMON EMBEDDED TREE WITH (h, d) $(\text{LCET}(h, d))$
INSTANCE: Two unordered trees T_1 and T_2.
SOLUTION: Find the largest common embedded tree of T_1 and T_2.

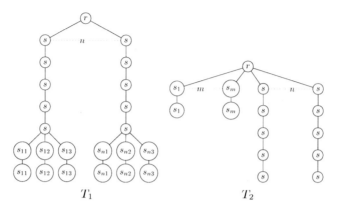

Fig. 2. Trees T_1 and T_2 [8]

3.1   The Problem of $\mathrm{UTED}(2, *, \iota_\mu)$

Zhang and Jiang [8] have shown that the problems of both $\mathrm{UTED}(7, *, \iota_\mu)$ and $\mathrm{LCET}(7, *)$ are MAX SNP-hard, by using trees T_1 and T_2 in Figure 2. In this section, we show that the problems of $\mathrm{UTED}(2, *, \iota_\mu)$ and $\mathrm{LCET}(2, *)$ are MAX SNP-hard.

First, we construct two trees T_3 and T_4 with height 2 in Figure 3, which are same trees given in [11] for proving NP-hardness of an unordered tree edit distance. We call the i-th child of r in T_3 a subtree of C_i, where $1 \le i \le n$ and $C_i \in \mathcal{C}$. Also we call the first m leaves in T_4 left leaves of T_4, and the last n children of r in T_4 dummy subtrees of T_4. We call this transformation from an instance of MAX 3SC-3 to unordered trees T_3 and T_4 f_1.

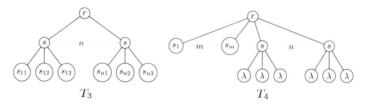

Fig. 3. Trees T_3 and T_4 (cf. [11])

Lemma 1. $\tau(T_3, T_4) = 6n + m - 4|\mathcal{C}^*|.$

Proof. Let M be the minimum cost mapping between T_3 and T_4, and $k = |\mathcal{C}^*|$. Then, without loss of generality, we can assume that $(r, r) \in M$.

For every $C \in \mathcal{C}^*$, M maps all of the leaves of a subtree of C in T_3 to three left leaves in T_4. In this case, M does not touch a node labeled by s in T_3. Then, the cost of M concerned with $\mathcal{C}^*$ is k. On the other hand, for every $C \in \mathcal{C} - \mathcal{C}^*$, M maps every subtree of C to some dummy subtree of T_4. Then, the cost of M concerned with $\mathcal{C} - \mathcal{C}^*$ is $6(n - k)$.

Since M does not touch $m - 3k$ nodes labeled by s and $n - (n - k) = k$ dummy subtrees in T_4, this cost of M is $m - 3k + 4k = m + k$. Hence, it holds that $\tau(T_3, T_4) = \iota_\mu(M) = k + 6(n - k) + m + k = 6n + m - 4k$. □

For an arbitrary mapping M between T_3 and T_4, let M' be a mapping between T_3 and T_4 such that $\{(u, v) \in M \mid l(u) = l(v)\}$. By the definition of ι_μ, it holds that $\iota_\mu(M) = \iota_\mu(M')$. Furthermore, let $\mathcal{C}'$ be a covering of $\mathcal{C}$ obtained from M' as follows: If M' contains (u_1, v_1), (u_2, v_2) and (u_3, v_3) such that $l(u_j) = l(v_j) = s_{ij}$ $(j = 1, 2, 3)$ for $1 \le i \le n$, then add C_i to $\mathcal{C}'$. We call the algorithm to construct from M to $\mathcal{C}'$ g_1, which is the same algorithm introduced by [8].

Theorem 1. *The problem of* $\mathrm{UTED}(2, *, \iota_\mu)$ *is MAX SNP-hard.*

Proof. Let I be an instance of MAX 3SC-3. By using the same discussion in the proof of Theorem 7 in [8], it holds that $n/7 \le opt(I)$. Then, by Lemma 1 and since $m \le 3n$, the following two inequalities hold.

$$opt(f_1(I)) = 6n + m - 4 \cdot opt(I) \le 9n - 4 \cdot opt(I) \le 59 \cdot opt(I),$$
$$\begin{aligned} s_2 - opt(f_1(I)) &= \iota_\mu(M) - \tau(T_3, T_4) = \iota_\mu(M') - \tau(T_3, T_4) \\ &\ge 6n + m - 4|\mathcal{C}'| - (6n + m - 4 \cdot opt(I)) = 4(opt(I) - s_1). \end{aligned}$$

Hence, (f_1, g_1) is an L-reduction from MAX 3SC-3 to $\mathrm{UTED}(2, *, \iota_\mu)$. □

Corollary 1. *The problem of* $\mathrm{LCET}(2, *)$ *is MAX SNP-hard.*

Proof. Consider the same algorithm f_1 in the proof of Theorem 1. Let T^* be the largest common embedded tree of T_3 and T_4 in Figure 3. Then, it holds that $|T^*| = 1 + 3|\mathcal{C}^*| + (n - |\mathcal{C}^*|) = 1 + n + 2|\mathcal{C}^*|$.

Consider the same algorithm g_2 introduced by [8] to construct a covering $\mathcal{C}'$ of $\mathcal{C}$ from a common embedded tree T' of T_3 and T_4 as follows. If all of the leaves of v_1, v_2 and v_3 such that $l(v_j) = s_{ij}$ $(j = 1, 2, 3)$ are in T', then add C_i to $\mathcal{C}'$. It is obvious that $|T'| \le 1 + n + 2|\mathcal{C}'|$, that is, $s_2 \le 1 + n + 2s_1$. Since $n/7 \le opt(I)$, we obtain the following two inequalities.

$$opt(f_1(I)) = 1 + n + 2 \cdot opt(I) \le 2n + 2 \cdot opt(I) \le 16 \cdot opt(I),$$
$$2(opt(I) - s_1) = (1 + n + 2 \cdot opt(I)) - (1 + n + 2s_1) \le opt(f_1(I)) - s_2.$$

Hence, (f_1, g_2) is an L-reduction from MAX 3SC-3 to $\mathrm{LCET}(2, *)$. □

3.2   The Problem of $\mathrm{UTED}(*, 2, \iota_\mu)$

In this section, we show that the problem of $\mathrm{UTED}(*, 2, \iota_\mu)$ is MAX SNP-hard.

First, we construct two trees T_5 and T_6 in Figure 4. We call the subtree of the form $s(s_{i1}, s(s_{i2}, s_{i3}))$ *a subtree of* C_i, where $1 \le i \le n$ and $C_i \in \mathcal{C}$. Also we call m leaves labeled by s_j in T_6 *left leaves of* T_6, where $1 \le j \le m$, and the children of the form $s(s)$ of r_2 in T_6 *dummy subtrees of* T_6. We call this transformation from an instance of MAX 3SC-3 to unordered binary trees T_5 and T_6 f_2.

Lemma 2. $\tau(T_5, T_6) = 7n + 3m - 2|\mathcal{C}^*| - 4$.

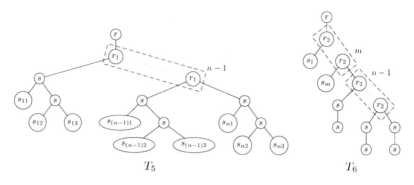

Fig. 4. Trees T'_5 and T'_6

Proof. Let M be the minimum cost mapping between T_5 and T_6, and $k = |\mathcal{C}^*|$. Without loss of generality, we can assume that $(r, r) \in M$ but M does not contain a pair (u, v) such that $l(u) = r_1$ and $l(v) = r_2$.

For every $C \in \mathcal{C}^*$, M maps all of the leaves of a subtree of C in T_5 to three left leaves in T_6. In this case, M does not touch nodes labeled by s in T_5. Then, the cost of M concerned with $\mathcal{C}^*$ is $2k$. On the other hand, for every $C \in \mathcal{C} - \mathcal{C}^*$, M maps every subtree of C to some dummy subtree of T_6. Then, the cost of M concerned with $\mathcal{C} - \mathcal{C}^*$ is $3(n - k)$.

Furthermore, M does not touch $m - 3k$ left leaves and $n - (n - k) = k$ dummy subtrees in T_6. Also M does not touch the nodes labeled by r_1 in T_5 and by r_2 in T_6. Then, this cost of M is $m - 3k + 2k + 2(n - 1 + m + n - 1) = 3m + 4n - k - 4$. Hence, it holds that $\tau(T_5, T_6) = \iota_\mu(M) = 2k + 3(n - k) + 3m + 4n - k - 4 = 7n + 3m - 2k - 2$. □

Theorem 2. *The problem of* $\mathrm{UTED}(*, 2, \iota_\mu)$ *is MAX SNP-hard.*

Proof. By Lemma 2 and since $n/7 \leq opt(I)$ [8] and $m \leq 3n$, the following two inequalities hold for an instance I of MAX 3SC-3 and the same algorithm g_1 in Theorem 1.

$$opt(f_2(I)) = 7n + 3m - 2 \cdot opt(I) - 4 \leq 16n - 2 \cdot opt(I) \leq 110 \cdot opt(I),$$
$$\begin{aligned} s_2 - opt(f_2(I)) &= \iota_\mu(M) - \tau(T_5, T_6) = \iota_\mu(M') - \tau(T_5, T_6) \\ &\geq 7n + 3m - 2|\mathcal{C}'| - 4 - (7n + 3m - 2 \cdot opt(I) - 4) \\ &= 2(opt(I) - s_1). \end{aligned}$$

Hence, (f_2, g_1) is an L-reduction from MAX 3SC-3 to $\mathrm{UTED}(*, 2, \iota_\mu)$. □

Corollary 2. *The problem of* $\mathrm{LCET}(*, 2)$ *is MAX SNP-hard.*

Proof. Consider the same algorithm f_2 in the proof of Theorem 2. Let T^* be the largest common embedded tree of T_5 and T_6 in Figure 4. Then, it holds that $|T^*| = 1 + 3|\mathcal{C}^*| + 2(n - |\mathcal{C}^*|) = 1 + 2n + |\mathcal{C}^*|$. Also, for a common embedded tree T' of T_5 and T_6, consider the same algorithm g_2 in Corollary 1. It is obvious that $|T'| \leq 1 + 2n + |\mathcal{C}'|$, that is, $s_2 \leq 1 + 2n + s_1$. Since $n/7 \leq opt(I)$, we obtain the following two inequalities.

$$opt(f_2(I)) = 1 + 2n + opt(I) \le 3n + opt(I) \le 22 \cdot opt(I),$$
$$opt(I) - s_1 = (1 + 2n + opt(I)) - (1 + 2n + s_1) \le opt(f_2(I)) - s_2.$$

Hence, (f_2, g_2) is an L-reduction from MAX 3SC-3 to LCET$(*, 2)$.   □

3.3   The Problem of UTED$(3, *, \mu)$

In this section, we show that the problem of UTED$(3, *, \mu)$ is MAX SNP-hard.

First, we construct two trees T_7 and T_8 in Figure 5. We call the i-th child of r in T_7 a *subtree* of C_i, where $1 \le i \le n$ and $C_i \in \mathcal{C}$, the first m children of r in T_8 *left subtrees* of T_8 and the last n children of r in T_8 *dummy subtrees* of T_8. We call this transformation from an instance of MAX 3SC-3 to unordered trees T_7 and T_8 f_3.

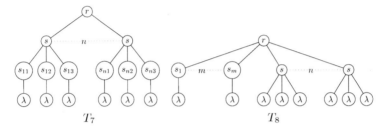

Fig. 5. Trees T_7 and T_8

Lemma 3. $\tau(T_7, T_8) = 3n + 2m - 4|\mathcal{C}^*|.$

Proof. Let M be the minimum cost mapping between T_7 and T_8, and $k = |\mathcal{C}^*|$. Without loss of generality, we can assume that $(r, r) \in M$.

For every $C \in \mathcal{C}^*$, M maps all of the nodes of a subtree of C except the root labeled by s in T_7 to three left subtrees in T_8. In this case, M does not touch a node labeled by s. Then, the cost of M concerned with $\mathcal{C}^*$ is k. On the other hand, for every $C \in \mathcal{C} - \mathcal{C}^*$, M maps every subtree of C to some dummy subtree of T_8. In this case, M contains (u, v) such that $l(u) = l(v) = s$ or $l(u) = l(v) = \lambda$ but not $l(u) = s_j$ and $l(v) = \lambda$, because M is the minimum cost. Then, the cost of M concerned with $\mathcal{C} - \mathcal{C}^*$ is $3(n - k)$.

Since M does not touch $m - 3k$ left subtrees and $n - (n - k) = k$ dummy subtrees in T_8, this cost of M is $2(m - 3k) + 4k = 2m - 2k$. Hence, it holds that $\tau(T_7, T_8) = \mu(M) = |\mathcal{C}^*| + 3(n - |\mathcal{C}^*|) + 2m - 2|\mathcal{C}^*| = 3n + 2m - 4|\mathcal{C}^*|.$ □

In contrast to the previous sections, we cannot use the algorithm g_1 [8] to construct a mapping $M' = \{(u, v) \in M \mid l(u) = l(v)\}$ from an arbitrary mapping M, because such an M' satisfies that $\mu(M) \le \mu(M')$, not $\mu(M) \ge \mu(M')$, for a unit cost μ, while it holds that $\iota_\mu(M) = \iota_\mu(M')$ for an indel unit cost ι_μ. Hence, in this paper, we introduce a *mapping rearrangement* of M.

Let M be an arbitrary mapping between T_7 and T_8. Note first that, if M contains either (r, v) or (u, r), where r is the root but u and v are not roots, then we can construct a new M_1 from M such that $\mu(M) \geq \mu(M_1)$, by deleting (r, v) or (u, r) from M and adding (r, r) to M_1. Hence, we can assume that $(r, r) \in M$.

After setting a mapping $M' = \{(u, v) \in M \mid l(u) = l(v)\}$, we apply the following *mapping rearrangement* of M. (See Figure 6.)

Case 1.

Case 2.

Case 3.   Case 4.

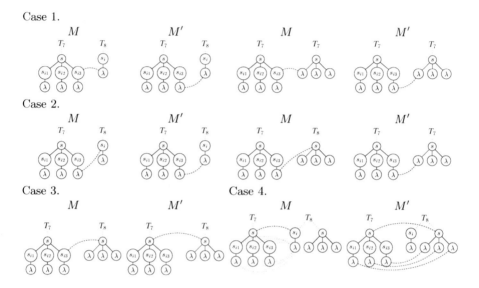

Fig. 6. A mapping rearrangement between T_7 and T_8

1. For a pair $(u, v) \in M$ such that $l(u) = s_i$ and $l(v) = \lambda$, add (u', v) to M', where u' is the unique child of u labeled by λ and not touched by M.
2. For a pair $(u, v) \in M$ such that $l(u) = \lambda$ and $l(v) = s_i$ or $l(v) = s$, add (u, v') to M', where v' is a child of v labeled by λ and not touched by M.
3. For a pair $(u, v) \in M$ such that $l(u) = s_i$ and $l(v) = s$, add (u', v) to M', where u' is the parent of u labeled by s and not touched by M.
4. For a pair $(u, v) \in M$ such that $l(u) = s$ and $l(v) = s_i$, first select a dummy subtree not touched by M or M', and then add (u, v_0), (u_1, v_1), (u_2, v_2) and (u_3, v_3) to M', where u_i is the leaf of the subtree rooted by u, v_0 is the root of the dummy subtree and v_i is the leaf of the dummy subtree for $i = 1, 2, 3$.

Lemma 4. $\mu(M) \geq \mu(M')$.

Proof. Since mapping rearrangements for Cases 1, 2 and 3 change $(u, v) \in M$ such that $l(u) \neq l(v)$ to $(u', v') \in M'$ such that $l(u') \neq l(v')$, it is obvious that $\mu(M) \geq \mu(M')$.

For Case 4, note that there always exists a dummy subtree not touched by M'. Then, for subsets $S \subseteq M$ and $S' \subseteq M'$ concerned with Case 4, it holds that $10 = 6 + 4 \leq \mu(S) \leq 8 + 4 = 12$ and $\mu(S') = 3 + 2 = 5$, which implies that $\mu(S) \geq \mu(S')$. Hence, it holds that $\mu(M) \geq \mu(M')$.   □

Let $\mathcal{C}'$ be a covering of $\mathcal{C}$ obtained from M' as same as g_1. We call the algorithm to construct from M to $\mathcal{C}'$ g_3.

Theorem 3. *The problem of* $\mathrm{UTED}(3, *, \mu)$ *is MAX SNP-hard.*

Proof. By Lemma 3 and 4, and since $n/7 \leq opt(I)$ [8] and $m \leq 3n$, the following two inequalities hold for an instance I of MAX 3SC-3.

$$opt(f_3(I)) = 3n + 2m - 4 \cdot opt(I) \leq 9n - 4 \cdot opt(I) = 59 \cdot opt(I),$$
$$s_2 - opt(f_3(I)) = \mu(M) - \tau(T_7, T_8) \geq \mu(M') - \tau(T_7, T_8)$$
$$\geq 3n + 2m - 4|\mathcal{C}'| - (3n + 2m - 4 \cdot opt(I)) = 4(opt(I) - s_1).$$

Hence, (f_3, g_3) is an L-reduction from MAX 3SC-3 to $\mathrm{UTED}(3, *, \mu)$. □

3.4   The Problem of $\mathrm{UTED}(*, 2, \mu)$

In this section, we show that the problem of $\mathrm{UTED}(*, 2, \mu)$ is MAX SNP-hard.

First, we construct two trees T_9 and T_{10} in Figure 7, where the number of nodes labeled by ρ in T_9 and T_{10} (enclosed by a solid line) is $8n + 2m + 2$, and we call it a ρ-*subtree*. We call the left child of q with height i in T_9 a *subtree of* C_i, where $1 \leq i \leq n$ and $C_i \in \mathcal{C}$. Also we call the left child of q with height j in T_{10} a *subtree of* s_j, where $1 \leq j \leq m$ and $s_j \in S$, and the left child of q with height k in T_{10} *dummy subtrees of* T_{10}, where $m + 1 \leq k \leq n + m$. We call this transformation from an instance of MAX 3SC-3 to unordered trees T_9 and T_{10} f_4. In particular, let M_0 be the mapping between ρ-subtrees of T_9 and T_{10} that touches all of the nodes labeled by ρ.

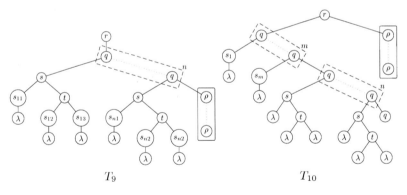

T_9 T_{10}

Fig. 7. Trees T_9 and T_{10}

Lemma 5. *Every mapping M between T_9 and T_{10} that does not touch ρ-subtrees of T_9 and T_{10} satisfies that* $\mu(M) > \mu(M_0)$.

Proof. For M_0, it holds that $\mu(M_0)$ is the total number of nodes in T_9 and T_{10} except ρ-subtrees, that is, $(9n + 1) + (6n + 3m + 2) = 15n + 3m + 3$. On the other hand, for every mapping M between T_9 and T_{10} that does not touch ρ-subtrees of T_9 and T_{10}, it holds that $\mu(M) \geq 2(8n + 2m + 2) = 16n + 4m + 4 > 15n + 3m + 3 = \mu(M_0)$. □

Lemma 6. $\tau(T_1, T_2) = 5n + 3m - 2|\mathcal{C}^*|$.

Proof. Let M be the minimum cost mapping between T_9 and T_{10}, and $k = |\mathcal{C}^*|$. Without loss of generality, we can assume that $(r, r) \in M$. Also, since it holds that $M_0 \subseteq M$ by Lemma 5, M touches no nodes labeled by q in T_9 and T_{10}.

For every $C \in \mathcal{C}^*$, M maps all of the nodes of a subtree of C except the nodes labeled by s and t in T_9 to three subtrees of s_i, s_j and s_k in T_{10}, where $1 \leq i, j, k \leq m$. In this case, M does not touch the nodes labeled by s and t. Then, the cost of M concerned with $\mathcal{C}^*$ is $2k$. On the other hand, for every $C \in \mathcal{C} - \mathcal{C}^*$, M maps every subtree of C to some dummy subtree of T_8. In this case, M contains (u, v) such that $l(u) = l(v) = s$, $l(u) = l(v) = t$ or $l(u) = l(v) = \lambda$ but not $l(u) = s_j$ and $l(v) = \lambda$, because M is the minimum cost. Then, the cost of M concerned with $\mathcal{C} - \mathcal{C}^*$ is $3(n - k)$.

Furthermore, M does not touch n nodes labeled by q in T_9 and $n + m$ nodes labeled by q in T_{10}. Also M does not touch $m - 3k$ subtrees of s_i in T_{10} and $n - (n - k) = k$ dummy subtrees in T_{10}. Then, this cost of M is $n + n + m + 2(m - 3k) + 5k = 2n + 3m - k$. Hence, it holds that $\tau(T_9, T_{10}) = \mu(M) = 2|\mathcal{C}^*| + 3(n - |\mathcal{C}^*|) + 2n + 3m - |\mathcal{C}^*| = 5n + 3m - 2|\mathcal{C}^*|$. □

Let M be an arbitrary mapping between T_9 and T_{10}. Without loss of generality, we can assume that $(r, r) \in M$. Then, we start a *mapping rearrangement* of M by constructing the following mapping M^* from M:

$$M^* = (M - (M_1 \cup M_2 \cup M_3)) \cup M_0, \quad M_2 = \{(u, v) \in M \mid l(u) = \rho\},$$
$$M_1 = \{(u, v) \in M \mid l(u) = l(v) = q\}, \quad M_3 = \{(u, v) \in M \mid l(v) = \rho\}.$$

Lemma 7. $\mu(M) > \mu(M^*)$.

Proof. If $(u, v) \in M$ such that $l(u) = l(v) = q$, then M contains no pair (u', v') such that $l(u') = l(v') = \rho$, which implies that $\mu(M) > \mu(M_0) \geq \mu((M - M_1) \cup M_0)$ by Lemma 5.

If $(u, v) \in M$ such that $l(u) = \rho$ and $l(v) \neq \rho$, then there exists a node v' in T_{10} such that $l(v') = \rho$ and M does not touch v' for such a node u, which implies that $\mu(M) \geq \mu((M - M_2) \cup M_0)$.

If $(u, v) \in M$ such that $l(u) \neq \rho$ and $l(v) = \rho$, then there exists a node u' in T_9 such that $l(u') = \rho$ and M does not touch u' for such a node v, which implies that $\mu(M) \geq \mu((M - M_3) \cup M_0)$. □

After setting a mapping $M' = \{(u, v) \in M^* \mid l(u) = l(v)\}$, we apply the following *mapping rearrangement* of M through M^*. (See Figure 8 and 9.)

1. For a pair $(u, v) \in M^*$ such that $l(u) = s_i$ and $l(v) = \lambda$, add (u', v) to M', where u' is the unique child of u labeled by λ and not touched by M^*.
2. For a pair $(u, v) \in M^*$ such that $l(u) = \lambda$ and $l(v) = s_i$, $l(v) = t$ or $l(v) = s$, add (u, v') to M', where v' is a descendant of v labeled by λ and not touched by M^*.
3. For a pair $(u, v) \in M^*$ such that $l(u) = s_i$ and $l(v) = s$, add (u', v) to M', where u' is the ancestor of u such that $l(u') = s$ and not touched by M^*.

Case 1.

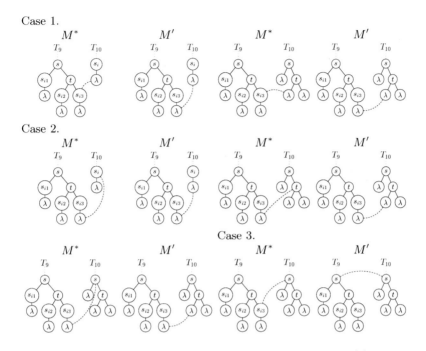

Case 2.

Case 3.

Fig. 8. A mapping rearrangement between T_9 and T_{10} (1)

4. Consider a pair $(u, v) \in M^*$ such that $l(u) = s_i$ and $l(v) = t$.
   Suppose that the label of the parent u_1 of u is t. If $(u_1, v_1) \in M^*$ such that
   v_1 is a parent of v labeled by s, then M^* does not touch the parent u_2 of
   u_1, so add (u_2, v_1) and (u_1, v) to M'. Otherwise, add (u_1, v) to M' simply.
   Suppose that the label of the parent u_1 of u is s. If $(u_2, v') \in M^*$ such that
   u_2 is a sibling of u_1 labeled by t and v' is some node in T_9, then add (u_2, v)
   and (u_3, v_1) to M', where u_3 is the unique child of u labeled by λ and v_1 is
   a sibling of v labeled by λ. Otherwise, add (u_2, v) to M' simply.

5. For a pair $(u, v) \in M^*$ such that $l(u) = s$ and $l(v) = s_i$, first select a
   dummy subtree not touched by M^* or M' and rooted by v_0, and then add
   (u, v_0), (u_1, v_1), (u_2, v_2), (u_3, v_3) and (u_4, v_4) to M', where u_i is the leaf of
   the subtree rooted by u, v_i is the leaf of the dummy subtree $(i = 1, 2, 3)$, u_4
   is the child of u labeled by t and v_4 is the child of v_0 labeled by t.

6. For a pair $(u, v) \in M^*$ such that $l(u) = t$ and $l(v) = s_i$, first select a dummy
   subtree not touched by M^* or M' and rooted by v_0, and then add (u, v_1),
   (u_1, v_0), (u_2, v_2), (u_3, v_3) and (u_4, v_4), where v_1 is the child of v_0 labeled by
   t, u_1 is the parent of u, u_2 and u_3 are the leaves of the subtree rooted by u,
   v_2 and v_3 are the leaves of the right subtree of the dummy subtree, u_4 is the
   leaf of the left child of u_3, and v_4 is the left child of v_3.

Lemma 8. $\mu(M^*) \geq \mu(M')$.

Case 4.

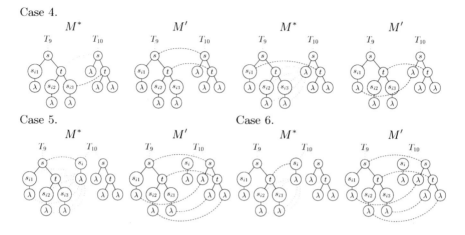

Case 5.   Case 6.

Fig. 9. A mapping rearrangement between T_9 and T_{10} (2)

Proof. Since mapping rearrangements for Cases 1, 2 and 3 change $(u, v) \in M^*$ such that $l(u) \neq l(v)$ to $(u', v') \in M'$ such that $l(u') = l(v')$, it is obvious that $\mu(M^*) \geq \mu(M')$. Since a mapping rearrangement for Case 4 changes $(u, v) \in M^*$ such that $l(u) \neq l(v)$ to $(u', v') \in M'$ such that $l(u') = l(v')$ and possibly adds $(u'', v'') \in M'$ such that $l(u'') = l(v'')$ to M', it holds that $\mu(M^*) \geq \mu(M')$.

For Cases 5 and 6, note that there always exists a dummy subtree not touched by M'. Then, for subsets $S^* \subseteq M^*$ and $S' \subseteq M'$ concerned with Cases 5 or 6, it holds that $12 = 7 + 5 \leq \mu(S^*) \leq 9 + 5 = 14$ and $\mu(S') = 3 + 2 = 5$, which implies that $\mu(S^*) \geq \mu(S')$. Hence, it holds that $\mu(M^*) \geq \mu(M')$. □

Let C' be a covering of C obtained from M' as same as g_1. We call the algorithm to construct from M to C' through M^* g_4.

Theorem 4. *The problem of* $\mathrm{UTED}(*, 2, \mu)$ *is MAX SNP-hard.*

Proof. By Lemma 6, 7 and 8, and since $n/7 \leq opt(I)$ [8] and $m \leq 3n$, the following two inequalities hold for an instance I of MAX 3SC-3.

$$opt(f_4(I)) = 5n + 3m - 2 \cdot opt(I) \leq 14n - 2 \cdot opt(I) = 96 \cdot opt(I),$$
$$\begin{aligned} s_2 - opt(f_4(I)) &= \mu(M) - \tau(T_9, T_{10}) \geq \mu(M') - \tau(T_9, T_{10}) \\ &\geq 5n + 3m - 2|C'| - (5n + 3m - 2 \cdot opt(I)) = 2(opt(I) - s_1). \end{aligned}$$

Hence, (f_4, g_4) is an L-reduction from MAX 3SC-3 to $\mathrm{UTED}(*, 2, \mu)$. □

4   Conclusion

In this paper, for two unordered trees with the maximum height h and the maximum degree d, we have investigated two optimization problems of $\mathrm{UTED}(h, d, \mu)$ to find an edit distance between given trees under a cost γ and $\mathrm{LCET}(h, d)$ to find the largest common embedded tree of given trees. Then, we have shown that $UTED(2, *, \iota_\mu)$, $LCET(2, *)$, $UTED(*, 2, \iota_\mu)$, $LCET(*, 2)$, $UTED(3, *, \mu)$ and $UTED(*, 2, \mu)$ are MAX SNP-hard, where ι_μ and μ are an indel unit cost and an unit cost, respectively.

Concerned with an edit distance for unordered trees, Jiang *et al.* [3] have shown that the problem of finding an *alignment* between unordered trees is MAX SNP-hard under an indel unit cost ι_μ, even if one tree is binary. It is a future work to discuss whether or not this problem is MAX SNP-hard under a unit cost μ.

References

1. Bille, P.: A survey on tree edit distance and related problems. Theoret. Comput. Sci. 337, 217–239 (2005)
2. Demaine, E.D., Mozes, S., Rossman, B., Weimann, O.: An optimal decomposition algorithm for tree edit distance. ACM Trans. Algorithms 6 (2009)
3. Jiang, T., Wang, L., Zhang, K.: Alignment of trees – an alternative to tree edit. Theoret. Comput. Sci. 143, 137–148 (1995)
4. Kann, V.: Maximum bounded 3-demensional matching is MAX SNP-complete. Inform. Process. Let. 37, 27–35 (1991)
5. Papadimitriou, C.H., Yannakakis, M.: Optimization, approximation and complexity. J. Comput. System Sci. 43, 425–440 (1991)
6. Shasha, D., Wang, J.T.-L., Zhang, K., Shih, F.Y.: Exact and approximate algorithms for unordered tree matching. IEEE Trans. Sys. Man. and Cybernet. 24, 668–678 (1994)
7. Tai, K.-C.: The tree-to-tree correction problem. J. ACM 26, 422–433 (1979)
8. Zhang, K., Jiang, T.: Some MAX SNP-hard results concerning unordered labeled trees. Inform. Process. Let. 49, 249–254 (1994)
9. Zhang, K., Shasha, D.: Simple fast algorithms for the editing distance between trees and related problems. SIAM J. Comput. 18, 1245–1262 (1989)
10. Zhang, K., Statman, R., Shasha, D.: On the editing distance between unordered labeled trees. Inform. Process. Let. 42, 133–139 (1992)
11. Zhang, K., Wang, J., Shasha, D.: On the editing distance between undirected acyclic graphs. Int. J. Found. Comput. Sci. 7, 43–58 (1995)

Approximation Algorithms for Orienting Mixed Graphs

Michael Elberfeld[1,*], Danny Segev[2,*], Colin R. Davidson[3],
Dana Silverbush[4], and Roded Sharan[4]

[1] Institute of Theoretical Computer Science,
University of Lübeck, 23538 Lübeck, Germany
elberfeld@tcs.uni-luebeck.de
[2] Department of Statistics, University of Haifa, Haifa 31905, Israel
segevd@stat.haifa.ac.il
[3] Faculty of Mathematics, University of Waterloo, Waterloo, Canada, N2L 3G1
colinrdavidson@gmail.com
[4] Blavatnik School of Computer Science, Tel Aviv University, Tel Aviv 69978, Israel
{danasilv,roded}@post.tau.ac.il

Abstract. Graph orientation is a fundamental problem in graph theory that has recently arisen in the study of signaling-regulatory pathways in protein networks. Given a graph and a list of ordered source-target vertex pairs, it calls for assigning directions to the edges of the graph so as to maximize the number of pairs that admit a directed source-to-target path. When the input graph is undirected, a sub-logarithmic approximation is known for the problem. However, the approximability of the biologically-relevant variant, in which the input graph has both directed and undirected edges, was left open. Here we give the first approximation algorithm to this problem. Our algorithm provides a sub-linear guarantee in the general case, and logarithmic guarantees for structured instances.

Keywords: protein-protein interaction network, mixed graph, graph orientation, approximation algorithm.

1 Introduction

Protein-protein interactions (PPIs) form the skeleton of signal transduction in the cell. While many of these interactions carry directed signaling information, current PPI measurement technologies, such as yeast two hybrid [10] and co-immunoprecipitation [14], cannot reveal the direction in which the signal flows. The problem of inferring this hidden directionality information is fundamental to our understanding of how these networks function. Previous work on this problem has relied on information from perturbation experiments [23], in which a gene is perturbed (cause) and as a result other genes change their expression levels (effects), to guide the orientation inference. Specifically, it is assumed that for an effect to take place, there must be a directed path in the network from

---

[*] These authors contributed equally to this work.

R. Giancarlo and G. Manzini (Eds.): CPM 2011, LNCS 6661, pp. 416–428, 2011.
© Springer-Verlag Berlin Heidelberg 2011

the causal gene to the affected gene. The arising combinatorial problem is to orient the edges of the network such that a maximum number of cause-effect pairs admit a directed path from the causal to the affected gene. When studying a PPI network in isolation, the input network is undirected. However, the more biologically relevant variant considers also protein-DNA interactions as these are necessary to explain the expression changes. Moreover, the directionality of some PPIs, like kinase-substrate interactions, is known in advance. Thus, in general, the input network is a mixed graph containing both directed and undirected edges.

The optimization problem that we study draws its recent interest from applications in network biology, but is rooted at practical applications from already a century ago: In 1939, Robbins [20], who was motivated by applications in street network design, showed that an undirected graph has a strongly connected orientation if and only if it has no bridge edge. The corresponding decision problem can be solved in linear time [22]. The characterization of Robbins was extended to mixed graphs by Boesch et al. [5]; linear time algorithms for deciding whether a mixed graph admits a strongly connected orientation were presented by Chung et al. [7]. Hakimi et al. [15] presented a polynomial algorithm for the problem of orienting an undirected graph so as to maximize the number of source-target pairs out of all possible ordered vertex pairs that admit a directed source-to-target path. A recent work by Dorn et al. [9] studies the parameterized complexity of orienting graphs. We refer to the textbook of Bang-Jensen and Gutin [3] for a comprehensive discussion of various graph orientation problems.

More recently, the problem of network orientation has been motivated by applications in network biology. Medvedovsky et al. [18] who formulated the problem that we study here, focused on restricted instances where the input graph is undirected, providing a logarithmic approximation algorithm for the problem. The approximation guarantee was later improved to $\Omega(\log \log n / \log n)$ by Gamzu et al. [13], where n denotes the number of vertices in the input graph. Gamzu et al. also showed that the orientation problem on mixed graphs can be approximated to within a poly-logarithmic ratio of $\Omega(1/\log^l n)$ where l is the maximum number of alternations between undirected and directed edges on a source-to-target path. Silverbush et al. [21] developed an ILP-based algorithm to optimally orient mixed networks, but the approximability of the problem (for non-constant l) was left open.

In this work, we study the approximability of the orientation problem on mixed graphs. We show that the problem is NP-hard to approximate to within a factor of 7/8. We then reduce the problem to orienting acyclic mixed graphs. We provide logarithmic approximation guarantees for tree-like reduced instances and use those to develop a sub-linear approximation algorithm for general instances.

The paper is organized as follows: In the next section we formally define the orientation problem, discuss its complexity and describe a generic reduction to acyclic mixed graphs. In Section 3 we present logarithmic factor approximation algorithms for tree-like instances. Section 4 presents the sub-linear approximation algorithm for the general case.

2   Preliminaries

Notation and terminology. We focus on simple graphs with no loops or parallel edges. A *mixed graph* is a triple $G = (V, E_U, E_D)$ that consists of a vertex set V, a set of *undirected edges* $E_U \subseteq \{e \subseteq V \mid |e| = 2\}$, and a set of *directed edges* $E_D \subseteq V \times V$. We assume that every pair of vertices is either connected by a single edge of a specific type (directed or undirected) or not connected at all. We also write $V(G)$, $E_U(G)$, and $E_D(G)$ to refer to the sets V, E_U, and E_D, respectively. When G is clear from the context, we will denote $n = |V|$.

Let G_1 and G_2 be two mixed graphs. The graph G_1 is a *subgraph* of G_2 when the relations $V(G_1) \subseteq V(G_2)$, $E_U(G_1) \subseteq E_U(G_2)$, and $E_D(G_1) \subseteq E_D(G_2)$ hold. A *path* of length ℓ in a mixed graph G is a sequence $p = \langle v_1, v_2, \ldots, v_\ell, v_{\ell+1} \rangle$ of distinct vertices such that for every $1 \le i \le \ell$, we have $\{v_i, v_{i+1}\} \in E_U(G)$ or $(v_i, v_{i+1}) \in E_D(G)$. It *crosses* a vertex $v \in V(G)$ when $v = v_i$ for some $i \in \{1, \ldots, \ell+1\}$. It is a *cycle* when $v_1 = v_{\ell+1}$. Given $s \in V(G)$ and $t \in V(G)$, we say that t *is reachable from* s when there exists a path in G that goes from s to t. In this case we also say that G *satisfies* the pair (s, t). A mixed graph with no cycles is called a *mixed acyclic graph* (MAG).

Let G be a mixed graph. An *orientation* of G is a directed graph over the same vertex set, whose edge set contains all the directed edges of G and a single directed instance of every undirected edge, but nothing more. When only a subset of the undirected edges have been oriented, we obtain a *partial orientation*.

Problem statement. The MAXIMUM-MIXED-GRAPH-ORIENTATION problem is defined as follows:

Input: A mixed graph G, and a collection of source-target vertex pairs $P \subseteq V(G) \times V(G)$.
Output: An orientation of G that satisfies a maximum number of pairs from P.

Hardness result. Arkin and Hassin [1] showed that it is NP-complete to decide whether, for a given mixed graph G and a collection of source-target pairs P, the graph G can be oriented to satisfy all pairs in P. Their reduction, based on the 3-SATISFIABILITY problem, guarantees that for every $k \in \mathbb{N}$ there exists an assignment with k satisfied clauses if and only if there exists an orientation with k satisfied pairs. Thus, the inapproximability of MAXIMUM-3-SATISFIABILITY [16] directly transfers to MAXIMUM-MIXED-GRAPH-ORIENTATION, implying that it is NP-hard to approximate it to within a factor of $7/8$. We note that this bound is slightly lower than the $12/13$-bound known for the special case where the input graph is undirected [18].

Reduction to mixed acyclic graphs. Given an orientation instance (G, P), we can orient the undirected edges of any mixed cycle in a consistent direction without affecting the maximum number of source-target pairs that are satisfied by an optimal orientation. This observation gives rise to a polynomial-time reduction from mixed graphs to MAGs: First, we iteratively orient mixed cycles in the input graph. Then, we contract strongly connected components into single vertices, and connect two components by an undirected (directed) edge when some vertex in the first component is connected by such an edge to a vertex in the second component (note that there cannot be more than one edge type as otherwise the two strongly connected components would have been merged). The pairs from P are adjusted accordingly from vertices of G to component vertices. A formal correctness proof of this reduction is given by Silverbush et al. [21].

Given a MAG G, the components of the undirected graph $(V(G), E_U(G))$ are called the *undirected components* of G; they must be trees that are connected by directed edges from $E_D(G)$ without producing cycles. The *graph of undirected components of* G is the directed acyclic graph G_{UCC} with $V(G_{UCC}) = \{G_i \mid G_i$ is an undirected component of $G\}$, and there is a directed edge from a node G_i to a node G_j when there is an edge from some vertex $v \in G_i$ to some vertex $w \in G_j$.

By the reduction above, we may focus our attention on treating MAGs. In addition, we may assume that each of the input pairs can be satisfied by some orientation; otherwise, it can be eliminated without affecting the optimum solution. Thus, throughout the paper, all instances considered will be assumed to satisfy these two properties.

3 Logarithmic Approximations for Tree-Like Instances

In this section we provide logarithmic approximations that apply to orientation instances where the graph is "similar" to a tree, as formally defined in the sequel. In the remainder of this section, we make use of the following result about orienting undirected trees due to Medvedovsky et al. [18]:

Lemma 3.1. *Let (G, P) be an orientation instance where G is an undirected tree. There is a polynomial-time algorithm that computes an orientation satisfying at least $|P|/(4\lceil \log n \rceil)$ pairs.*

3.1 Orienting Mixed Trees

The above lemma guarantees that a logarithmic fraction of the input pairs can always be satisfied, and since it is constructive, we immediately derive an $\Omega(1/\log n)$ approximation algorithm for undirected trees. The following sequences of claims are of similar nature: We prove the existence of orientations satisfying a certain fraction of all input pairs, and this leads to approximation algorithms with the corresponding ratio. We start with orientations for mixed trees.

Lemma 3.2. *Let (G, P) be an orientation instance with a mixed tree G. There is a polynomial-time algorithm that computes an orientation satisfying at least $|P|/(4\lceil \log n \rceil)$ of the pairs.*

Proof. First contract all directed edges of the tree into single vertices and update the end vertices of the input pairs accordingly. The resulting graph is an undirected tree with source-to-target paths for every pair in P. By Lemma 3.1, there exists a polynomial-time algorithm that finds an orientation of the resulting tree satisfying at least $1/(4\lceil \log n \rceil)$ of the pairs. Carrying over the edge directions to the initial mixed tree produces an orientation that satisfies exactly the same collection of pairs in the original orientation instance.   □

3.2   Crossings through a Junction Component

Let (G, P) be an orientation instance and let $T_1, T_2, \ldots$ be the undirected components of G. We construct a subgraph of G, called the *skeleton* $S = S(G)$ of G by deleting all but one directed edge between any pair of trees T_i and T_j. Note that the exact structure of a skeleton graph depends on the (polynomial-time) procedure used for its construction; we choose any fixed procedure to define S unambiguously. It is not difficult to verify that the skeleton S contains source-to-target paths for the pairs P, and that every orientation of S satisfying certain pairs directly translates into an orientation for G satisfying at least the same pairs. Figure 1 shows an example of a graph G and a skeleton for it. Note that for any MAG G and its skeleton $S = S(G)$, we have $G_{\text{vcc}} = S_{\text{vcc}}$.

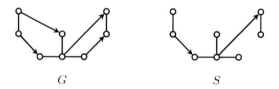

$$G \qquad\qquad\qquad S$$

Fig. 1. An example MAG G and a skeleton S of it

The next lemma is crucial to establish the remaining results of this section, as well as the sub-linear approximation algorithm described in Section 4.

Lemma 3.3. *Let (G, P) be an orientation instance and T be an undirected component of G. If each pair in P admits a source-to-target path that crosses a vertex from T then there is a polynomial-time algorithm that computes an orientation satisfying at least $|P|/(4\lceil \log n \rceil)$ of the pairs.*

Proof. Since the skeleton $S = S(G)$ is a MAG and, therefore, S_{vcc} is a directed acyclic graph, for every undirected component $T' \neq T$ of S_{vcc}, exactly one of the following options holds: (1) T' is reachable from T in S_{vcc}; (2) T is reachable from T' in S_{vcc}; or (3) there is no path between T and T' in either direction. Consequently, we can consider two subtrees that are rooted at T: The first subtree spans the vertices of S_{vcc} that are reachable from T, and the second subtree spans the vertices of S_{vcc} that can reach T. We merge both subtrees at T and call the resulting directed tree T_{vcc}. To compute an orientation for G, we consider the subtree of S that is constructed by taking all undirected components from T_{UCC}, and connect two vertices in different components by a directed edge if this edge is already present in S and the components are connected in T_{UCC}. This subtree of S contains a source-to-target path for each pair in P. Therefore, by Lemma 3.2, we can construct (in polynomial time) an orientation satisfying at least $|P|/(4\lceil \log n \rceil)$ pairs in S and, thus, also in the original graph G.   □

Lemma 3.3 implies an $\Omega(1/\log n)$ approximation for a special case of the orientation problem, which we call MAXIMUM-JUNCTION-TREE-ORIENTATION. In Section 4 we shall apply the algorithm to instances where all pairs have source-to-target paths crossing a distinguished vertex r.

3.3   Orientations for Small Feedback Vertex Sets or Treewidth

We end this section by providing logarithmic approximations to the orientation problem on tree-like instances. Precisely, we consider two graph parameters: *feedback vertex number* and *treewidth*, showing that whenever either one of these is bounded by a constant, it is possible to compute an orientation that satisfies a poly-logarithmic fraction of the input pairs.

Lemma 3.4. *Let (G, P) be an orientation instance where the underlying undirected subgraph of G_{UCC} can be turned into a tree by deleting at most k vertices. There is a polynomial-time algorithm that computes an orientation satisfying at least $|P|/(4(2k+1)\lceil \log n \rceil)$ pairs.*

Proof. We begin by detecting a small-sized feedback vertex set $F = \{T_1, \ldots, T_\ell\}$, consisting of ℓ vertices whose removal turns the underlying undirected subgraph of G_{UCC} into a tree. Even though finding a minimum cardinality vertex set of this type is NP-hard [17], this problem can be approximated to within a factor of 2 in undirected graphs [2], implying that we can assume $\ell \leq 2k$. We now partition P into two subsets, the collection of pairs P^+ for which we can find source-to-target paths in G that cross undirected components from F, and the collection $P^- = P \setminus P^+$. We further partition P^+ into ℓ subsets $P_1^+, \ldots, P_\ell^+$, where a pair $(s, t) \in P^+$ lies in P_i^+ if i is the minimal index for which there exists a source-to-target path for this pair that crosses the undirected component T_i. With these definitions at hand, note that by deleting the undirected components F from G, we can use Lemma 3.2 to efficiently compute an orientation of G satisfying at least $|P^-|/(4\lceil \log n \rceil)$ pairs; after deleting F the skeleton of the resulting graph is a tree and all pairs in P^- remain connected since they are only connected through paths that not visit vertices from F. On the other hand, for each collection P_i^+ we can satisfy at least $|P_i^+|/(4\lceil \log n \rceil)$ pairs by applying Lemma 3.3. Picking the option that generates the highest number of satisfied pairs results in an orientation satisfying at least $|P|/(4(2k+1)\lceil \log n \rceil)$ of the pairs in P.   □

We note that the above approximation result can be improved by a factor of 2 if the feedback vertex set has bounded size. For such instances we can invoke an exact fixed parameter algorithm [6] to find an optimal feedback set, rather than using the 2-approximation algorithm.

Lemma 3.5. *Let (G, P) be an orientation instance where the underlying undirected subgraph of G_{UCC} has treewidth k. There is a polynomial-time algorithm that computes an orientation satisfying at least $|P|/(4(k+1)\lceil \log n \rceil^2)$ pairs.*

Proof. We first compute a tree decomposition of width k for the undirected underlying graph of G_{UCC}. A tree decomposition $(\mathcal{T}, \{B_t\}_{t \in V(\mathcal{T})})$ consists of a tree $\mathcal{T}$ whose nodes are labeled with possibly-overlapping subsets B_t of vertices, called *bags*, such that: (1) the incident vertices of every edge are both contained in some bag; and (2) for every original vertex, the nodes of the bags that contain it make up a connected subtree. Its width is defined as the maximum number of vertices in a bag minus 1. For a comprehensive discussion on tree decompositions and their polynomial-time computability in the case of bounded tree width, we refer to the book of Flum and Grohe [11].

Based on the tree decomposition $(\mathcal{T}, \{B_t\}_{t \in V(\mathcal{T})})$, we partition P into subsets $P_1, P_2, \ldots, P_L$ with $L \leq \lceil \log n \rceil$ such that for every subset we can efficiently find an orientation that satisfies a fraction of at least $1/(4(k+1)\lceil \log n \rceil)$ of its pairs. By picking the largest subset of pairs and its corresponding orientation, we obtain an orientation satisfying at least $|P|/(4(k+1)\lceil \log n \rceil^2)$ pairs.

For the purpose of constructing P_1, consider a *centroid* node t of $\mathcal{T}$ whose removal breaks this tree into subtrees of cardinality at most $|V(\mathcal{T})|/2$, noting that any tree necessarily contains a centroid (see, for instance, [12]). Let P_1 be the pairs in P with source-to-target paths that cross vertices from undirected components of $B_t = \{T_1, \ldots, T_l\}$, where $l \leq k+1$. We further partition P_1 into l collections $P_1^1, \ldots, P_1^l$ such that a pair $(s, t) \in P_1$ lies in P_1^i if there exists an s-t path that crosses vertices from T_i but no s-t paths that cross vertices from components T_j with $j < i$. By Lemma 3.3, we can compute an orientation that satisfies at least $|P_1^i|/(4\lceil \log n \rceil)$ of the pairs in P_1^i, for every $1 \leq i \leq l$. By taking the largest collection, we can satisfy at least $|P_1|/(4(k+1)\lceil \log n \rceil)$ pairs in P_1.

To construct P_2, we proceed with the pair collection $P \setminus P_1$ that contains exactly the pairs from P with no source-to-target paths that cross vertices from the components of P_1. We delete the node t from $\mathcal{T}$, as well as the components in B_t from G. This results in a graph that contains source-to-target paths for all pairs from $P \setminus P_1$ and a forest of tree decompositions for the graph. For each tree decomposition we compute a centroid bag and, in the same way as above, the collection P_2 of pairs in $P \setminus P_1$ with source-to-target paths that cross components from these centroid bags. Using the same arguments as above, we can compute an orientation that satisfies at least $|P_2|/(4(k+1)\lceil \log n \rceil)$ of the pairs in P_2. We proceed recursively in the same way to construct $P_3, P_4, \ldots, P_L$ as long as each tree decomposition (and the corresponding subgraph of G) is not empty. Since the maximal size of a subtree decreases by a factor of at least 2 in each level of the recursion, this process terminates within $\lceil \log n \rceil$ steps.   □

4   Sub-linear Approximations for General Instances

In what follows, we focus our attention on approximating the orientation problem in its utmost generality, that is, without making simplifying structural assumptions on the underlying (mixed-acyclic) graph G and on the collection of input pairs P. The main result of this section can be briefly stated as follows.

Theorem 4.1. *The* MAXIMUM-MIXED-GRAPH-ORIENTATION *problem can be approximated within a factor of* $\Omega(1/(M^{1/\sqrt{2}} \log n))$, *where* $M = \max\{n, |P|\}$.

In addition, we provide an improved approximation guarantee for input instances with bounded-distance pairs. This result is described in Section 4.3.

4.1   The Algorithm

For each pair $(s_i, t_i) \in P$, let p_i be a shortest path from s_i to t_i in G, and let $\mathcal{P}$ be the set of all shortest paths, i.e., $\mathcal{P} = \{p_i : (s_i, t_i) \in P\}$. Our algorithm is based on a greedy framework where paths in $\mathcal{P}$ are oriented (from source to target) one after the other, trying not to interfere with future orientations of too many other paths by picking the shortest path in each step. Somewhat informally, this process concludes as soon as one of the following termination conditions is met:

The greedy step. At any point in time, we will be holding a partial orientation G_ℓ of G and a subset $\mathcal{P}_\ell \subseteq \mathcal{P}$ of shortest paths, where these sets are indexed according to the step number that has just been completed. In other words, at the conclusion of step ℓ we have G_ℓ and $\mathcal{P}_\ell$, where initially $G_0 = G$ and $\mathcal{P}_0 = \mathcal{P}$. Now, as long as none of the termination conditions described below is met, we proceed as follows:

1. Let $\hat{p} = <s, \ldots, t>$ be a shortest path in $\mathcal{P}_\ell$.
2. Orient $\hat{p}$ in the direction from s to t to obtain $G_{\ell+1}$.
3. Discard from $\mathcal{P}_\ell$ the path $\hat{p}$ as well as any path that has a non-empty edge intersection with $\hat{p}$. This way, we obtain $\mathcal{P}_{\ell+1}$.

Termination conditions. There are two conditions that will cause the greedy iterations to terminate. For now, we state both conditions in terms of two parameters $\alpha \geq 0$ and $\beta \geq 0$, whose values will be optimized later on.

Condition 1: $|\mathcal{P}_\ell| \leq n^\alpha$. In this case, we will orient an arbitrary path from $\mathcal{P}_\ell$, and update the current orientation to G_ℓ, as in the preceding greedy iterations. We then complete the orientation by arbitrarily orienting all yet-unoriented edges.

Condition 2: There exists a vertex r such that at least $|\mathcal{P}_\ell|^\beta$ paths in $\mathcal{P}_\ell$ go through r. We construct a MAXIMUM-JUNCTION-TREE-ORIENTATION instance with input graph G_ℓ, junction vertex r, and pairs $\{(s_i, t_i) : p_i \in \mathcal{P}_\ell$ goes through $r\}$. We then apply the algorithm described in Section 3.2 for this special case, and return its output as our final orientation.

4.2   Analysis

To establish a lower bound on the number of satisfied pairs, we break the analysis into two cases, depending on the condition that caused the greedy iterations to terminate. In the remainder of this section, we assume that L greedy iterations have been completed prior to satisfying one of the termination conditions.

Connections due to condition 1: In this case we satisfy a single pair out of $\{(s_i, t_i) : p_i \in \mathcal{P}_L\}$, noting that $|\mathcal{P}_L| \leq n^{\alpha}$.

Connections due to condition 2: Following Lemma 3.3, the number of pairs satisfied out of $\{(s_i, t_i) : p_i \in \mathcal{P}_L\}$ is $\Omega(1/\log n) \cdot |\mathcal{P}_L|^{\beta}$.

We proceed by arguing that an $\Omega(1/n^{1-\alpha(1-2\beta)})$ fraction of the pairs in $\{(s_i, t_i) : p_i \notin \mathcal{P}_L\}$ are already satisfied by the partial orientation G_L. To this end, note that in each iteration $1 \leq \ell \leq L$ we satisfy a single pair by orienting the shortest path $\hat{p} \in \mathcal{P}_{\ell-1}$, and eliminating several others to obtain $\mathcal{P}_{\ell}$. To prove the claim above, it is sufficient to show that the number of eliminated paths satisfies $|\mathcal{P}_{\ell-1} \setminus \mathcal{P}_{\ell}| \leq n^{1-\alpha(1-2\beta)}$. Denote by $E(p)$ the set of edges of a path p, so that $|E(p)|$ is its length. We begin by observing that, since condition 2 has not been met in iteration ℓ, each edge can have at most $|\mathcal{P}_{\ell-1}|^{\beta}$ paths from $\mathcal{P}_{\ell-1}$ going through it, implying that $|\mathcal{P}_{\ell-1} \setminus \mathcal{P}_{\ell}| \leq |E(\hat{p})| \cdot |\mathcal{P}_{\ell-1}|^{\beta}$. Since $|E(\hat{p})|$ is upper bounded by the average length of the paths in $\mathcal{P}_{\ell-1}$, we have

$$|E(\hat{p})| \leq \frac{1}{|\mathcal{P}_{\ell-1}|} \sum_{p_i \in \mathcal{P}_{\ell-1}} |E(p_i)| \leq \frac{1}{|\mathcal{P}_{\ell-1}|} \sum_{p_i \in \mathcal{P}_{\ell-1}} |V(p_i)|$$

$$= \frac{1}{|\mathcal{P}_{\ell-1}|} \sum_{v \in V} |\{p_i \in \mathcal{P}_{\ell-1} : v \in V(p_i)\}|$$

$$\leq \frac{1}{|\mathcal{P}_{\ell-1}|} \cdot n \cdot |\mathcal{P}_{\ell-1}|^{\beta} = \frac{n}{|\mathcal{P}_{\ell-1}|^{1-\beta}} ,$$

where the third inequality holds since condition 2 has not been met. Hence,

$$|\mathcal{P}_{\ell-1} \setminus \mathcal{P}_{\ell}| \leq \frac{n}{|\mathcal{P}_{\ell-1}|^{1-2\beta}} \leq \frac{n}{n^{\alpha(1-2\beta)}} = n^{1-\alpha(1-2\beta)} ,$$

where the second inequality follows from $|\mathcal{P}_{\ell-1}| > n^{\alpha}$, as condition 1 has not been met.

Putting it all together. Based on the above discussion, it follows that the number of satisfied pairs when we terminate the algorithm due to condition 1 is

$$\Omega\left(\frac{1}{n^{1-\alpha(1-2\beta)}}\right)(|P| - |\mathcal{P}_L|) + 1 = \Omega\left(\frac{1}{n^{1-\alpha(1-2\beta)}}\right)(|P| - n^{\alpha}) + \frac{1}{n^{\alpha}}n^{\alpha}$$

$$= \Omega\left(\frac{1}{\max\{n^{1-\alpha(1-2\beta)}, n^{\alpha}\}}\right)|P|$$

$$= \Omega\left(\frac{1}{n^{\max\{1-\alpha(1-2\beta),\alpha\}}}\right)|P| .$$

Similarly, the number of satisfied pairs when the algorithm is terminated due to condition 2 is

$$\Omega\left(\frac{1}{n^{1-\alpha(1-2\beta)}}\right)(|P|-|\mathcal{P}_L|) + \Omega\left(\frac{1}{\log n}\right)|\mathcal{P}_L|^\beta$$

$$= \Omega\left(\frac{1}{n^{1-\alpha(1-2\beta)}}\right)(|P|-|\mathcal{P}_L|) + \Omega\left(\frac{1}{|\mathcal{P}_L|^{1-\beta}\log n}\right)|\mathcal{P}_L|$$

$$= \Omega\left(\frac{1}{\max\{n^{1-\alpha(1-2\beta)}, |P|^{1-\beta}\log n\}}\right)|P|$$

$$= \Omega\left(\frac{1}{M^{\max\{1-\alpha(1-2\beta),1-\beta\}}}\right)\frac{1}{\log n}|P| \ .$$

To obtain the best-possible performance guarantee, we pick values for α and β so as to minimize $\max\{\alpha, 1-\beta, 1-\alpha(1-2\beta)\}$. As explained below, the last term is optimized for $\alpha^* = \sqrt{1/2}$ and $\beta^* = \sqrt{1/2}/(2\sqrt{1/2}+1) = 1 - \sqrt{1/2}$, in which case its value is $\sqrt{1/2} \approx 0.707$.

Optimizing α and β. Suppose we know the value of α^*. In this case, β^* should be picked so as to minimize $\max\{1-\beta, 1-\alpha^*(1-2\beta)\}$. Since $1-\beta$ is a decreasing linear function of β and $1-\alpha^*(1-2\beta)$ is an increasing linear function, this minimum is attained when $1-\beta = 1-\alpha^*(1-2\beta)$, that is, $\beta^* = \alpha^*/(2\alpha^*+1)$. For this value, we have $\min\max\{1-\beta, 1-\alpha^*(1-2\beta)\} = (\alpha^*+1)/(2\alpha^*+1)$. It remains to find a value of α that minimizes $\max\{\alpha, (\alpha+1)/(2\alpha+1)\}$. Using similar arguments, it is not difficult to verify that the right value to pick is $\alpha^* = \sqrt{1/2}$.

4.3 An Improved Approximation for Bounded-Distance Pairs

In practice, the diameter of biological networks is sub-logarithmic due to their scale-free property [4,8,19]. For example, in the yeast physical network described in [21], the maximum source-target distance is 14. This motivates examining the approximation guarantee in terms of the maximum length of a shortest source-target path in the reduced mixed acyclic graph, which we denote by $\Delta = \Delta(G, P)$. In the following we present an $\Omega(1/\sqrt{\Delta|P|\log n})$ approximation to the orientation problem.

Our algorithm remains essentially unchanged, except for its termination conditions. Unlike the more general procedure, we ignore condition 1, and terminate the greedy iterations as soon as condition 2 is met, i.e., when there exists a vertex $r \in V$ such that at least $|\mathcal{P}_\ell|^\beta$ paths in $\mathcal{P}_\ell$ go through r. In this case, we construct a MAXIMUM-JUNCTION-TREE-ORIENTATION instance as before, with input graph G_ℓ, junction vertex r, and pairs $\{(s_i, t_i) : p_i \in \mathcal{P}_\ell \text{ goes through } r\}$. Our logarithmic approximation for this particular setting is then applied.

Similarly to the analysis in Section 4.2, we can prove the next two claims:

Connections due to termination condition 2: The number of pairs satisfied out of $\{(s_i, t_i) : p_i \in \mathcal{P}_L\}$ is $\Omega(1/\log n) \cdot |\mathcal{P}_L|^\beta$.

Connections due to greedy iterations: A fraction of $\Omega(1/(\Delta|P|^\beta))$ of the pairs in $\{(s_i, t_i) : p_i \notin \mathcal{P}_L\}$ are already satisfied by the partial orientation $\mathcal{G}_L$. This follows by observing that the number of paths that are eliminated from $\mathcal{P}_{\ell-1}$ in iteration ℓ is at most $\Delta|\mathcal{P}_{\ell-1}|^\beta \leq \Delta|P|^\beta$.

Consequently, the number of satisfied pairs upon termination is:

$$\Omega\left(\frac{1}{\Delta|P|^\beta}\right)(|P| - |\mathcal{P}_L|) + \Omega\left(\frac{1}{\log n}\right)|\mathcal{P}_L|^\beta$$

$$= \Omega\left(\frac{1}{\Delta|P|^\beta}\right)(|P| - |\mathcal{P}_L|) + \Omega\left(\frac{1}{|\mathcal{P}_L|^{1-\beta}\log n}\right)|\mathcal{P}_L|$$

$$= \Omega\left(\frac{1}{\max\{\Delta|P|^\beta, |P|^{1-\beta}\log n\}}\right)|P| .$$

By choosing $\beta = \frac{1}{2}(1 + \log_{|P|}(\frac{\log n}{\Delta}))$, we obtain an approximation ratio of $\Omega(1/\sqrt{\Delta|P|\log n})$.

In this section we used the usual definition of path lengths: the length of a path is the number of its edges. The above analyses work in a similar way if we measure the length of a path by the number of its undirected edges or even by the number of undirected components the path visits. This yields the same asymptotic bounds with respect to the size of the input, but highlights the increasing performance of the algorithm for structured inputs where these path length measures are small.

5   Conclusions

In this paper we presented approximation algorithms for the MAXIMUM-MIXED-GRAPH-ORIENTATION problem, which has recently arisen in the study of biological networks. We first showed that tree-like instances admit orientations (that can be computed in polynomial time) satisfying a poly-logarithmic fraction of the input pairs. Then we extended these algorithms to develop the first approximation algorithm for the problem whose ratio depends only on the size of the input instance, where no structural properties are assumed. The algorithm has a sub-linear approximation ratio, which can be improved when the input pairs are connected by short paths. The known upper and lower bounds for the approximation ratio of MAXIMUM-MIXED-GRAPH-ORIENTATION are far from being tight. Closing this gap, both in the undirected and mixed cases, remains an open problem.

Acknowledgments

M.E. was supported by a research grant from the Dr. Alexander und Rita Besser-Stiftung. C.R.D. would like to thank Gerry Schwartz, Heather Reisman, and the University of Waterloo-Haifa International Experience Program for funding his visit to the University of Haifa, during which part of this work was done. R.S. was supported by a research grant from the Israel Science Foundation (grant no. 385/06).

References

1. Arkin, E.M., Hassin, R.: A note on orientations of mixed graphs. Discrete Applied Mathematics 116(3), 271–278 (2002)
2. Bafna, V., Berman, P., Fujito, T.: A 2-approximation algorithm for the undirected feedback vertex set problem. SIAM Journal on Discrete Mathematics 12(3), 289–297 (1999)
3. Bang-Jensen, J., Gutin, G.: Digraphs: Theory, Algorithms and Applications, 2nd edn. Springer, Heidelberg (2008)
4. Barabási, A.-L., Oltvai, Z.N.: Network biology: understanding the cell's functional organization. Nature Reviews Genetics 5(2), 101–113 (2004)
5. Boesch, F., Tindell, R.: Robbins's theorem for mixed multigraphs. The American Mathematical Monthly 87(9), 716–719 (1980)
6. Chen, J., Fomin, F.V., Liu, Y., Lu, S., Villanger, Y.: Improved algorithms for feedback vertex set problems. Journal of Computer and System Sciences 74(7), 1188–1198 (2008)
7. Chung, F.R.K., Garey, M.R., Tarjan, R.E.: Strongly connected orientations of mixed multigraphs. Networks 15(4), 477–484 (1985)
8. Cohen, R., Havlin, S., ben-Avraham, D.: Structural properties of scale-free networks. In: Handbook of Graphs and Networks: From the Genome to the Internet. Wiley-VCH, Weinheim (2002)
9. Dorn, B., Hüffner, F., Krüger, D., Niedermeier, R., Uhlmann, J.: Exploiting bounded signal flow for graph orientation based on cause–effect pairs (To appear). In: Marchetti-Spaccamela, A., Segal, M. (eds.) TAPAS 2011. LNCS, vol. 6595, pp. 104–115. Springer, Heidelberg (2011)
10. Fields, S.: High-throughput two-hybrid analysis. The promise and the peril. The FEB Journal 272(21), 5391–5399 (2005)
11. Flum, J., Grohe, M.: Parameterized Complexity Theory. Springer, Heidelberg (2006)
12. Frederickson, G.N., Johnson, D.B.: Generating and searching sets induced by networks. In: ICALP 1980. LNCS, vol. 85, pp. 221–233. Springer, Heidelberg (1980)
13. Gamzu, I., Segev, D., Sharan, R.: Improved orientations of physical networks. In: Moulton, V., Singh, M. (eds.) WABI 2010. LNCS, vol. 6293, pp. 215–225. Springer, Heidelberg (2010)
14. Gavin, A., Bösche, M., Krause, R., Grandi, P., Marzioch, M., Bauer, A., Schultz, J., Rick, J.M., Michon, A.-M., Cruciat, C.-M., Remor, M., Höfert, C., Schelder, M., Brajenovic, M., Ruffner, H., Merino, A., Klein, K., Hudak, M., Dickson, D., Rudi, T., Gnau, V., Bauch, A., Bastuck, S., Huhse, B., Leutwein, C., Heurtier, M.-A., Copley, R.R., Edelmann, A., Querfurth, E., Rybin, V., Drewes, G., Raida, M., Bouwmeester, T., Bork, P., Seraphin, B., Kuster, B., Neubauer, G., Superti-Furga, G.: Functional organization of the yeast proteome by systematic analysis of protein complexes. Nature 415(6868), 141–147 (2002)
15. Hakimi, S.L., Schmeichel, E.F., Young, N.E.: Orienting graphs to optimize reachability. Information Processing Letters 63(5), 229–235 (1997)
16. Håstad, J.: Some optimal inapproximability results. Journal of the ACM 48(4), 798–859 (2001)
17. Karp, R.M.: Reducibility among combinatorial problems. In: Complexity of Computer Computations, pp. 85–103. Plenum Press, New York (1972)

18. Medvedovsky, A., Bafna, V., Zwick, U., Sharan, R.: An algorithm for orienting graphs based on cause-effect pairs and its applications to orienting protein networks. In: Crandall, K.A., Lagergren, J. (eds.) WABI 2008. LNCS (LNBI), vol. 5251, pp. 222–232. Springer, Heidelberg (2008)
19. Newman, M.E.J.: The structure and function of complex networks. SIAM Review 45(2), 167–256 (2003)
20. Robbins, H.E.: A theorem on graphs, with an application to a problem of traffic control. The American Mathematical Monthly 46(5), 281–283 (1939)
21. Silverbush, D., Elberfeld, M., Sharan, R.: Optimally orienting physical networks. In: Bafna, V., Sahinalp, S.C. (eds.) RECOMB 2011. LNCS, vol. 6577, pp. 424–436. Springer, Heidelberg (2011)
22. Tarjan, R.E.: A note on finding the bridges of a graph. Information Processing Letters 2(6), 160–161 (1974)
23. Yeang, C., Ideker, T., Jaakkola, T.: Physical network models. Journal of Computational Biology 11(2-3), 243–262 (2004)

Frequent Submap Discovery

Stéphane Gosselin, Guillaume Damiand, and Christine Solnon*

Université de Lyon, CNRS
Université Lyon 1, LIRIS, UMR5205, F-69622, France

Abstract. Combinatorial maps are nice data structures for modeling the topology of nD objects subdivided in cells (*e.g.*, vertices, edges, faces, volumes, ...) by means of incidence and adjacency relationships between these cells. In particular, they can be used to model the topology of plane graphs. In this paper, we describe an algorithm, called mSpan, for extracting patterns which occur frequently in a database of maps. We experimentally compare mSpan with gSpan on a synthetic database of randomly generated $2D$ and $3D$ maps. We show that gSpan does not extract the same patterns, as it only considers adjacency relationships between cells. We also show that mSpan exhibits nicer scale-up properties when increasing map sizes or when decreasing frequency.

1 Introduction

Combinatorial maps are nice data structures for modeling the topology of nD objects subdivided in cells (*e.g.*, vertices, edges, faces, volumes, ...) by means of incidence and adjacency relationships between these cells. First defined in $2D$ [9,19,12,4], they have been extended to nD [2,14,15]. Combinatorial maps are often used to model the partition of an image in regions and to describe the topology of this partition (e.g., [1] for $2D$ images and [5] for $3D$ images). There exist efficient image processing algorithms using this topological information. However, there exist few algorithms for analyzing or comparing combinatorial maps, which are key issues in image processing.

In this paper, we describe an algorithm for extracting patterns which occur frequently in a database of maps. This algorithm is a first step for analyzing and characterizing sets of maps. Finding frequent patterns in databases is a classical data mining problem, the tractability of which highly depends on the existency of efficient algorithms for deciding if two patterns are actually different or if they are two occurrences of a same object. Hence, if finding frequent subgraphs is intractable in the general case, it may be solved in incremental polynomial time when considering classes of graphs for which subgraph isomorphism may be solved in polynomial time, such as trees [3] or outerplanar graphs [11]. We have introduced efficient polynomial-time algorithms to decide of submap isomorphism in [7], and to search for a map into a database of maps in [10]. These algorithms allow us to design an incremental polynomial time algorithm for extracting frequent patterns from a database of maps.

---

* The authors acknowledge an ANR grant BLANC 07-1_184534: this work was done in the context of project SATTIC.

R. Giancarlo and G. Manzini (Eds.): CPM 2011, LNCS 6661, pp. 429–440, 2011.

Outline. Basic definitions on combinatorial maps are recalled in section 2. The algorithm for extracting frequent submaps from $2D$ maps is described in section 3, and its extension to nD maps is described in section 4. First experimental results on a synthetic database of randomly generated $2D$ and $3D$ maps and on a database of maps extracted from images are reported in section 5 and 6.

2   Recalls on Combinatorial Maps

Combinatorial maps describe the subdivision of nD objects into cells of dimensions lower or equal to n ($0D$ vertices, $1D$ edges, $2D$ faces, $3D$ volumes, ...), and describe the topology of these cells by means of incidence and adjacency relationships between these cells. For sake of simplicity, we first introduce maps in $2D$ and describe our algorithm within this $2D$ context. The extension to nD maps is rather straightforward and is described in section 4.

In $2D$, a combinatorial map models a plane graph *i.e.*, the embedding of a planar graph into a plane, as illustrated in Fig. 1. It is defined by a set of darts and two functions β_1 and β_2 as follows.

Definition 1 (Combinatorial map [15]). *A $2D$ combinatorial map (or map) is defined by a tuple $M = (D, \beta_1, \beta_2)$ where D is a finite set of darts; β_1 is a permutation on D (i.e., a one-to-one mapping from D to D); and β_2 is an involution on D (i.e., a one-to-one mapping from D to D such that $\beta_2 = \beta_2^{-1}$).*

A dart d is said to be i-sewn with another dart d' if $d' = \beta_i(d)$. β_1 is a permutation which models dart successions when turning around faces with respect to some given order. β_2 models adjacency relations between faces.

In some cases, it may be useful to allow some β_i to be partially defined, thus leading to open combinatorial maps. The basic idea is to add a new element ϵ to the set of darts, and to allow darts to be i-sewn with ϵ. By definition, $\beta_1(\epsilon) = \beta_2(\epsilon) = \epsilon$. Fig. 2 gives an example of open map (see [8,17] for precise definitions).

In this paper, we extract patterns from maps, where patterns are maps which are isomorphic to submaps of these maps. More precisely, map isomorphism has been defined e.g. in [16] as follows.

Definition 2 (Map isomorphism). *Two maps $M = (D, \beta_1, \beta_2)$ and $M' = (D', \beta_1', \beta_2')$ are isomorphic if there exists a bijection $f : D \rightarrow D'$ such that $\forall d \in D, f(\beta_1(d)) = \beta_1'(f(d))$ and $f(\beta_2(d)) = \beta_2'(f(d))$.*

This definition has been extended to open maps in [7] by adding that $f(\epsilon) = \epsilon$, thus enforcing that, when a dart is i-sewn with ϵ, then the dart matched to it by f is i-sewn with ϵ. Submap isomorphism simply derives from the definition of map isomorphism: there is a submap isomorphism from a map M to a map M' if there exists a submap of M' which is isomorphic to M', where a submap is basically obtained by removing some darts (and free-ing darts that were i-sewn

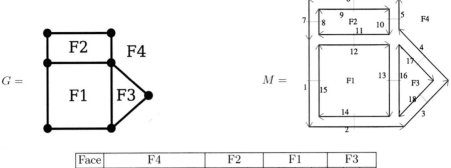

Face	F4							F2				F1				F3		
Dart	1	2	3	4	5	6	7	8	9	10	11	12	13	14	15	16	17	18
β_1	2	3	4	5	6	7	1	9	10	11	8	13	14	15	12	17	18	16
β_2	15	14	18	17	10	9	8	7	6	5	12	11	16	2	1	13	4	3

Fig. 1. The map M describes the topology of the plane graph G. Darts are represented by numbered arrows. 1-sewn darts are drawn consecutively, and 2-sewn darts are concurrently drawn and in reverse orientation, with a little grey segment between the two darts. Darts 1 to 7 correspond to face F4, darts 8 to 11 to face F2 and so on.

with the removed darts). For example, there is a submap isomorphism from the map of Fig. 2 to the map of Fig. 1 as it is isomorphic to the submap of Fig. 1 obtained by removing darts 1 to 11.

In [7], we have described an algorithm which decides of submap isomorphism from a map $M = (D, \beta_1, \beta_2)$ to a map $M' = (D', \beta_1', \beta_2')$ in $\mathcal{O}(|D| \cdot |D'|)$, provided that M is connected, *i.e.*, there must exist a path of sewn darts between every pair of darts of M.

In [10], we have introduced a signature which allows us to efficiently search for a map M in a database B containing k maps such that the largest map has t darts: the time complexity for building the signature of the database is $\mathcal{O}(k \cdot t^2)$; the space complexity of this signature is $\mathcal{O}(k \cdot t)$, and the time complexity of searching for all maps of B which are isomorphic to M is $\mathcal{O}(n \cdot t^2)$.

3   Frequent Submap Discovery

When considering $2D$ maps, the basic cell is the face. Therefore, a pattern is a connected set of faces. We can then define the problem of frequent submap

Face	F1				F2		
Dart	a	b	c	d	e	f	g
β_1	b	c	d	a	f	g	e
β_2	ϵ	ϵ	e	ϵ	c	ϵ	ϵ

Fig. 2. Open combinatorial map example. Darts a, b, d, f and g are not 2-sewn.

Algorithm 1. $mSpan(S, \sigma)$

Input: a set of maps S and a real number $\sigma \in]0; 1]$
Output: the set F of all maps which are submaps of at least $\sigma \cdot |S|$ maps of S
1 $F_1 \leftarrow$ all patterns composed of 1 face and occurring in at least $\sigma \cdot |S|$ maps of S
2 $F \leftarrow F_1$
3 **while** $F_1 \neq \emptyset$ **do**
4   choose a pattern f in F_1
5       $Cand \leftarrow \{f\}$
6   **while** $Cand \neq \emptyset$ **do**
7   remove a pattern p from $Cand$
8           $F_p \leftarrow grow(p, F_1)$
9           $Cand \leftarrow Cand \cup F_p$
10          $F \leftarrow F \cup F_p$

       /* All frequent patterns which contain face f belong to F */
11   remove f from F_1

12 **return** F

---

discovery in a similar way as [13] has defined the problem of frequent subgraph discovery: given a set of maps S and a parameter σ such that $0 < \sigma \leq 1$, the goal is to find all patterns M such that $freq(M, S) \geq \sigma \cdot |S|$, where $freq(M, S)$ is the frequency of M in S, *i.e.*, the number of maps $M' \in S$ such that there is (at least) one submap isomorphism from M to M'.

A map may have an exponential number of different submaps so that a naive representation of the search space for this problem has exponential size in the length of the input. To reduce the set of candidate patterns to be explored, we exploit the fact that the frequency constraint is anti-monotone with respect to the submap isomorphism partial order relation: if a pattern p is not frequent, then any pattern p' such that p is subisomorphic to p' cannot be frequent.

Algorithm 1 describes our frequent submap mining algorithm, called mSpan for <u>M</u>ap-based <u>S</u>ubstructure <u>Pattern</u> mining. mSpan follows the same basic principle as gSpan [20] which extracts frequent subgraphs: it constructs patterns with a depth-first search algorithm and exploits the frequency constraint to prune parts of the search space which do not contain frequent patterns.

More precisely, we first compute the set F_1 of all frequent patterns composed of a single face, and we initialize the set F of all frequent patterns with F_1. Then, for each face f of F_1, we build all frequent patterns which contain f plus some faces of F_1 and we add these frequent patterns to F (lines 4-10). Finally, we remove f from F_1 (line 11) in order to prevent us from re-building frequent patterns containing f in the next iterations of the while loop of lines 3-11. The set of all frequent patterns which contain f plus some faces of F_1 is built iteratively by using a set $Cand$ of frequent patterns which are candidate to be extended by sewing to them one face of F_1: at each iteration (lines 6-10), we remove a pattern p from $Cand$ (line 7) and the *grow* function computes all frequent patterns that

---

Algorithm 2. $grow(p, F_1)$

Input: a frequent pattern p and a set of frequent 1-face patterns F_1
Output: a set F_p of all frequent patterns built by adding a face of F_1 to p

1 $L_p \leftarrow \emptyset$
2 **for** *each occurrence o of the pattern p in a map of S* **do**
3 　**for** *each dart d which belongs to the boundary of this occurrence o of p* **do**
4 　　**if** $\beta_2(d) \neq \epsilon$ *so that there exists a face which is 2-sewn with d* **then**
5 　　　let f be the face 2-sewn with d
6 　　　**if** $f \in F_1$ **then**
7 　　　　let p_f be the pattern obtained by 2-sewing face f to dart d of o
8 　　　　**if** $p_f \notin L_p$ **then** add p_f to L_p and initialize $freq(p_f)$ to 1
9 　　　　**else** update $freq(p_f)$

10 return $\{p_i \in L_p \mid freq(p_i) \geq \sigma \cdot |S|\}$

---

may be built by sewing a face of F_1 to p (line 8); these frequent patterns are added to the set *Cand* (line 9) in order to further build new patterns which contain them.

The *grow* function is described in algorithm 2. Given a frequent pattern p and a set of frequent 1-face patterns F_1, it returns all frequent patterns obtained by sewing one face of F_1 to p. This is done by traversing the boundary of every occurrence o of p in a map of S: for each dart d of this boundary, if the face which is 2-sewn to d belongs to F_1 then the pattern p_f obtained by 2-sewing this face to dart d of o is a candidate frequent pattern which is added to L_p if it does not already belong to it (line 8). Once all candidate patterns have been computed in L_p, we return all patterns of L_p which are frequent (line 10).

Data structure used to memorize pattern occurrences (line 2 of Algo. 2). When trying to grow pattern p by adding a new face to it, we do not compute all occurrences of p in a map of S. This information is incrementally stored: each time an occurrence of a pattern p_f is found (line 7), we keep track of it in an occurrence list $occ(p_f)$ which contains one dart for every pattern occurrence of p_f as illustrated in Fig. 3.

Traversing the boundary of a pattern occurrence (line 3 of Algo. 2). The darts which belong to the boundary of an occurrence o of a pattern p are found by performing a traversal of o, guided by the pattern p, starting in parallel from dart 1 of p and from the initial dart associated with o in the occurrence list $occ(p)$, as illustrated in Fig. 3 (see [8] for more details). This is done in linear time with respect to the number of darts of the pattern p.

Data structure used to decide if a pattern p_f belongs to L_p (line 8 of Algo. 2). Each time a new pattern p_f is found (line 7), we compute its signature and we add this signature to a signature tree. If the pattern p_f has k darts, then the space complexity of the signature of p_f is $\mathcal{O}(k)$ and the time complexity to

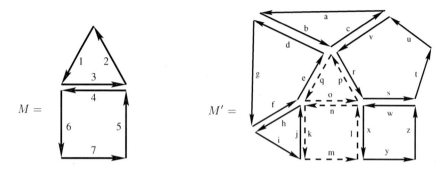

Fig. 3. Example of pattern occurrence list. Pattern M occurs 5 times in map M'. For each occurrence o, we memorize the dart of M' which corresponds to dart 1 of pattern M. Hence, the occurrence list associated with pattern M in map M' is $occ(M) = < q, a, j, p, i >$. To find the boundary of an occurrence of M, we search for the darts of M' which correspond to the 2-free darts of M (i.e., 1, 6, 7, 5, 2). For example, the boundary of the occurrence of M which starts at dart q contains darts q, k, m, l, p.

compute the signature and to add it to the signature tree is $\mathcal{O}(k^2)$ in the worst case. Using this tree signature allows us to check if p_f already belongs to L_p (line 8) in $\mathcal{O}(k)$, whatever the size of L_p is (see [10] for more details).

Frequency update (line 9 of Algo. 2). A pattern may appear several times in the same map, however, its frequency is increased by 1 at most once for each map. We explore occurrences map by map, so it is sufficient to use a flag to know if the frequency of a pattern for a given map has already been increased.

4   Generalization to nD Combinatorial Maps

For sake of simplicity, we have described our frequent submap mining algorithm for $2D$ combinatorial maps. However, it can be extended to nD maps in a very straightforward way. Actually, we have implemented it for the nD case and we report experimental results on $2D$ and $3D$ maps in the next section.

If $2D$ maps are described by two functions β_1 and β_2 which respectively describe adjacency relations between edges and faces, nD maps are described by n functions, β_1 to β_n, such that each β_i function describes adjacency relations between cells of dimension i, called i-cells (1-cells are edges, 2-cells faces, 3-cells volumes, ...). We have extended submap isomorphism to nD maps in [8].

In nD, mined patterns are connected n-cells (*i.e.*, connected faces in $2D$, connected volumes in $3D$, ...). Algorithms 1 and 2 are extended to the nD case by replacing faces with n-cells: we first search for all frequent patterns composed of one n-cell and, for each of these patterns, we iteratively compute all frequent patterns which contain it. The *grow* function builds new frequent patterns by n-sewing a frequent n-cell with a frequent pattern.

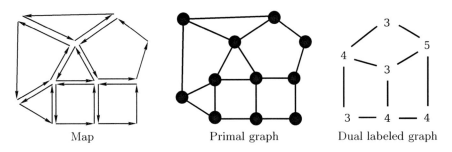

Map   Primal graph   Dual labeled graph

Fig. 4. Example of primal graph and dual labeled graph associated with a map

5   Experimental Evaluation on Synthetic Databases

Using synthetic databases allows us to evaluate scale-up properties when decreasing the frequency threshold σ, and when increasing the size of the maps, *i.e.*, the number of faces in $2D$ and the number of volumes in $3D$.

Considered datasets. We have generated different databases. Each database $D(n, k)$ contains 1000 connected maps such that $n \in \{2, 3\}$ corresponds to the dimension of the map, and k to the number of n-cells (faces for $n = 2$ and volumes for $n = 3$). Maps are randomly generated in such a way that they are connected and their faces (resp. volumes) have degrees varying between 3 and 10 (resp. 4 and 10). When $n = 2$ (resp. $n = 3$), each n-map is generated by first building k closed n-cells such that the degree of each n-cell is randomly chosen within $[3; 10]$ (resp. $[4; 10]$) according to a uniform distribution, and then randomly n-sewing these n-cells until we obtain a connected map. Note that generated maps may have holes and do not have outer (infinite) n-cell.

Maps vs graphs. We compare mSpan with gSpan[1], which is a state-of-the-art algorithm for extracting frequent connected subgraphs from a database of graphs [20]. Let us first note that mSpan and gSpan solve different problems which have different theoretical complexities: if submap isomorphism has a polynomial-time complexity, subgraph isomorphism is NP-complete. Therefore, it is not surprising if gSpan and mSpan exhibit different scale-up properties. Given a $2D$ map, we can generate a primal graph in a very straightforward way (see Fig. 4). However, mining the primal graph is not really meaningful and the extracted patterns cannot be compared with those extracted by mSpan. Indeed, mSpan extracts connected sets of faces whereas patterns extracted by gSpan are connected subgraphs which may not correspond to connected sets of faces at all (*e.g.*, trees). For a fair comparison, we consider the dual graph which associates a vertex with every face of the $2D$ map and which connects two vertices iff the corresponding faces in the map are adjacent. We also label each vertex of the dual graph with the degree of the corresponding face (*i.e.*, its number of edges). Patterns extracted by gSpan from the labeled dual graph are connected

---

[1] Implementation found in http://www.cs.ucsb.edu/~xyan/software/gSpan.htm

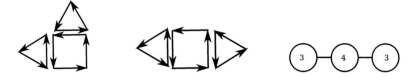

Fig. 5. Two different submaps which correspond to the same dual labeled graph

subgraphs and, therefore, correspond to connected sets of faces in the $2D$ map. Labels associated with vertices allow gSpan to discriminate faces which have different degrees and greatly improve performances of gSpan. However, gSpan does not consider the topology of the graph (*i.e.*, the order in which faces are encountered when turning around one face) so that two different submaps may correspond to the same subgraph in the dual labeled graph, as illustrated in Fig. 5. Therefore, mSpan and gSpan do not extract the same frequent patterns.

For $3D$ maps, we also generate dual labeled graphs: we associate a vertex with every volume of the $3D$ map; we connect two vertices iff the corresponding volumes are adjacent; and we label each vertex with the degree of the corresponding volume. This way, connected subgraphs of dual labeled graphs correspond to connected sets of volumes. However, like in $2D$, different connected sets of volumes may correspond to a same connected graph.

Note that labeled dual graphs are much smaller than the corresponding maps: a $2D$ (resp. $3D$) map which has 350 faces (resp. 80 volumes) has 1800 (resp. 1200) darts or so, whereas the corresponding dual graph has 800 (resp. 160) edges or so and 350 (resp. 80) vertices.

Scale-up properties when increasing map sizes. Top and middle curves of Fig. 6 display results of mSpan and gSpan on $D(n, k)$ databases with $\sigma = 0.9$ when increasing the number of faces k from 4 to 350 for $n = 2$, and when increasing the number of volumes k from 2 to 80 for $n = 3$. Each run has been limited to 3600 seconds of CPU time. mSpan is able to extract all frequent patterns within this time limit, even for the largest values of k. gSpan is faster than mSpan when $k < 50$ in $2D$, and when $k < 30$ in $3D$. However, for larger values of k it becomes slower, and it is not able to compute all frequent patterns within the CPU time limit of 3600 seconds when $k > 120$ in $2D$ and when $k > 50$ in $3D$. Actually, gSpan extracts much more frequent patterns than mSpan, and the greater k, the larger the difference. This comes from the fact that graphs do not model the topology so that different map patterns (which may not be frequent) correspond to the same graph pattern (which may become frequent).

Scale-up properties when increasing σ. Bottom curves of Fig. 6 displays results of mSpan and gSpan on the $D(2, 60)$ database when increasing the frequency threshold from 0.1 to 1. It shows us that gSpan is faster than mSpan when

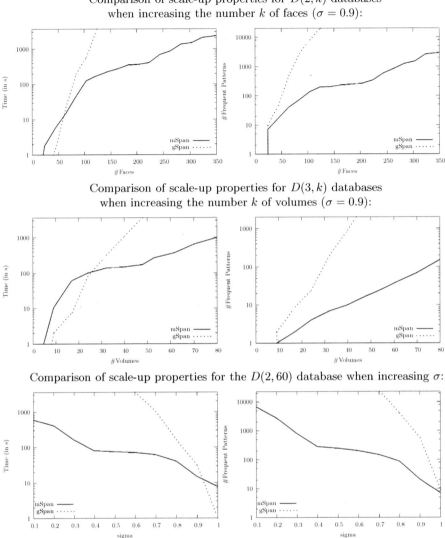

Fig. 6. Comparison of mSpan (bold lines) and gSpan (dashed lines) scale-up properties: curves on the left (resp. right) plot the evolution of CPU-time in seconds (resp. number of extracted patterns).

$\sigma > 0.9$, but for smaller values of σ, mSpan becomes faster and gSpan is not able to compute all frequent patterns within the CPU time limit of 3600 seconds when $\sigma < 0.7$. Actually, the number of extracted patterns grows much quicker for gSpan than for mSpan when decreasing the frequency threshold σ. We have performed similar experiments on $3D$ maps, and observed very similar results.

6 Application to Image Classification

The goal of this section is not to define a new approach for classifying images, but to show that frequent patterns may be used to describe images by numerical vectors, thus allowing one to use the numerous tools defined on vector spaces for searching, classifying or clustering purposes. We have considered a supervised classification problem, which involves deciding the class of a new image knowing the classes of a sample learning set of images, and we have used the C4.5 classification method [18].

We have considered a database of 4 classes of images such that each class contains 40 images (see a sample in Table 1). Each image of the database has been segmented into a 2D combinatorial map, using the algorithm described in[6]. These maps have 98 faces on average (minimum 10 and maximum 253).

Table 1. Sample of the database composed of 4 classes with 40 images per classe

We have considered a leave-k-out experimental protocol, with $k = 10\%$: we have selected 10 different learning sets, such that each learning set contains 144 images (36 images of each class) and, for each of these learning sets, we have classified the 16 remaining images; we report average results obtained over the 10 different learning sets.

For each learning set, we have used mSpan to extract frequent patterns with the frequency threshold σ set to 0.1. On average over the 10 learning sets, mSpan has extracted 854 frequent patterns, whose sizes are ranging between 1 and 8 faces, in less than one second of CPU-time on an Intel Core 2 Duo with

Table 2. Confusion Matrix. Each cell on line i and column j gives the percentage of images which belong to class i and have been classified in class j.

classified as / real class	Cherry	Football	Sea	Greenland
Cherry	77.5	5	5	12.5
Football	2.5	80	17.5	0
Sea	0	15	85	0
GreenLand	2.5	0	15	82.5

4GB RAM. Then, each image i has been represented by a numerical vector V_i whose dimension is equal to the number of frequent patterns and such that the jth element of V_i is equal to the number of occurrences of the jth frequent pattern in the map associated with i.

Table 2 displays the confusion matrix of a C4.5 classification of these numerical vectors. The average classification rate is equal to 81% or so. This result is very promising as it has been obtained with frequent patterns only. Indeed, this kind of topological information is only a small part of the information contained in an image, and it could be easily combined with any other classical features such as colour or texture to improve the classification process, thus bridging the gap between traditional pattern recognition techniques based on feature vectors, and structural pattern recognition techniques based on structured representations of images such as graphs.

7   Conclusion

We have introduced an algorithm called mSpan for extracting frequent patterns from combinatorial maps. This algorithm uses efficient polynomial time procedures for deciding of submap isomorphism [7], and for searching for isomorphic occurrences of a given map in the signature of a base of maps [10].

Combinatorial maps model the topology of nD objects subdivided in cells (*e.g.*, vertices, edges, faces, volumes, ...) by means of incidence and adjacency relationships between these cells. We have shown that we can use dual labeled graphs to model adjacency relationships between cells, but these graphs do not model the topology of these cells (*i.e.*, the order in which they are encountered when turning around a given cell). Therefore, different map patterns (which may not be frequent) may be modeled by a same dual labeled graph (which may become frequent) so that a graph mining algorithm extracts much more patterns. Of course, the relevancy of extracted patterns depends on the application. We have already applied mSpan to an aperiodic tiling application, the goal of which is to find the largest pattern occurring frequently in a given aperiodic tiling. Clearly, on this kind of application, the topology is of uppermost importance and patterns extracted from dual labeled graphs are not relevant.

References

1. Braquelaire, J.-P., Brun, L.: Image segmentation with topological maps and inter-pixel representation 9(1), 62–79 (March 1998)
2. Brisson, E.: Representing geometric structures in d dimensions: topology and order. In: SCG, Saarbrücken, Germany, pp. 218–227 (1989)
3. Chi, Y., Muntz, R.R., Nijssen, S., Kok, J.N.: Frequent subtree mining - an overview. Fundam. Inf. 66, 161–198 (2004)
4. Cori, R.: Un code pour les graphes planaires et ses applications. In: Astérisque, Soc. Math. de France, Paris, France, vol. 27 (1975)
5. Damiand, G.: Topological model for 3d image representation: Definition and incremental extraction algorithm. CVIU 109(3), 260–289 (2008)
6. Damiand, G., Bertrand, Y., Fiorio, C.: Topological model for two-dimensional image representation: definition and optimal extraction algorithm. CVIU 93(2), 111–154 (2004)
7. Damiand, G., De La Higuera, C., Janodet, J.-C., Samuel, E., Solnon, C.: Polynomial Algorithm for Submap Isomorphism: Application to searching patterns in images. In: GbR. LNCS, pp. 102–112. Springer, Heidelberg (2009)
8. Damiand, G., Solnon, C., De La Higuera, C., Janodet, J.-C., Samuel, E.: Polynomial Algorithms for Subisomorphism of nD Open Combinatorial Maps. In: Computer Vision and Image Understanding (CVIU) (December 2011)
9. Edmonds, J.: A combinatorial representation for polyhedral surfaces. Notices of the American Mathematical Society 7 (1960)
10. Gosselin, S., Damiand, G., Solnon, C.: Efficient search of combinatorial maps using signatures. Theoretical Computer Science 412(15), 1392–1405 (2011); Theoretical Computer Science Issues in Image Analysis and Processing
11. Horvath, T., Ramon, J., Wrobel, S.: Frequent subgraph mining in outerplanar graphs. In: KDD 2006, pp. 197–206 (2006)
12. Jacques, A.: Constellations et graphes topologiques. Combinatorial Theory and Applications 2, 657–673 (1970)
13. Kuramochi, M., Karypis, G.: Frequent subgraph discovery. In: IEEE International Conference on Data Mining, vol. 0, p. 313 (2001)
14. Lienhardt, P.: Subdivision of n-dimensional spaces and n-dimensional generalized maps. In: SCG, Saarbrücken, Germany, pp. 228–236 (1989)
15. Lienhardt, P.: Topological models for boundary representation: a comparison with n-dimensional generalized maps. Computer-Aided Design 23(1), 59–82 (1991)
16. Lienhardt, P.: N-dimensional generalized combinatorial maps and cellular quasi-manifolds. IJCGA 4(3), 275–324 (1994)
17. Poudret, M., Arnould, A., Bertrand, Y., Lienhardt, P.: Cartes combinatoires ouvertes. Research Notes 2007-1, Laboratoire SIC E.A. 4103 (October 2007)
18. Salzberg, S.L.: C4.5: Programs for machine learning by j. ross quinlan. In: Machine Learning 1994, vol. 16, pp. 235–240. Morgan Kaufmann Publishers, Inc., San Francisco (1993), doi:10.1007/BF00993309
19. Tutte, W.T.: A census of planar maps. Canad. J. Math. 15, 249–271 (1963)
20. Yan, X., Han, J.: gspan: Graph-based substructure pattern mining. In: Proceedings of the 2002 IEEE International Conference on Data Mining, ICDM 2002, pages 721. IEEE Computer Society, Washington, DC (2002)

Edit Distance with Duplications and Contractions Revisited

Tamar Pinhas, Dekel Tsur, Shay Zakov, and Michal Ziv-Ukelson

Department of Computer Science Ben-Gurion University of the Negev, Israel
{matuskat,dekelts,zakovs,michaluz}@cs.bgu.ac.il

Abstract. In this paper, we propose three algorithms for the problem of string edit distance with duplication and contraction operations, which improve the time complexity of previous algorithms for this problem. These include a faster algorithm for the general case of the problem, and two improvements which apply under certain assumptions on the cost function. The general algorithm is based on fast min-plus multiplication of square matrices, and obtains the running time of $O\left(\frac{|\Sigma|n^3 \log^3 \log n}{\log^2 n}\right)$, where n is the length of the input strings and $|\Sigma|$ is the alphabet size. This algorithm is further accelerated, under some assumption on the cost function, to $O\left(|\Sigma|\left(n^2 + \frac{nn'^2 \log^3 \log n'}{\log^2 n'}\right)\right)$ time, where n' is the length of the run-length encoding of the input. Another improvement is based on a new fast matrix-vector min-plus multiplication under a certain discreteness assumption, and yields an $O\left(|\Sigma|\frac{n^3}{\log^2 n}\right)$ time algorithm. Furthermore, this algorithm is online, in the sense that one of the strings may be given letter by letter. As part of this algorithm we present the currently fastest online algorithm for weighted CFG parsing for discrete weighted grammars. This result is useful on its own.

1 Introduction

Comparing strings is a well-studied problem in computer science as well as in bioinformatics. In this paper we address the problem of string edit distance, with the additional operations of duplication and contraction. Such algorithms are motivated by the study of minisatellites and their comparisons in the context of population genetics [11].

Traditionally, string similarity is measured in terms of edit distance, which reflects the minimum-cost edit of one string to the other based on the edit operations of substitutions (including matches) and deletions/insertions (indels). However, in comparing minisatellite maps, one has to also consider that regions of the map have arisen as a result of duplication events from the neighboring units. A minisatellite is a section of DNA that consists of tandem repetitions of short (6–100 bp) sequence motifs spanning 0.5 kilobases to several kilobases. The repeated motifs also vary in sequence through base substitutions. A *minisatellite map* represents a minisatellite region, where each motif (denoted *unit*) is encoded by a character and handled as one entity. For one minisatellite locus,

R. Giancarlo and G. Manzini (Eds.): CPM 2011, LNCS 6661, pp. 441–454, 2011.

both the type and the number of units in the map vary between individuals in a population. Therefore, pairwise comparisons of minisatellite maps are typically applied in studying the evolution of populations.

The single copy duplication model, where only one unit can duplicate at a time, is the most popular, and its biological validation was asserted for the MSY1 minisatellites [8,11]. According to this model, one unit can mutate into another unit via a mutation of a single nucleotide within it. Also, a unit can be duplicated, that is, an additional copy of the unit may appear next to the original one in one of the strings (tandem repeat). Thus, when comparing minisatellite maps, two additional operations are considered: duplication and contraction. As we assume the single copy duplication model, duplications (and contractions) add (and subtract, respectively) a single letter at a time.

The problem of comparing two minisatellite maps under the single copy duplication model was first defined and studied by Bérard and Rivals [8]. Bérard and Rivals suggested an $O(n^4)$ time and $O(n^3)$ space algorithm, where n is the length of the two input strings (for the sake of simplicity, we assume that both strings are of the same length). This was followed by the work of Behzadi and Steyaert [5], who gave an $O(|\Sigma|n^3)$ time and $O(|\Sigma|n^2)$ space algorithm for the problem, where $|\Sigma|$ is the alphabet size (typically a few tens of unique units). Behzadi and Steyaert [5] improved the running time of their algorithm based on run length encoding, to $O(n^2 + |\Sigma|nn'^2)$, where n' is the length of the run-length encoding of the input strings [4]. Run length encoding was also used by Bérard et al. [7], who presented an $O(n^3 + |\Sigma|n'^3)$ time algorithm. Recently, Abouelhoda et al. [1] presented an alphabet-independent algorithm, with time complexity $O(n^2 + nn'^2)$. The algorithms mentioned above need some restrictions on the edit scripts they consider in order to guarantee the optimality of the solution. For example, the algorithm of Behzadi and Steyaert [5] cannot find an optimal edit script from $s = ab$ to $t = cd$ if the unique optimal script is $ab \to bb \to b \to d \to dd \to cd$. The algorithm of Abouelhoda et al. [1] works correctly on the instance above, but it cannot find an optimal edit script from $s = a$ to $t = bc$ if the unique optimal script is $a \to d \to dd \to bd \to bc$. It is easy to design cost functions in which the example scripts are optimal. Thus, the two algorithms above are correct only when the cost function forbids such scripts from being optimal. Therefore, the algorithms presented in this paper are currently the fastest algorithms for the problem under general cost functions, even though their time complexity contains a $|\Sigma|$ factor.

1.1   Our Contribution and Roadmap

We propose three algorithms for the problem of edit distance with duplications and contractions, based on fast matrix multiplication. We start, in Section 2, with the problem definition and its basic solution.

In Section 3, we present an algorithm for general cost functions that is based on min-plus square matrix multiplication. Using the matrix multiplication algorithm of Chan [9], this algorithm runs in $O\left(\frac{|\Sigma|n^3 \log^3 \log n}{\log^2 n}\right)$ time.

In Section 4, we extend our approach to the algorithm given in [4] which exploits run length encodings of the input strings, assuming some restrictions on the cost functions. This yields an $O\left(n^2 + \frac{|\Sigma|nn'^2 \log^3 \log n'}{\log^2 n'}\right)$ time algorithm.

In Section 5, we describe another improvement to the general algorithm for the case of discrete cost functions. This algorithm is based on fast min-plus matrix-vector multiplication and is online, in the sense that one of the strings may be given letter by letter. For this purpose, we adapt Williams' matrix-vector multiplication algorithm over a finite semiring [14], and modify it to efficiently handle the min-plus matrix-vector multiplication, for the case where the differences between adjacent cells are taken from a finite integer interval. This yields an overall time complexity of $O\left(\frac{|\Sigma|n^3}{\log^2 n}\right)$. An additional result obtained along the way is the adaptation of the algorithm for weighted context-free grammar (CFG) parsing under discrete cost functions to be online as well.

Due to space restrictions, additional materials, including proofs, figures and algorithm details, are on the Internet at
http://www.cs.bgu.ac.il/~negevcb/publications.php.

2   Edit Distance with Duplications and Contractions

In this section, we define the *Edit Distance with Duplications and Contractions* (EDDC) problem, show some of its properties and give a basic solution for it.

2.1   Problem Definition

In the edit distance model, one is given a source string s and a target string t, and it is assumed that t was obtained by applying a sequence of local edit operations, called an *edit script*, over s. We denote such a script by $s = u^0 \to u^1 \to u^2 \to \ldots \to u^l = t$, where each intermediate string u^k is obtained by applying a single edit operation on u^{k-1}. Each edit operation has a *cost*, and the cost of an edit script is defined as the sum of costs of its operations. The *edit distance* from s to t, denoted by $ed\,(s, t)$, is the minimum cost of an edit script from s into t. The goal of the problem is to compute the edit distance between the two given strings. An edit script whose cost is equal to the edit distance is called an *optimal* script from s into t.

In the standard problem definition [8], the allowed edit operations include *insertion* (inserting a letter in some position in the string), *deletion* (deleting a letter from some position in the string), and *mutation* (replacing one letter in the string by another). The EDDC variant adds two operations: *duplication* and *contraction* [8]. The duplication operation replaces a single-letter substring α with the substring $\alpha\alpha$, where contraction is its symmetric operation.

Throughout the rest of this paper, fix the following entities. Let Σ be a finite alphabet. For a letter $\alpha \in \Sigma$, denote by $ins(\alpha), dup(\alpha)$, and $del(\alpha)$ the costs of the insertion, duplication, and deletion operations applied on α, respectively, by

$cont(\alpha)$ the cost of contracting a string of the form $\alpha\alpha$ into α, and by $mut(\alpha, \beta)$ the cost of mutating α into some letter $\beta \in \Sigma$. We assume that all operation costs are nonnegative. Denote by s_i the ith letter in the string s, starting at 0. We denote $s_{i,j}$ the substring $s_i s_{i+1} \ldots s_{j-1}$ of s.

2.2   The Basic Recursion

We next show how to solve the EDDC problem recursively. It is possible to show that there is an optimal edit script from s into t of the following form: s is first transformed into a string $w = w_0 w_1 \ldots w_r$ (the *common ancestor* of s and t), where each letter w_i is obtained by reducing some substring s^i of s, and $s = s^0 s^1 \ldots s^r$. Then, w is transformed into t, where every letter w_i generates some substring t^i of t through the application of some edit script, and $t = t^0 t^1 \ldots t^r$. This is expressed in the following recursive formula:

$$ed\,(s_{0,i}, t_{0,j}) = \min_{\substack{0 \le k < i, \\ 0 \le l < j, \\ \alpha \in \Sigma}} \{ed\,(s_{0,k}, t_{0,l}) + ed\,(s_{k,i}, \alpha) + ed\,(\alpha, t_{l,j})\}, \qquad (2.1)$$

with the boundary cases $ed\,(s_{0,i}, t_{0,0}) = ed\,(s_{0,0}, t_{0,j}) = \infty$ for $i, j > 0$, and $ed\,(s_{0,0}, t_{0,0}) = 0$.

In previous works [5,1], the EDDC between a letter and a string, $ed\,(s_{k,i}, \alpha)$ and $ed\,(\alpha, t_{l,j})$, was computed separately preceding the computation of the recursive formula above. We keep this approach and show, in the next section, how to compute these values by applying a CFG parsing algorithm. While being slightly different, our recursive formula resembles previous formulations [5,1]. Thus, we defer the assertion of its correctness to the supplementary materials.

2.3   Context-Free Grammar Representation

An *optimal generating edit script* is an edit script that contains only insertion, duplication and mutation operations. Consider the application of a generating edit script to a single-letter string $s = \gamma$ that results in some target string t. Such a script may be described in terms of weighted CFG parsing, as follows. The set of terminals in the grammar is Σ. The set of non-terminals is $N = \{\tilde{\alpha} : \alpha \in \Sigma\} \cup \{I\}$, where I is a unique non-terminal, representing a letter insertion placeholder. The start non-terminal is $\tilde{\gamma}$, corresponding to the source letter γ. The parse rules are:

1. Mutations: $\forall \tilde{\alpha}, \tilde{\beta} \in N, \ \tilde{\alpha} \to \tilde{\beta}$. The cost of such rules is $mut(\alpha, \beta)$.
2. Duplications: $\forall \tilde{\alpha} \in N, \ \tilde{\alpha} \to \tilde{\alpha}\tilde{\alpha}$. The cost of such rules is $dup(\alpha)$.
3. Elongations (a preliminary step to allow insertions): $\forall \tilde{\alpha} \in N, \ \tilde{\alpha} \to \tilde{\alpha}I$ and $\tilde{\alpha} \to I\tilde{\alpha}$. The cost of such rules is 0 (note that an application of such rules must be followed by an application of an insertion rule of type 4).
4. Insertions: $\forall \tilde{\alpha} \in N, \ I \to \tilde{\alpha}$. The cost of such rules is $ins(\alpha)$.
5. Terminal achievement: $\forall \tilde{\alpha} \in N, \ \tilde{\alpha} \to \alpha$, where α is the terminal corresponding to non-terminal $\tilde{\alpha}$. The cost of such rules is 0.

A generating edit script from a letter γ into t corresponds to a minimal cost parse tree of t via the grammar above, which can we achieved by applying the CKY algorithm [12,10,2]. A similar grammar can be used to generate a single letter string from a non-empty string using contraction and deletion operations.

2.4   A Basic Dynamic Programming Algorithm for EDDC

In this section we describe a dynamic programming (DP) algorithm that implements the recursion of Eq. 2.1. The algorithm consists of two stages. Given a pair of strings s and t, the first stage of the algorithm computes all letter-to-substring edit distances $ed\,(s_{i,j},\alpha)$ and $ed\,(\alpha,t_{i,j})$. In the second stage, all prefix-to-prefix edit distances $ed\,(s_{0,i},t_{0,j})$, and in particular the edit distance between the two complete strings, are computed. The algorithm maintains the following matrices:

- Matrices S^α, for every $\alpha \in \Sigma$. An entry $S_{i,j}^\alpha$ holds the value $ed\,(s_{i,j},\alpha)$.
- Matrices T^α, for every $\alpha \in \Sigma$. An entry $T_{i,j}^\alpha$ holds the value $ed\,(\alpha,t_{i,j})$.
- Matrices TD^α, for every $\alpha \in \Sigma$. An entry $TD_{i,j}^\alpha$ holds the minimum of $ed\,(s_{0,i},t_{0,l}) + ed\,(\alpha,t_{l,j})$, for $0 \leq l < j$.
- A matrix TD, where an entry $TD_{i,j}$ holds the value $ed\,(s_{0,i},t_{0,j})$.

Stage 1. In this stage the matrices S^α and T^α, for all $\alpha \in \Sigma$, are computed using a weighted CFG parsing algorithm [12,10,2], as explained in Section 2.3.

Stage 2. This stage takes as input the matrices S^α and T^α which were computed in the first stage, in addition to the strings s and t, and computes the matrix TD according to Eq. 2.1. We point out that a naive implementation of Eq. 2.1 would yield an $O(|\Sigma| n^4)$ running time. However, as done in [1], it is possible to split the computation into two independent parts by replacing Eq. 2.1 with the two interleaved recursions (Eq. 2.2 and Eq. 2.3 below), in order to obtain a more efficient, $O(|\Sigma| n^3)$ algorithm. This is achieved by utilizing the auxiliary matrices TD^α:

$$TD_{k,j}^\alpha = \min_{0 \leq l < j} \{TD_{k,l} + T_{l,j}^\alpha\}, \tag{2.2}$$

$$TD_{i,j} = \min_{\substack{0 \leq k < i \\ \alpha \in \Sigma}} \{TD_{k,j}^\alpha + S_{k,i}^\alpha\}. \tag{2.3}$$

The algorithm interleaves the simultaneous computation of matrices of the form TD^α (Eq. 2.2), with the computation of the matrix TD (Eq. 2.3), by computing their entries column by column (ordered by increasing column indices), alternating between the matrices. Note that the computation of Eq. 2.2 requires values $TD_{k,l}$ for $l < j$, which were already computed in TD due to the column-by-column interleaved computation order, and similarly for the computation of Eq. 2.3. All queried values of the form $T_{l,j}^\alpha$ and $S_{k,i}^\alpha$, for any α, were already computed in the first stage and are available via direct lookup.

Time Complexity Analysis. As explained in Section 2.3, the computation of all tables S^α and T^α can be implemented by running a CFG parsing algorithm over s and t, with respect to a grammar with $O(|\Sigma|)$ non-terminals and derivation rules. The CKY algorithm for this problem takes $O(|\Sigma|n^3)$ running time [12,10,2]. The second stage also takes $O(|\Sigma|n^3)$ time. Thus, the overall running time of the basic algorithm is $O(|\Sigma|n^3)$.

3   Accelerating the Algorithm Using Fast Square Matrix Multiplication

In this section, we show how to accelerate the basic algorithm which was presented in the previous section. Stage 1 of the basic algorithm, is accelerated using Valiant's algorithm for CFG parsing [13,6]. The DP algorithm of Stage 2 of the basic algorithm, is accelerated as follows.

The key observation utilized here is that the computations in Eq. 2.2 and Eq. 2.3 can be expressed in terms of *min-plus vector multiplications*. While naively, the running time for computing the min-plus multiplication of a pair of vectors is linear in the length of the vectors, there are fast algorithms for min-plus matrix multiplications which perform vector multiplications in sub-linear (amortized) time [9]. We adapt and utilize such algorithms here to improve the worst-case bounds of EDDC. For this we need some notations and observations, which we give in the next section.

3.1   Matrix Multiplication Preliminaries

We use the notation $A_{n \times m}$ to imply that the matrix A has n rows and m columns, where indices start at 0. For matrices $A_{n \times m}$ and $B_{n \times m}$, the *entry-wise min* operation $A \oplus B$ yields a matrix $C_{n \times m}$, where the entries of C are defined by $C_{i,j} = \min\{A_{i,j}, B_{i,j}\}$. For matrices $A_{n \times k}$ and $B_{k \times m}$, the *min-plus multiplication* operation $A \otimes B$ yields a matrix $C_{n \times m}$, where the entries of C are defined by $C_{i,j} = \min_{0 \le r < k}\{A_{i,r} + B_{r,j}\}$.

An entry $A_{i',j'}$ in A is said to be *adjacent* to an entry $A_{i,j}$ if either $i' = i$ and $j' = j + 1$, or $i' = i + 1$ and $j' = j$.

Denote by $D = [a, b]$ the interval of integers $a, a + 1, \ldots, b$, and by $|D|$ the number of integers contained in D. Say that a matrix A is a *D-discrete matrix* if for every pair of adjacent entries $A_{i,j}$ and $A_{i',j'}$ in A, the value $A_{i',j'} - A_{i,j}$ is in D.

For a subset of row indices I and a subset of column indices J, define $A_{I,J}$ to be the submatrix of A which is induced by the rows in I and the columns in J. When I contains a single row i or J contains a single column j, we simplify the notation and write $A_{i,J}$ or $A_{I,j}$.

A matrix consisting of a single column is called a *vector*. For a vector x, denote by x_i the ith entry of x, and by x_I the sub-vector of x induced by the row indices I.

The following observation is immediately implied by the min-plus multiplication definition and it generalizes the notion of computing a single entry in the multiplication result matrix to that of computing a complete sub-matrix.

Observation 1. *Let $A_{n \times k}, B_{k \times m}$ and $C_{n \times m}$ be matrices such that $C = A \otimes B$. Let $K = [0, k - 1]$, and let $K_1, K_2, \ldots, K_r$ be a partition of K into sub-intervals. Then, for every subset I of $[0, n - 1]$ and every subset J of $[0, m - 1]$, $C_{I,J} = A_{I,K} \otimes B_{K,J} = (A_{I,K_1} \otimes B_{K_1,J}) \oplus (A_{I,K_2} \otimes B_{K_2,J}) \oplus \cdots \oplus (A_{I,K_r} \otimes B_{K_r,J})$.*

3.2   The General Case Algorithm

Observe that it is possible to formulate Eqs. 2.2 and 2.3 in terms of min-plus vector multiplications, as follows:

$$TD^{\alpha}_{k,j} = TD_{k,[0,j-1]} \otimes T^{\alpha}_{[0,j-1],j} \tag{3.1}$$

$$TD_{i,j} = \min_{\alpha \in \Sigma} \{ (S^{\alpha})^T_{i,[0,i-1]} \otimes TD^{\alpha}_{[0,i-1],j} \} \tag{3.2}$$

where $(S^{\alpha})^T$ denotes the transposed matrix S^{α}.

Our algorithm is based on a general framework presented in [15], which improves the time complexity of the computation. This framework generalizes Valiant's algorithm for CFG parsing [13] to a family of algorithms, and organizes the computations of vector multiplications in a way that allows to exploit fast square matrix multiplication algorithms. It is simple to verify that the computation of Eqs. 3.1 and 3.2 can be done efficiently, using the techniques of [15]. This yields an $O\left(\frac{|\Sigma| n^3 \log^3 \log n}{\log^2 n}\right)$ time algorithm. While we do not give here a detailed description of the algorithm, it will become evident in the next section, where we describe its extension to exploit run-length encoding of the input. Also, the reader is referred to the supplementary materials for a detailed description of the algorithm.

4   Fast Square Matrix Multiplication on Top of Run Length Encoding

In this section we extend the recursive fast-matrix multiplication approach of Section 3 to the algorithm given in [4], which exploits run length encodings of the maps, assuming some restrictions on the cost function.

4.1   Sparsifying the Computation Using Run Length Encoding

A *Run Length Encoding* (RLE) of a string is a representation of the string, in which each maximal substring, composed of r repeats of the same letter α, is encoded by the letter α and the number r (denoted α^r). For example, the RLE of the string $aaabbacccc$ is $a^3 b^2 a^1 c^4$. Such a maximal single-lettered substring is called a *run*, and the length of an RLE is the number of runs in the encoding (e.g. the length of the RLE in the previous example is 4). In [4], RLE was used in order to accelerate the computation of the EDDC problem, under certain restrictions to the cost function. Specifically, it was required that the costs of duplications and contractions are less than the costs of all other operations. Actually, this requirement can be relaxed, and the following assumption suffices:

Assumption 1. *For every $\alpha, \beta \in \Sigma$: $dup(\alpha) \leq \min\{dup(\beta)+mut(\beta,\alpha), ins(\alpha)\}$, and $cont(\alpha) \leq \min\{mut(\alpha,\beta) + cont(\beta), del(\alpha)\}$.*

Under the assumption above, it is possible to show the following observation:

Observation 2 (Fact 1 in [4]). *There exists an optimal generating script of a string t from a letter α, in which for every run of size $k > 1$ in t, the $k - 1$ right letters of the run are generated by duplications of the leftmost letter of the run. This property holds symmetrically for reducing scripts from a string s into a letter α, replacing duplications by contractions.*

Due to Obs. 2, the edit distance between a letter and a string, required by Stage 1 of the basic algorithm, can be directly computed as follows. We compute the edit distance between the letter and the run length encoded string, considering each run as a single letter. Then, we add the cost of $k-1$ duplications $dup(\alpha)$ for every run α^k. Let n' denote the RLE length of the input strings (we assume for simplicity that both input strings have the same RLE length). The complexity of this algorithm is dominated by the complexity of CFG parsing on a string of length n'. This gives the time complexity of $O\left(\frac{|\Sigma|n'^3 \log^3 \log n'}{\log^2 n'}\right)$ for computing all matrices S^α and T^α [15].

Also, regarding Stage 2 of the basic algorithm, and due to Obs. 2, it was noted in [4] that some operations in the recursive computation of $ed\,(s,t)$ are redundant. In general, instead of examining $O(|\Sigma|\,n)$ solutions to sub-problems (as implied by Eq. 2.3 here), it is sufficient to examine only $O(|\Sigma|\,n')$ expressions. We next adopt the refined computation due to this observation, with respect to our formulation of the recurrence. The proof of correctness is similar to that presented in [4]. Denote the following subsets of indices in a string w:

- $E_l(w)$ is the set of all indices that are the start of a run in w.
- $E_r(w)$ is the set of all indices that are the end of a run in w.
- $E(w, i) = \{j < i | j = i - 1 \text{ or } j \in E_l(w) \cup E_r(w)\}$.

Then, Eqs. 2.2 and 2.3 can be refined as follows.

$$TD^\alpha_{k,j} = \min_{l \in E(t,j)} \{TD_{k,l} + T^\alpha_{l,j}\} \tag{4.1}$$

$$TD_{i,j} = \begin{cases} \min\limits_{k \in E(s,i), \alpha \in \Sigma} \left(TD^\alpha_{k,j} + S^\alpha_{k,i}\right), & \text{if } i \in E_l(s), j \in E_l(t), \\[2mm] \min \left(\begin{array}{c} TD_{i,j-1} + dup(t_{j-1}), \\ \min\limits_{k \in E(s,i), \alpha \in \Sigma} (TD_{k,j-1} + S^\alpha_{k,i} + T^\alpha_{j-1,j}) \end{array} \right), & \text{if } i \in E_l(s), j \notin E_l(t), \\[4mm] \min \left(\begin{array}{c} TD_{i-1,j} + cont(s_{i-1}), \\ \min\limits_{l \in E(t,j), \alpha \in \Sigma} (TD_{i-1,l} + T^\alpha_{l,j} + S^\alpha_{i-1,i}) \end{array} \right), & \text{if } i \notin E_l(s), j \in E_l(t), \\[4mm] \min \left(\begin{array}{c} TD_{i,j-1} + dup(t_{j-1}), \\ TD_{i-1,j} + cont(s_{i-1}), \\ \min\limits_{\alpha \in \Sigma} (TD_{i-1,j-1} + S^\alpha_{i-1,i} + T^\alpha_{j-1,j}) \end{array} \right), & \text{otherwise.} \end{cases} \tag{4.2}$$

The $TD^\alpha_{k,j}$ entries are computed only for $k \in E(s,n)$ and $j \in E_l(t)$. An illustration of the four cases of Eq.4.2 can be found in the supplementary materials.

4.2 The Algorithm

The algorithm is based on forward dynamic programming: during the computation, each $TD_{i,j}$ or $TD^\alpha_{i,j}$ entry is either in a *finished* or *active* state. Each finished entry contains the correct value (namely, $TD_{i,j} = ed\,(s_{0,i}, t_{0,j})$). The value of an active entry $TD_{i,j}$ is updated according to some $TD_{k,l}$ and $TD^\alpha_{k,l}$ entries it depends on, as described in Eq. 4.2, and each of these entries is in a finished state. After the state of $TD_{k,l}$ changes from active to finished, the algorithm updates the $TD_{i,j}$ and $TD^\alpha_{i,j}$ entries whose values depend on $TD_{k,l}$.

The algorithm works using a recursion, similarly to Valiant's algorithm [13] (see also [15]). The recursive procedure COMPUTE(R) receives a region $R = [i_1, i_2] \times [j_1, j_2]$, where $i_1 \in E_l(s)$ and $j_1 \in E_l(t)$, and computes the entries $TD_{i,j}$ and $TD^\alpha_{i,j}$ for $(i,j) \in R$. When COMPUTE(R) is called, each entry $TD_{i,j}$ (or $TD^\alpha_{i,j}$) for $(i,j) \in R$ is already updated with the intermediate results obtained by the application of Eqs. 4.1 and 4.2 with respect to all entries that are outside of the region R. If there is only one index pair i, j in the region in which both i and j start a run (e.g. $|E_l(s) \cap [i_1, i_2]| = 1$ and $|E_l(t) \cap [j_1, j_2]| = 1$), then the entries inside the region R are computed directly by applying Eqs. 4.1 and 4.2. Otherwise, procedure COMPUTE performs the following steps.

1. Partition the region R into 2 disjoint regions R_1 and R_2.
2. Call COMPUTE(R_1).
3. Update the $TD_{i,j}$ and $TD^\alpha_{i,j}$ entries for $(i,j) \in R_2$ using entries inside the region R_1 of the TD and TD^α matrices that were computed in step 3.
4. Call COMPUTE(R_2).

Partition stage The partitioning of R into subregions is done as follows. The procedure alternates between partitioning R vertically and horizontally, depending on the recursion level (on odd levels the partition is vertical, and on even levels the partition is horizontal). When the region cannot be partitioned vertically (this occurs when $|E_l(s) \cap [i_1, i_2]| = 1$), all following partitions are performed horizontally, and vice versa.

To partition $R = [i_1, i_2] \times [j_1, j_2]$ vertically, the procedure finds the index $i_3 \in E_l(s) \cap [i_1, i_2]$ such that $|E_l(s) \cap [i_1, i_3 - 1]| = \lfloor |E_l(s) \cap [i_1, i_2]|/2 \rfloor$. Then, the subregions are $R_1 = [i_1, i_3 - 1] \times [j_1, j_2]$ and $R_2 = [i_3, i_2] \times [j_1, j_2]$. The horizontal partitioning is performed analogously.

Update stage After a recursive call COMPUTE(R_1) for a subregion $R_1 = [i_1, i_3 - 1] \times [j_1, j_2]$, procedure COMPUTE updates values in the region $R_2 = [i_3, i_2] \times [j_1, j_2]$. By Eq. 4.2, the following updates need to be performed.

1. For every $(i,j) \in R_2$ with $i \in E_l(s)$ and $j \in E_l(t)$, the value of $TD_{i,j}$ is changed to the minimum of its current value, and

$$\min_{k \in E(s,i) \cap [i_1, i_3 - 1], \alpha \in \Sigma} \left(TD^\alpha_{k,j} + S^\alpha_{k,i} \right).$$

2. For every $(i,j) \in R_2$ with $i \in E_1(s)$ and $j \notin E_1(t)$, the value of $TD_{i,j}$ is changed to the minimum of its current value, and

$$\min_{k \in E(s,i) \cap [i_1, i_3 - 1], \alpha \in \Sigma} \left(TD_{k,j-1} + S_{k,i}^{\alpha} + T_{j-1,j}^{\alpha} \right).$$

The first update above is done by performing for every $\alpha \in \Sigma$, a min-plus multiplication of sub-matrices of S^{α} and TD^{α}: the sub-matrix of S^{α} is induced by the rows in $E(s,i) \cap [i_1, i_3 - 1]$ and the columns in $E_1(s) \cap [i_3, i_2]$ (this sub-matrix is transposed when performing the multiplication), and the sub-matrix of TD^{α} is induced by the rows in $E(s,i) \cap [i_1, i_3-1]$ and the columns in $E_1(t) \cap [j_1, j_2]$. The second update is also done using min-plus matrix multiplication. In addition, in the update stage we handle updates to adjacent cells according to Eq. 4.2.

The update above applies to vertical partitioning. Symmetrically, a similar update is applied to horizontal partitioning (here, the update is done on the TD matrix and the TD^{α} matrices).

Time Complexity Analysis. Applying min-plus square matrix multiplication to run-length encoded EDDC yields an $O\left(|\Sigma|\left(n^2 + \frac{nn'^2(\log^3 \log n')}{\log^2 n'}\right)\right)$ time algorithm. A detailed proof appears in the supplementary materials.

5   A Faster, Online Algorithm Using Fast Matrix-Vector Multiplication

In this section we describe another acceleration of the general algorithm presented in Section 3.2, intended for cost functions in which all operation costs are integers[1]. Furthermore, this algorithm is online in the sense that one of the strings may be given letter by letter.

5.1   A High-Level Overview of the Algorithm

Consider the case where all edit operation costs are integers. Observe that for $a = \max_{\alpha \in \Sigma}(del(\alpha))$, $b = \max_{\alpha \in \Sigma}(ins(\alpha))$, and $D = [-a,b]$, all matrices used by the algorithm (i.e. T^{α}, S^{α}, TD^{α} and TD) can be regarded as D-discrete matrices (see supplementary materials). This observation is exploited here, as it allows fast min-plus D-discrete matrix-vector multiplications, by employing an algorithm which is described in Section 5.2.

We first sketch a high-level overview of a faster EDDC algorithm for discrete cost functions. Based on Obs. 1 in Section 3.1, complete columns of TD^{α} and TD can be computed via matrix-vector min-plus multiplications:

$$TD_{[0,n-1],j}^{\alpha} = TD_{[0,n-1],[0,j-1]} \otimes T_{[0,j-1],j}^{\alpha} \tag{5.1}$$

$$TD_{[0,n-1],j} = \oplus_{\alpha \in \Sigma} \{ (S^{\alpha})_{[0,n-1],[0,n-1]}^{T} \otimes TD_{[0,n-1],j}^{\alpha} \} \tag{5.2}$$

---

[1] Costs which are rational numbers can be scaled to integers in a straightforward manner.

Let s be the string that is known in advance and let t be the string which is received online, letter by letter. The algorithm starts by computing S^α for every non-terminal $\tilde{\alpha}$, using Valiant's algorithm for CFG parsing [13], as in Section 3. The algorithm continues by updating all other matrices incrementally column by column, as additional letters of t are received. For the jth letter of t, the jth columns in all matrices T^α, TD^α and TD, are computed as follows.

First, we apply our algorithm for online weighted CFG parsing with discrete weights, to fill the added column in matrices T^α for every $\alpha \in \Sigma$. This algorithm is explained in the next paragraph. Second, the new columns in matrices TD^α and TD are computed according to Eq. 5.1 and Eq. 5.2, respectively. Note that all column computations are of the form of min-plus matrix-vector multiplications, which are achieved via the matrix-vector multiplication algorithm described in Section 5.2.

We give here a short description of our algorithm for online weighted CFG parsing with discrete weights: The algorithm computes a DP table column-by-column, using min-plus vector multiplications, in a similar manner to the formulations given in Eqs 5.1 and 5.2 and [15]. It applies the algorithm of Section 5.2, in an increasing order of the result sub-vectors. Thus, each of the $O(n)$ columns in the DP tables is computed in $O(\frac{n^2}{\log^2 n})$ time. The details of the algorithm appear in the supplementary materials.

Time Complexity Analysis. The online weighted CFG parsing algorithm (described above) is first executed once on the string s of size n and on the grammar G with $O(|\Sigma|)$ non terminals and rules. Then, the same algorithm is incrementally applied again for each of the n letter increments in the accumulating string t, yielding altogether an overall time complexity of $O\left(\frac{|\Sigma|n^3}{\log^2 n}\right)$. Summing the work of computing new columns of TD^α and TD over all n iterations, the algorithm performs $O(|\Sigma|n)$ fast matrix-vector min-plus multiplications (see Section 5.2), each taking $O\left(\frac{n^2}{\log^2 n}\right)$ time, and computes the minimum of $|\Sigma|$ n-length vectors, for each column in the matrix TD. Hence, the total time complexity of the algorithm is $O\left(\frac{|\Sigma|n^3}{\log^2 n}\right)$.

5.2 Fast D-discrete Matrix-Vector Multiplication

We now turn to describe an algorithm for min-plus D-discrete matrix-vector multiplication, whose running time is $O(\frac{n^2}{\log^2 n})$ (after preprocessing the matrix), under the RAM computational model assumptions. Our algorithm is based on Williams' algorithm [14] for finite semiring matrix-vector multiplications. The main acceleration technique follows the concept of the *Four-Russians algorithm* [3], i.e. pre-computing reoccurring computations, tabulating their results in lookup tables, and querying these tables in order to efficiently retrieve required results. The following lemma implies that the set of D-discrete matrices is closed under the $\otimes$ and $\oplus$ operations.

Lemma 1. *Let X and Y be two D-discrete matrices. Then, if the operations $X \otimes Y$ and $X \oplus Y$ are defined for the dimensions of X and Y, then their results are also D-discrete matrices. (The proof appears in the supplementary materials.)*

We use the following representation for D-discrete vectors. A Δ-*encoding* of a D-discrete vector x is a pair $(x_0, \Delta(x))$, where x_0 is the value of the first entry in x and $\Delta(x)$ is the sequence of differences between adjacent entries in x. The notation $x = (x_0, \Delta(x))$ is used to imply that $(x_0, \Delta(x))$ is the Δ-encoding of x. Call a vector of the form $x = (0, \Delta(x))$ a *canonical D-discrete vector.*

Let q be an integer whose exact value will be determined later (where $q = O(\log n)$). We apply Δ-encodings on q-length D-discrete vectors. Note that in such encodings $(x_0, \Delta(x))$, the sequence of differences $\Delta(x)$ contains $q-1$ elements from D. Therefore, we can index all possible sequences $\Delta(x)$ by integers between 0 and $|D|^{q-1} - 1$, and assume that the $\Delta(x)$ element in Δ-encodings is represented by the corresponding integer in this range.

The algorithm tabulates results of two kinds of computations: matrix-vector min-plus multiplications $B \otimes x$, and vector entry-wise min $x \oplus y$.

1. Tabulation for efficient matrix-vector $\otimes$ computations. Due to Lemma 1, the result of a D-discrete matrix-vector min-plus multiplication is a D-discrete vector. The next observation states an additional property of such a multiplication.

Observation 3. *Let B be a matrix, $x = (0, \Delta(x))$ a canonical vector, and $y = (y_0, \Delta(y)) = B \otimes x$. Then, for any vector $x' = (x'_0, \Delta(x))$, the vector $y' = B \otimes x'$ satisfies $y' = (y_0 + x'_0, \Delta(y))$.*

Obs. 3 implies that matrices $B_{q \times q}$ can be processed as follows. A lookup table MUL_B is computed, containing the results of all multiplications of the form $B \otimes x$, where x is a canonical q-length D-discrete vector. The results in the table are represented by their Δ-encodings. Let $x' = (x'_0, \Delta(x'))$ be a q-length D-discrete vector, for which we wish to compute the value of $B \otimes x'$. One could alternatively retrieve the result $(y_0, \Delta(y))$ of the multiplication between B and the corresponding canonical vector $(0, \Delta(x'))$, and return the result of the operation $B \otimes x'$ by the Δ-encoding $(y_0 + x'_0, \Delta(y))$.

2. Tabulation for efficient vector $\oplus$ computations. The following lemma implies that when the absolute difference between the first entries of a given pair of q-length D-discrete vectors x and y is sufficiently large, one of the vectors can be immediately taken as the result of the $x \oplus y$ operation.

Lemma 2. *Let $x = (x_0, \Delta(x))$ and $y = (y_0, \Delta(y))$ be two D-discrete vectors of length q. If $y_0 - x_0 \geq (|D| - 1)(q - 1)$, then $x \oplus y = x$. (The proof appears in the supplementary materials.)*

The next observation explains how to apply constant shifts in order to use a canonical representation of one of the vectors when computing $\oplus$ operations.

Observation 4. *Let* $x = (x_0, \Delta(x))$ *and* $y = (y_0, \Delta(y))$ *be two vectors of the same length, and let* $z = (z_0, \Delta(z)) = x \oplus y$. *For every number* μ, *it holds that* $(z_0 + \mu, \Delta(z)) = (x_0 + \mu, \Delta(x)) \oplus (y_0 + \mu, \Delta(y))$.

In order to efficiently compute $\oplus$ operations on q-length D-discrete vectors, the algorithm computes a lookup table MIN. For every pair of $(q-1)$-length difference sequences $\Delta(x)$ and $\Delta(y)$ and for every integer y_0 in the interval $[-(|D|-1)(q-1) + 1, (|D|-1)(q-1) - 1]$, the table MIN stores the Δ-encoding of the result of $(0, \Delta(x)) \oplus (y_0, \Delta(y))$. Due to Lemma 2 and Obs. 4, we get the following routine for computing $x \oplus y$ for q-length D-discrete vectors $x = (x_0, \Delta(x))$ and $y = (y_0, \Delta(y))$:

case 1: if $y_0 - x_0 \geq (|D|-1)(q-1)$ return x.
case 2: if $x_0 - y_0 \geq (|D|-1)(q-1)$ return y.
case 3: else let $(z_0, \Delta(z)) = MIN[\Delta(x), \Delta(y), y_0 - x_0]$, return $(z_0 + x_0, \Delta(z))$.

We next show how to process the input matrix A using the two kinds of tabulation techniques presented above, and how to utilize the obtained lookup tables to efficiently compute operations of the form $A \otimes x$.

1. Processing the input matrix. For an integer r, let Q_r denote the interval $Q_r = [rq, (r+1)q - 1]$ of length q. The algorithm decomposes the matrix $A_{n \times m}$ into blocks of the form $B = A_{Q_i, Q_j}$. Each block B is processed independently by computing the lookup table MUL_B. In addition, the algorithm constructs the lookup table MIN, as described above.

2. Computing matrix-vector multiplications. Given an m-length D-discrete vector x, the algorithm computes the Δ-encoding of all sub-vectors x_{Q_j}, for $0 \leq j < m/q$. Then, the algorithm computes $y = A \otimes x$ by independently computing all sub-vectors y_{Q_i} of y, for $0 \leq i < n/q$, according to Obs. 1:

$$y_{Q_i} = (A_{Q_i, Q_0} \otimes x_{Q_0}) \oplus (A_{Q_i, Q_1} \otimes x_{Q_1}) \oplus \cdots \oplus \left(A_{Q_i, Q_{m/q-1}} \otimes x_{Q_{m/q-1}}\right).$$

Each one of the computations of the form $A_{Q_i, Q_j} \otimes x_{Q_j}$ is implemented as follows. Let $x_{Q_j} = (x_0, \Delta(x))$, and let $B = A_{Q_i, Q_j}$. The algorithm performs a single query to the table MUL_B with the canonical vector $(0, \Delta(x))$. For the query result $(y_0, \Delta(y))$, the algorithm returns $(y_0 + x_0, \Delta(y))$. Then, the algorithm performs the efficient $\oplus$ computations required for y_{Q_i}. All intermediate vectors, of this computation are q-length D-discrete vectors (by Lemma 1).

Time Complexity Analysis. A D-discrete matrix $A_{n \times m}$ can be processed in $O(mn^{1.5})$ time so that min-plus multiplications $A \otimes x$, where x is an m-length D-discrete vector, take $O(\frac{nm}{\log^2 n})$ time. Details appear in the supplementary materials.

Acknowledgments. We would like to thank Prof. Yefim Dinitz for kindly pointing us to some relevant references. This research was partially supported by ISF grant 478/10 and by the Frankel Center for Computer Science at Ben Gurion University of the Negev.

References

1. Abouelhoda, M.I., Giegerich, R., Behzadi, B., Steyaert, J.-M.: Alignment of minisatellite maps based on run-length encoding scheme. J. of Bioinformatics and Computational Biology 7(2), 287–308 (2009)
2. Akutsu, T.: Approximation and exact algorithms for RNA secondary structure prediction and recognition of stochastic context-free languages. J. of Combinatorial Optimization 3(2), 321–336 (1999)
3. Arlazarov, V.L., Dinic, E.A., Kronod, M.A., Faradzev, I.A.: On economical construction of the transitive closure of an oriented graph. Soviet Math. Dokl. 11, 1209–1210 (1970)
4. Behzadi, B., Steyaert, J.M.: The minisatellite transformation problem revisited: A run length encoded approach. Algorithms in Bioinformatics, 290–301 (2004)
5. Behzadi, B., Steyaert, J.M.: An improved algorithm for generalized comparison of minisatellites. J. of Discrete Algorithms 3(2-4), 375–389 (2005)
6. Benedí, J.M., Sánchez, J.A.: Fast stochastic context-free parsing: A stochastic version of the Valiant algorithm. Pattern Recognition and Image Analysis, 80–88 (2007)
7. Bérard, S., Nicolas, F., Buard, J., Gascuel, O., Rivals, E.: A fast and specific alignment method for minisatellite maps. Evolutionary bioinformatics online 2, 303 (2006)
8. Bérard, S., Rivals, E.: Comparison of minisatellites. J. of Computational biology 10(3-4), 357–372 (2003)
9. Chan, T.M.: More algorithms for all-pairs shortest paths in weighted graphs. In: Proc. 39th ACM Symposium on Theory of Computing (STOC), pp. 590–598 (2007)
10. Chappelier, J.C., Rajman, M.: A generalized CYK algorithm for parsing stochastic CFG. In: Tabulation en analyse syntaxique et déduction. Journées, pp. 133–137 (1998)
11. Jobling, M.A., Heyer, E., Dieltjes, P., de Knijff, P.: Y-chromosome-specific microsatellite mutation rates re-examined using a minisatellite, MSY1. Human Molecular Genetics 8(11), 2117–2120 (1999)
12. Kasami, T.: An efficient recognition and syntax-analysis algorithm for context-free languages. Defense Technical Information Center (1965)
13. Valiant, L.G.: General context-free recognition in less than cubic time. J. of Computer and System Sciences 10(2), 308–314 (1975)
14. Williams, R.: Matrix-vector multiplication in sub-quadratic time(some preprocessing required). In: Proc. 18th ACM-SIAM Symposium on Discrete Algorithms (SODA), pp. 995–1001 (2007)
15. Zakov, S., Tsur, D., Ziv-Ukelson, M.: Reducing the worst case running times of a family of RNA and CFG problems, using Valiant's approach. In: Algorithms in Bioinformatics, pp. 65–77 (2010)

Polynomial-Time Approximation Algorithms for Weighted LCS Problem[*]

Marek Cygan[1], Marcin Kubica[1], Jakub Radoszewski[1],
Wojciech Rytter[1,2], and Tomasz Waleń[1]

[1] Dept. of Mathematics, Computer Science and Mechanics,
University of Warsaw, Warsaw, Poland
{cygan,kubica,jrad,rytter,walen}@mimuw.edu.pl
[2] Dept. of Math. and Informatics,
Copernicus University, Toruń, Poland

Abstract. We deal with a variant of the well-known Longest Common Subsequence (LCS) problem for weighted sequences. A (biological) weighted sequence determines the probability for each symbol to occur at a given position of the sequence (such sequences are also called Position Weighted Matrices, PWM). Two possible such versions of the problem were proposed by (Amir et al., 2009 and 2010), they are called LCWS and LCWS2 (Longest Common Weighted Subsequence 1 and 2 Problem). We solve an open problem, stated in conclusions of the paper by Amir et al., of the tractability of a log-probability version of LCWS2 problem for bounded alphabets, showing that it is NP-hard already for an alphabet of size 2. We also improve the $(1/|\Sigma|)$-approximation algorithm given by Amir et al. (where Σ is the alphabet): we show a polynomial-time approximation scheme (PTAS) for the LCWS2 problem using $O(n^5)$ space. We also give a simpler $(1/2)$-approximation algorithm for the same problem using only $O(n^2)$ space.

1 Introduction

We consider (biological) *weighted sequences*, in which for each position we know the probability of an occurrence of any symbol from the alphabet Σ (more formal definition follows). Weighted sequences are also referred to in the literature as p-weighted sequences or Position Weighted Matrices (PWM) [2,16]. The notion of weighted sequence was introduced as a tool for motif discovery and local alignment, and is extensively used in computational molecular biology. In particular, binding sites, profiles of protein families and complete chromosome sequences, that have been obtained using a whole-genome shotgun strategy [15,17] can be modelled as weighted sequences [10].

[*] The first author is supported by grant no. N206 355636 of the Polish Ministry of Science and Higher Education. The third author is supported by grant no. N206 568540 of the National Science Centre. The fourth author is supported by grant no. N206 566740 of the National Science Centre.

R. Giancarlo and G. Manzini (Eds.): CPM 2011, LNCS 6661, pp. 455–466, 2011.

Multiple algorithmic results related to combinatorics of weighted sequences, i.e., repetitions, regularities and pattern matching, have already been presented. The basic concepts (including pattern matching, repeats discovery and cover computation) were studied using three different approaches: weighted suffix trees [11], Crochemore partition [7] utilized in [13] and Karp-Miller-Rabin algorithm [8] utilized in [6]. There are also results dealing with: approximate and gapped pattern matching [3,20], property matching [1], swapped matching [19], all-covers and all-seeds problem [18,21], and extracting motifs (repeated motifs, common motifs and all maximal pairs) from weighted sequences [14]. On the practical side, there are recent results concerning massive exact and approximate pattern matching for mapping short weighted sequences to a reference genome [4], also in the parallel setting [12].

Recently Amir et al. [2] extended another well-known string problem, the Longest Common Subsequence problem [5], to weighted sequences. They introduced two versions of the Longest Common Weighted Subsequence problem, LCWS and LCWS2. Despite their similarity the complexity status of these problems is dramatically different, the first has a polynomial time solution, while the latter one is NP-hard. The results from [1,3] are also related to the LCS problem for weighted sequences, however none of the papers considers the LCS problem explicitly (the first one defines and considers weighted Hamming and edit distances, while the second considers a general property matching setting). Moreover, all these papers are limited by the assumption that patterns are ordinary strings.

The main problem considered in this paper is LCWS2. We solve an open problem stated by Amir et al. [2] and show that a log-probability version of LCWS2 is NP-hard for a bounded alphabet, moreover, even for alphabet of size 2 — the proof can be found in Section 3. We also improve the $(1/|\Sigma|)$-approximation algorithm for LCWS2 proposed in [2] by providing a polynomial-time approximation scheme (PTAS) for LCWS2. Note that obtaining a fully polynomial-time approximation scheme (FPTAS) for the LCWS2 problem is not possible, since this would imply tractability of LCWS2. Additionally we give a simpler $(1/2)$-approximation algorithm with smaller space requirements — it uses $O(n^2)$ space instead of $O(n^5)$ space needed by PTAS. Both algorithms are described in Section 4. We start by recalling the definitions of the respective problems, for which we propose a novel, more abstract formulation.

2 Preliminaries

Let Σ be a finite alphabet, $\Sigma = \{\sigma_1, \sigma_2, \ldots, \sigma_K\}$. We will be assuming for simplicity, as in most applications, that $K = |\Sigma| = O(1)$. By Σ^* we denote the set of all words over Σ. By Σ^d we denote the set of words of length d.

Definition 1 (Weighted sequence). *A weighted sequence $X = x_1 x_2 \ldots x_n$ of length $|X| = n$ over an alphabet $\Sigma = \{\sigma_1, \sigma_2, \ldots, \sigma_K\}$ is a sequence of sets of pairs of the form:*

$$x_i = \{(\sigma_j, \ p_i^{(X)}(\sigma_j)) \ : \ j = 1, 2, \ldots, K\}.$$

x_1	x_2	x_3	x_4
$p_1(\mathsf{a}) = 1/3$	$p_2(\mathsf{a}) = 1$	$p_3(\mathsf{a}) = 0$	$p_4(\mathsf{a}) = 1/2$
$p_1(\mathsf{b}) = 1/3$	$p_2(\mathsf{b}) = 0$	$p_3(\mathsf{b}) = 1/2$	$p_4(\mathsf{b}) = 1/4$
$p_1(\mathsf{c}) = 1/3$	$p_2(\mathsf{c}) = 0$	$p_3(\mathsf{c}) = 1/2$	$p_4(\mathsf{c}) = 1/4$

Fig. 1. A weighted sequence $X = x_1 x_2 x_3 x_4$ over the alphabet $\Sigma = \{\mathsf{a}, \mathsf{b}, \mathsf{c}\}$

If the considered weighted sequence is unambiguous, we will write p_i instead of $p_i^{(X)}$. Here $p_i(\sigma_j)$ is the occurrence probability of the character σ_j at the position i, these values are non-negative and sum up to 1 for a given i.
By $\mathcal{WS}(\Sigma)$ we denote the set of all weighted sequences over the alphabet Σ.

Now we recall the definitions of two versions of the Longest Common Weighted Subsequence problem [2]. We simplify their formulation by introducing an auxiliary notion of α-*subsequence*.

Let $X \in \mathcal{WS}(\Sigma)$. Let $\mathrm{Seq}_d^{|X|}$ be the set of all increasing sequences of d positions in X. For a string $s \in \Sigma^d$ and $\pi \in \mathrm{Seq}_d^{|X|}$, define $\mathcal{P}_X(\pi, s)$ as the probability that the substring on positions corresponding to π in X equals s. More formally, if $\pi = (i_1, i_2, \ldots, i_d)$ then

$$\mathcal{P}_X(\pi, s) = \prod_{k=1}^{d} p_{i_k}^{(X)}(s_k). \qquad (1)$$

Denote $SUBS(X, \alpha) = \left\{ s \in \Sigma^* : \exists \left(\pi \in \mathrm{Seq}_{|s|}^{|X|} \right) \ \mathcal{P}_X(\pi, s) \geq \alpha \right\}. \qquad (2)$

In other words $SUBS(X, \alpha)$ is the set of deterministic strings which match a subsequence of X with probability at least α. Every $s \in SUBS(X, \alpha)$ is called an α-*subsequence* of X. An α-subsequence of length d is also called an (α, d)-subsequence.

Example 1. Consider the weighted sequence from Fig. 1. The string a is its α-subsequence for any $\alpha \in (0, 1]$, the string aba is its $\frac{1}{4}$-subsequence, while the string aaaa is *not* its α-subsequence for any $\alpha > 0$.

Definition 2 (α-LCWS problem)
Input: *Two weighted sequences $X, Y \in \mathcal{WS}(\Sigma)$ and a cut-off probability α.*
Output: *The longest string $s \in \Sigma^*$ such that*

$$\exists \left(\pi \in \mathrm{Seq}_{|s|}^{|X|}, \ \pi' \in \mathrm{Seq}_{|s|}^{|Y|} \right) \ \mathcal{P}_X(\pi, s) \cdot \mathcal{P}_Y(\pi', s) \geq \alpha.$$

Equivalently, s is the longest string in $SUBS(X, \alpha_1) \cap SUBS(Y, \alpha_2)$ for some $\alpha_1 \cdot \alpha_2 \geq \alpha$.

Definition 3 ((α_1, α_2)-LCWS2 problem)
Input: *Two weighted sequences X, Y and two cut-off probabilities α_1, α_2.*
Output: *The longest string $s \in SUBS(X, \alpha_1) \cap SUBS(Y, \alpha_2)$.*

The complexity status of these problems is dramatically different, the α-LCWS problem has a polynomial time solution, while a log-probability version of the (α_1, α_2)-LCWS2 problem is NP-hard.

Theorem 1. *[2]*
(a) *The α-LCWS problem can be solved in $O(n^3)$ time and $O(n^2)$ space. If we are only interested in the length of the output, the problem can be solved in $O(Ln^2)$ time, where L is the length of the solution.*
(b) *The log-probability version of the (α_1, α_2)-LCWS2 problem is NP-hard over unbounded alphabets and admits a $(1/|\Sigma|)$-approximation algorithm (thus the problem itself admits the same approximation algorithm).*

The main problem considered in this paper is LCWS2 (however, we also utilize the polynomial time solution of LCWS problem). We tackle its version with a single cut-off probability, stated in the following Definition 4, which is, by Lemma 1, equivalent to the general version with two parameters.

Definition 4 (α-LCWS2 problem)
Input: *Two weighted sequences $X, Y \in \mathcal{WS}(\Sigma)$ and a cut-off probability α.*
Output: *The longest string $s \in SUBS(X, \alpha) \cap SUBS(Y, \alpha)$.*

The following lemma shows that the (α_1, α_2)-LCWS2 and α-LCWS2 problems are equivalent.

Lemma 1. *The (α_1, α_2)-LCWS2 problem can be reduced in linear time to the $\min(\alpha_1, \alpha_2)$-LCWS2 problem.*

Lemma 1 is a consequence of the following claim.

Claim 2. *Let $X, Y \in \mathcal{WS}(\Sigma)$ and let $\alpha_1, \alpha_2 \in (0, 1]$, $\alpha_1 < \alpha_2$. Then there exist $X', Y' \in \mathcal{WS}(\Sigma')$, where $\Sigma' = \Sigma \cup \{\#\}$ is the original alphabet extended by a symbol $\# \notin \Sigma$, such that for any string s:*

$$s \text{ is a solution to the } (\alpha_1, \alpha_2)\text{-LCWS2 problem for } X \text{ and } Y \Leftrightarrow$$
$$s \text{ is a solution to the } \alpha_1\text{-LCWS2 problem for } X' \text{ and } Y'.$$

In particular, no solution to the α_1-LCWS2 problem for X' and Y' contains the symbol $\#$.
   Moreover, $|X'| = |X|$, $|Y'| = |Y|$ and both weighted sequences can be constructed from X and Y in $O(n)$ time.

Proof. First assume that $\alpha_2 < 1$. Let $\gamma = \log_{\alpha_2} \alpha_1$. Recall that $\Sigma = \{\sigma_1, \ldots, \sigma_K\}$. We define weighted sequences X' and Y' over Σ' by the following probabilities:

$$p_i^{(X')}(\sigma_j) = p_i^{(X)}(\sigma_j), \quad p_i^{(X')}(\#) = 0$$

$$p_i^{(Y')}(\sigma_j) = p_i^{(Y)}(\sigma_j)^{\gamma}, \quad p_i^{(Y')}(\#) = 1 - \sum_{j=1}^{k} p_i^{(Y')}(\sigma_j).$$

It is easy to see that the conditions on the probabilities imposed by Definition 1 are satisfied — observe that $\gamma > 1$ (since $1 > \alpha_2 \geq \alpha_1$). We will prove that the following equality holds for this definition of X' and Y':

$$SUBS(X, \alpha_1) \cap SUBS(Y, \alpha_2) \;=\; SUBS(X', \alpha_1) \cap SUBS(Y', \alpha_1). \qquad (3)$$

Note that the left side of the equality (3) is a subset of Σ^*, while the right side is a subset of $(\Sigma')^*$. We prove (3) by showing two inclusions.

($\subseteq$) Let $s \in SUBS(X, \alpha_1) \cap SUBS(Y, \alpha_2)$, $s = s_1 s_2 \ldots s_d$. Then obviously $s \in SUBS(X', \alpha_1)$. Let $\pi \in Seq_d^{|Y|}$, $\pi = (i_1, i_2, \ldots, i_d)$, be a sequence of positions for which $\mathcal{P}_Y(\pi, s) \geq \alpha_2$. Then the same sequence of positions shows that $s \in SUBS(Y', \alpha_1)$:

$$\mathcal{P}_{Y'}(\pi, s) \;=\; \prod_{j=1}^{d} p_{i_j}^{(Y)}(s_j)^{\gamma} \;=\; \mathcal{P}_Y(\pi, s)^{\gamma} \;\geq\; \alpha_2^{\gamma} \;=\; \alpha_1.$$

($\supseteq$) Let $s \in SUBS(X', \alpha_1) \cap SUBS(Y', \alpha_1)$, $s = s_1 s_2 \ldots s_d$. First note that $s \in \Sigma^*$, otherwise s would not be an α_1-subsequence of X' (since $p_i^{(X')}(\#) = 0$ for all i). Hence, $s \in SUBS(X, \alpha_1)$. Let $\pi \in Seq_d^{|Y'|}$, $\pi = (i_1, i_2, \ldots, i_d)$, be a sequence of positions for which $\mathcal{P}_{Y'}(\pi, s) \geq \alpha_1$. Then the same sequence of positions shows that $s \in SUBS(Y, \alpha_2)$:

$$\mathcal{P}_Y(\pi, s) \;=\; \prod_{j=1}^{d} p_{i_j}^{(Y')}(s_j)^{1/\gamma} \;=\; \mathcal{P}_{Y'}(\pi, s)^{1/\gamma} \;\geq\; \alpha_2.$$

We are left with the case $\alpha_2 = 1$. If a string $s \in \Sigma^*$ is a 1-subsequence of the sequence Y, it cannot use any position i_j for a letter σ_j such that $p_{i_j}^{(Y)}(s_j) < 1$, hence we set:

$$p_i^{(X')}(\sigma_j) = p_i^{(X)}(\sigma_j), \quad p_i^{(X')}(\#) = 0,$$

$$p_i^{(Y')}(\sigma_j) = 1 \quad \text{for } p_i^{(Y)}(\sigma_j) = 1 \text{ and } p_i^{(Y')}(\sigma_j) = 0 \text{ otherwise,}$$

$$p_i^{(Y')}(\#) = 1 - \sum_{j=1}^{k} p_i^{(Y')}(\sigma_j).$$

It is easy to check that $s \in \Sigma^*$ is a 1-subsequence of the sequence Y iff s is an α_1-subsequence of the sequence Y'. This concludes the proof of the claim. ☐

3   Integer LCWS2 over a Bounded Alphabet Is NP-Hard

We recall the definition of the integer log-probability version of the LCWS2 problem as given by Amir et al. [2]. Define an *I-weighted sequence* X over the alphabet $\Sigma = \{\sigma_1, \sigma_2, \ldots, \sigma_K\}$ as a sequence of sets of pairs of the form:

$$x_i = \{(\sigma_j, \; w_i^{(X)}(\sigma_j)) \;:\; j = 1, 2, \ldots, K\}, \quad \text{where } w_i^{(X)}(\sigma_j) \in \mathbb{Z}_+.$$

Let us introduce notations similar to (1) and (2). For an I-weighted sequence X and $s \in \Sigma^d$, define:

$$\mathcal{W}_X(\pi, s) = \sum_{k=1}^{d} w_{i_k}^{(X)}(s_k) \qquad \text{for } \pi = (i_1, \ldots, i_d) \in \text{Seq}_d^{|X|}.$$

For an I-weighted sequence X and $\alpha \in \mathbb{Z}_+$, denote:

$$SUBS(X, \alpha) = \left\{ s \in \Sigma^* : \exists \left(\pi \in \text{Seq}_{|s|}^{|X|} \right) \ \mathcal{W}_X(\pi, s) \leq \alpha \right\}.$$

Using these notations, the LCIWS2 problem can be stated as follows:

Definition 5 (α-LCIWS2 problem)
Input: *Two I-weighted sequences X, Y and a cut-off value $\alpha \in \mathbb{Z}_+$.*
Output: *The longest string $s \in SUBS(X, \alpha) \cap SUBS(Y, \alpha)$.*

The previously known proof [2] of NP-hardness of the α-LCIWS2 problem depended on the assumption of an unbounded alphabet Σ. We show NP-hardness of α-LCIWS2 over the alphabet $\Sigma = \{a, b\}$.

For this, we perform a reduction of α-LCIWS2 to the following NP-complete problem [9] (the same NP-complete problem was utilized in [2]).

Definition 6 (Partition problem)
Input: *A finite set S, $S \subseteq \mathbb{Z}_+$.*
Binary output: *Is there a subset $S' \subseteq S$ such that $\sum S' = \sum S \setminus S'$.*

We make the reduction from the Partition problem to the LCIWS2 problem as follows. Let $S = \{q_1, q_2, \ldots, q_n\}$ be an instance of the Partition problem. We construct I-weighted sequences $X = x_1 x_2 \ldots x_n$ and $Y = y_1 y_2 \ldots y_n$ over the alphabet $\Sigma = \{a, b\}$ with the following weights of letters from Σ:

$$w_i^{(X)}(a) = q_i + c, \quad w_i^{(X)}(b) = c, \qquad w_i^{(Y)}(a) = c, \quad w_i^{(Y)}(b) = q_i + c.$$

Here $c > 0$ is an arbitrary positive integer. Finally let $\alpha = \frac{1}{2} \sum S + nc$.

Lemma 3. *The Partition problem for an instance S has a positive answer iff the length of the solution to α-LCIWS2 for X and Y is n.*

Proof. ($\Rightarrow$) Let $\pi = (i_1, i_2, \ldots, i_k)$ be an increasing sequence of positions such that $S' = \{q_{i_1}, q_{i_2}, \ldots, q_{i_k}\}$ is a solution to the Partition problem for S, let $\pi' = (i_1', i_2', \ldots, i_{n-k}')$ be the sequence of all remaining positions in S. Then the string $s \in \Sigma^n$ such that $s_{i_j} = a$ for i_j in π and $s_{i_j'} = b$ for i_j' in π' is a solution to the α-LCIWS2 problem for X and Y. Indeed, let $id_n = (1, 2, \ldots, n)$. Then:

$$\mathcal{W}_X(id_n, s) = \sum_i w_i^{(X)}(s_i) = \sum_{i_j} (q_{i_j} + c) + \sum_{i_j'} c = \sum S' + nc = \alpha,$$

$$\mathcal{W}_Y(id_n, s) = \sum_i w_i^{(Y)}(s_i) = \sum_{i_j} c + \sum_{i_j'} (q_{i_j'} + c) = \sum (S \setminus S') + nc = \alpha,$$

thus $s \in SUBS(X, \alpha) \cap SUBS(Y, \alpha)$. Hence, s is a solution to the α-LCIWS2 problem for X and Y, since it is the longest string in this set.

($\Leftarrow$) Let $s \in \Sigma^n$ be a solution to α-LCIWS2 problem for X and Y. Denote by $\pi = (i_1, i_2, \ldots, i_k)$ and $\pi' = (i'_1, i'_2, \ldots, i'_{n-k})$ the increasing sequences of positions within s containing letters a and b respectively. Then, for $id_n = (1, 2, \ldots, n)$, the following inequalities must hold:

$$\alpha \geq \mathcal{W}_X(id_n, s) = \sum_i w_i^{(X)}(s_i) = \sum_{i_j} (q_{i_j} + c) + \sum_{i'_j} c = \sum_{i_j} q_{i_j} + nc,$$

$$\alpha \geq \mathcal{W}_Y(id_n, s) = \sum_i w_i^{(Y)}(s_i) = \sum_{i_j} c + \sum_{i'_j} (q_{i'_j} + c) = \sum_{i'_j} q_{i'_j} + nc.$$

By recalling the definition of α and reducing equal addends, we obtain

$$\sum_{i_j} q_{i_j} \leq \frac{1}{2} \sum S \quad \text{and} \quad \sum_{i'_j} q_{i'_j} \leq \frac{1}{2} \sum S. \tag{4}$$

Note that both left sides of the inequalities (4) are non-negative and sum up to $\sum S$. Hence, both inequalities (4) are equalities, and therefore the set S' defined as $S' = \{q_{i_1}, q_{i_2}, \ldots, q_{i_k}\}$ is a solution to the Partition problem for the set S. □

Due to Lemma 3, we have a reduction from the Partition problem to the LCIWS2 problem, and even more, to a decision version of LCIWS2 (asking whether there exists a common subsequence of a given length). We conclude that the decision version of LCIWS2 problem is NP-complete, and moreover:

Theorem 2. *LCIWS2 problem over a* binary *alphabet is NP-hard.*

4   Approximating LCWS2

Previous work on the α-LCWS2 problem [2] contained a $(1/|\Sigma|)$-approximation algorithm. We start this section by presenting a $(1/2)$-approximation algorithm for α-LCWS2 and then proceed to a polynomial-time approximation scheme (PTAS) for this problem. The first algorithm is more space-efficient than the presented general approximation scheme.

4.1   (1/2)-Approximation Algorithm for LCWS2

Let $X, Y \in \mathcal{WS}(\Sigma)$, $n = \max(|X|, |Y|)$, and $\alpha \in (0, 1]$. By $OPT(X, Y, \alpha)$ we denote the length of the solution to the α-LCWS2 problem for X and Y. In case the length of the solution to α^2-LCWS problem for X, Y is even, we obtain a $(1/2)$-approximation of $OPT(X, Y, \alpha)$ simply by taking one half of the solution to the α^2-LCWS problem. In the case of odd length, we need to use the solutions to the α'-LCWS problem for all prefixes of X and Y, for different values of α', as shown in the proof of the following theorem.

Theorem 3
(a) *We can compute a solution to the α-LCWS2 problem for $X, Y \in \mathcal{WS}(\Sigma)$ of length at least $\lfloor OPT(X, Y, \alpha)/2 \rfloor$ in $O(n^3)$ time and $O(n^2)$ space.*
(b) *There exists a $(1/2)$-approximation algorithm for the α-LCWS2 problem which runs in $O(n^5)$ time and $O(n^2)$ space.*

Proof (a) Let L be the size of the solution to the α-LCWS2 problem for X and Y and let d be the size of the solution to the α^2-LCWS problem for X and Y. We will show that $\lfloor \frac{d}{2} \rfloor \leq L \leq d$. This suffices to prove point (a), since d can be computed in $O(n^3)$ time and $O(n^2)$ space, see Theorem 1.

The inequality $d \geq L$ is a consequence of the following trivial inclusion:

$$SUBS(X, \alpha) \cap SUBS(Y, \alpha) \subseteq \{SUBS(X, \alpha_1) \cap SUBS(Y, \alpha_2) \; : \; \alpha_1 \cdot \alpha_2 \geq \alpha^2\},$$

in which the left side is the set of candidates for the solution of α-LCWS2 problem, while the right side are candidates for the solution to α^2-LCWS problem, both for the weighted sequences X and Y.

As for the other inequality, $L \geq \lfloor \frac{d}{2} \rfloor$, let $\pi \in \mathrm{Seq}_d^{|X|}$, $\pi = (i_1, \dots, i_d)$, and $\pi' \in \mathrm{Seq}_d^{|Y|}$, $\pi' = (i'_1, \dots, i'_d)$, be the sequences of positions corresponding to the solution $s = s_1 \dots s_d$ of α^2-LCWS. Thus s satisfies:

$$\mathcal{P}_X(\pi, s) \cdot \mathcal{P}_Y(\pi', s) = \prod_{j=1}^{d} p_{i_j}^{(X)}(s_j) \cdot \prod_{j=1}^{d} p_{i'_j}^{(Y)}(s_j) \geq \alpha^2. \qquad (5)$$

Let $g = \lfloor \frac{d}{2} \rfloor$. The left side of the inequality (5) can be written as $A \cdot B \cdot C \cdot D$, where:

$$A = \prod_{j=1}^{g} p_{i_j}^{(X)}(s_j), \quad B = \prod_{j=1}^{g} p_{i'_j}^{(Y)}(s_j), \quad C = \prod_{j=g+1}^{d} p_{i_j}^{(X)}(s_j), \quad D = \prod_{j=g+1}^{d} p_{i'_j}^{(Y)}(s_j).$$

Note that at most one of the numbers A, B, C, D can be less than α. Indeed, otherwise the product of these numbers would certainly be less than α^2, since all of them are at most 1. Hence:

$$(A \geq \alpha \wedge B \geq \alpha) \quad \vee \quad (C \geq \alpha \wedge D \geq \alpha).$$

Consequently, at least one of the strings $s_1 \dots s_g$ or $s_{g+1} \dots s_d$ is a solution to the α-LCWS2 problem, therefore $L \geq \lfloor \frac{d}{2} \rfloor$.

(b) If d is odd we need to additionally check if $L \geq \lceil \frac{d}{2} \rceil$. For this, we iterate over all possibilities of the last positions of the α-subsequence of length L within X and Y and all letters that could be the last letter of the resulting string. For every such choice we obtain an instance of the (α_1, α_2)-LCWS2 problem for some α_1, α_2, which we transform into an instance with a single cut-off probability using Lemma 1. Recall that the alphabet is of a constant size. We have $O(n^2)$ pairs of last positions, so the complexity grows by a quadratic factor. $\square$

4.2   Polynomial-Time Approximation Scheme for LCWS2

Let $X, Y \in \mathcal{WS}(\Sigma)$, $n = \max(|X|, |Y|)$, and $\alpha \in (0, 1]$. We say that an instance (X, Y, α) of the α-LCWS2 problem is a (γ, T)-*power* if all the weights in the sequence X are powers of γ, where $0 < \gamma < 1$ and $\gamma^{T-1} \geq \alpha > \gamma^T$.

Lemma 4. *The α-LCWS2 problem for (γ, T)-power instances can be solved in $O(n^3 T)$ time and space.*

Proof. Without the loss of generality, we can assume, that $m = |Y| \leq |X| = n$. We use the dynamic programming technique. Our approach is a generalisation of the standard LCS algorithm. We have $O(n^3 T)$ states, each described by a tuple (a, b, ℓ, t), where:

- a is the position in the sequence X, $1 \leq a \leq n$;
- b is the position in the sequence Y, $1 \leq b \leq m$;
- ℓ is the length of the subsequence already chosen, $0 \leq \ell \leq m$;
- t is a γ-based logarithm of the product of $p_i(\sigma_j)$ values of the chosen subsequence of X; by the definition of the (γ, T)-power, we only consider integral values of t from the interval $[0, T - 1]$.

For a state (a, b, ℓ, t) we store, as $\mathrm{val}(a, b, \ell, t)$, the greatest value β such that there exists a string $z = z_1 z_2 \dots z_\ell \in \Sigma^\ell$ that is a (β, ℓ)-subsequence of $y_1, \dots, y_b$, and there exists a sequence of positions $\pi \in \mathrm{Seq}_\ell^a$ for which

$$\log_\gamma \mathcal{P}_X(\pi, z) = t.$$

If no such (β, ℓ)-subsequence exists, we set $\mathrm{val}(a, b, \ell, t) = 0$.

Intuitively, in a state (a, b, ℓ, t) besides two positions a, b used in the classical LCS algorithm we control the length of the subsequence using the parameter ℓ and with the parameter t we count the number of times the value γ is multiplied in the product of probabilities used in the subsequence of the sequence X.

When computing the value for a state (a, b, ℓ, t), it suffices to consider three options: either drop the a-th position in the sequence X, or drop the b-th position in the sequence Y, or use one of the $|\Sigma|$ letters at the positions a, b in the sequence X and Y respectively. Formally, we have the following recursive formula:

$$\mathrm{val}(a, b, \ell, t) = \max\Big(\mathrm{val}(a-1, b, \ell, t), \ \mathrm{val}(a, b-1, \ell, t),$$

$$\max_j \Big\{ p_b^{(Y)}(\sigma_j) \cdot \mathrm{val}(a-1, b-1, \ell-1, t - \log_\gamma(p_a^{(X)}(\sigma_j))) \Big\}\Big).$$

We set $\mathrm{val}(a, b, 0, t) = 1$, and for all remaining border cases $\mathrm{val}(a, b, \ell, t) = 0$. Thus we compute all the values $\mathrm{val}(a, b, \ell, t)$ in $O(n^3 T)$ time.

After computing values for all states we look for the greatest value of d for which there exists $t \in [0, T-1]$ such that $\mathrm{val}(|X|, |Y|, d, t) \geq \alpha$. By extending the dynamic programming algorithm in a standard manner we can construct a string $s \in \Sigma^d$ and two sequences of positions $\pi \in \mathrm{Seq}_d^{|X|}, \pi' \in \mathrm{Seq}_d^{|Y|}$ such that $\mathcal{P}_X(\pi, s), \mathcal{P}_Y(\pi', s) \geq \alpha$.   □

Lemma 5. *For any $\epsilon > 0$ we can compute in $O(n^4/\epsilon)$ time and space a string which is an $\alpha^{1+\epsilon}$-subsequence of X and an α-subsequence of Y of length at least* $\mathrm{OPT}(X, Y, \alpha)$.

Proof. We assume that $\alpha < 1$, if $\alpha = 1$ we can solve the problem using a reduction to the standard LCS problem in $O(n^2)$ time and space.

We start by scaling and rounding probability distributions of the weighted sequence X. Let $T = \frac{n}{\epsilon}$ and $\gamma = \alpha^{1/T}$. For all i, j we set:

$$p_i'(\sigma_j) = \gamma^{\lfloor \log_\gamma (p_i^{(X)}(\sigma_j)) \rfloor}.$$

Observe that:

$$p_i^{(X)}(\sigma_j) = \gamma^{\log_\gamma (p_i^{(X)}(\sigma_j))}, \quad p_i'(\sigma_j) \geq p_i^{(X)}(\sigma_j) \geq p_i'(\sigma_j)\alpha^{\epsilon/n}.$$

Hence, $p_i'(\sigma_j)$ is greater than $p_i^{(X)}(\sigma_j)$ at most by a factor of $\alpha^{-\epsilon/n}$. Now the conclusion follows from Lemma 4, since with the new weight p' for the sequence X the instance (X, Y, α) is a (γ, T)-power and multiplication of at most n weights differs from α by at most a factor of α^ϵ. Note that the new weight p' is not a probability distribution, as it does not satisfy the equality $\sum_{j=1}^K p'(\sigma_j) = 1$, however the algorithm from Lemma 4 does not use this assumption.   □

Lemma 6. *Let (X, Y, α) be an instance of the LCWS2 problem. In $O(n^5)$ time and space one can find a string s which is an $(\alpha, d - 1)$-subsequence of both X and Y such that no $(\alpha, d + 1)$-subsequence of both X and Y exists.*

Proof. We set $\epsilon = 1/n$ and use the algorithm from Lemma 5 for the instance (X, Y, α, ϵ). Thus in $O(n^5)$ time and space we obtain a string $z \in \Sigma^d$ which is an $(\alpha^{1+1/n}, d)$-subsequence of X and an (α, d)-subsequence of Y. Note that no $(\alpha, d + 1)$-subsequence of both X and Y exists. Now we remove exactly one character of the string z which has the smallest value of $p_{i_k}^{(X)}(z_k)$. Thus we obtain the final string $s \in \Sigma^{d-1}$ which is an $(\alpha, d - 1)$-subsequence of both X and Y, since:

$$\alpha^{(1+1/n)(1-1/d)} \geq \alpha^{(1+1/n)(1-1/n)} = \alpha^{1-1/n^2} \geq \alpha.   □$$

Theorem 4. *For any real value $\epsilon \in (0, 1]$ there exists a $(1 - \epsilon)$-approximation algorithm for the LCWS2 problem which runs in polynomial time and uses $O(n^5)$ space. Consequently the LCWS2 problem admits a PTAS.*

Proof. The algorithm works as follows. Using the algorithm from Lemma 6 we find a positive integer d and an $(\alpha, d - 1)$-subsequence.

If $d \geq 1/\epsilon$ then we are done since in that case we have $(d-1)/d = 1-1/d \geq 1-\epsilon$ which means that we have found a $(1 - \epsilon)$-approximation.

If $d < 1/\epsilon$ then we search for an (α, d)-subsequence using a brute-force approach, i.e., we try all $\binom{|X|}{d}$, $\binom{|Y|}{d}$ subsets of positions in each sequence and all $|\Sigma|^d$ possible strings of length d, which leads to $n^{O(1/\epsilon)}$ running time. If we find an (α, d)-subsequence we simply return it as the final answer. Otherwise we return the $(\alpha, d - 1)$-subsequence found by the algorithm from Lemma 6. Either way when $d < 1/\epsilon$ our answer is optimal.   □

Acknowledgements. We thank M. Christodoulakis, M. Crochemore, C. Iliopoulos and G. Tischler for helpful discussions about the weighted LCS problem.

References

1. Amir, A., Chencinski, E., Iliopoulos, C.S., Kopelowitz, T., Zhang, H.: Property matching and weighted matching. Theor. Comput. Sci. 395(2-3), 298–310 (2008)
2. Amir, A., Gotthilf, Z., Shalom, B.R.: Weighted LCS. J. Discrete Algorithms 8, 273–281 (2010)
3. Amir, A., Iliopoulos, C.S., Kapah, O., Porat, E.: Approximate matching in weighted sequences. In: Lewenstein, M., Valiente, G. (eds.) CPM 2006. LNCS, vol. 4009, pp. 365–376. Springer, Heidelberg (2006)
4. Antoniou, P., Iliopoulos, C.S., Mouchard, L., Pissis, S.P.: Algorithms for mapping short degenerate and weighted sequences to a reference genome. I. J. Computational Biology and Drug Design 2(4), 385–397 (2009)
5. Bergroth, L., Hakonen, H., Raita, T.: A survey of longest common subsequence algorithms. In: SPIRE, pp. 39–48 (2000)
6. Christodoulakis, M., Iliopoulos, C.S., Mouchard, L., Perdikuri, K., Tsakalidis, A.K., Tsichlas, K.: Computation of repetitions and regularities of biologically weighted sequences. Journal of Computational Biology 13(6), 1214–1231 (2006)
7. Crochemore, M.: An optimal algorithm for computing the repetitions in a word. Inf. Process. Lett. 12(5), 244–250 (1981)
8. Crochemore, M., Rytter, W.: Jewels of Stringology. World Scientific, Singapore (2003)
9. Garey, M.R., Johnson, D.S.: Computers and Intractability: A Guide to the Theory of NP-Completeness. W. H. Freeman, New York (1979)
10. Gusfield, D.: Algorithms on Strings, Trees, and Sequences – Computer Science and Computational Biology. Cambridge University Press, Cambridge (1997)
11. Iliopoulos, C.S., Makris, C., Panagis, Y., Perdikuri, K., Theodoridis, E., Tsakalidis, A.K.: The weighted suffix tree: An efficient data structure for handling molecular weighted sequences and its applications. Fundam. Inform. 71(2-3), 259–277 (2006)
12. Iliopoulos, C.S., Miller, M., Pissis, S.P.: Parallel algorithms for degenerate and weighted sequences derived from high throughput sequencing technologies. In: Holub, J., Zdárek, J. (eds.) Stringology, pp. 249–262. Prague Stringology Club, Department of Computer Science and Engineering, Faculty of Electrical Engineering, Czech Technical University in Prague (2009)
13. Iliopoulos, C.S., Mouchard, L., Perdikuri, K., Tsakalidis, A.K.: Computing the repetitions in a biological weighted sequence. Journal of Automata, Languages and Combinatorics 10(5/6), 687–696 (2005)
14. Iliopoulos, C.S., Perdikuri, K., Theodoridis, E., Tsakalidis, A., Tsichlas, K.: Motif extraction from weighted sequences. In: Apostolico, A., Melucci, M. (eds.) SPIRE 2004. LNCS, vol. 3246, pp. 286–297. Springer, Heidelberg (2004)
15. Myers, E.W., Celera Genomics Corporation: A whole-genome assembly of drosophila 287(5461), 2196–2204 (2000)
16. Thompson, J.D., Higgins, D.G., Gibson, T.J.: CLUSTAL W: improving the sensitivity of progressive multiple sequence alignment through sequence weighting, positions-specific gap penalties and weight matrix choice. Nucleic Acids Res. 22, 4673–4680 (1994)

17. Venter, J.C., Celera Genomics Corporation: The sequence of the human genome. Science 291, 1304–1351 (2001)
18. Zhang, H., Guo, Q., Fan, J., Iliopoulos, C.S.: Loose and strict repeats in weighted sequences of proteins. Protein and Peptide Letters 17(9), 1136–1142(7) (2010)
19. Zhang, H., Guo, Q., Iliopoulos, C.S.: String matching with swaps in a weighted sequence. In: Zhang, J., He, J.-H., Fu, Y. (eds.) CIS 2004. LNCS, vol. 3314, pp. 698–704. Springer, Heidelberg (2004)
20. Zhang, H., Guo, Q., Iliopoulos, C.S.: An algorithmic framework for motif discovery problems in weighted sequences. In: Calamoneri, T., Diaz, J. (eds.) CIAC 2010. LNCS, vol. 6078, pp. 335–346. Springer, Heidelberg (2010)
21. Zhang, H., Guo, Q., Iliopoulos, C.S.: Varieties of regularities in weighted sequences. In: Chen, B. (ed.) AAIM 2010. LNCS, vol. 6124, pp. 271–280. Springer, Heidelberg (2010)

Restricted Common Superstring and Restricted Common Supersequence

Raphaël Clifford[1], Zvi Gotthilf[2],
Moshe Lewenstein[2,*], and Alexandru Popa[1]

[1] Department of Computer Science, University of Bristol, UK
{clifford,popa}@cs.bris.ac.uk
[2] Department of Computer Science, Bar-Ilan University,
Ramat Gan 52900, Israel
{gotthiz,moshe}@cs.biu.ac.il

Abstract. The *shortest common superstring* and the *shortest common supersequence* are two well studied problems having a wide range of applications. In this paper we consider both problems with resource constraints, denoted as the Restricted Common Superstring (shortly *RCSstr*) problem and the Restricted Common Supersequence (shortly *RCSseq*). In the *RCSstr* (*RCSseq*) problem we are given a set S of n strings, $s_1, s_2, \ldots, s_n$, and a multiset $t = \{t_1, t_2, \ldots, t_m\}$, and the goal is to find a permutation $\pi : \{1, \ldots, m\} \to \{1, \ldots, m\}$ to maximize the number of strings in S that are substrings (subsequences) of $\pi(t) = t_{\pi(1)} t_{\pi(2)} \cdots t_{\pi(m)}$ (we call this ordering of the multiset, $\pi(t)$, a permutation of t). We first show that in its most general setting the *RCSstr* problem is *NP-complete* and hard to approximate within a factor of $n^{1-\epsilon}$, for any $\epsilon > 0$, unless P = NP. Afterwards, we present two separate reductions to show that the *RCSstr* problem remains NP-Hard even in the case where the elements of t are drawn from a binary alphabet or for the case where all input strings are of length two. We then present some approximation results for several variants of the *RCSstr* problem. In the second part of this paper, we turn to the *RCSseq* problem, where we present some hardness results, tight lower bounds and approximation algorithms.

1 Introduction

1.1 Motivation

In AI planning research it is very important to exploit the interactions between different parts of plans. This was observed early in the area [21,26,30]. One very important type of interaction is the *merging* of different actions to make the total plan more efficient.

In the general setting we have a set of goals (or tasks) which have to be accomplished and we want to find the most cost efficient plan which achieves

* This work was partially supported by the Israel Science Foundation (ISF) grant # 1484/08.

R. Giancarlo and G. Manzini (Eds.): CPM 2011, LNCS 6661, pp. 467–478, 2011.
© Springer-Verlag Berlin Heidelberg 2011

all the goals. This problem is also known as the *shortest common superstring* in the case that every goal has to be done continuously or the *shortest common supersequence* if we can abandon a task and resume its process later. In both problems we assume that we have an unlimited set of resources and we want to achieve all our goals. Of course, in real life this is never the case: our resources are always limited.

Therefore, a more realistic question is: given a fixed set of resources, how many goals can be achieved (continuously or not)?

It seems that most of the applications of the shortest common superstring and the shortest common supersequence problem, are more suitable for the case of limited resources. The main challenge for such applications is to find the best arrangement that will lead us to accomplish the maximum number of goals.

As an example, Wilensky [29] gives the scenario where John is planning to go camping for a week. He goes to the supermarket to buy a week's worth of groceries. John has to achieve a set of goals (i.e., to buy food for meals during the camping weekend) and he is able to merge some goals (i.e., to buy different products during a single trip to a supermarket) in order to make the plan more efficient.

Another application, from the computational biology area, is the case where only the set of amino acids can be determined and not their precise ordering. Here we want to know which ordering would maximize the number of short strings which can be substrings or subsequences of some ordering of the symbols in a given text.

1.2   Previous Work

In the shortest common supersequence we are given a set S of n strings, $s_1, s_2, \ldots, s_n$ and we want to find the shortest string that is a supersequence of every string in S. For arbitrary n the problem is known to be NP-Hard [14] even in the case of a binary alphabet [19]. However for fixed n a dynamic programming approach takes polynomial time and space. The shortest common supersequence problem has been studied extensively both from a theoretical point of view [9,12,15,18,20,27], from an experimental point of view [1,6] and from the perspective of its wide range of applications in data compression [24], query optimization in database systems [23] and text editing [22].

In the shortest common superstring problem we are given a set S of n strings, $s_1, s_2, \ldots, s_n$ and we want to find the shortest string that is a superstring of every string in S. For arbitrary n the problem is known to be *NP-Complete* [8] and APX-hard [4]. Even for the case of binary alphabet Ott [16] presented lower bounds for the achievable approximation ratio. The best known approximation ratio so far is 2.5 [13,25]. In [10], we considered two variants of the Restricted Common Superstring problem with swap permutations.

1.3   Our Contributions

We consider the complexity and the approximability of two problems which are closely related to the well-known shortest common superstring and shortest common supersequence problems.

Problem 1. (Restricted Common Superstring (Supersequence)) The input consists of a set $S = \{s_1, s_2, \ldots, s_n\}$ of n strings over an alphabet Σ and a multiset $t = \{t_1, t_2, \ldots, t_m\}$ over the same alphabet. The goal is to find an ordering of the multiset t that maximizes the number of strings in S that are a substring (subsequence) of the ordered multiset. We denote this ordering by $\pi(t) = t_{\pi(1)} t_{\pi(2)} \cdots t_{\pi(m)}$ (and we say that $\pi(t)$ is a permutation of t). If all the strings in S have length at most ℓ, we refer to the problem as *RCSstr[ℓ] (RCSseq[ℓ])*. For simplicity of presentation, we assume throughout that all the input strings are distinct and every string $s_i \in S$ is a substring of at least one permutation $\pi(t)$.

Example 1. Let multiset $t = \{a, a, b, b, c, c\}$ and set $S = \{abb, bbc, cba, aca\}$ be an instance of *RCSstr* (and also of *RCSstr[3]*). In this example the maximum number of strings from S that can be a substring of a permutation of t is 3. One such possible permutation is $\pi(t) = acabbc$ which contains the strings aca, abb, bbc as substrings.

Example 2. Let multiset $t = \{a, a, b, c\}$ and set $S = \{ab, bc, cb, ca\}$ be an instance of *RCSseq* and also *RCSseq[2]*. In this example the maximum number of strings from S that can be a subsequence of a permutation of t is 3. One such possible permutation is $\pi(t) = abca$ which contains the strings ab, bc, ca as a subsequence.

The paper is organized as follows. In Section 2.1 we study the hardness of the *RCSstr* problem. We show first that in its most general setting the *RCS* problem is *NP-complete* and hard to approximate within a factor of less than $n^{1-\epsilon}$, for any $\epsilon > 0$, unless P = NP. Then, we show that even if all input strings are of length two (*RCSstr[2]*) and t is a set, i.e. no symbols are repeated, then the *RCSstr* problem is *APX-Hard*. Afterwards, we prove that the *RCSstr* problem remains NP-Hard even in the case of a binary alphabet.

In Section 2.2, we design approximation algorithms for several restricted variants of the *RCSstr* problem. We first present a 3/4 approximation algorithm for the *RCSstr[2]* problem where t is a set. Moreover, we give a $1/(\ell(\ell(\ell+1)/2-1))$-approximation algorithm for *RCSstr[ℓ]*, when ℓ is the length of the longest input string.

The *RCSseq* problem is studied in Section 3. In Section 3.1 we show that the hardness results for *RCSstr* hold also for *RCSseq*. Moreover, we show an approximation lower bound of $1/\ell!$ when ℓ is the length of the longest input string.

In Section 3.2, we present approximation algorithms for two variants of the *RCSseq* problem. The first is a $(1 + \Omega(1/\sqrt{\Delta}))/2$ approximation algorithm for *RCSstr[2]*, where Δ is the number of occurrences of the most frequent character in S. Then, for *RCSseq* we show that a selection of an arbitrary permutation, $\pi(t)$, yields a $1/\ell!$ randomized approximation algorithm, thus matching the lower bound presented in Section 3.1.

2   *RCSstr*

2.1   Hardness of the *RCSstr*

In this section we present hardness results for several variants of the *RCSstr* problem.

We show here that *RCSstr* problem is *NP-complete* and hard to approximate within a factor better than $n^{1-\epsilon}$, for any $\epsilon > 0$, unless P = NP. To do so, we present an approximation-preserving reduction from the classical *maximum clique* problem.

Definition 1. *(Maximum Clique) Given an undirected graph $G = (V, E)$ the maximum clique problem is to find a vertex set $V' \subseteq V$ of maximum cardinality, such that for every two vertices in V', there exists an edge connecting the two.*

The following seminal hardness result will be useful.

Theorem 1. *[31] The maximum clique problem does not have an $n^{1-\epsilon}$ approximation, for any $\epsilon > 0$, unless $P = NP$.*

We can now present our main hardness result of the *RCSstr* problem.

Theorem 2. *RCSstr is NP-complete and hard to approximate within a factor of $n^{1-\epsilon}$, for any $\epsilon > 0$, unless $P = NP$.*

Proof. We present an approximation-preserving reduction from the maximum clique problem to the *RCSstr* problem. Given an undirected graph $G = (V, E)$, where $V = \{v_1, v_2, \dots, v_n\}$, we construct an instance (S, t) of the *RCSstr* problem in the following way.

Set t to be $\{v_1^n, v_2^n, \dots, v_n^n\}$ and for each vertex $v_i \in V$ define a string $s_i \in S$ as follows. Set $d(v_i)$ to be the ordered sequence of the vertices not adjacent to v_i. Set s_i to be $v_i^n \cdot d(v_i)$, where $\cdot$ denotes concatenation.

We now prove that the optimal solution of the *RCSstr* instance (S, t) has size x if and only if the optimal solution of maximum clique problem on the graph G has size x.

Let π be a permutation on the multiset t and let $A \subseteq S$ be all the strings that are substrings of $\pi(t)$. Denote by A' the set of vertices in G corresponding to the set of strings A. We prove that the vertices in A' form a clique. Suppose that this is not true and there exist two vertices $v_i, v_j \in A'$ such that $(v_i, v_j) \notin E$. Note that, in any common superstring of the strings s_i and s_j either v_i or v_j must have at least $n + 1$ occurrences, since v_i is not present in the neighbors list of v_j and vice versa. This is a contradiction since the multiset t has only n copies of each character. Therefore the set of vertices A' forms a clique.

On the other hand, let $A' = \{v_1, \dots, v_k\} \subseteq V$ be a clique and let $A = \{s_1, \dots, s_k\} \subseteq S$ be the set of corresponding strings. We can find a permutation of t which contains all the strings in A as a substring by concatenating $s_1, \dots, s_k$

and appending the remaining characters arbitrarily at the end. No character is used more than n times since the vertices from A' form a clique and, therefore, $v_i \notin d(v_j)$ for any $v_i, v_j \in A'$.

Thus, the $RCSstr$ problem is *NP-complete* and hard to approximate within a factor $n^{1-\epsilon}$, for any $\epsilon > 0$, unless P = NP.   □

We now show that the $RCSstr[2]$ problem is APX-Hard even if t is a set, i.e. each character in t is unique. To do so, we present an approximation-preserving reduction from the classical *Asymmetric maximum TSP* problem with edge weights of 0 and 1.

Definition 2. *(Maximum Asymmetric Travelling Salesman Problem)*
Given a complete weighted directed graph $G = (V, E)$ the Maximum Travelling Salesman Problem *is to find a closed tour of maximum weight visiting all vertices exactly once.*

Theorem 3. *[7] For any constant $\epsilon > 0$, it is NP-Hard to approximate the Maximum Asymmetric Travelling Salesman with 0, 1 edge weights within $320/321+\epsilon$.*

The hardness result for the $RCSstr[2]$ problem is stated in the following theorem.

Theorem 4. *There exists a constant $\beta > 0$, such that the RCSstr problem is NP-Hard to approximate within a factor of $1 - \beta$, even if all the strings in S have length two and t is a set.*

Proof. We present a gap-preserving reduction from the maximum asymmetric TSP to the $RCSstr[2]$ problem where t is a set.

Given a complete directed graph $G = (V, E)$, with $|V| = n$, $|E| = n(n-1)$ and edge weights of 0 and 1, we construct an instance (S, t) of the $RCSstr[2]$ problem in the following way.

Set $t = V$ and for each arc $(a, b) \in E$ with weight 1 set a string ab in S. Let $OPT(G)$ be the length of the optimal tour on the graph G and let $OPT(S, t)$ be the maximum number of strings from S which can be substrings of a permutation of t. In order to have an inapproximability factor less than 1, we also assume that $n > 322$.

We now prove that the reduction presented is a gap-preserving reduction. Specifically, we prove that:

$$OPT(G) = n \Rightarrow OPT(S, t) = n - 1$$

$$OPT(G) < (1 - \alpha)n \Rightarrow OPT(S, t) < (1 - \beta)(n - 1)$$

where $\alpha > 0$ and $\beta > 0$ are constants which are defined later. The permutation $v_1 v_2 \ldots v_n$ corresponding to a tour of length n contains $n - 1$ strings from S as substrings: $v_1 v_2, v_2 v_3, \ldots, v_{n-1} v_n$. Therefore, the first implication is true.

Suppose now that $OPT(G) < (1 - \alpha)n$. Then, $OPT(S, t) < (1 - \alpha)n$, since a permutation of t defines a path in the graph, which is shorter than a tour. We want to find a constant β such that $(1 - \alpha)n \leq (1 - \beta)(n - 1)$. The following inequality gives the desired.

$$\beta \leq 1 - \frac{1 - \alpha}{1 - \frac{1}{n}}$$

Therefore, if the maximum ATSP problem does not admit a $1 - \alpha$ approximation, then the *RCSstr[2]* problem (even in case that t is a set) does not admit a $1 - \beta$ approximation (the reader may refer to [28] for a more detailed argument of this claim). From Theorem 3, we know that is hard to approximate the Maximum Asymmetric Travelling Salesman with 0, 1 edge weights within $320/321 + \epsilon$, for any $\epsilon > 0$. Therefore, our problem is inapproximable within $1 - \beta \geq n(320/321 + \epsilon)/(n - 1)$, for any $\epsilon > 0$.

□

We now show that even over a binary alphabet the *RCSstr* problem remains NP-Hard.

Theorem 5. *If $|\Sigma| = 2$, then the RCSstr problem is NP-Hard.*

Proof. Let $\Sigma = \{0, 1\}$. We prove that if we can solve the *RCSstr* problem on the alphabet Σ in polynomial time, then we can solve in polynomial time the shortest common superstring problem on the alphabet Σ.

Consider a shortest common superstring instance S, where the longest string has length ℓ. It is easy to see that $s_1 \cdot s_2 \cdots \cdots s_n$ is a superstring of all the strings in S. Hence, the solution is no longer than $n\ell$. We show that $O(n^2\ell^2)$ calls to *RCSstr* are sufficient to find the shortest common superstring of the given strings.

We name an *RCSstr* instance (S, t) *complete*, if all the strings of S are substrings of the optimal solution $\pi(t)$.

Note that there exists a string x with i 0's and j 1's that is a common superstring of all the strings in S if and only if the *RCSstr* instance $(S, 0^i 1^j)$ is *complete*. Therefore, we want to find the shortest string t such that the *RCSstr* instance (S, t) is *complete*. The shortest common superstring is given by the permutation $\pi(t)$ returned by calling the *RCSstr* on the instance (S, t). The number of multisets $0^i 1^j$ where $i + j \leq n\ell$ is $O(n^2\ell^2)$. Therefore we can call the *RCSstr* on all of them and we can find the shortest common superstring on the given strings in polynomial time (note that this time can be improved somewhat by employing a binary search). The shortest common superstring problem is NP-Hard and the theorem follows.

□

2.2   Approximating *RCSstr*

In this section we present approximation algorithms for two variants of the *RCSstr* problem.

We first present a 3/4-approximation algorithm for the *RCSstr[2]* problem where each character of t is unique. Our algorithm follows immediately from the NP-Hardness reduction presented in the previous section. Since each character in t is unique we can construct a complete directed graph $G = (V, E)$, with $V = \Sigma$

as in the proof of Theorem 4. We then apply the $3/4$ approximation algorithm for the *Maximum ATSP* [3] and we obtain a cycle $t_{\pi(1)}, t_{\pi(2)}, \ldots, t_{\pi(n)}, t_{\pi(1)}$ of total weight k, where $\pi : \{1, \ldots, n\} \to \{1, \ldots, n\}$ is a permutation.

If, for some $i < n$, $t_{\pi(i)} t_{\pi(i+1)} \notin S$, we output $t_{\pi(i+1)} t_{\pi(i+2)} \cdots t_{\pi(n-1)} t_{\pi(n)} t_{\pi(1)}$ $\cdots t_{\pi(i)}$, that contains k strings from S as substrings (and yields an approximation ratio of $3/4$). Otherwise, we output $t_{\pi(1)} t_{\pi(2)} \cdots t_{\pi(n-1)} t_{\pi(n)}$ that contains exactly $n - 1$ strings from S as substrings, which is optimal.

Here we present a simple $1/(\ell(\ell(\ell + 1)/2 - 1))$-approximation algorithm for *RCSstr[ℓ]*.

The idea is output a concatenation of a maximal collection of strings from S. One can observe that each of the ℓ characters of a string in our solution cannot be used by more than $\ell(\ell + 1)/2 - 1$ strings in the optimal solution. Therefore, the algorithm yields a $1/(\ell(\ell(\ell + 1)/2 - 1))$-approximation ratio. Formally, the algorithm is presented below.

Algorithm 1. A $1/(\ell(\ell(\ell + 1)/2 - 1))$ approximation algorithm for *RCSstr[ℓ]*

Find a maximal subset $S' = s'_1, s'_2, \ldots, s'_q \subset S$ of strings under the following constraint: there exists a permutation $\pi(t)$ of the multiset such that $s'_1 \cdot s'_2 \cdot \cdots \cdot s'_q$ is a prefix of $\pi(t)$.

Output: $\pi(t)$

Theorem 6. *Algorithm 1 is a $1/(\ell(\ell(\ell+1)/2-1))$-approximation algorithm for RCSstr[ℓ].*

Proof. Note that, a single character can be used simultaneously in at most $\ell(\ell + 1)/2 - 1$ non-unit-length strings of the optimal solution. Since for every $s_i \in S$, $|s_i| \leq \ell$, we can conclude that a single string in our solution can cause at most $\ell(\ell(\ell+1)/2-1)$ other strings of the optimal solution not to be chosen. Thus, the size of the optimal solution is at most $q(\ell(\ell(\ell+1)/2-1))$ and the approximation ratio follows. $\square$

One tight example for the above analysis of Algorithm 1 is the following: $t = \{a, b, c, q, q, q, z, z, z, w, w, w, x, x, x\}$, and $S = \{abc, qa, az, wqa, qaz, azx, qb, bz, wqb, qbz, bzx, qc, cz, wqc, qcz, czx\}$. If we first select into the maximal collection the string abc, then we cannot add any other string to our solution. The optimal solution has size 15 and consists of all the other strings.

Observation 1. *Given an RCSstr[ℓ] instance, if for every $s_i \in S$, s_i is not a substring of any other $s_j \in S$, then Algorithm 1 is an ℓ^2-approximation algorithm.*

Proof. Note that, a single character can be used simultaneously in at most ℓ strings of the optimal solution, thus, a single string in our solution can stop at most ℓ^2 other strings of the optimal solution from being placed. $\square$

One can notice that, in case that all input strings are of length ℓ the above observation must holds.

3   *RCSseq*

We now turn to the *RCSseq* problem. We first present hardness results and
lower bound for several variants of the *RCSseq* problem and then we present
two approximation algorithms.

3.1   Hardness of the *RCSseq* problem

In the following theorem we show that the hardness result for the general *RCSstr*
holds also to the *RCSseq*.

Theorem 7. *RCSseq is NP-complete and hard to approximate within a factor*
$n^{1-\epsilon}$, *for any* $\epsilon > 0$, *unless* $P = NP$.

Proof. Omitted (similar to the proof of Theorem 2).

Moreover, we state that even over a binary alphabet the *RCSseq* problem remains
NP-Hard.

Theorem 8. *If* $|\Sigma| = 2$, *then the RCSseq problem is NP-Hard.*

Proof. Omitted (similar to the proof of Theorem 5).

We now prove that *RCSseq* is APX-Hard even if all the input strings are of
length two and t is a set. To do so, we present an approximation-preserving
reduction from the classical *maximum acyclic subgraph* problem.

Definition 3. *(Maximum Acyclic Subgraph) Given a directed graph* $G = (V, E)$
the maximum acyclic subgraph *problem is to find a subset* A *of the arcs such
that* $G' = (V, A)$ *is acyclic and* A *has maximum cardinality.*

Theorem 9. *[17] The Maximum Acyclic Subgraph problem is APX-Complete.*

We can now present our hardness result.

Theorem 10. *RCSseq is APX-Hard even if all the strings in S have length two
and t is a set.*

Proof. We present an approximation-preserving reduction from the maximum
acyclic subgraph problem. Given a directed graph $G = (V, E)$ we construct an
instance (S, t) of the *RCSseq* problem as follows. Set $t = V$ and for every arc
$(a, b) \in E$ we add a string ab to S.

Let π be a permutation of the set t and let $A \subseteq S$ be all the strings that are
subsequences of $\pi(t)$. The corresponding edge set A defines an acyclic subgraph of
G. On the other hand, let $A \subseteq E$ be an acyclic subgraph. Consider a topological
ordering of (V, A). All strings corresponding to edges A are subsequences of $\pi(t)$
that corresponds to the topological ordering.

Note that the optimal solution of the *RCSseq* instance (S, t) has size x if and
only if the optimal solution of maximum acyclic subgraph problem on the graph
G has size x. Thus, the *RCSseq* problem is APX-Hard.   □

In [11] the following result is proven.

Theorem 11. *The maximum acyclic subgraph problem is Unique-Games hard to approximate within a factor better than the trivial 1/2 achieved by a random ordering.*

The maximum acyclic subgraph is a special case of permutation constraint satisfaction problem (permCSP). A permCSP of arity k is specified by a subset S of permutations on $\{1, 2, \ldots, k\}$. An instance of such a permCSP consists of a set of variables V and a collection of constraints each of which is an ordered k-tuple of V. The objective is to find a global ordering σ of the variables that maximizes the number of constraint tuples whose ordering (under σ) follows a permutation in S. In [5] Charikar, Guruswami and Manokaran prove the following result.

Theorem 12. *For every permCSP of arity 3, beating the random ordering is Unique-Games hard.*

Our problem corresponds a *permCSP* where S contains only the identical permutation. Therefore we can conclude the following.

Theorem 13. *RCSseq[2] is Unique-Games hard to approximate within a factor better than 1/2.*

Theorem 14. *RCSseq[3] is Unique-Games hard to approximate within a factor better than 1/6.*

Currently there is an unpublished result by Charikar, Håstad and Guruswami stating that every k-ary *permCSP* is approximation resistant. This implies that $RCSseq[\ell]$ cannot have an approximation algorithm better than $1/\ell!$.

3.2 Approximating *RCSseq*

In this subsection we present a $(1 + \Omega(1/\sqrt{\Delta}))/2$ approximation algorithm for the $RCSseq[2]$ problem where Δ is the number of occurrences of the most frequent character in S. We also present a simple randomized approximation algorithm which achieves an approximation ratio of $1/\ell!$.

Theorem 15. *[2] The maximum acyclic subgraph problem is approximable within $(1 + \Omega(1/\sqrt{\Delta}))/2$, where Δ is the maximum degree of a node in the graph.*

Given a multiset t, let P' be the set of characters that have a single occurrence in t and let P be $\Sigma \backslash P'$, where Σ is the alphabet of t. We define Q to be the following multiset. For every $\sigma \in P$, if σ has r occurrences in t, then σ has $r - 2$ occurrences Q.

Figure 1 is an example of Algorithm 2. In the first stage we construct a graph according to the first two steps, note that $P = \{e\}$, $P' = \{a, b, c, d\}$ and $Q = \emptyset$. Then we present an acyclic directed subgraph and we output $F \cdot F' \cdot F \cdot F''$, where $F = e$ and $F' = cadb$.

Algorithm 2. A $(1 + \Omega(1/\sqrt{\Delta}))/2$ approximation algorithm for $RCSseq[2]$

---

1. Given a multiset t, construct a graph $G = (V, E)$ such that:
   $v_i \in V$ iff $v_i \in P'$ and $(a, b) \in E$ iff $a, b \in P'$ and $ab \in S$.
2. Apply the $(1 + \Omega(1/\sqrt{\Delta}))/2$ approximation algorithm for the maximum acyclic subgraph to the graph G.
   Denote the output subgraph by $G'(V, E')$.
3. Let F' be a topological order of the vertices of G'.
   Let F and F'' be an arbitrary ordering of P and Q respectively.
4. Output $F \cdot F' \cdot F \cdot F''$.

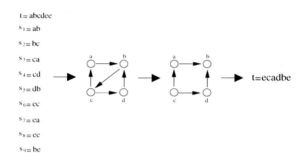

Fig. 1. Algorithm 2 example

Theorem 16. *Algorithm 2 is a $(1 + \Omega(1/\sqrt{\Delta}))/2$ approximation algorithm for the RCSseq[2] problem, where Δ is the maximum number of occurrences of a character in the set S.*

Proof. Given a string $ab \in S$. If $a \in P$ or $b \in P$ (or both), then ab is always a subsequence of $F \cdot F' \cdot F$. Otherwise, if both a and b appear only once in t, then ab is a subsequence of $F \cdot F' \cdot F$ if only if the edge (a, b) is selected in the arc set of the maximum acyclic subgraph. Since the maximum acyclic subgraph problem has an approximation ratio of $(1 + \Omega(1/\sqrt{\Delta}))/2$, the same approximation ratio holds for $RCSseq[2]$ problem. □

We now deal with $RCSseq[\ell]$ instances. We show that selecting an arbitrary permutation $\pi(t)$ achieves an expected approximation ratio of $\frac{1}{\ell!}$.

We define by $P(s_i, \pi(t))$ the probability that a string $s_i \in S$ is a subsequence of a permutation $\pi(t)$.

Note that, $P(s_i, \pi(t)) \geq \frac{\binom{|t|}{\ell}(|t|-\ell)!}{|t|!} = \frac{1}{\ell!}$. Therefore, the expected number of strings from S to be subsequences of an arbitrary permutation $\pi(t) \geq \frac{|S|}{\ell!}$. Thus, selecting an arbitrary permutation $\pi(t)$ achieves an expected approximation ratio of at least $\frac{|S|}{|S|\ell!} = \frac{1}{\ell!}$.

References

1. Barone, P., Bonizzoni, P., Vedova, G.D., Mauri, G.: An approximation algorithm for the shortest common supersequence problem: an experimental analysis. In: Symposium on Applied Computing, pp. 56–60 (2001)
2. Berger, B., Peter, W.S.: Approximation algorithms for the maximum acyclic subgraph problem. In: SODA, pp. 236–243 (1990)
3. Bläser, M.: A 3/4-approximation algorithm for maximum atsp with weights zero and one. In: APPROX-RANDOM, pp. 61–71 (2004)
4. Blum, A., Jiang, T., Li, M., Tromp, J., Yannakakis, M.: Linear approximation of shortest superstrings. J. ACM 41(4), 630–647 (1994)
5. Charikar, M., Guruswami, V., Manokaran, R.: Every permutation csp of arity 3 is approximation resistant. In: IEEE Conference on Computational Complexity, pp. 62–73 (2009)
6. Cotta, C.: Memetic algorithms with partial lamarckism for the shortest common supersequence problem. In: IWINAC, vol. (2), pp. 84–91 (2005)
7. Engebretsen, L., Karpinski, M.: Approximation hardness of tsp with bounded metrics. In: ICALP, pp. 201–212 (2001)
8. Garey, M.R., Johnson, D.S.: Computers and intractability. A guide to the theory of NP-completeness. W. H. Freeman, New York (1979)
9. Gotthilf, Z., Lewenstein, M.: Improved approximation results on the shortest common supersequence problem. In: SPIRE, pp. 277–284 (2009)
10. Gotthilf, Z., Lewenstein, M., Popa, A.: On shortest common superstring and swap permutations. In: SPIRE, pp. 270–278 (2010)
11. Guruswami, V., Manokaran, R., Raghavendra, P.: Beating the random ordering is hard: Inapproximability of maximum acyclic subgraph. In: FOCS, pp. 573–582 (2008)
12. Jiang, T., Li, M.: On the approximation of shortest common supersequences and longest common subsequences. SIAM J. Comput. 24(5), 1122–1139 (1995)
13. Kaplan, H., Lewenstein, M., Shafrir, N., Sviridenko, M.: Approximation algorithms for asymmetric tsp by decomposing directed regular multigraphs. J. ACM 52(4), 602–626 (2005)
14. Maier, D.: The complexity of some problems on subsequences and supersequences. J. ACM 25(2), 322–336 (1978)
15. Middendorf, M.: The shortest common nonsubsequence problem is NP-complete. Theor. Comput. Sci. 108(2), 365–369 (1993)
16. Ott, S.: Lower bounds for approximating shortest superstrings over an alphabet of size 2. In: WG, pp. 55–64 (1999)
17. Papadimitriou, C.H., Yannakakis, M.: Optimization, approximation, and complexity classes. J. Comput. Syst. Sci. 43(3), 425–440 (1991)
18. Pevzner, P.A.: Multiple alignment, communication cost, and graph matching. SIAM Journal of Applied Mathematics 52(6), 1763–1779 (1992)
19. Räihä, K.-J., Ukkonen, E.: The shortest common supersequence problem over binary alphabet is NP-complete. Theor. Comput. Sci. 16, 187–198 (1981)
20. Rubinov, A.R., Timkovsky, V.G.: String noninclusion optimization problems. SIAM J. Discrete Math. 11(3), 456–467 (1998)
21. Sacerdoti, E.D.: A structure for plans and behavior. Elsevier, Amsterdam (1977)
22. Sankoff, D., Kruskal, J.: Time Warps, String Edits, and Macromolecules: The Theory and Practice of Sequence Comparison. CSLI Publications, Stanford (1983)

23. Sellis, T.K.: Multiple-query optimization. ACM Trans. Database Syst. 13(1), 23–52 (1988)
24. Storer, J.A.: Data Compression: Methods and Theory. Computer Science Press, Rockville (1988)
25. Sweedyk, Z.: A $2\frac{1}{2}$-approximation algorithm for shortest superstring. SIAM J. Comput. 29(3), 954–986 (1999)
26. Tate, A.: Generating project networks. In: IJCAI, pp. 888–893 (1977)
27. Timkovsky, V.G.: Some approximations for shortest common nonsubsequences and supersequences. In: SPIRE, pp. 257–268 (2008)
28. Vazirani, V.V.: Approximation Algorithms. Springer, Heidelberg (2004)
29. Wilensky, R.: Planning and understanding. Addison-Wesley, Reading (1983)
30. Wilkins, D.E.: Practical planning: Extending the classical AI planning paradigm. Morgan Kaufmann, San Francisco (1988)
31. Zuckerman, D.: Linear degree extractors and the inapproximability of max clique and chromatic number. Theory of Computing 3(1), 103–128 (2007)

Author Index